The Encyclopedia of Medical and Veterinary Entomology

The Encyclopedia of Medical and Veterinary Entomology

Richard C. Russell
University of Sydney, Australia

Domenico Otranto
University of Bari, Italy

Richard L. Wall
University of Bristol, United Kingdom

www.cabi.org

CABI is a trading name of CAB International

CABI
Nosworthy Way
Wallingford
Oxfordshire, OX10 8DE
UK

CABI
38 Chauncey Street
Suite 1002
Boston, MA 02111
USA

Tel: +44 (0)1491 832111
Fax: +44 (0)1491 833508
E-mail: info@cabi.org
Website: www.cabi.org

Tel: +1 800 552 3083 (toll free)
Tel: +1 (0)617 395 4051
E-mail: cabi-nao@cabi.org

A catalogue record for this book is available from the British Library, London, UK.

Library of Congress Cataloging-in-Publication Data

Russell, Richard C. (Richard Charles), 1947-
   The encyclopedia of medical and veterinary entomology / by Richard C. Russell, Domenico Otranto, Richard L. Wall.
    p. cm.
Includes bibliographical references and index.
ISBN 978-1-78064-037-2 (hbk: alk. paper) 1. Arthropod vectors--Encyclopedias.
2. Veterinary entomology--Encyclopedias. 3. Insects as carriers of disease--
Encyclopedias. I. Otranto, Domenico. II. Wall, Richard, Ph. D. III Title.

RA641.A7R87 2013
614.4'303--dc23
                          2013006880
ISBN-13: 978 1 78064 037 2

Commissioning editor: Sarah Hulbert
Editorial assistant: Emma McCann
Production editor: Simon Hill

Typeset by Columns Design XML Ltd.
Printed and bound by Grutenberg Press Ltd, Tarxien, Malta.

Contents

Preface and Acknowledgements

This Encyclopedia was based on and adapted from the original text *Medical and Veterinary Entomology*, written by Professor Douglas S. Kettle. This comprehensive and authoritative book, first published in 1984, has for many years been a standard reference source in its field, being reprinted many times in two different editions, the second in 1995. We are greatly indebted to Doug for allowing us the privilege of updating and restructuring his seminal work, and although the book has been transformed to an encyclopedia format with listing by common name, we have retained much of his content and style as a tribute. Overall, while the subject matter we have included is necessarily selective in its detail, we have endeavoured to provide an extensive overview of the various topics without trying to present a comprehensive account. Sadly, Doug died in October 2012, when this book was in its final stages of publication, so he was not able to see the final product.

In producing this volume, we would like to thank Sarah Hulbert, Commissioning Editor at CABI, for her support and encouragement and Emma McCann at CABI, who acted as editorial assistant for the final production. We are grateful also to Carly Au (University of Bristol, UK), who provided early editorial assistance, and to a number of colleagues for reading and commenting on various sections: Cinzia Cantacessi (James Cook University, Australia), Aleksandra Cupina (University of Novi Sad, Serbia), Filipe Dantas-Torres (Fundação Oswaldo Cruz, Brazil and University of Bari, Italy), Stephen Doggett (Westmead Hospital, Australia), Durland Fish (Yale University, USA), Annunziata Giangaspero (University of Foggia, Italy), Alessio Giannelli (University of Bari, Italy), Vincenzo Lorusso (University of Edinburgh, UK), Dusan Petric (University of Novi Sad, Serbia), Scott Ritchie (James Cook University, Australia), Cameron Webb (Westmead Hospital and University of Sydney, Australia) and Peter Whelan (Northern Territory Department of Health, Australia). We are grateful to a number of colleagues who contributed colour images, but we wish to thank in particular Alan Walker (University of Edinburgh, UK) and Stephen Doggett (Westmead Hospital, Australia), who generously provided many of the photographs.

Richard Russell, Domenico Otranto and Richard Wall
April 2013

Introduction –
Part I: The Creatures

1. WHAT IS MEDICAL AND VETERINARY ENTOMOLOGY?

The science of entomology should, strictly speaking, be restricted to 'six-legged animals' or insects, but the applied entomologist is expected to cover a wider field. Medical and veterinary entomologists are expected to deal with other terrestrial creatures within the **Phylum Arthropoda**, such as scorpions and spiders, ticks and mites and also centipedes and millipedes. In this book, therefore, the terms 'entomology' and 'insects' should be taken to include the wider range of arthropods of medical and veterinary importance.

Medical and veterinary entomology is concerned with the role of arthropods in the causation of disease or discomfort in animals and humans. This concern is paramount, and hence the focus of interest of medical and veterinary entomologists must be the incidence and control of the affliction, not necessarily the control of the causative insect. Of course, insect control is one means of controlling and/or preventing the disease or discomfort, but in a particular setting may not be the most appropriate method. Where disease incidence is low and the insect widespread, it may be more realistic to treat cases as they arise rather than attempt to control the pest or vector. The medical or veterinary entomologist who forgets this primary involvement with the affliction caused by the insect and focuses attention solely on the insect itself will often fail to effect appropriate management, to the detriment of the patient.

2. WHAT IS THE PHYLUM ARTHROPODA?

The largest phylum in the animal kingdom is the Arthropoda, which contains about 80% of the known species of animals. They are bilaterally symmetrical, segmented animals with jointed legs. Each segment consists of a dorsal sclerotized plate, the tergum, and a similar ventral plate, the sternum, the two being joined together laterally by membranous pleura (singular pleuron). The terga and sterna of successive segments are separated by intersegmental membranes, which, together with the pleura, provide flexibility. The exoskeleton provides a limit to growth, and periodically arthropods have to develop a new skin under the existing one and then cast the old skin. The process is known as ecdysis or moulting and, in the insects, the interval between ecdyses is a stadium or stage (described as an instar where there is morphological change between moults). Internally, arthropods have the typical invertebrate arrangement of ventral nerve cord and dorsal heart, the reverse of the vertebrate arrangement.

3. THE CLASSIFICATION OF ARTHROPODA

The Phylum Arthropoda may be divided into those with antennae and mandibles as mouth parts, and those creatures lacking the mandibulata antennae but possessing chelicerate mouthparts, the Chelicerata. Two classes of Arthropoda are especially important to the medical and veterinary entomologist – the **Class Insecta** (insects) in the Mandibulata and the **Class Arachnida** (scorpions, spiders, ticks and mites) in the Chelicerata. Two other classes are also of minor significance within the Subphylum Myriapoda: the **Class Chilopoda** (centipedes) and the **Class Diplopoda** (millipedes). Given the diversity of the arthropods and the lack of clarity about the taxonomic relationships between existing groups, some debate inevitably exists about the precise classification of the various taxa.

© R.C. Russell 2013. *The Encyclopedia of Medical and Veterinary Entomology* (R.C. Russell *et al.*)

3.1 The Class Chilopoda

The Chilopoda are long, soft-bodied terrestrial arthropods which are dorsoventrally flattened. They have one pair of antennae on the head and three pairs of appendages associated with the mouth. Behind the head, the body is metamerically segmented and composed of at least 17 segments, each of which, with the exception of the last, bears a pair of legs. The most significant order is the Scolopendromorpha, which includes some of the larger centipedes that can inflict venomous bites (Fig. I.1).

3.2 The Class Diplopoda

The Diplopoda are metamerically segmented terrestrial arthropods which feed on decaying vegetable matter. They have one pair of antennae and two pairs of appendages associated with the mouth. Each apparent segment bears two pairs of legs and two pairs of spiracles, the respiratory openings. The orders Polydesmida and Spirobolida are of medical interest because some of these can produce chemical secretions when disturbed that can irritate and discolour human skin (Fig. I.2).

Fig. I.1. Dorsal view of a chilopod, *Scolopendron morsitans*. Redrawn from Lewis (1981).

3.3 The Class Arachnida

The Arachnida are carnivorous terrestrial chelicerate arthropods which have no antennae. They have one pair of chelicerae for feeding, a pair of pedipalps whose function varies from order to order and four pairs of walking legs. The Arachnida is a large and diverse class, containing several subclasses and orders of importance.

3.3.1 Subclass Acari (mites and ticks)

The Acari are often small arthropods, varying widely in shape but in which the prosoma (anterior portion of the body) and opisthosoma (hind portion) are broadly fused and abdominal segmentation is inconspicuous or absent (Fig. I.3). The pedipalps are short sensory structures associated with the chelicerae in a discrete

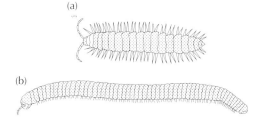

Fig. I.2. (a) Dorsal view of a flat-back diplopod, *Polydesmus* sp.; (b) lateral view of a cylindrical diplopod, *Cylindroiulus* sp. Redrawn from Hopkin and Read (1992).

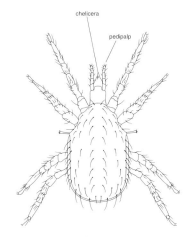

Fig. I.3. Dorsal view of an acarine, *Zercoseius ometes*. Reproduced with permission from Savoury, T.H. (1977) Arachnida, Academic Press, London.

gnathosoma. Acari are extremely important agents of disease and vectors of pathogens. The Subclass Acari is usually considered to be composed of two superorders: Parasitiformes and Acariformes. The Parasitiformes includes, for example, ticks and bird mites; the Acariformes includes, for example, chiggers, plant mites, scabies and stored product and oribatid mites.

3.3.2 Order Scorpiones (scorpions)

Scorpions have powerful chelate pedipalps and abdomens ending in a large, globular sting, and some species can cause severe envenomation and occasionally death in humans (Fig. I.4).

3.3.3 Order Araneae (spiders)

Spiders are characterized by having a uniform prosoma (anterior portion of the body) joined by a narrow pedicel to an unsegmented opisthosoma (hind portion) (Fig. I.5a and b). The pedipalps are tactile, leg-like structures, shorter than the ambulatory legs. In the male, they are modified as intromittent organs, and male spiders are readily recognized by the terminal swelling on the pedipalp. The chelicerae are two-segmented, but not chelate. The distal segment is sharply pointed and bears at its tip the opening of a poison duct. Some spiders are highly venomous, with bites that can be life threatening.

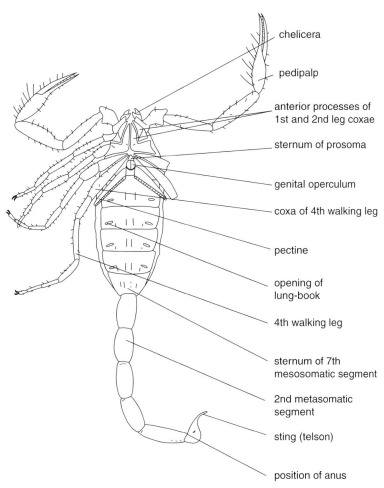

chelicera

pedipalp

anterior processes of
1st and 2nd leg coxae

sternum of prosoma

genital operculum

coxa of 4th walking leg

pectine

opening of
lung-book

4th walking leg

sternum of 7th
mesosomatic segment

2nd metasomatic
segment

sting (telson)

position of anus

Fig. I.4. Ventral view of a scorpion. *Source*: Snow, K.R. (1970) *The Arachnids*. Routledge and Kegan Paul, London.

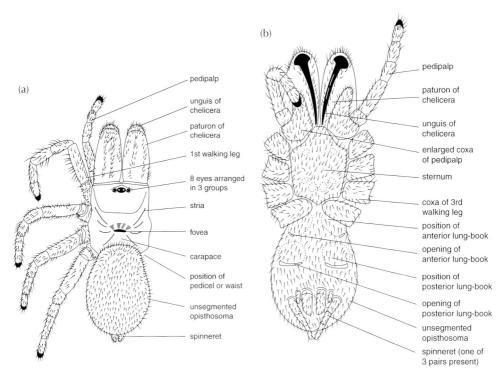

Fig. I.5. (a) Dorsal and (b) ventral views of a mygalomorph spider. *Source*: Snow, K.R. (1970) *The Arachnids*. Routledge and Kegan Paul, London.

3.3.4 Other arachnid orders of minor or incidental interest (Fig. I.6)

ORDER AMBLYPYGI (WHIPSPIDERS)

Whipspiders are flat-bodied creatures found in tropical and subtropical regions worldwide. They are often nocturnal predators with powerful pedipalps, but lacking poison glands. Only the posterior three pairs of legs are used for walking and those of the first pair are tactile and, in action, are stretched out in front of the animal. Amblypygids range in size from 4 to 45 mm in length and the first pair of legs is excessively long, more than twice the length of the walking legs.

ORDER OPILIONES (HARVESTMEN)

Harvestmen have a widespread global distribution and are characterized by their exceptionally long walking legs compared with their body size (although there are also short-legged species). They are mostly nocturnally active omnivores. They do not have venom glands but do have a pair of defensive glands that can discharge a peculiar smelling fluid containing noxious quinones to repel enemies when disturbed.

ORDER PSEUDOSCORPIONES (PSEUDOSCORPIONS)

The Pseudoscorpiones are a widely distributed order of small arachnids (<8 mm in length) with large chelate pedipalps, superficially resembling small scorpions but they lack a sting. They are common in soil and decaying vegetation and a few species are to be found in food stores and among books, presumably feeding on book lice (psocids). They produce injectable poisons from glands in their pedipalps.

ORDER SOLIFUGAE (SUN SPIDERS, CAMEL SPIDERS)

The Solifugae are large, hairy, nocturnal carnivorous arachnids of desert areas in the tropics and subtropics. Solifugids are recognized easily by their possession of large,

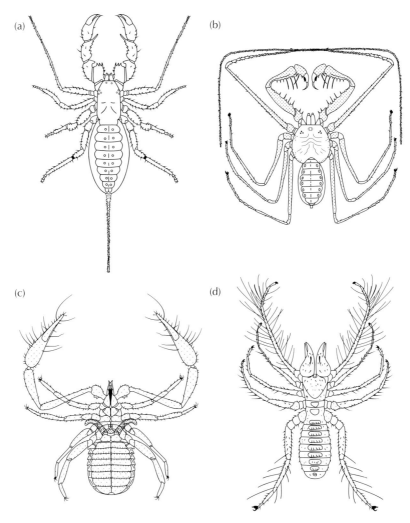

Fig. I.6. (a) A uropygid, *Thelyphonus insularis*; (b) an amblypygid, *Stegophrynus dammermani*; (c) a pseudoscorpion, *Chelifer cancroides*; (d) a solifugid, *Galeodes arabs*. Reproduced with permission from Savoury, T.H. (1977) *Arachnida*. Academic Press, London.

powerful chelate chelicerae with which they seize their prey. They will attack any suitable prey, including small vertebrates and each other. They have no poison glands and rely solely on the crushing power of the chelicerae, which can cause ragged wounds, are often contaminated and can result in secondary infections.

ORDER UROPYGI (WHIP SCORPIONS)

The Uropygi are flat-bodied creatures found in many tropical and subtropical dry regions, but not in Europe, Australia or Africa (except for an introduced species). They are nocturnal predators with powerful pedipalps but lacking poison glands. Only the posterior three pairs of legs are used for walking and those of the first pair are tactile and, in action, are stretched out in front of the animal. Uropygids range in size from 25 to 85 mm and the abdomen terminates in a segmented flagellum as long as the rest of the abdomen. They do not have venom glands, but do have defensive glands at the rear of the abdomen which can spray a fluid containing acetic and octanoic acids to repel enemies.

3.4 The Class Insecta

The Class Insecta are ectognathous and usually hypognathous arthropods, i.e. with the mouthparts exposed and directed ventrally (Figs I.7 and I.8). The body of an insect is organized into three regions – head, thorax and abdomen. The head bears a pair of sensory antennae, large compound eyes and three pairs of mouthparts (mandibles, 1st

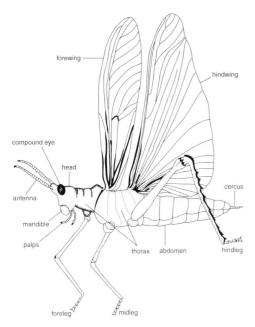

Fig. I.7. Lateral view of a generalized insect.

maxillae and labium or fused 2nd maxillae). Within the Insecta, a number of orders are of major or significant importance with respect to medical and veterinary entomology.

3.4.1 Order Blattodea (cockroaches)

Cockroaches are dorsoventrally flattened, exopterygote terrestrial insects with long antennae and wings, when present, folded flat over the body (Fig. I.9). The mouthparts are mandibulate and the legs cursorial, i.e. adapted for running, at which cockroaches are very adept. The hardened forewings, called tegmina (singular tegmen), overlap and cover much of the dorsal surface of the body, protecting the more delicate hindwings, which are the effective flying organs. The abdomen ends in paired, jointed cerci. The eggs are enclosed in a hardened, purse-like case, the ootheca.

3.4.2 Order Phthiraptera (lice)

Lice are small, dorsoventrally flattened, exopterygote, wingless, obligatory ecto-parasites of birds and mammals (Fig. I.10). Two different forms of lice have evolved – the Mallophaga retain the primitive insectan mandibulate mouthparts and feed on epidermal structures of birds and mammals, while the Anoplura have evolved specialized mouthparts for blood feeding and are found only on mammals. The classification of the Phthiraptera is complex. The order is divided into four sub-orders: Anoplura, Amblycera, Ischnocera and Rhynchophthirina. The Anoplura are known as sucking lice. The suborders Amblycera and

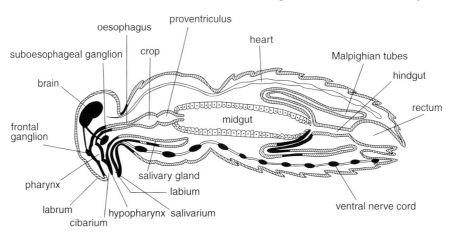

Fig. I.8. Diagrammatic longitudinal section of a generalized insect.

Ischnocera are known as chewing lice and are usually discussed together and described as the Mallophaga. The latter is a very small suborder, including just two species known from the elephant and the wart hog.

SUBORDER ANOPLURA (SUCKING LICE)
Sucking lice have relatively long, narrow heads with retracted mouthparts that are not discernible externally (Fig. I.10a). The three thoracic segments are fused to form a single

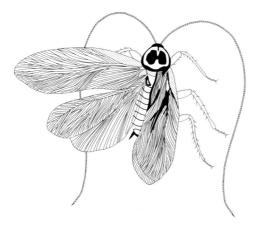

Fig. I.9. Dorsal view of a cockroach, *Periplaneta australasiae*.

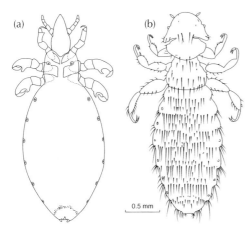

Fig. I.10. (a) Dorsal view of an anopluran, *Linognathus vituli*, the long-nosed cattle louse (Phthiraptera, Anoplura); (b) dorsal view of *Paraheterodoxus insignis* (Phthiraptera, Mallophaga).

structure. The most important sucking louse is *Pediculus humanus*, a parasite of humans, which has had a major impact on human history and social development by being the biological vector of epidemic typhus and epidemic relapsing fever.

SUBORDERS AMBLYCERA AND ISCHNOCERA (MALLOPHAGA) (CHEWING LICE)
Chewing lice have broad heads to accommodate mandibulate mouthparts and their associated muscles (Fig. I.10b). These mouthparts are obvious externally. The prothorax is free from the mesothorax, which may be fused to, or separate from, the metathorax. Chewing lice are mainly ectoparasites of birds, but some species occur on mammals.

3.4.3 Order Hemiptera (bugs)
The Hemiptera are exopterygote insects with highly specialized mouthparts produced into a ventrally reflected proboscis (Fig. I.11). Most Hemiptera feed on the fluid contents of plants, either by tapping the phloem or by piercing the cells of the mesophyll. This habit of piercing plants makes them ideal vectors of plant pathogens and of viruses which cause disease among crops, being of considerable importance to agriculturalists and horticulturalists. Some Hemiptera are predacious and a few, including the Cimicidae (bed bugs) and Triatomidae (kissing bugs), feed on blood, and some reduviids are vectors of Chagas' disease.

3.4.4 Order Coleoptera (beetles)
Beetles are endopterygote, mandibulate insects with thickened forewings known as elytra (singular elytron), which meet edge to edge in the mid-dorsal line (Fig. I.12). The

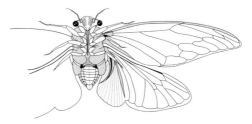

Fig. I.11. Ventral view of a hemipteran, the double drummer, *Thopha saccata*.

Coleoptera is the largest order of the Insecta, but it is only of very minor importance in medical or veterinary entomology. When some beetles (e.g. members of the Families Meloidae (Fig. I.12a) and Staphylinidae (Fig. I.12b) are disturbed, they produce a vesicating fluid.

(a)

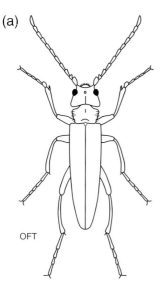

OFT

(b)

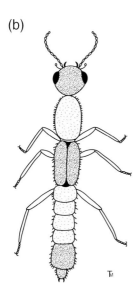

Fig. I.12. Coleoptera: dorsal views of (a) *Lytta vesicatoria* and (b) *Paederus cruenticollis*. *Source*: *L. vesicatoria* from Patton, W.S. (1931) *Insects, Ticks and Venomous, Animals Part II, Public Health*. H.R. Grubb, Croydon, UK.

3.4.5 Order Hymenoptera (ants, bees, wasps)

The Hymenoptera are endopterygote, mandibulate insects in which the forewings are larger than the hindwings and both fore- and hindwings are coupled mechanically to function as a single entity (Fig. I.13). Many Hymenoptera have a complex social organization. The female ovipositor is often modified into a sting, which is used to immobilize prey, as in the hunting wasps, or in defence of the colony, as in ants and bees. Although a wasp or bee sting is painful, the main danger lies in an allergic response, which may lead to anaphylactic shock and rapid collapse of the person stung.

3.4.6 Order Lepidoptera (butterflies, moths)

The Lepidoptera are endopterygote insects whose body, legs and wings are covered with detachable scales which are morphologically flattened hairs (Fig. I.14). The scales produce colourful patterns, which make members of this order so attractive and the object of collectors. The larval stage is a caterpillar with chewing mandibulate mouthparts, while the adult has a coiled proboscis for feeding on nectar. In West Africa, a few moths have been found to feed on secretions of the eyes of cattle, and one Asian species actually pierces the skin and feeds on the blood of large mammals. Many caterpillars are covered with

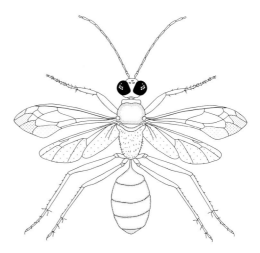

Fig. I.13. Dorsal view of a hymenopteran (Sphecidae).

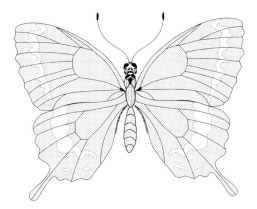

Fig. I.14. Dorsal view of a lepidopteran, *Papilio canopus.*

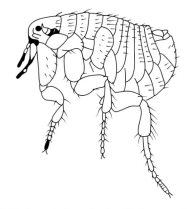

Fig. I.15. Lateral view of a flea (Siphonaptera).

long, urticating hairs (setae), which may or may not have a poison gland at the base and can cause severe dermatitis.

3.4.7 Order Siphonaptera (fleas)

Fleas are laterally flattened, wingless, endopterygote, blood-sucking ectoparasites of birds and mammals (Fig. I.15). The larva is worm-like and apodous (legless), while in the adult the hindlegs are saltatorial, i.e. adapted for leaping, and the source of the familiar escape response of adult fleas. Fleas are important vectors of disease, of which the most important is plague.

3.4.8 Order Diptera (two-winged flies)

Diptera are endopterygote insects with only one pair of functional wings (Fig. I.16). Dipterous larvae exploit both aquatic and terrestrial environments, are invariably apodous and those of the higher Diptera, which have very reduced heads, are known as maggots. From a medical point of view, the Diptera is the most important order of insects and contains families notorious for being vectors of disease organisms and endo- and ectoparasites of various animals, such as the Chironomidae (non-biting midges), Culicidae (mosquitoes), Ceratopogonidae (biting midges), Psychodidae (sand flies), Simuliidae (black flies), Chloropidae (eye flies), Glossinidae (tsetse flies), Muscidae (house flies and stable flies), Calliphoridae (blow flies and screw-worm flies), Sarcophagidae (flesh flies) and the Oestridae and Hippoboscidae (bot and warble flies).

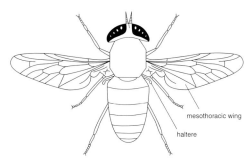

Fig. I.16. Dorsal view of a dipteran.

3.4.9 Other insect orders of minor or incidental interest

ORDER COLLEMBOLA (SPRINGTAILS)
Springtails have occasionally been found to be responsible for human infestations, causing an itching dermatitis.

ORDER DERMAPTERA (EARWIGS)
Some species of Dermaptera are known to be able to pierce human skin with their hind pincers and (although they have chewing mouthparts) there is a record of them imbibing blood.

ORDER EPHEMEROPTERA (MAYFLIES)
Mayflies can cause inhalation allergies when very abundant following large emergences.

ORDER PHASMATODEA (STICK INSECTS, WALKING STICKS)
The Phasmatodea include a number of species with prothoracic glands which can produce

odorous and irritating defensive sprays that have been known to cause stinging in the eyes. Also, some species are equipped with strong spines on their legs, which they use in defensive actions, and these can inflict considerable pain and draw blood.

ORDER PSOCOPTERA (BOOK LICE, BARK LICE)
The Psocoptera have been known to cause skin irritations when populations are abundant in domestic or storage situations.

ORDER THYSANOPTERA (THRIPS)
Thrips are occasionally incriminated in causing skin irritation, as they probe skin for moisture (and occasionally take blood); they can be particularly attracted to light-coloured clothing (including laundry drying on a line after washing) and so come into skin contact.

SELECTED BIBLIOGRAPHY

Beaty, B.J. and Marquadt, W.C. (1996) *The Biology of Disease Vectors*. University Press of Colorado, Niwot, Colorado.

Bettini, S. (ed.) (1978) *Arthropod Venoms*. Springer-Verlag, Berlin.

Burgess, N.R.H. and Cowan, G.O. (1993) *A Colour Atlas of Medical Entomology*. Chapman and Hall, London.

Chapman, R.F. (1998) *The Insects: Structure and Function*, 4th edn. Cambridge University Press, Cambridge, UK.

Goddard, J. (2000) *Physician's Guide to Arthropods of Medical Importance*, 3rd edn. CRC Press, Boca Raton, Florida.

Gullan, P.J. and Cranston, P.S. (1995) *The Insects: An Outline of Entomology*, 4th edn. Wiley-Blackwell, Chichester, UK.

Grimaldi, D. and Engel, M.S. (2004) *Evolution of the Insects*. Cambridge University Press, Cambridge, UK.

Hopkin, S.P. and Read, H.J. (1992) *The Biology of Millipedes*. Oxford University Press, Oxford, UK, 233 pp.

Kettle, D.S. (1995) *Medical and Veterinary Entomology*, 2nd edn. CAB International, Wallingford, UK.

Klowden, M.J. (2007) *Physiological Systems in Insects*, 2nd edn. Academic Press, Burlington, Maryland.

Lane, R.P. and Crosskey, R.W. (eds) (1993) *Insects and Acarines of Medical Importance*. Chapman and Hall, London.

Lewis, J.G.E. (1981) *The Biology of Centipedes*. Cambridge University Press, Cambridge, UK.

Mullen, G. and Durden, L. (eds) (2002) *Medical and Veterinary Entomology*. Academic Press, San Diego, California.

Peters, W. (1992) *A Colour Atlas of Arthropods in Clinical Medicine*. Wolfe Publishing Ltd, London.

Savory, T.H. (1977) *Arachnida*. Academic Press, London.

Service, M.W. (1978) A short history of early medical entomology. *Journal of Medical Entomology* 14, 603–626.

Service, M.W. (1980) *A Guide to Medical Entomology*. Macmillan Press, London.

Snow, K.R. (1970) *The Arachnids*. Routledge and Kegan Paul, London.

Taylor, M.A., Coop, R.L. and Wall, R.L. (2007) *Veterinary Parasitology*, 3rd edn. Blackwell Science Ltd, Oxford, UK.

Walker, A. (1994) *The Arthropods of Humans and Domestic Animals, A Guide to Preliminary Identification*. Chapman and Hall, London.

Introduction –
Part II: The Context

1. MEDICAL AND VETERINARY ISSUES

Arthropods of medical and veterinary importance have significant health, welfare and economic impacts on humans and domestic animals. The main cause of this is the transmission of disease-causing pathogens. In both humans and domestic animals, a large range of arthropods may also cause mild to severe afflictions through their capacity to bite, in some cases resulting in significant levels of blood loss. They may cause myiasis (the infestation of living tissues with fly larvae), sting (envenomate), carry urticating hairs and discharge vesicating fluids and have body parts, secretions or excreta which can induce allergic, dermal or respiratory reactions. When present in high numbers, their activity may cause nuisance and disturbance to humans and domestic animals, including livestock, and in the latter case, this may disrupt normal patterns of grazing or resting behaviour; in some cases, violent avoidance responses can lead to self-wounding. Another medical issue known as delusionary parasitosis is outlined further below, as is the application of an understanding of insect colonization of cadavers to the forensic sciences.

2. DISEASE TRANSMISSION

Animal and human diseases arise fundamentally from two causes: the presence of an introduced agent or pathogen upsetting the normal functioning of the organism, or a breakdown of the organism's integrating system leading to the development of organic disease. Medical and veterinary entomology is concerned only with diseases caused by pathogens. Sometimes, the insect itself may be the pathogen, as in scabies, a skin disease due to the presence of the mite, *Sarcoptes scabiei*, or pediculosis, due to infestation with the human body louse, *Pediculus humanus humanus*, and the head louse, *Pediculus humanus capitis*. More commonly, the role of the insect is as a vector (carrier) of the pathogen from one host to another. Most important in this category are the biting and other flies, although some fleas, lice, bugs, ticks and mites are also relevant.

Insects function as vectors in one of two ways, either mechanically or biologically. In mechanical transmission, the insect acquires the pathogen from one source and deposits it in other locations, where it may infect a new host. For example, the role of house flies in the transmission of enteric disease organisms is mechanical. House flies are attracted equally to faeces and food, on both of which they feed. Consequently, organisms picked up on and in the body of a house fly when it is feeding on faeces are carried away and may be deposited on human food or cooking utensils when the fly visits there.

In biological transmission, the pathogen normally is transmitted from host to host through the body of the insect vector. Thus, while it is possible to transmit malaria by blood transfusion, in nature the only way the malarial organism is passed from person to person is through the bite of a susceptible mosquito, such as the species *Anopheles gambiae*. Similarly, viral diseases such as yellow fever are maintained in their natural vertebrate hosts (monkeys) through the bites of infected mosquitoes and can reach humans only through the bite of other infective mosquitoes such as the species *Aedes aegypti*. The role of the insect is critical in biological transmission, and insect control is a major weapon in the armoury of disease control.

Biological transmission cycles vary with the nature of the pathogen. For example, some pathogens go through a cycle which involves maturing through a series of developmental stages with or without any multiplication in the numbers of pathogens, while others undergo multiplication without development. For example, in mosquitoes, the malaria protozoan, *Plasmodium*, undergoes development in the gut with the ingested sexual gametocytes combining to form a zygote that becomes an ookinete, which forms an oocyst in the gut wall in which a vast number of sporozoites form to then infect the salivary glands (from where they can be transmitted during the next blood feed) – with an exponential increase in numbers. However, with the filarial nematodes, the microfilarial stages that are ingested by competent mosquitoes undergo development and growth through three larval stages via various internal tissues, before they arrive in the head and mouthparts (from where they can be transmitted) – but there is no multiplication of the filaria and the numbers of infective third-stage larvae that eventually arrive are invariably fewer than the number of microfilaria ingested (due to attrition during various stages of development). With viruses, in competent mosquito species, the ingested virions invade cells in the midgut wall, proliferate and escape to the haemocoel, where they invade and multiply in various tissues and eventually within the salivary glands (from where they are able to be transmitted) – with an enormous increase in numbers but with no biological development of the virus itself.

Incidentally, the viruses transmitted to vertebrates by insects and acarines are known as arboviruses. The term simply means arthropod- (ar) borne (bo) viruses. Arboviruses are defined as viruses that multiply in both their vertebrate and invertebrate hosts, and the term is therefore restricted to viruses which are transmitted biologically. The criteria for a virus being an arbovirus are based on its biology and ecology, and not on its morphology. Consequently, arboviruses are found in five families: the Bunyaviridae, Flaviviridae, Reoviridae, Rhabdoviridae and Togaviridae. The greatest number of arboviruses is in the Bunyaviridae, but the most important mosquito-borne arboviruses are in the Flaviviridae and Togaviridae. The arboviruses in the Flaviviridae are assigned to the genus *Flavivirus* and in the Togaviridae to the genus *Alphavirus*. Within each genus, the species are serologically related but quite distinct from species in other genera. Entomologically, the most important feature is that the majority of species of *Alphavirus* are transmitted by mosquitoes, while the vectors of *Flavivirus* species are mostly mosquitoes or ticks.

3. ALLERGIC REACTIONS

A range of arthropods produce allergic reactions in humans and companion and domestic animals, because some species contain potent allergens in their saliva (e.g. fleas, biting midges, ticks), body parts and faecal wastes (e.g. mange mites, caterpillars, cockroaches) or venom (e.g. ants, wasps). Saliva, faeces and venoms contain a range of immunogenic, immunomodulating and other pharmacologically active molecules, and the various disintegrating body parts also contain molecules that may be contacted, injected or inhaled and sensitize people or animals by inducing IgE-mediated allergic reactions.

The allergens can produce various local and systemic reactions, such as dermal rashes with itching and swelling, rhinitis and asthma, headaches and conjunctivitis and, in extreme cases, life-threatening anaphylaxis. In most cases, these allergens are proteins, or molecules with protein components, which induce an IgE-type immunoglobulin (antibody) response. When these allergens can be identified, isolated and characterized, they can be used to assist in diagnosing the exact cause of suffering in the patients and in immunotherapy treatments that can relieve their condition.

Mosquitoes are well known to produce inflammatory skin reactions with their bites, and anaphylactic reactions are rare but have been recorded. Other biting flies, such as horse flies, black flies, mosquitoes, sand flies and biting midges, cause various degrees of local and systemic allergic reactions to their saliva. Fleas have allergens in their saliva that can produce 'hives' in sensitive humans and severe flea allergic dermatitis in domestic dogs and cats. In sensitized individuals, allergic responses may also be triggered in response to flea environment detritus (e.g. their faecal matter, moulting enzymes, larval and pupal skins or adult exoskeletons). Saliva associated with bites from lice, bed bugs, kissing bugs, ticks and mites also causes allergic reactions.

The feeding activity of mange mites, such as *Sarcoptes scabiei* or *Psoroptes ovis*, produces an inflammatory response of the skin. This is stimulated by the feeding, burrowing or production of antigenic material by the mite. The keratinocytes release cytokines (especially IL-1) in response to non-specific damage, which diffuses into the dermis, leading to cutaneous inflammation. In addition, mite antigens *de novo* or processed by antigen-presenting cells can be carried in the dermal lymphatics to local lymph nodes, where an immunological response occurs. This response can be humoral or cell mediated, resulting in either a protective immune response or hypersensitivity. Mange mite faecal antigens in particular are thought to be important in the production of hypersensitivity, which amplifies the innate inflammatory response.

Dust mites and storage mites are also domestic sources of potent allergens that cause dermal and respiratory symptoms. House flies are the source of allergens for some persons, with derived particles and secretions having been implicated. Some non-biting chironomid midges contain haemoglobin molecules that are potent allergens, and persons who are sensitized react (e.g. with respiratory, conjunctival and dermal responses) to exposure to adult and larval debris through skin contact with and inhalation of aerial-borne particles. Allergic responses to cockroaches have become much more recognized in recent years; prevalence as high as 60% has been reported from some populations that have been studied and the allergens are associated with cockroach bodies, secretions and wastes.

Ant, bee and wasp venoms contain proteins that induce IgE-mediated allergic reaction, both local and systemic, and there can be a degree of cross-reactivity to similar allergens from different species that can result in a broad sensitivity to multiple species.

Caterpillars of some butterflies and moths (and a few spiders) have so-called urticating hairs; they can be either true setae (as detachable hairs with some caterpillars and butterflies, and some tarantula spiders) or modified setae (as stiffened barbed hairs in caterpillars) or spines (with attached venom glands in caterpillars), and they have a protective function against predators. The true setae-type urticating hairs of caterpillars contain proteins and other molecules that bring on IgE-mediated allergic reactions in sensitized persons (even anaphylactic reactions). Dermal exposure to these hairs brings symptoms that are generally non-specific and local skin reactions vary between individuals, although itching, local swelling and erythema are common features; more serious systemic symptoms can occur and life-threatening anaphylactic reactions have been reported. Oral exposure can lead to swelling problems within the mouth and the oesophagus and possible inhalation that can bring on asthma-like symptoms (domestic animals rather than humans are more commonly affected, as they are more likely to ingest the caterpillars or their hairs), and there can be dramatic and life-threatening consequences. Ocular exposure in humans (and other animals) can be serious, with keratitis, conjunctivitis and damage to the inner eye having been reported.

4. DIRECT (ECTO- AND ENDOPARASITE) DAMAGE

As a result of their feeding and biting activity, insects may have a variety of direct and indirect parasitic effects on their hosts. The blood-feeding activity of insects such as horse flies or fleas may result in significant levels of blood loss and when these insects are present in large numbers, the blood loss may be directly debilitating and anaemia may occur in heavily infested hosts.

The larvae of many fly species are adapted to feed directly on the tissues of living human and animal hosts. This is known as myiasis and may be subclassified as specific (obligatory), semi-specific (facultative) or accidental (pseudo-myiasis) – depending on the need of the parasite to complete part of its life cycle in/on the host. The vertebrate hosts are usually mammals, occasionally birds and, less commonly, amphibians or reptiles. A wide range of tissues may be infested and the myiasis may be dermal, subdermal or cutaneous, nasopharyngeal, ocular, intestinal/enteric or urinogenital. The impact of this feeding activity depends on the site of infestation and the number of larvae. In some cases, for example small numbers of bot larvae in the stomach, a horse may have almost no clinical signs, whereas the infestation of humans or cattle by the larvae of screw-worm flies may be fatal if left untreated.

The physical presence of insects, particularly flies, as they attempt to feed or oviposit and when they are present at high density may cause appreciable levels of irritation and disturbance. In livestock, this may result in head shaking, stamping, skin twitching, tail switching or scratching, and these activities may result in reduced growth and loss of condition because the time spent in avoidance is lost from grazing or resting. In extreme cases, persistent fly activity may result in serious self-injury following collision with fences and other objects.

Large populations of flies may breed in animal dung, particularly in and around intensive husbandry units. Their activity may result in considerable social nuisance and legal problems, especially where suburban developments have encroached on previously rural areas. Adult flies and their faeces may also decrease the aesthetic appearance and value of houses, farm facilities and produce, such as hens' eggs, and cause irritation and annoyance to employees.

5. ENTOMOPHOBIA AND DELUSIONAL PARASITOSIS

In addition to the harm that insects can do to humans and their livestock by stinging, biting and transmitting pathogens, there is the very real but irrational response that some individuals can have to the presence of insects, called 'entomophobia'. It is a response out of all proportion to the real or imagined danger involved, often induced by spiders, bees or wasps, and psychological treatment might be required to desensitize the sufferer.

In comparison, 'delusionary parasitosis' is a much more difficult issue to deal with because the insects and their effects are imaginary. This condition is synonymous with Ekbom Syndrome – someone believing that his or her body is infested by small creatures that are invisible or that only he or she can see. There is some overlap with what is known as Morgellons (also known as 'fibre disease'); a similar delusional condition in which individuals believe they have fibres emerging from their skin. The person with delusional parasitosis experiences tactile (itching, crawling, prickling) sensations and often 'sees' little 'bugs' coming out of their skin but cannot manage to collect them. The condition is often characterized by the fact that the 'bugs' appear to change shape from time to time. The sufferers collect myriad specimens from themselves (often with serious excoriations and lacerations of their skin) and their personal environment and refer them to a physician or entomologist, convinced they have the causative 'bugs'; typically, though, there is nothing in the submitted samples but skin and scabs, lint and household debris, and the person refuses to believe that the cause cannot be found to be a live 'bug' of some sort. Overall, there can be a range of possible causes (apart from an actual insect infestation) that include physical, physiological and psychological factors, but psychiatric counselling is usually required to eradicate the truly imaginary infestations. There are reports that early psychiatric intervention can be successful, but many (possibly most) sufferers refuse to believe their condition is psychological and cannot be convinced to consult a psychologist or psychiatrist. Also, while several neuropharmacologic drugs appear to be useful for treatment, long-term maintenance often appears to be required. There is little that the medical entomologist can do to resolve the situation for sufferers of delusional parasitosis, other than try to convince them that the cause of their problem is not entomological and that they should seek psychotherapy assistance as the delusion almost never resolves spontaneously. Despite the fact that as a delusional disorder it is not well understood by the psychological community, there appears to be a good outcome from an early psychological intervention. From the perspective of the medical entomologist, it has been said that although delusional parasitosis is not an entomological problem, it is most certainly a problem for entomologists.

6. FORENSIC ENTOMOLOGY

Medicolegal forensic entomology is now a recognized discipline concerned with the application of the study of insects and other arthropods to legal issues. It has three categories – urban, stored products and medicolegal, and it is the last which is related to medical (and in some cases veterinary) entomology. The most common application is in determining the minimum time since death (post-mortem interval), but also it may be used to detect cases of neglect or the movement of a corpse, manner of death and other aspects of forensic investigations (e.g. the detection of illegal drugs taken up by the individual before death).

Insects occurring in carrion can be assigned to one of four categories. There are those that feed on the carrion, of which blow flies are usually the first to appear, arriving within a few hours of death. They are followed by predators and parasites of the necrophagous species, omnivorous species, which feed on both the corpse and its inhabitants, and adventive species, which use the corpse as an extension of their environment. As a result of seepage, changes also occur in the fauna under the corpse.

The entomologist has to deduce, from the arthropods present on a corpse, the time at which the insect attack began. The arthropod specimens must be identified to species and stage of development. The time taken to reach a particular stage is temperature dependent, but allowance must be made for the fact that the temperature in the cadaver will be higher than the air temperature due to the production of metabolic heat. Account must be taken of the location of the cadaver; for example, whether it has been exposed to the sun or sheltered from the wind. Time for the development of a species must be known in degree-hours above a threshold temperature. Then, it should be possible to calculate when the first insect eggs were laid. However, account must be taken of the variations in the composition of the local insect populations. To confirm findings, it may be necessary to simulate conditions using a freshly killed pig as a surrogate human cadaver. However, insect attack will be delayed when a corpse has been covered, hidden in an enclosed space, burned or covered by soil. A covering of 2.5 cm of soil is enough to exclude calliphorid blow flies from a corpse, but the muscid fly, *Muscina stabulans*, lays its eggs on the surface and the larvae are able to penetrate soil to a depth of 10 cm in search of carrion. Decomposition of a corpse in water will be slow because of loss of heat to the surrounding medium. Indications that a corpse has been moved from the place of death may be given by the fauna of the corpse differing from that of the locality in which it has been found, and the fauna below the corpse being less advanced than would be expected from the stage of decomposition. Further, insects can assist in determining the cause of death by the accumulation in the insect of drugs, heavy metals and toxic compounds, or their degradation products.

7. EMERGING ISSUES

During the last decades of the 20th century and the early years of the 21st century, a global emergence and resurgence of many insect vectors and vector-borne diseases has been evident in human and animal populations. There has been a global resurgence in bed bug populations. North America has seen the importation of and subsequent rapid spread of West Nile virus disease, where birds are the primary reservoir. Europe has witnessed unexpected outbreaks of bluetongue virus in sheep in northern countries and the appearance of a previously unknown arbovirus in livestock called Schmallenberg virus. There has also been a continued spread and establishment of the exotic mosquito, *Aedes albopictus* (an Asian vector of dengue fever and other viruses), throughout the world, including Europe (facilitated by the global trade in used tyres, which act as carriers of mosquito eggs), and the emergence of chikungunya virus in Italy following outbreaks in islands off East Africa and subsequently India. Crimean-Congo haemorrhagic fever virus and its *Hyalomma* tick vector have also emerged in parts of Turkey and for the first time caused clinical disease in humans in both Turkey and Greece. Usutu virus has emerged in Austria, Hungary and Spain. Outbreaks of West Nile virus continue to appear in France, Hungary and Romania and the virus has emerged in Italy and Greece. Tick-borne encephalitis has been found recently for the first time in mountainous regions of the Czech Republic and has expanded its range in Scandinavia, and many new tick-transmitted rickettsial pathogens have been identified of concern to both human and veterinary health. Such changes are the result of a complex interaction of factors, each of which may carry a different weight and play a different role under specific local circumstances.

Change in social and economic conditions is often an important key influence on disease incidence patterns. High levels of population growth, along with economic collapse or civil unrest, may often result in major movements of people and their animals, primarily to urban centres. Unplanned urbanization, in combination with inadequate housing and poor-quality water, sewage and waste management systems produce conditions in which high densities of hosts and poor animal and public health allow for increased transmission of vector-borne diseases in and between human and animal populations.

The increased movement of goods, humans and companion animals worldwide is also an important mechanism for the introduction of exotic vectors and disease agents. Container shipping and air transport have had major impacts on the rate of vector introduction. However, while novel vector introductions probably occur relatively frequently, for this to be of significance the introduced vectors must either be carrying pathogens or find a source of endemic pathogens at their new location. They must also be able to survive long enough to become infectious and be able to disperse and breed under the climatic conditions of their new location. This chain of probabilities is usually sufficiently low that, in most instances, introduced arthropod vectors quickly die out or are identified and eliminated. The alternative route is for the introduction of companion animals infected with pathogens for which endemic arthropods are competent vectors. Again, this is a relatively rare phenomenon. There are notable exceptions, some mentioned above, such as the introduction and spread of the invasive mosquito *Ae. albopictus* throughout most of the Mediterranean region, the incursion of bluetongue into northern Europe from sub-Saharan Africa and the introduction of West Nile virus into New York in 1999. In the case of invasive mosquitoes, their arrival and establishment in Italy, coupled with the international travel of humans infected with tropical pathogens, led to the first documented outbreak of chikungunya virus in humans in Italy in 2007. In 2010, these same invasive mosquitoes were responsible for autochthonous cases of dengue in humans. There is also clear evidence now that *Ae. albopictus* acts as an effective vector of canine dirofilariosis in Italy and is contributing to the re-emergence of this disease in dogs and the emergence of human dirofilariasis.

Habitat change may have a significant impact; large-scale irrigation and flood prevention systems for example, such as the south-eastern Anatolia irrigation project, have been built in the past 50 years without regard to their effect on vector-borne diseases. In parts of Europe, there are large-scale plans for wetland creation to mitigate the effects of climate change (i.e. adaptation) on biodiversity loss, as well as coastal wetland creation to mitigate sea-level rise. The potential exists, therefore, for new aquatic habitats for mosquitoes, often in locations where land previously had been drained for agriculture. Ironically, some of these sites were the main mosquito and malaria grounds of the 19th century. Conversely, tropical forests are being cleared at an increasing rate, and conversion of land to intensive agricultural production, such as rice, has also increased – increasing breeding sites for mosquito vectors, as well as increasing the exposure rate between the mosquitoes and reservoir hosts for pathogens. Movement and importation of livestock have been important in contributing to and initiating arbovirus outbreaks such as Crimean-Congo haemorrhagic fever and Rift Valley fever. On the other hand, in some areas of the world, initiatives are being undertaken to promote biodiversity enhancement, to increase woodland management for biodiversity and to link fragments of existing biodiverse-rich habitat under the banner of 'habitat connectivity'. These developments have considerable significance for arthropod-borne disease, since they increase the interaction between humans and domestic and wild animals, such as the deer which support adult tick populations and the rodents which are important reservoirs of a range of disease pathogens, such as lyme borreliosis and babesiosis.

Considerable concerns are associated with the reliance on insecticides and chemotherapy, since this may lead to problems such as the development of resistance. Growing insecticide and drug resistance, and the slowed rate of new product development, make it harder to combat vectors and vector-borne diseases. In addition, a change in public health emphasis in some parts of the world from proactive 'disease surveillance and prevention' to reactive 'response and treatment' makes outbreaks more likely.

Finally, climate change may well promote the emergence of vector-borne disease, at least in some regions. Undoubtedly, increased or decreased temperature and rainfall at certain times of the year will impact invertebrate vectors and the pathogens they might transmit, since the rates of physiological processes in arthropods are highly dependent on ambient temperature. Hence, viewed at least superficially, long-term changes in climate would be expected to have a direct effect on their distribution and abundance, in addition to their ability to mature pathogen infection.

SELECTED BIBLIOGRAPHY

Alexander, J.O'D. (1984) *Arthropods and Human Skin*. Springer-Verlag, Berlin.

Arlian, L.G. (2002) Arthropod allergens and human health. *Annual Review of Entomology* 47, 395–433.

Battisti, A., Holm, G., Bengt, F. and Larsson, S. (2011) Urticating hairs in arthropods: their nature and medical significance. *Annual Review of Entomology* 56, 203–220.

Beaty, B.J. and Marquadt, W.C. (1996) *The Biology of Disease Vectors*. University Press of Colorado, Niwot, Colorado.

Bettini, S. (ed) (1978) *Arthropod Venoms*. Springer-Verlag, Berlin.

Blazar, J.M., Lienau, E.K. and Allard, M.W. (2011) Insects as vectors of foodborne pathogenic bacteria. *Terrestrial Arthropod Reviews* 4, 5–16.

Brouqui, P. (2011) Arthropod-borne diseases associated with political and social disorder. *Annual Review of Entomology* 56, 357–374.

Burgess, N.R.H. and Cowan, G.O. (1993) *A Colour Atlas of Medical Entomology*. Chapman and Hall, London.

Burns, D.A. (2009) Diseases caused by arthropods and other noxious animals. In: Burns, D.A., Breathnach, S.M. and Griffiths, C.E.M. (eds) *Rook's Textbook of Dermatology*. Blackwell Publishing, London, pp. 38.1–38.61.

Byrd, J.H. and Castner, J.L. (eds) (2001) *Forensic Entomology*. CRC Press, Boca Raton, Florida.

Carn, V.M. (1996) The role of dipterous insects in the mechanical transmission of animal viruses. *British Veterinary Journal* 152, 377–393.

Catts, E.P. and Goff, M.L. (1992) Forensic entomology in criminal investigations. *Annual Review of Entomology* 37, 253–272.

Catts, E.P. and Haskell, N.H. (eds) (1990) *Entomology and Death: A Procedural Guide*. Joyce's Print Shop, Clemson, South Carolina.

Colwell, D.D., Dantas-Torres, F. and Otranto, D. (2011) Vector-borne parasitic zoonoses: emerging scenarios and new perspectives. *Veterinary Parasitology* 24(182), 14–21.

Ebeling, W. (1975) *Urban Entomology*. University of Califiornia Press, Berkeley, California.

Eldridge, B.F. and Edman, J.D. (eds) (2000) *Medical Entomology: A Textbook on Public Health and Veterinary Problems Caused by Arthropods*. Kluwer Academic, Dordrecht/Norwell, Massachusetts.

Fallis, A.M. (1980) Arthropods as pests and vectors of disease. *Veterinary Parasitology* 6, 47–73.

Francesconi, F. and Lupi, O. (2012) Myiasis. *Clinical Microbiology Reviews* 25, 79–105.

Geary, M.J., Smith, A. and Russell, R.C. (2009) Maggots Down Under. *Wound Practice and Research* 17, 36–42.

Goddard, J. (2000) *Physician's Guide to Arthropods of Medical Importance*, 3rd edn. CRC Press, Boca Raton, Florida.

Gratz, N.G. (1999) Emerging and resurging vector-borne diseases. *Annual Review of Entomology* 44, 51–75.

Gratz, N.G. (2006) *The Vector- and Rodent-borne Diseases of Europe and North America: Their Distribution and Public Health Burden*. Cambridge University Press, Cambridge, UK.

Gubler, D.J. (1998) Resurgent vector-borne diseases as a global health problem. *Emerging Infectious Diseases* 4, 442–450.

Hardy, T.N. (1988) Entomophobia: the case for Miss Muffet. *Bulletin of the Entomological Society of America* 34, 64–69.

Harwood, R.F. and James, M.T. (1979) *Entomology in Human and Animal Health*, 7th edn. Macmillan Co, New York.

Hinkle, N.C. (2010) Ekbom Syndrome: the challenge of the 'invisible bug' infestations. *Annual Review of Entomology* 55, 77–94.

Hoffman, D.R. (1987) Allergy to biting insects. *Clinical Reviews of Allergy* 5, 177–190.

Kettle, D.S. (1995) *Medical and Veterinary Entomology*, 2nd edn. CAB International, Wallingford, UK.

King, T.P. and Monsalve, R.I. (2008) Allergens from stinging insects: ants, bees, and vespids. In: Kay, A.B., Kaplan, A.P., Bousquet, J. and Holt, P.G. (eds) *Allergy and Allergic Diseases. Volume 1: The Scientific Basis of Allergy*. Wiley-Blackwell, Oxford, UK, pp. 1123–1130.

Klotz, J.H., Klotz, S.A. and Pinnas, J.L. (2009) Animal bites and stings with anaphylactic potential. *The Journal of Emergency Medicine* 36, 148–156.

Lane, R.P. and Crosskey, R.W. (eds) (1993) *Insects and Acarines of Medical Importance*. Chapman and Hall, London.

Lehane, M.J. (1991) *Biology of Blood-sucking Insects*. Harper Collins Academic, London.

Monath, T.P. (ed.) (1988) *The Arboviruses: Ecology and Epidemiology*,Vols I–IV. CRC Press, Boca Raton, Florida.

Monath, T.P. (ed) (1989) *The Arboviruses: Ecology and Epidemiology*, Vol V. CRC Press, Boca Raton, Florida.

Müller, U.R., Haeberli, G. and Helbling, A. (2008) Allergic reactions to stinging and biting insects. In: Rich, R.R., Fleisher, T.A., Shearer, W.T., Schroeder, H.W., Frew, A.J. and Weyand, C.M. (eds) *Clinical Immunology*, 3rd edn. Elsevier, London, pp. 657–666.

Natural History Museum (2012) Introduction to myiasis (http://www.nhm.ac.uk/research-curation/research/projects/myiasis-larvae/intro-myiasis/index.html, accessed 1 February 2012).

Pearson, M.L., Selby, J.V., Katz, K.A., Cantrell, V., Braden, C.R., Parise, M.E., *et al.* (2012) Clinical, epidemiological, histopathologic and molecular features of an unexplained dermopathy. *PLoS ONE* 7(1), e29908.

Peters, W. (1992) *A Colour Atlas of Arthropods in Clinical Medicine*. Wolfe Publishing Ltd, London.

Reiter, P. (2001) Climate change and mosquito-borne disease. *Environmental Health Perspectives* 109, 141–161.

Robinson, W.H. (1996) *Urban Entomology: Insect and Mite Pests in the Human Environment*. Chapman and Hall, London.

Rogers, D.J. and Randolph, S.E. (2006) Climate change and vector-borne diseases. *Advances in Parasitology* 62, 345–381.

Russell, R.C. (2001) The medical significance of Acari in Australia. In: Halliday, R.B., Walter, D.E., Proctor, H.C., Norton, R.A. and Colloff, M.J. (eds) *Proceedings of the 10th International Congress of Acarology*. CSIRO Publishing, Melbourne, pp. 535–546.

Russell, R.C. (2009) Mosquito-borne disease and climate change in Australia: time for a reality check. *Australian Journal of Entomology* 48, 1–7.

Sellers, R.F. (1980) Weather, host and vector – their interplay in the spread of insect borne animal virus diseases. *Journal of Hygiene* 81, 65–102.

Service, M.W. (1980) *A Guide to Medical Entomology*. Macmillan Press, London.

Service, M.W. (ed.) (2001) *Encyclopedia of Arthropod-transmitted Infections*. CAB International, Wallingford, UK.

Smith, K.G.V. (1986) *A Manual of Forensic Entomology*. Trustees of the British Museum (Natural History), London.

Steelman, D.C. (1976) Effects of external and internal arthropod parasites on domestic livestock production. *Annual Review of Entomology* 21, 155–178.

Steen, C.J., Carbonaro, P.A. and Schwartz, R.A. (2004) Arthropods in dermatology. *Journal of the American Academy of Dermatology* 50, 819–842.

Walker, A. (1994) *The Arthropods of Humans and Domestic Animals, A Guide to Preliminary Identification*. Chapman and Hall, London.

Wall, R. and Shearer, D. (1997) *Veterinary Entomology*. Chapman and Hall, London.

Wall, R. and Shearer, D. (2000) *Veterinary Ectoparasites: Biology, Pathology and Control*, 2nd edn. Blackwell Science Ltd, Abingdon, UK.

Webber, R. (2009) *Communicable Disease Epidemiology and Control*, 3rd edn. CAB International, Wallingford, UK.

Wirtz, R.A. (1984) Allergic and toxic reactions to non-stinging arthropods. *Annual Review of Entomology* 29, 47–69.

Wykoff, R.F. (1987) Delusions of parasitosis: a review. *Reviews of Infectious Diseases* 9, 433–437.

Ants (Hymenoptera: Formicidae)

1. INTRODUCTION

The ants are members of the order Hymenoptera (which also contains bees and wasps). They are endopterygote, mandibulate insects in which the forewings (when present) are larger than the hindwings and both fore- and hindwings are coupled mechanically to function as a single entity. They exist throughout most of the world. Some species can inflict serious stings and bites, usually in defence of their colony, but often very aggressively, and can inflict pain and cause allergic reactions (often severe in sensitized persons). Some ants act as intermediate hosts for liver flukes and tapeworms affecting wild and domestic animals (and, at least occasionally, humans).

2. TAXONOMY

It has been estimated there may be more than 15,000 ant species worldwide, with at least 11 subfamilies. A number of ant subfamilies are responsible for health-related problems, but species of the Dorylinae (driver ants and safari ants), Ecitoninae (army ants and legionary ants), Myrmeciinae (bulldog ants), Myrmicinae (including harvester ants, fire ants, pharaoh ants), Ponerinae (Ponerine ants) and Pseudomyrmecinae (acacia ants) are generally the most important.

3. MORPHOLOGY

Typically, ants have a narrow petiole of one or two segments, with a dorsal lobe at the beginning of the abdomen. However, many ant species have a caste system of queens, workers (sterile females) and males, with morphological variation between the castes (which complicates identification) and wings present only in the reproductive forms.

4. LIFE CYCLE

The life cycle of different ants varies greatly. Development is holometabolous, with egg, larval, pupal and adult stages, often all associated with colonies, but there is great variation in colony structure and behaviour.

5. BEHAVIOUR AND BIONOMICS

Colonies often consist of one or more queens 'surrounded' by many workers. Reproductive males and females may be produced only once a year in temperate regions; these are winged and they usually leave the colony for mating, with the males typically dying soon after the mating season and the inseminated queen losing its wings and founding a new nest site. Colonies of many ant species are perennial rather than annual establishments, although inseminated queens will often start a new colony each spring.

Nesting by ants can be found in a wide variety of situations, from aerial nests in trees, to surface mounds, to subterranean excavations (specific examples below), but the behaviour of greatest significance in this context is that of potential contact with ants and resultant stinging. Pheromones are often involved in alerting members of ant colonies to defence reactions. Ant stinging often involves multiple plunges of the stinging apparatus (modified ovipositor) into the skin, with the venom glands contracting sim- ultaneously and pumping venom down the channel of the apparatus. With a few ants (e.g. harvester ants, *Pogonomyrmex* spp.), the stinging results in the apparatus and venom sac remaining in the wound when the insect leaves; thus, the ant is eviscerated and will die soon after stinging.

5.1 Ants of particular importance

5.1.1 Green tree (weaver) ants (*Oecophylla* spp.)

Two species are of note: *Oecophylla smaragdina* is found in tropical areas of South-east Asia and Australia and *Oecophylla longinoda* is in sub-Saharan tropical Africa. These usually pale green or orange-brown ants (up to 10 mm long) live in tree nests, often above head height, composed of leaves drawn together by worker ants and bound with silk produced by larvae held by other workers. There can be colonies of nests, even more than a hundred, spanning adjacent groups of trees and containing many hundreds of thousands of workers. These ants are highly territorial and if their nest is disturbed or threatened, they attack *en masse*; they do not sting but inflict wounds with their mandibles and spray venom (formic acid) from their terminal abdominal glands into the wounds, resulting in a burning sensation.

5.1.2 Army ants (legionary ants, safari ants)

This name is applied to many species (e.g. *Dorylus laevigatus*, *Dorylus molestus*, *Eciton hamatum*, *Eciton burchelli*) of ants in different subfamilies characterized by the same basic pattern of aggressive predatory behaviour. Unlike other ants, army ants do not construct permanent nests and the colony (with tens of millions of individuals) moves continuously. Huge numbers of ants forage simultaneously over a certain area and attack prey (virtually anything, including small vertebrates) en masse. Because there are various species included under the name, few features are discussed in detail here. Some individuals can be as large as 12 mm in length, with the mandibles being approximately one-third of the overall length. Although some species are capable of stinging, most rely on their large powerful mandibles.

5.1.3 Jumper and bull ants

These include different species (e.g. *Myrmecia pilosula*, *Myrmecia pyriformis* and other *Myrmecia* spp.) of typically large ants (up to 3 cm or more in length, with the smallest species >6 mm). They can be a major problem in some areas of Australia. They nest in soil, often with a low mound covered in small pebbles, and colonies are relatively small (mostly hundreds to a few thousand workers). They feed on nectars, plant juices and small prey, can run rapidly and are very aggressive; they have long mandibles (up to 5 mm in length) and a potent sting that can cause severe pain and life-threatening allergic reactions in sensitive individuals.

5.1.4 Fire ants

These ants, belonging to the genus *Solenopsis*, are a particular problem in parts of the USA; however, they have become established elsewhere as well. The workers are approximately 2–5 mm long and are soil nesters in rural, urban and suburban localities, and they are opportunistic feeders on soil arthropods and seeds. Their established colonies often contain >100,000 individuals and they respond quickly to disturbances at or near their nest, swarming over intruders and biting and stinging *en masse* in response to alarm pheromones. Their sting results in a burning sensation (hence the name) and the most important species are *Solenopsis invicta* and *Solenopsis richteri*, both introduced from South America and known as 'imported fire ants'.

5.1.5 European fire ants

Found across Europe from the UK to Central Asia, these ants (e.g. *Myrmica rubra* and *Myrmica ruginodes*) are 4–6 mm long and yellowish- to reddish-brown in colour, with a slightly darker pigmentation on the head. They live under stones, fallen trees and in soil in meadows and gardens, feeding on honeydew from aphids and preying on many species of insects and other invertebrates. Although they are not as aggressive as the imported fire ant *Solenopsis* species mentioned above, they will attack and sting anything that disturbs their nest.

5.1.6 Harvester ants

These ants of the genus *Pogonomyrmex* are relatively large (5–10 mm) and are a particular problem in the USA, where a number of species are known for their aggressiveness and stinging, which causes intense pain. They build their nests in dry sandy to harder soils and with a characteristic crater in the centre of a low mound. The nests can be up to 10 m in diameter and colonies may contain up to 10,000 workers.

5.1.7 Pharaoh ants

This ant (*Monomorium pharaonsis*) is a tiny species (~2 mm long) that can be found throughout most of the world. It nests in a wide range of habitats, both inside and outside buildings. It is omnivorous and its colonies can comprise >1,000,000 individuals. Capable of stinging, it is also of concern because it has been implicated in the mechanical transmission of pathogens within hospitals, where it has been found infesting surgical dressings and other contaminated materials and biting new-born babies.

6. MEDICAL AND VETERINARY IMPORTANCE

Most ants possess a sting and their venoms can contain chemicals that can cause pain and, occasionally, more serious systemic effects. Many species which do not possess stings squirt defensive secretions containing pain-causing chemicals (e.g. formic acid) from the end of the abdomen on to wounds made by their biting mouthparts. Venoms comprise mostly mixtures of simple proteinaceous organic compounds (active allergenic antigens) including enzymes (lipases, esterases and proteases that cause pain). Harvester ants have high amounts of hyaluronidase, haemolysins and histamines; bull ant and jumper ant venoms are rich in phospholipases, hyaluronidase, phosphatases and histamine, and fire ant venom is high in piperidine alkaloids.

While the sting can be painful, the main danger lies in an allergic response stimulated by venom components, which may lead to anaphylactic shock and rapid collapse of the person stung. While the fear engendered by stinging insects is out of all proportion to the risks involved, as it has been estimated that the annual mortality from hymenopteran stings throughout the world is from one to six deaths per 10 million people, death can occur in sensitized individuals within an hour of being stung, with respiratory failure brought on by anaphylactic shock.

Fire ants cause a burning sensation, with some swelling, and a characteristic fluid-filled vesicle, which develops to a pustule, usually forms at the sting site. Most ant stings are relatively painful, but those of *Paraponera* and *Pogonomyrmex* species are particularly noteworthy. Stings from harvester ants (*Pogonomyrmex* spp.) are extremely intense, can last several hours, spread along the lymph vessels to axillary nodes and cause piloerection and sweating around the sting site. *Paraponera clavata* of Central and South America is reported to have the most painful sting of all hymenoptera – an intense debilitating pain that lasts several hours, with the affected area expanding up to 30 cm in the first hour. Bulldog and jack jumper ants (*Myrmecia* spp.) of Australia sting repeatedly while holding with their mandibles: the venom is highly allergenic and anaphylactic reactions have resulted in deaths.

From a veterinary perspective, various species can cause problems for some wild and domestic animals and fire ants are notorious for attacking smaller, newborn animals and ground-dwelling birds. But, some ants also play a role as intermediate hosts for *Dicrocoelium* flukes and *Raillietina* tapeworms.

Dicrocoelium dendriticum, the 'small liver fluke', infects the bile ducts and gall bladder of domestic and wild ruminants (e.g. sheep, goats, cattle, etc.), but occasionally also dogs, pigs, horses and humans. While infections in the vertebrate host are often asymptomatic or relatively mild, they can lead to weight loss and reduced milk production. It is widely distributed in dry lowland or mountain pastures of the northern hemisphere (Europe, Asia, Africa, North America), and it has been also reported from South America and Australia (although there is no evidence that infestation is endemic in the latter). Many mollusc species have been shown to act as the first intermediate host for the fluke when they ingest the eggs and support the development of the miracidium to the cercarial stage, which is exuded from the snail in slime balls. These are ingested by ants (e.g. *Formica fusca*, *Formica pratensis* and *Formica rufibarbis*), which become the second intermediate host. In the ants, the cercariae develop to metacercariae, which localize in the ant's suboesophageal ganglion and control the movement of the ant to and from the tips of vegetation in the morning and evening (at temperatures of ~15–20°C or below, the metacercariae cause a cateleptic cramp that locks the ant's jaws on to the vegetation), thus making the ant more available for ingestion when the ruminants are grazing. In addition, this phenomenon protects the ants from the higher temperatures of the midday hours (when the jaws are released and the ant descends the vegetation for protection from the sun).

Various species of *Raillietina* tapeworms infest vertebrates, mostly birds and rarely humans, with *Raillietina echinobothrida* being an important species in terms of prevalence and pathogenicity among domestic birds, particularly the domestic fowl (in which the tapeworm is responsible for stunted growth, emaciation and decreased egg production). The adult tapeworm lives in the intestine of birds, which are the definitive hosts, and ants (particularly species of *Tetramorium*) are the intermediate host. Eggs are passed with the bird's faeces and immature stages are ingested by the ants, entering the alimentary canal and migrating into the abdominal cavity, where they develop into cysticercoids, which infect the birds that ingest the ants.

6.1 Treatments

Cold packs can provide relief and baking soda and meat tenderizer have been recommended to neutralize and denature the venom com-ponents. Antihistamines and topical corticosteroids may be helpful in relieving swelling and other allergic reactions. If there is a serious allergic or other systemic reaction (e.g. anaphylaxis), medical assistance should be sought immediately; an injection of adrenaline (epinephrine) is the usual initial response. It is recommended that people with known hypersensitivity should carry antihistamines and an adrenalin syringe; such individuals can undergo desensitization therapy. Immunotherapy against fire ant stings is common in the USA, where the ants are abundant.

For treating dicrocoeliosis, various benzimidazole and pro-benzimidazole derivatives, and other fluke treatments, have been used with various degrees of effectiveness in animals, while praziquantel has been successful in treating human infections. *Raillietina* infestations are often treated with albendazole.

7. PREVENTION AND CONTROL

Typically, avoiding contact with ants is the best way to prevent envenomation from stings. If required, insecticidal treatments of nests should be administered at night, when the adults are least active and all individuals are in the nest. Although direct application of an insecticide to a nest can aid control of ants, professional programmes usually involve the use of baits (appropriate to the target species) containing a slow-acting insecticide that spreads throughout the colony in the nest.

Prevention of dicrocoeliosis involves animal husbandry management practices relating to grazing times; administering molluscicides or insecticides is usually not feasible on a large scale. Habitat management against the intermediate host is the most appropriate method for reducing risk of infestation with *Raillietina* tapeworms in domestic poultry, although insecticides may be useful in some circumstances.

SELECTED BIBLIOGRAPHY

Alexander, J.O'D. (1984) *Arthropods and Human Skin*. Springer-Verlag, Berlin, pp. 150–158.

Arlian, L.G. (2002) Arthropod allergens and human health. *Annual Review of Entomology* 47, 95–433.

Bettini, S. (ed.) (1978) *Arthropod Venoms*. Springer-Verlag, Berlin.

Blum, M.S. (1984) Poisonous ants and their venoms. In: Tu, A.T. (ed.) *Insect Poison, Allergens, and Other Invertebrate Venoms. Handbook of Natural Toxins*,Vol 2, Chapter 7. Dekker, New York, pp. 225–242.

Bolton, B. (1994) *Identification Guide to the Ant Genera of the World*. Harvard University Press, Cambridge, Massachusetts.

Brown, S.G.A., van Eeden, P., Wiese, M.D., Mullins, R.J., Solley, G.O., Puy, R., *et al*. (2011) Causes of ant sting anaphylaxis in Australia: the Australian Ant Venom Allergy Study. *Medical Journal of Australia* 195, 69–73.

Camazine, S. (1988) Hymenopteran stings: reactions, mechanisms, and medical treatment. *Bulletin of the Entomological Society of America* 34, 17–21.

Charpin, D., Birnbaum, J. and Vervolet, D. (1994) Epidemiology of hymenopteran allergy. *Clinical and Experimental Allergy* 24, 1010–1015.

Friedman, L.S., Modi, P., Liang, S. and Hryhorczuk, D. (2010) Analysis of Hymenoptera stings reported to the Illinois Poison Center. *Journal of Medical Entomology* 47, 907–912.

Gauld, I.D. and Bolton, B. (1988) *The Hymenoptera*. Oxford University, Press, London.

Goulet, H. and Huber, J.T. (1993) *Hymenoptera of the World: An Identification Guide to Families*. Research Branch, Agriculture Canada, Ottawa.

Holldobler, B. and Wilson, E.O. (1990) *The Ants*. Belknap/Harvard University Press, Cambridge, Massachusetts.

King, T.P. and Monsalve, R.I. (2008) Allergens from stinging insects: ants, bees, and vespids. In: Kay, A.B., Kaplan, A.P., Bousquet, J. and Holt, P.G. (eds) *Allergy and Allergic Diseases. Volume 1: The Scientific Basis of Allergy*. Wiley-Blackwell, Oxford, UK, pp. 1123–1130.

McGain, F. and Winkel, K.D. (2002) Ant sting mortality in Australia. *Toxicon* 40, 1095–1100.

Meier, J. (1995) Biology and distribution of hymenopterans of medical importance, their venom apparatus and venom composition. In: Meier, J. and White, J. (eds) *Handbook of Clinical Toxicology of Animal Venoms and Poisons*, Chapter 21. CRC Press, Boca Raton, Florida, pp. 331–348.

Moffitt, J.E., Barker, J.R. and Stafford, C.T. (1997) Management of imported fire ant allergy: results of a survey. *Annals of Allergy and Asthma Immunology* 79, 125–130.

Müller, U.R., Haeberli, G. and Helbling, A. (2008) Allergic reactions to stinging and biting insects. In: Rich, R.R., Fleisher, T.A., Shearer, W.T., Schroeder, H.W., Frew, A.J. and Weyand, C.M. (eds) *Clinical Immunology*, 3rd edn. Elsevier, Amsterdam, pp. 657–666.

Otranto, D. and Traversa, D. (2002) A review of dicrocoeliosis of ruminants including recent advances in the diagnosis and treatment. *Veterinary Parasitology* 107, 317–335.

Prahlow, J.A. and Barnard, J.J. (1998) Fatal anaphylaxis due to fire ant stings. *American Journal of Forensic Medicine and Pathology* 19, 137–142.

Piek, T. (1986) *Venoms of the Hymenoptera, Biochemical, Pharmacological and Behavioural Aspects*. Academic Press, Orlando, Florida.

Reisman, R.E. (1994a) Insect stings. *New England Journal of Medicine* 331, 523–527.

Reisman, R.E. (1994b) Venom hypersensitivity. *Journal of Allergy and Clinical Immunology* 94, 651–658.

Bed Bugs (Hemiptera: Cimicidae)

1. INTRODUCTION

Bed bugs are ectoparasites or blood seekers ranked within the Order Hemiptera, Family Cimicidae. Species of two families of Hemiptera, the Cimicidae and Polyctenidae, and of the Subfamily Triatominae of the Family Reduviidae (assassin bugs) are bloodsucking. The Polyctenidae are small ectoparasites of bats of no medical importance. The Triatomidae are dealt with elsewhere under the entry 'Kissing Bugs'. The Cimicidae are blood-feeding, temporary ectoparasites of birds and mammals. Species ranked within the Cimicidae are widespread in the northern hemisphere. No bird-feeding cimicids occur in tropical Africa or Central America and no native cimicids occur in Australia.

The 21 species of *Cimex* are mainly parasites of bats, with one species associated with birds. The bed bugs *Cimex lectularius* and *C. hemipterus* feed principally on humans. While *C. lectularius* is a cosmopolitan species of temperate and subtropical regions feeding on humans, bats, chickens and other domestic animals, *C. hemipterus* is tropicopolitan and subtropical and feeds on humans and chickens, and rarely bats. *Leptocimex boueti* is a parasite of bats and humans in West Africa.

Bed bugs, once rife all over the world, had been made rare in many countries by the domestic (and commercial) use of chlorinated hydrocarbon pesticides such as DDT from the 1950s. After a dramatic global resurgence (starting in the 1990s), both *C. lectularius* and *C. hemipterus* are now distributed around the world once again, and they are becoming particularly abundant in poor socio-economic settings. Increasing insecticide resistance, international travel, the lack of availability of highly residual insecticides (modern insecticides such as the pyrethroids that have an excito-repellent effect can result in insect dispersal and poor pest management) are thought to be key factors responsible for the growth and spread of bed bug populations.

2. TAXONOMY

The 91 species within the Cimicidae are arranged in 23 genera, most of which are parasites of bats or birds. Twelve genera are found only in the New World, nine only in the Old World and two, *Cimex* and *Oeciacus*, in both.

The genus *Cimex* is associated with both mammals and birds, and the two main species that bite humans are the common bed bug, *C. lectularius*, and the tropical bed bug, *C. hemipterus*. These species are mostly allopatric in their geographical distributions, but where they are sympatric, cross-mating occurs. Male *C. hemipterus* and female *C. lectularius* mate as readily with each other as they do with their conspecific partners; however, the few eggs produced from such matings are infertile. The reverse mating (male *C. lectularius* and female *C. hemipterus*) occurs readily in the laboratory and results in a high proportion of abnormal eggs; however, occasional hybrids are produced.

3. MORPHOLOGY

The two bed bug species will be considered together and specific differences indicated. They have oval, flattened bodies and, at first sight appear to be wingless but they are micropterous, with the forewings reduced to hemelytral pads and the hindwings absent (Fig. 1). The adult bed bug measures 5–7 mm when unfed, with females being slightly larger

© R.C. Russell 2013. *The Encyclopedia of Medical and Veterinary Entomology* (R.C. Russell *et al.*)

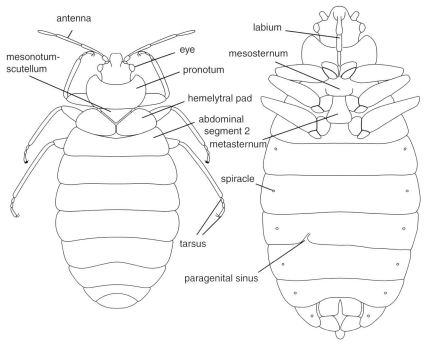

Fig. 1. Female *Cimex lectulariu*s. Dorsal view (left). Ventral view (right). Source: Usinger, 1966.

than males and *C. hemipterus* being about 25% longer than *C. lectularius*. They are generally red-brown in colour, although they appear darker following a blood meal. The distinguishing morphological difference between the two species is the broader prothorax (located behind the head) of *C. lectularius* compared with *C. hemipterus*.

The head bears long four-segmented antennae, of which the last three segments are long and slender, and a pair of widely separated compound eyes, laterally placed at the sides of the head; there are no ocelli. The labium has three obvious segments and is reflected backwards under the head, reaching as far as the coxae of the first pair of legs.

The prothorax is recessed anteriorly and its sides surround the posterior part of the head. In *C. lectularius*, the breadth of the pronotum is more than 2.5 times the length of the prothorax in the midline, and it is less in *C. hemipterus*. The mesonotum-scutellum is triangular in shape, with the base adjoining the pronotum and the apex backwardly directed. Laterally, there are the hemelytral pads. The

tarsus has three segments, the last segment bearing a pair of simple claws with no aerolium. The metasternum is a more or less square, flat plate between the coxae, with rounded posterior corners. The mesosternum is rectangular, being wider than it is long.

The abdomen is 11-segmented, with segments 2–9 being easily recognizable dorsally. When the bed bug engorges, the abdomen increases greatly in volume, thus exposing the intersegmental membranes and expanding the membranous areas in the mid-ventral line of the second to fifth abdominal segments (hunger folds). There are seven pairs of spiracles located ventrally on abdominal segments 2–8. Ventrally, on the right side of the female, there is a notch or paragenital sinus on the posterior margin of the fifth segment. It opens into the ectospermalege. There is no corresponding structure on the left side of the female. The male is similarly asymmetrical and only the left paramere is developed at the posterior end of the abdomen, directed towards the left side (Fig. 2); there is no right paramere.

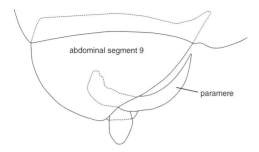

Fig. 2. Ventral view of the terminalia of a male *Cimex lectularius*.

4. LIFE CYCLE

There are some minor differences between the two species, but for *C. lectularius*, eggs are laid on rough rather than smooth surfaces and generally inserted into cracks and crevices. They are laid individually and held in place by transparent cement. The eggs are cream in colour and ~1 mm in length and less than 0.5 mm in breadth. The eggs are fertilized while still in the ovary and the embryos undergo some development before being laid. At 22°C, eggs take 10–12 days to hatch and the shortest time for development is 4–5 days at 30–35°C. Hatching does not occur at 37°C or above, or at temperatures below 13°C, although eggs can remain viable for short periods (<3 months) as the temperature approaches 0°C. In temperate climates, eggs laid in the autumn are likely to have died before the temperature rises above the threshold in spring, except in homes with heating.

There are five juvenile stages, known as nymphs, with each stage requiring at least one blood meal (often two) to moult to the next stage. Nymphs will feed within 24 h of emergence or of moulting to the next instar. When feeding, the bug grasps the skin with its forelegs as a prelude to piercing it. Saliva injected during feeding contains anticoagulants to prevent the blood clotting. Feeding takes 5–10 min and nymphs take in 2–6 times their own body weight of blood. Time for development through the instars is very similar for the first four, but the fifth is usually somewhat longer. The length of the life cycle is

very dependent on ambient temperature. The entire nymphal development can take 6–8 weeks at 22°C, after which the adults can live for up to around 6 months, but at 30°C development from egg to adult can be completed in 3 weeks and overall lifespan is shorter.

5. BEHAVIOUR AND BIONOMICS

Bed bugs are obligate blood feeders and will seek out humans to acquire a blood meal, but they do not live on humans or burrow into their skin. They are nocturnal and their activity peaks before dawn. They are negatively phototactic, which, combined with positive thigmotaxis, ensures that they hide away in cracks and crevices during the day, including under and within the seams of mattresses, bed frames and other furniture, floorboards, paintings and carpets, behind skirting boards, in various cracks and crevices of walls and behind loose wallpaper. Bed bugs are attracted by the body heat and carbon dioxide (and perhaps skin odours) of a host, mainly coming out at night to bite the sleeping victim; temperature receptors are probably located on the basal segments of the antennae.

Adult bed bugs have a scent gland that opens ventrally on the metathorax on to an evaporative area; the main components of the secretion are two aldehydes (octenal and hexenal), which function as alarm pheromones and cause dispersal of aggregated bugs; the receptors for the pheromones are found on the terminal segment of the antenna.

Bed bugs also produce an aggregation pheromone, which brings them together, and thigmotaxis ensures that they stay grouped; the origin of this pheromone is unknown and it does not appear to be a sex pheromone, because the scent of males or females attracts both sexes equally. Juvenile bed bugs respond more strongly to juvenile-produced aggregation pheromones, while adults respond more strongly to adult male-produced pheromones.

The method of insemination in *Cimex* is most unusual. The male climbs on to the back of the female, with his head on the left side of the pronotum and the abdomen tucked under the right side of the female. There are three stages in the insemination of *C. lectularius*. In the spermalege stage, the male penetrates the ectospermalege and injects a mass of sperm into the adjacent mesospermalege. The spermatozoa become mobile in about 30 min and after 3–4 h move into the haemocoel to begin the second phase. The spermatozoa concentrate at the base of the genital apparatus near the junction of the paired and median oviducts. They penetrate into the seminal receptacle, which has no direct communication with the lumen of the oviduct. The spermatozoa move up in the wall of the oviduct to the pedicel of the ovariole and concentrate in the syncitial tissue at the distal end of the ovariole. The egg is fertilized in the ovariole before the chorion is formed. This is the intragenital phase. This traumatic method of insemination allows transfer of nutrient materials from the male to the female, which could be of value in the survival of the species under adverse conditions; however, repeated inseminations can be harmful to the female.

Adult bugs may feed twice a week, and the frequency of oviposition and the number of eggs laid are dependent on the availability of blood meals. At 23°C and 75% RH (relative humidity), newly moulted female *C. lectularius* can take on average twice their own body weight of blood in their first meal and begin to oviposit eggs 5–6 days later for up to 6 days, producing 6–10 eggs. Females kept with males and fed once a week produce on average nearly 3 eggs in the first week, between 7 and 8 eggs in the second and third weeks and more than 8 eggs in the fourth week. Thereafter, egg production can average 6–7 eggs/week for the next 13 weeks, after which it may decline rapidly. In heated premises and with an adequate food supply, a small starting population of bed bugs can develop into several thousand within a year.

The lifespan of a bed bug is usually about 6 months, depending on climate. In longevity experiments, the 1st instar of both species survives for the shortest time. At all temperatures, adult *C. lectularius* tend to survive much longer than adult *C. hemipterus*. There are reports of bed bugs living up to almost 2 years in cold climates without heating; however, at 22°C, such as in heated premises, then the maximum lifespan is 6 months.

Although their own powers of dispersal are relatively limited, bed bugs are carried throughout the world by humans, in their personal (clothing and baggage) or domestic (furniture and furnishings) possessions. However, infestations can spread widely through apartment complexes in a period of several months to years.

6. MEDICAL AND VETERINARY IMPORTANCE

The nocturnal biting of bed bugs can be debilitating to humans when sleep is disturbed. Reactions resulting from the bite of bed bugs can vary greatly, with the severity of irritation often dependent on both the sensitivity of the individual as well as the number of bed bugs. For many, the bites can be painless, but general discomfort may be experienced by others. Allergic reactions can range from the local to systemic. For many people, symptoms appear usually within 7–11 days of the initial bite(s); however, repeated exposure usually decreases the time between bite and the onset of pronounced cutaneous manifestations, such as small clusters of extremely pruritic, erythematous papules. Systemic hypersensitivity reactions have been reported, including anaphylaxis. In contrast, some people can become desensitized with prolonged untreated infestations. Relative to other bloodsucking insects, bed bugs take a very large blood meal and there are reports of humans associated with large infestations suffering from anaemia.

The most commonly affected areas of the body are the exposed arms and shoulders, where bites can often be arranged in an irregular linear pattern. Common allergic reactions (which may appear several days after the bites) include the development of large wheals, often 1–6 cm in diameter, accompanied by itching and inflammation, and usually subsiding to red spots that can last for several days to even months. Bullous eruptions have also been reported and these are thought to be IgE-mediated type 1 immune responses (similar to envenomizing hymenoptera). Anaphylaxis is rare, but may occur in patients with severe allergies and/or when exposed to

RUSSELL, RICHARD C. (RICHARD CHARLES), 1947-

ENCYCLOPEDIA OF MEDICAL AND VETERINARY ENTOMOLOGY.

                    Cloth   429 P.
WALLINGFORD: CABI PUBLISHING, 2013

AUTH: UNIV. OF SYDNEY. ENCYCLOPEDIA W/ B&W &
COLOR ILLUS.
LCCN 2013-6880
  ISBN 1780640374   **Library PO#** FIRM REFERENCE

	List	240.00	USD
5395 NATIONAL UNIVERSITY LIBRAR	**Disc**	14.0%	
App. Date 4/09/14 REFERENCE 8214-15	**Net**	206.40	USD

SUBJ: 1. ARTHROPOD VECTORS--ENCYCL. 2. VETERINARY
ENTOMOLOGY--ENCYCL.

CLASS RA641   DEWEY# 614.4303   LEVEL ADV-AC

YBP Library Services

RUSSELL, RICHARD C. (RICHARD CHARLES), 1947-

ENCYCLOPEDIA OF MEDICAL AND VETERINARY ENTOMOLOGY.

                    Cloth   429 P.
WALLINGFORD: CABI PUBLISHING, 2013

AUTH: UNIV. OF SYDNEY. ENCYCLOPEDIA W/ B&W &
COLOR ILLUS.
LCCN 2013-6880
  ISBN 1780640374   **Library PO#** FIRM REFERENCE

	List	240.00	USD
5395 NATIONAL UNIVERSITY LIBRAR	**Disc**	14.0%	
App. Date 4/09/14 REFERENCE 8214-15	**Net**	206.40	USD

SUBJ: 1. ARTHROPOD VECTORS--ENCYCL. 2. VETERINARY
ENTOMOLOGY--ENCYCL.

CLASS RA641   DEWEY# 614.4303   LEVEL ADV-AC

high densities of bed bugs. There are also rare reports of bed bugs triggering asthmatic reactions similar to that of dust mites.

There is no evidence of any transmission of disease-causing agents to humans. As with the bite of any arthropod, scratching of the site can lead to secondary bacterial infection. Although bed bugs have been suspected in the transmission of many disease organisms of man and bats, in most cases conclusive evidence is lacking or experimental data have demonstrated that bed bugs are incompetent vectors. Hepatitis B surface antigen has been recovered from unengorged *C. hemipterus* and *C. lectularius*. The antigen has been shown to persist in both species for up to 6 weeks after a blood meal and to be excreted in the faeces of *C. hemipterus*, from which it might infect a susceptible person by contamination of skin lesions or mucosal surfaces, or by inhalation of dust. It was found that HIV could survive for 1 h in *C. lectularius*, offering the possibility that mechanical transmission could occur; however, bed bugs are not a major route for the transmission of hepatitis B virus and, therefore, are even less likely to transmit other blood-borne viruses (such as HIV) which are less infectious than HBV. However, although there may not be a real disease concern for their infestations, the physical, emotional and psychological impacts of nuisance biting resulting from heavy infestations may be significant.

7. PREVENTION AND CONTROL

There is no specific treatment available for bed bug bites but care should be taken to avoid secondary infection of the bite site. Most bed bug bite reactions are self-limited and require little specific treatment other than corticosteroid creams and oral/dermal antihistamines for local allergic reactions, and topical antiseptic to minimize or antibiotics to treat for secondary infections.

Measures to protect residential dwellings from infestations of bed bugs require the maintenance of good levels of cleanliness and hygiene, especially minimizing clutter. However, there is a risk of introduction of bed bugs by visitors or homeowners travelling domestically or internationally. Indeed, the two most common routes of bed bug introduction appear to be via baggage and secondhand furniture. Infested clothing can be washed in hot water and dried on the hot cycle of the clothes drier, while delicate materials can be placed into the freezer. On the other hand, many infestations in apartment complexes are derived from adjoining or nearby infested units. Poor insect management processes can lead to the spread of active infestations. It is imperative that pest controllers and accommodation managers are trained in appropriate bed bug eradication options. Industry standards on bed bug management have been developed in Australia, Europe and the USA, and these should be followed not only for advice on how to control active infestations but also on how to minimize potential infestations.

When travelling, care should be taken to inspect the bed thoroughly for signs of bed bugs. Longer-term infestations are often reported to be accompanied by a distinctive 'sweet sickly smell' (a complex mixture of octenal, hexenal and minor amounts of other components), but this tends to be noticeable only if the infestation is extremely heavy or has been disturbed during eradication processes. It can take some months for an infestation to be noticed but the most common evidence of infestation is the presence of blood spotting, bug exuviae (cast skins) or the insects themselves on the bed or bedding – particularly in the beading at the edges of mattresses. Avoiding bed bug infestations will reduce the risk of nuisance biting and transfer of insects. The use of insect repellents or insecticide-treated bed nets have been suggested as possible methods to protect against bites when staying in accommodation suspected of containing bed bugs, although the widespread prevalence of pyrethroid resistance lessens protection from the latter. The effectiveness of repellents at preventing bed bug bites requires further investigation.

Bed bugs are considered one of the most difficult domestic pests to eradicate, due to a high level of insecticide resistance widespread among modern bed bug strains. While synthetic pyrethroids may be the first choice in domestic situations, these may not be effective (because of widespread resistance) and may even repel the bugs. While carbamates and organophosphates may be more effective, they may not be acceptable for treating mattresses because of their odours, and (in addition) they are banned in many countries. The neonicotinoids are now becoming widely employed in bed bug management as a tropical treatment, because bed bugs are not resistant to this class of compounds; however, these products have poor residual effect and need to be applied with other compounds such as diatomaceous earth.

Unlike many arthropod pests where population reduction is usually sufficient to alleviate pest impacts, in the case of bed bugs, complete eradication is the only acceptable objective. A careful inspection must be undertaken in not only the infested room but also all adjoining rooms to identify all likely sources. Insecticides with residual activity are required.

Non-chemical approaches to control depend on the articles to be treated but include the use of vacuuming, freezing, heating and/or steaming (although attempting to heat-kill bugs in bedding by exposing mattresses enclosed in plastic to the sun has been shown to be ineffective); in general, insecticides need to be applied in conjunction with any non-chemical methods. As bed bugs are cryptic in their habits and insecticides do not kill the eggs, complete control is usually not possible with an initial treatment. A post-control treatment evaluation and re-treatment, approximately 10–12 days (dependent on ambient temperatures) following the initial treatment, is essential and, in some cases, more than two evaluations and treatments may be necessary. Scent-trained dogs are used in some places to detect new infestations or to assess the successfulness of treatment.

SELECTED BIBLIOGRAPHY

Araujo, R.N., Costa, F.S., Gontijo, N.F., Goncalves, T.C.M. and Pereira, M.H. (2009) The feeding process of *Cimex lectularius* (Linnaeus 1758) and *Cimex hemipterus* (Fabricius 1803) on different bloodmeal sources. *Journal of Insect Physiology* 55, 1151–1157.

Arlian, L.G. (2002) Arthropod allergens and human health. *Annual Review of Entomology* 47, 395–433.

Benoit, J.B., Lopez-Martinez, G., Teets, N.M., Phillips, S.A. and Denlinger, D.L. (2009) Responses of the bed bug, *Cimex lectularius*, to temperature extremes and dehydration: levels of tolerance, rapid cold hardening and expression of heat shock proteins. *Medical and Veterinary Entomology* 23, 418–425.

Delaunay, P., Blanc, V., Del Giudice, P., Levy-Bencheton, A., Chosidow, O., Marty, P., *et al.* (2011) Bedbugs and infectious diseases. *Clinical Infectious Diseases* 52, 200–210.

Doggett, S.L. (2011) *A Code of Practice for the Control of Bed Bug Infestations in Australia*, 4th edn. Department of Medical Entomology and The Australian Environmental Pest Managers Association, Westmead Hospital, Sydney, Australia.

Doggett, S.L. and Russell, R.C. (2009) Bed bugs: what the GP needs to know. *Australian Family Physician* 38, 880–884.

Doggett, S.L., Dwyer, D.E., Penas, P. and Russell, R.C. (2012) Bed bugs: clinical relevance and control options. *Clinical Microbiological Reviews* 25, 164–192.

Eddy, C. and Jones, S.C. (2011a) Bed bugs, public health, and social justice: Part 1, a call to action. *Journal of Environmental Health* 73, 8–14.

Eddy, E. and Jones, S.C. (2011b) Bed bugs, public health, and social justice: Part 2, an opinion survey. *Journal of Environmental Health* 73, 15–17.

Goddard, J. and de Shazo, R. (2009) Bed bugs (*Cimex lectularius*) and clinical consequences of their bites. *Journal of the American Medical Association* 301, 1358–1366.

Hamann, I.D. (2004) Insect bites and skin infestations. *Medicine Today* 5, 39–46.

How, Y.F. and Lee, C.Y. (2010a) Survey of bed bugs in infested premises in Malaysia and Singapore. *Journal of Vector Ecology* 35, 89–94.

How, Y.F. and Lee, C.Y. (2010b) Fecundity, nymphal development and longevity of field-collected tropical bedbugs, *Cimex hemipterus*. *Medical and Veterinary Entomology* 24, 108–116.

Hwang, S.W., Svoboda, T.J., De Jong, L.J., Kabasele, K.J. and Gogosis, E. (2005) Bed bug infestations in an urban environment. *Emerging Infectious Diseases* 11, 533–538.

Kolb, A., Needham, G.R., Neyman, K.M. and High, W.A. (2009) Bedbugs. *Dermatologic Therapy* 22, 347–352.

Myamba, J., Maxwell, C.A., Asidi, A. and Curtis, C.F. (2002) Pyrethroid resistance in tropical bedbugs, *Cimex hemipterus*, associated with use of treated bednets. *Medical and Veterinary Entomology* 16, 448–451.

Ogston, C.W. and London, W.T. (1980) Excretion of hepatitis B surface antigen by the bedbug *Cimex hemipterus* Fabr. *Transactions of the Royal Society of Tropical Medicine and Hygiene* 74, 823–825.

Pfiester, M., Koehler, P.G. and Pereira, R.M. (2009) Effect of population structure and size on aggregation behavior of *Cimex lectularius* (Hemiptera: Cimicidae). *Journal of Medical Entomology* 46, 1015–1020.

Potter, M.F. (2011) The history of bed bug management. *American Entomologist* 57, 14–25.

Reinhardt, K. and Siva-Jothy, M.T. (2007) Biology of the bed bugs (Cimicidae). *Annual Review of Entomology* 52, 351–374.

Reinhardt, K., Kempke, D., Naylor, R. and Siva-Jothy, M.T. (2009) Sensitivity to bites by the bedbug, *Cimex lectularius*. *Medical and Veterinary Entomology* 23, 163–166.

Richards, L., Boase, C.J., Gezan, S. and Cameron, M.M. (2009) Are bed bug infestations on the increase within greater London? *Journal of Environmental Health Research* 9, 17–24.

Romero, A., Potter, M.F., Potter, D.A. and Haynes, K.F. (2007) Insecticide resistance in the bed bug: a factor in the pest's sudden resurgence? *Journal of Medical Entomology* 44, 175–178.

Romero, A., Potter, M.F. and Haynes, K.F. (2009) Behavioral responses of the bed bug to insecticide residues. *Journal of Medical Entomology* 46, 51–57.

Schofield, C.J. and Dolling, W.R. (1993) Bedbugs and kissing-bugs (blood-sucking Hemiptera). In: Lane, R.P. and Crosskey, R.W. (eds) *Medical Insects and Arachnids*. Chapman and Hall, London, pp. 483–516.

Temu, E.A., Minjas, J.N., Shiff, C.J. and Majala, A. (1999) Bedbug control by permethrin-impregnated bednets in Tanzania. *Medical and Veterinary Entomology* 13, 457–459.

Thomas, I., Kihiczak, C.G. and Schwartz, R.A. (2004) Bedbug bites: a review. *International Journal of Dermatology* 43, 430–433.

Usinger, R.L. (1966) Monograph of Cimicidae (Hemiptera-Heteroptera). *The Thomas Say Foundation* 7, 1–585.

Walpole, D.E. (1988) Cross-mating studies between two species of bedbugs (Hemiptera: Cimicidae) with a description of a marker of interspecific mating. *South Africa Journal of Science* 84, 215–216.

Wang, C., Saltzmann, K., Chin, E., Bennett, G.W. and Gibb, T. (2010) Characteristics of *Cimex lectularius* (Hemiptera: Cimicidae), infestation and dispersal in a high-rise apartment building. *Journal of Economic Entomology* 103, 172–177.

Bees and Wasps (Hymenoptera: Apidae, Vespidae and Others)

1. INTRODUCTION

Bees and wasps (along with ants) belong to the order Hymenoptera. They are endopterygote, mandibulate insects in which the forewings are larger than the hindwings and both fore- and hindwings are coupled mechanically to function as a single entity (Fig. 3). For their capacity to inflict venomous stings, most often in defence of their nests, some bees and many wasps are significant for human and animal health.

2. TAXONOMY

A number of wasp species are responsible for health-related problems: in particular, many in the family Vespidae and some in the families Sphecidae, Mutillidae and Pompilidae. Within the Vespidae, the subfamilies Vespinae and Polistinae contain those known as hornets, yellow jackets and the paper wasps, respectively.

Bees are often regarded as a single family, the Apidae, but an alternative view has them as nine families; however, most of the stinging bees are social species and are in the family Apidae in either classification.

3. MORPHOLOGY

Wings are well developed in most wasps and bees. There is a narrow petiole (constriction of the second abdominal segment) more obvious in wasps, and bees are more hairy on the body and legs (which enables the gathering and transport of pollen). In all the species considered here, except in the stingless bees (Meliponini), the ovipositor is modified into a stinging apparatus in females and the accessory glands have been modified to form venom glands that secrete into a storage sac.

4. LIFE CYCLE

Development is holometabolous, with egg, larval, pupal and adult stages, often all associated with colonies, but there is great variation in colony structure and behaviour. Colonies often consist of one or more queens 'surrounded' by many workers, and reproductive males and females may be produced only once a year in temperate regions, leaving the colony for mating and with the males typically dying soon after the mating season. While colonies of some bees may be perennial, those of other bees and many wasps are annual (particularly in temperate regions).

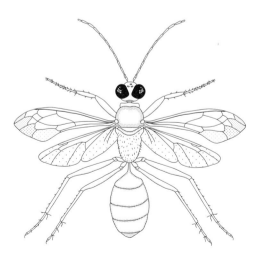

Fig. 3. Dorsal view of a hymenopteran (Sphecidae).

5. BEHAVIOUR AND BIONOMICS

The behaviour of greatest significance in this context is that of stinging. Bees and wasps that are foraging are less likely to sting (unless threatened) than those defending the nest/ colony, and their attack will be initiated by various visual (colour), chemical (alarm pheromone) and physical (vibration) cues. Many species react to vibrations of the substrate near the nest: vespid wasps often attack dark objects moving near their colonies and honeybees release volatile 'alarm chemicals' from their sting and mandibular glands when they feel the colony is threatened. Wasp stinging often involves multiple plunges of the apparatus into the skin, with the venom glands simultaneously contracting and pumping venom down the channel of the apparatus. With bees (and some social wasps and a few ants), the stinging results in the apparatus and venom sac remaining in the wound when the insect flies off (eviscerated and soon to die), because of large barbs on the apparatus that do not occur in most wasps and ants. In the tropics, some species of stingless bees can be obnoxious in their persistent gathering by the dozens on human skin to suck sweat. Three tiny South-east Asian species suck tears from human and animal eyes. In Central America, species of *Oxytrigona* release a caustic secretion from their mandibular glands, which together with their biting cause painful lesions and scars that can be permanent.

5.1 Bees and wasps of particular importance

5.1.1 Honeybees (*Apis* spp.)

These are widespread and the common honeybee (*Apis mellifera*) has been introduced to all parts of the world for its honey-making and plant-pollinating capabilities. Its nesting habit in hives is well known and intruders can be stung by large numbers of bees near colonies. An aggressive subspecies, the Africanized honeybee, *Apis mellifera scutellata*, has expanded its distribution to the Americas and it has a propensity for mass-stinging attacks on both humans and other animals (because of its lower threshold of response to alarm cues), and it is reportedly responsible for many fatalities.

5.1.2 Bumblebees (*Bombus* spp.)

These are large, hairy, black and yellow bees, which colonize subterranean cavities or structural voids in buildings. Most species are non-aggressive and typically pose little stinging risk for humans, although in some species the workers can be aggressive to nest intruders.

5.1.3 Other bees

In South-east Asia, sweat-sucking meliponine bees include species of *Lepidotrigona* (e.g. *Lepidotrigona terminata*) and *Tetragonula* (e.g. *Tetragonula laeviceps*, *T. testaceitarsis*). Tear drinkers belong to *Lisotrigona cacciae*, *Lisotrigona furva* and *Pariotrigona klossi*.

5.1.4 Hornets, yellow jackets and paper wasps

These are social wasps that are very defensive and many are likely to sting readily, both near their nest and also while foraging. Their colonies are usually annual, with nests of large fibrous combs, some aerial and some subterranean, with less than a hundred to more than a thousand individuals. Hornets (*Vespa* spp.) are a particular stinging problem in Asia. 'Yellow jacket' is an American term for *Vespula* spp. and *Dolichovespula* spp., which are generally patterned yellow and black and are typically subterranean fibre carton nesters in soil cavities or in wall spaces and structural voids in buildings; they forage for protein and liquid carbohydrates around human activity sites and are responsible for many stings – some of which can result in very serious reactions. The European species *Vespula germanica* has become widely established in both the northern and southern hemispheres; it is aggressive and prone to sting, and its attraction to cans of sugary drink has led to stings inside the mouth that become life-threatening when the air passages swell with the reaction. Paper wasps (*Polistes* spp. and *Ropalidia* spp.) are common stinging wasps in many parts of the world, constructing annual nests of a single, paper-like comb, usually with a relatively small number (<100) of cells, which is attached by a pedicle to both natural (e.g. tree) and built (e.g. house eaves, window and door frames) sites, where they come into close contact with humans and stinging is likely to occur, particularly if the nest is disturbed.

5.1.5 Sphecids (mud daubers)

These are solitary wasps, of which only a few are of concern for their stings (despite their large and often threatening appearance), that build mud nests around buildings and provision them with spiders or other arthropods such as cicadas.

5.1.6 Mutillids (velvet ants)

These are solitary wasps, often brightly coloured, in which the females are wingless and wander the ground looking for prey to parasitize. They can inflict a very painful sting.

5.1.7 Pompilids (spider wasps)

These are solitary wasps that are often metallic blue or black, live in burrows and prey on spiders, and some of the larger species can inflict a painful sting.

6. MEDICAL AND VETERINARY IMPORTANCE

Although a wasp or bee sting is painful, the main danger lies in an allergic response, which may lead to anaphylactic shock and rapid collapse of the person stung. Biochemical and pharmacological aspects of the venoms of Hymenoptera have been investigated and multiple chemical components identified; these include histamines, serotonin and cate-cholamines (which induce pain, itching and redness), peptides such as haemolysins (which destroy red blood cells and cause pain), neuro-toxins and hyaluronidase (which facilitates spread and activity of venom components).

Allergy to stinging insects is commonly caused. The major venom allergens from these families are proteins of 10–50 kDa. Nearly all have been cloned and their structures are known. Each hymenopteran family has unique venom allergens, as well as allergens homologous with those of other families. Different species within each family have the same set of homologous venom proteins. Several venom allergens are glycoproteins with similar carbohydrate side chains. Antigenic cross-reactivity of homologous venom allergens can be due to their common protein and/or carbohydrate epitopes. Recombinant venom allergens are available and can be useful reagents for diagnosis and treatment. Bee and wasp venoms are rich in cytotoxic peptides, melittin and mastoparan, respectively, and their allergenicity may be due to the combined action of these enzymes and bioactive peptides. Vespids typically cause immediate and sometimes intense and prolonged pain when they sting, with swelling, erythema and per-sistent (perhaps for days) itching at the site; their venoms comprise serotonin and kinins, histamines, tyramine and catecholamines (all of which cause pain), peptides and phospholipases inducing haemolysis, enzymes that act as specific allergens and even neurotoxic com-pounds. There is some cross-reactivity between hymenopteran venoms, particularly with the hyaluronidase allergens of the various species of vespids.

Bee venom is a complex mixture of proteins, peptides and many small organic molecules, the most dangerous being phospholipases and hyaluronidase, which can sensitize individuals and lead to serious allergic reactions and occasionally death, and there are large quantities of a potent melittin that enhances the effects of the phospholipases as well as causing pain, triggering lysis of red blood cells and enhancing the spread of the toxins.

The fear engendered by stinging insects is out of all proportion to the risks involved. The annual mortality from hymenopteran stings throughout the world has been estimated to be 1–6 deaths/10 million people. However, death from respiratory failure brought on by anaphylactic shock can occur in sensitized individuals within an hour of being stung. Individuals can be desensitized with graduated injections of venom and venom sac proteins.

From a veterinary perspective, bee and wasp stings on the head can sometimes be problematic for dogs.

6.1 Treatments

Embedded bee stings should be removed quickly and the usual advised method is by lateral scraping not pinch–pulling (but rapid removal should be the priority). For both bee and wasp stings, cold (ice) packs can provide relief and applications of baking soda and meat tenderizer (papain) have been recommended to neutralize and denature the venom components. Antihistamines are helpful in relieving swelling and other allergic reactions, and topical application of corticosteroids may also help. If there is a serious allergic or other systemic reaction (e.g. anaphylaxis), medical assistance should be sought immediately and injected adrenaline (epinephrine) is the usual initial response. People with known hypersensitivity are usually recommended to carry antihistamines and an adrenalin syringe, and such individuals can undergo desensitization immunotherapy.

7. PREVENTION AND CONTROL

Typically, evading contact with foraging bees and wasps is the best way to avoid envenomation from incidental stings, while wearing light-coloured clothing has been proposed as a way of reducing risks of mass attacks near colonies.

If required, insecticidal treatments of nests should be undertaken at night, when the adults are least active and all individuals are in the nest. While knock-down insecticides are generally used for this task, a follow-up with a residual insecticide will provide for killing adults that later emerge from capped cells. Control of honeybee nests inside wall voids of dwellings usually requires specialist professional expertise.

SELECTED BIBLIOGRAPHY

Akre, R.D. and Reed, H.C. (1984) Biology and distribution of social Hymenoptera. In: Tu, A.T. (ed.) *Insect Poison, Allergens, and Other Invertebrate Venoms. Handbook of Natural Toxins*, Vol 2, Chapter 1. Dekker, New York, pp. 3–47.

Alexander, J.O'D. (1984) *Arthropods and Human Skin*. Springer-Verlag, Berlin, pp. 135–150.

Arlian, L.G. (2002) Arthropod allergens and human health. *Annual Review of Entomology* 47, 395–433.

Bänziger, H. and Bänziger, S. (2010) Mammals, birds and reptiles as hosts of *Lisotrigona* bees, the tear drinkers with the broadest host range (Hymenoptera, Apidae). *Mitteilungen der Schweizerischen Entomologischen Gesellschaft* 83, 271–282.

Bänziger, H., Boongird, S., Sukumalanand, P. and Bänziger, S. (2009) Bees that drink human tears (Hymenoptera: Apidae). *Journal of the Kansas Entomological Society* 82, 135–150.

Barnard, J. (1973) Studies of 400 Hymenopteran sting deaths in the United States. *Journal of Allergy and Clinical Immunology* 52, 259–264.

Bettini, S. (ed.) (1978) *Arthropod Venoms*. Springer-Verlag, Berlin.

Camazine, S. (1988) Hymenopteran stings: reactions, mechanisms, and medical treatment. *Bulletin of the Entomological Society of America* 34, 17–21.

Charpin, D., Birnbaum, J. and Vervelot, D. (1994) Epidemiology of Hymenopteran allergy. *Clinical and Experimental Allergy* 24, 1010–1015.

Friedman, L.S., Modi, P., Liang, S. and Hryhorczuk, D. (2010) Analysis of Hymenoptera stings reported to the Illinois Poison Center. *Journal of Medical Entomology* 47, 907–912.

Gauld, I.D. and Bolton, B. (1988) *The Hymenoptera*. Oxford University Press, London.

Goulet, H. and Huber, J.T. (1993) *Hymenoptera of the World: An Identification Guide to Families*. Research Branch, Agriculture Canada, Ottawa.

Hamilton, R.G. (2002) Diagnosis of hymenoptera venom sensitivity. *Current Opinion in Allergy and Clinical Immunology* 2, 347–351.

Johansen, B., Eriksson, A. and Ornehult, L. (1991) Human fatalities caused by bee and wasp stings in Sweden. *International Journal of Legal Medicine* 104, 99–103.

King, T.P. and Monsalve, R.I. (2008) Allergens from stinging insects: ants, bees, and vespids. In: Kay, A.B., Kaplan, A.P., Bousquet, J. and Holt, P.G. (eds) *Allergy and Allergic Diseases. Volume 1: The Scientific Basis of Allergy*. Wiley-Blackwell, Oxford, UK, pp. 1123–1130.

King, T.P. and Spangfort, M.D. (2000) Structure and biology of stinging insect venom allergens. *International Archives of Allergy Immunology* 123, 99–106.

Langley, R. and Morrow, W. (1997) Deaths resulting from animal attacks in the United States. *Wild and Environmental Medicine* 8, 8–16.

McGain, F., Harrison, J. and Winkel, K.D. (2000) Wasp sting mortality in Australia. *Medical Journal of Australia* 173, 198–200.

Meier, J. (1995) Biology and distribution of hymenopterans of medical importance, their venom apparatus and venom composition. In: Meier, J. and White, J. (eds) *Handbook of Clinical Toxicology of Animal Venoms and Poisons*, Chapter 21. CRC Press, Boca Raton, Florida, pp. 331–348.

Müller, U.R., Haeberli, G. and Helbling, A. (2008) Allergic reactions to stinging and biting insects. In: Rich, R.R., Fleisher, T.A., Shearer, W.T., Schroeder, H.W., Frew, A.J. and Weyand, C.M. (eds) *Clinical Immunology*, 3rd edn. Elsevier, Amsterdam, pp. 657–666.

Piek, T. (1986) *Venoms of the Hymenoptera, Biochemical, Pharmacological and Behavioural Aspects.* Academic Press, Orlando, Florida.

Reisman, R.E. (1994a) Insect stings. *New England Journal of Medicine* 331, 523–527.

Reisman, R.E. (1994b) Venom hypersensitivity. *Journal of Allergy and Clinical Immunology* 94, 651–658.

Schumacher, M.J. and Egen, N.B. (1995) Significance of Africanized bees for public health. *Archives of Internal Medicine* 155, 2038–2043.

Beetles (Coleoptera: Meloidae, Oedemeridae, Staphylinidae and Others)

1. INTRODUCTION

Beetles (order Coleoptera) are of very minor importance in medical or veterinary entomology. When disturbed or crushed, some species exude a vesicating fluid; other species serve as intermediate hosts of helminths parasitizing domestic and wild animals; some beetles infest grains and other stored products and can cause respiratory allergies; and larvae of others are covered with sharp spines that can cause skin reactions.

2. TAXONOMY

The beetles make up about 40% of all known insects, with more than 300,000 species described. There are four suborders including more than 150 families, although their systematics is under debate.

3. MORPHOLOGY

Beetles vary in shape and size and, while most are terrestrial, some are aquatic. Overall, they are endopterygote insects with thickened forewings known as elytra, which meet edge to edge in the mid-dorsal line and cover the membranous hindwings used for flight. They are generally hard-bodied insects with the elytra covering the body, although some species have shortened elytra and softer exposed body parts. The mouthparts are of the biting and chewing type.

4. LIFE CYCLE

Development is holometabolous; eggs are laid variously, dependent on habitat preferences. Larvae are typically legless 'grubs' and are of diverse types but usually consistent within families; species may overwinter as pupae in temperate climates, with adults emerging in the summer and maturing through to autumn, when they may hibernate. The majority of beetles have a single generation per year, but adults may live up to and beyond a year (some undergoing diapause in adverse conditions).

5. BEHAVIOUR AND BIONOMICS

Adult behaviours vary; relevant behaviours associated with the particular beetles of medical and veterinary significance are addressed below.

6. MEDICAL AND VETERINARY IMPORTANCE

Some beetles, for example the histerid beetle (*Carcinops pumilio*), hairy fungus beetle (*Typhaea stercorea*), lesser mealworm (*Alphitobius diaperinus*) and rice weevil (*Sitophilus oryzae*), have been associated with the carriage of foodborne pathogens that affect humans. In veterinary medicine, various beetles can serve as intermediate hosts of helminth parasites, where beetles ingest parasite eggs present in faeces or faecally contaminated food and transmit the parasitic infective stage when they in turn are ingested accidentally by an animal host. Some of these parasite infestations by acanthocephalans, cestodes, nematodes and trematodes occasionally also affect humans.

However, the beetle families of most significance for human or animal health are those that can produce vesicating fluids – the Meloidae (blister beetles), Oedemeridae (false blister beetles), Staphylinidae (rove beetles) and Tenebrionidae (flour, grain beetles). These fluids may be secreted by glands (e.g. pygidial glands) and can be ejected as a defensive action towards a predator, as with some Tenebrionids, or simply as chemicals dispersed throughout the body. These can cause irritation to humans and animals when the insect is crushed, as with the blister and rove beetles, where direct contact with beetle body fluids can result in skin (and eye) irritations; inhalation of body parts may cause respiratory reactions. Contacts with humans are more common during the seasons when beetles are abundant and/or near bright lights at night. Incidental invasions of orifices (eyes and ears) by small crawling or windborne beetles can lead to distress in both humans and animals (e.g. the condition long known as 'Christmas eye' or 'harvester's eye' in South-eastern Australia that is caused by tiny, <1 mm, *Orthopterus* spp. of the family Corylophidae).

6.1 Meloidae (*Cylindrothorax, Epicauta, Lytta, Mylabris, Psalydolytta* spp.)

Vesicating beetles in this family have been reported from Europe, China, India, Africa and North America and are most often encountered in association with foliage and flowers (where they feed on pollen). At least 20 species are known to cause dermatitis via their body fluids (containing cantharidin), which cause stinging and burning sensations, and within 24 h a blistering that may lead to a vesicular dermatitis with itching and oozing lesions (and a potential for secondary infections). The best-known member of the 'blister beetles' is the southern European species *Lytta vesicatoria* (Fig. 4), known as Spanish fly, whose dried extract was once used as a counterirritant and illegally as an aphro-

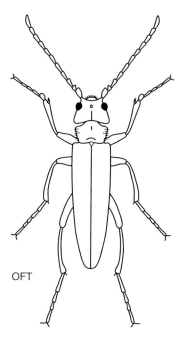

Fig. 4. Coleoptera: dorsal view of *Lytta vesicatoria*. *Source*: Patton, W.S. (1931) *Insects, Ticks and Venomous, Animals Part II, Public Health.* H.R. Grubb, Croydon UK.

disiac. Some meloids can present a problem for stock animals foraging where the beetles may be abundant, since accidental ingestion of a large number of beetles can lead to cantharidin poisoning. Dead beetles in harvested crops retain their toxin and also pose a similar hazard. Other species that have been reported to cause vesications include *Cylindrothorax melanocephala* in West Africa, species of *Epicauta* in North America, India, Sudan and Senegal, *Mylabris* in Nigeria and India and *Psalydolytta* in West Africa.

6.2 Oedemeridae (*Alloxacis, Eobia, Oxycopis, Sessinia* spp.)

Reactions caused by these beetles are most often reported from the USA, Central America and the Caribbean, but also from Australia, New Zealand and parts of the Pacific. Similar to the meloids, they are attracted to flowers and their body fluids contain cantharidin, which causes various dermal responses including blistering. *Oxycopis vittata* has been reported from Puerto Rico, *Sessinia* spp. from Kiribati, *Eobia apicifusca* from Australia, *Thelyphassa lineata* from New Zealand and *T. apicata* from Hawaii.

6.3 Staphylinidae (*Paederus* spp.)

Worldwide there are some 600 species of *Paederus*, the most significant genus of rove beetle in medical and veterinary entomology. Adults and larvae are beneficial predators, which frequent mostly leaf litter and debris and similarly moist habitats. Adults are active in daylight and are attracted to lights at night, which favours their contact with humans, especially in the tropics. They secrete a vesicating fluid (the active ingredient of which is pederin) when crushed or physically disturbed, as happens when *Paederus* lands on bare skin and is injudiciously brushed off or when it becomes trapped between clothing and skin. There is no immediate response to the fluid; however, in a day or two's time an angry weal appears, followed by the formation of blisters. The condition is known as dermatitis linearis or 'whiplash dermatitis' on the skin and it can cause conjunctivitis if rubbed in the eye. The erythema may persist for months and the conjunctivitis can produce temporary blindness (Nairobi eye). In East Africa, *Paederus sabaeus* has been reported, in the Middle East, *Paederus ilsae* and *Paederus iliensis* have been found responsible for these conditions, while in Japan and Australia, *Paederus fuscipes* and *Paederus australis* or *Paederus cruenticollis*, respectively, cause dermatitis linearis.

6.4 Tenebrionidae (*Blaps, Eleodes, Tribolium* spp.)

These beetles produce defensive secretions which function to repel small natural predators but which can cause darkening, burning and blistering of human skin because of the quinones that they contain. Most of the species are scavengers of dry or decaying plant material in dry environments, but some have become adapted to stored products such as cereals in

the human domestic environment (e.g. the common flour beetle, *Tribolium castaneum*).

6.5 Dermestidae (*Anthrenus, Dermestes, Trogoderma* spp.)

As a group, these beetles can be classified as allergenic rather than vesicating. They have been reported to cause urticaria, dermatitis, vasculitis and lymphadenopathy, but the effects appear to be purely mechanical and involve no toxin. The beetles are widely associated with human domestic and food storage environments and the adults generally are of little importance from a medical perspective (apart from their capacity to infest and 'spoil' various foodstuffs); however, their larvae (e.g. *Trogoderma inclusum*) are covered with very large numbers of barbed, spear-headed hastisetae and slender spicisetae that are capable of penetrating human skin and causing allergic responses.

6.6 Beetles as intermediate hosts of helminth parasites

Various beetles, including species of scarabids, carabids and tenebrionids, are intermediate hosts of poultry tapeworms (e.g. *Raillietina cesticillus* and *Choanotaenia infundibulum*) and birds are infested when they ingest beetles containing the encysted larvae arising from tapeworm eggs that were ingested by the beetles as larvae or adults.

Various other cestodes, nematodes and acanthocephalans infesting domestic companion animals and livestock are transmitted in a similar fashion. For example, the feeding mechanisms of coprophagous dung beetles (Scarabaeidae) make them efficient vectors for the transmission of *Spirocerca lupi* (Nematoda: Spirocercidae), which causes a potentially fatal disease to dogs in mostly warmer tropical and subtropical regions.

7. PREVENTION AND CONTROL

Considering that the vast majority of species of beetle are beneficial to the environment, while only small numbers of species may be a nuisance, any treatments must be planned carefully. In general, grain storage facilities, and the areas around them, should be cleaned thoroughly prior to storing grain. Storage facilities infested by beetles may be sprayed or dusted, if needed, with insecticides; however, the residual effect and toxicity of these chemicals must be considered. Each control strategy must take into consideration the use of preventative measures. The following strategies can assist in controlling beetle populations: pest-proof physical barriers, pheromone-based baits (for monitoring), mass-capture traps, spraying of larvicide insect growth inhibitors (e.g. diflubenzuron, pyriproxyfen), natural repellents or biocides (e.g. spinosad, *Cymbopogon nardus* or *Lavandula officinalis*) and/or pyrethroids (e.g. alfamethrin, cypermethrin, deltamethrin, etofenprox). The administration of control agents should be strictly regulated due to the potential ecological risks associated with their indiscriminate use. For effects from vesicating beetles, immediate washing of the affected skin is advisable.

SELECTED BIBLIOGRAPHY

Ahmed, A.R., Moy, R., Barr, A.R. and Price, Z. (1981) Carpet beetle dermatitis. *Journal of the American Academy of Dermatology* 5, 428–432.

Alexander, J.O'D. (1984) *Arthropods and Human Skin*. Springer-Verlag, Berlin, pp. 75–85.

Banney, L.A., Wood, D.J. and Francis, G.D. (2000) Whiplash rove beetle dermatitis in central Queensland. *Australasian Journal of Dermatology* 41, 162–167.

Bettini, S. (ed.) (1978) *Arthropod Venoms*. Springer-Verlag, Berlin.

Blazar, J.M., Lienau, E.K. and Allard, M.W. (2011) Insects as vectors of foodborne pathogenic bacteria. *Terrestrial Arthropod Reviews* 4, 5–16.

Davidson, S.A., Norton, S.A., Carder, M.C. and Debboun, M. (2009) Outbreak of dermatitis linearis caused by *Paederus ilsae* and *Paederus iliensis* (Coleoptera: Staphylinidae) at a military base in Iraq. *United States Army Medical Department Journal* 6–15.

Frank, J.H. and Kanamitsu, K. (1987) *Paederus*, sensu lato (Coleoptera: Staphylinidae): natural history and medical importance. *Journal of Medical Entomology* 24, 155–191.

Giglioli, M.E.C. (1965) Some observations on blister beetles, family Meloïdae, in Gambia, West Africa. *Transactions of the Royal Society of Tropical Medicine and Hygiene* 59, 657–663.

Hald, B., Olsen, A. and Madsen, M. (1998) *Typhaea stercorea* (Coleoptera: Mycetophagidae), a carrier of *Salmonella enterica* serovar Infantis in a Danish broiler house. *Journal of Economic Entomology* 91, 660–664.

Hazeleger, W.C., Bolder, N.M., Beumer, R.R. and Jacobs-Reitsma, W.F. (2008) Darkling beetles (*Alphitobius diasperinus*) and their larvae as potential vectors for the transfer of *Campylobacter jejuni* and *Salmonella enterica* Serovar Paratyphi B Variant Java between successive broiler flocks. *Applied and Environmental Microbiology* 74, 6887–6891.

Helman, R.G. and Edwards, W.C. (1997) Clinical features of blister beetle poisoning in equids: 70 cases (1983–1996). *Journal of the American Veterinary Medicine Association* 211, 1018–1021.

Karras, D.J., Farrell, S.E., Harrigan, R.A., Henretig, F.M. and Gealt, L. (1996) Poisoning from 'Spanish fly' (cantharidin). *American Journal of Emergency Medicine* 14, 478–483.

Mbonile, L. (2011) Acute haemorrhagic conjunctivitis epidemics and outbreaks of *Paederus* spp. keratoconjunctivitis ('Nairobi red eyes') and dermatitis. *South African Medical Journal* 101, 541–543.

McCallister, J.C., Steelman, C.D. and Skeeles, J.K. (1994) Reservoir competence of the Lesser Mealworm (Coleoptera: Tenebrionidae) for *Salmonella typhimurium* (Eubacteriales: Enterobacteriaceae). *Journal of Medical Entomology* 31, 369–372.

McCrae, A.W.R. and Visser, S.A. (1975) *Paederus* (Coleoptera: Staphylinidae) in Uganda. I. Outbreaks, clinical effects, extraction and bioassay of the vesicating toxin. *Annals of Tropical Medicine and Parasitology* 69, 109–120.

Nicholls, D.S.H., Christmas, T.I. and Greig, D.E. (1990) Oedemerid blister beetle dermatosis: a review. *Journal of the American Academy of Dermatology* 22, 815–819.

Okiwelu, S.N., Umeozor, O.C. and Akpan, A.J. (1996) An outbreak of the vesicating beetle *Paederus sabaeus* Er. (Coleoptera: Staphylinidae) in Rivers State, Nigeria. *Annals of Tropical Medicine and Parasitology* 90, 345–346.

Patton, W.S. (1931) *Insects, Ticks and Venomous Animals Part II, Public Health*. H.R. Grubb, Croydon UK.

Schmitz, D.G. (1989) Cantharidin toxicosis in horses. *Journal of Veterinary Internal Medicine* 3, 208–215.

Southcott, R.V. (1989) Injuries from Coleoptera. *Medical Journal of Australia* 151, 654–659.

Tagwireyi, D., Ball, D.E., Loga, P.J. and Moyo, S. (2000) Cantharidin poisoning due to 'Blister beetle' ingestion. *Toxicon* 38, 1865–1869.

Todd, R.E., Guthridge, S.L. and Montgomery, B.L. (1996) Evacuation of an aboriginal community in response to an outbreak of blistering dermatitis caused by a beetle (*Paederus australis*). *Medical Journal of Australia* 164, 238–240.

Weatherston, J. and Percy, J.E. (1978) Venoms of Coleoptera. In: Bettini, S. (ed.) *Arthropod Venoms*. Springer-Verlag, Berlin, pp. 511–554.

Biting Midges (Diptera: Ceratopogonidae)

1. INTRODUCTION

Ceratopogonids are referred to variously as biting midges (for example in Europe) and sand flies (for example in some parts of coastal Australia and the Caribbean), but 'biting midges' is the preferred common name and they should not be confused with the true sand flies (family Psychodidae, subfamily Phlebotominae – see entry for 'Sand Flies') or the black flies (family Simuliidae – see entry for 'Black Flies'), which are also incorrectly called 'sand flies' in some places. Members of the family Ceratopogonidae are important biting pests and vectors of minor pathogens to humans and vectors of major pathogens to livestock, including bluetongue virus to sheep, nodule-forming nematodes to cattle and a *Leucocytozoon* blood parasite to birds.

2. TAXONOMY

The Ceratopogonidae is a large family and is divided into four subfamilies: Leptoconopinae, Forcipomyiinae, Dasyheleinae and Ceratopogoninae, with more than 125 genera and more than 6000 species overall.

The Leptoconopinae includes only one genus, *Leptoconops*, but several subgenera: *Holoconops*, *Styloconops* and *Leptoconops*, including about 80 species. They are distinguished by possession of milky-white wings, which contrast sharply with the black head and thorax; there are no macrotrichia (obvious hairs) on the wings; no *r–m* cross-vein (Fig. 5); and the antenna has 12–14 segments in the female. In other ceratopogonids, the *r–m* cross-vein is present and the antennae are composed of 15 segments. *Leptoconops* species are largely restricted to the warmer areas of the world, with the subgenera

Holoconops and *Styloconops* being associated with sandy coasts; for example, *Leptoconops* (*Holoconops*) *becquaerti* in the Caribbean and *Leptoconops* (*Styloconops*) *spinosifrons* in the Indian Ocean, while the subgenus *Leptoconops* is more prevalent in inland locations, e.g. *Leptoconops* (*Leptoconops*) *torrens* in Utah (USA).

The Forcipomyiinae contains a single genus and a single subgenus, *Forcipomyia* (*Lasiohelea*), which is bloodsucking and represented by about 50 species. These have densely hairy wings, a well-developed empodium on the last tarsal segment and a long second radial cell (Fig. 5). *Lasiohelea* is associated with tropical and subtropical rainforests, although *Forcipomyia* (*Lasiohelea*) *sibirica* is a pest at Krasnoyarsk (Siberia).

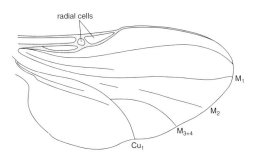

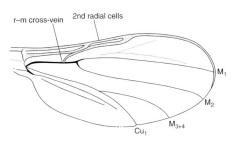

Fig. 5. Top: wing of female *Leptoconops* (*Styloconops*) *australiensis*. Bottom: wing of *Forcipomyia* (*Lasiohelea*) *townsvillensis*.

No member of the Dasyheleinae is of medical or veterinary importance. They have hairy wings, a short second radial cell, vestigial empodium and sculptured antennal segments.

The Ceratopogoninae contains the most important and largest genus, *Culicoides*, of which nearly 1400 species have been described, but also the smallest genus, *Austroconops*, with only one species. The genus *Culicoides* is widely distributed in the world from the tropics to the tundra, and from sea level to 4200 m above sea level in Tibet. *Austroconops* has only been recorded from near Perth in Western Australia. There is considerable variation within the Ceratopogoninae but all have vestigial empodia and their antennal segments are not sculptured. The wings of most species of *Culicoides* (Fig. 6) are patterned dark and light (Fig. 7). The pattern is due to pigmentation in the wing membrane and therefore cannot be rubbed off, but the pigment fades in specimens stored in alcohol and exposed to light. *Culicoides* wings have a petiolate media vein, i.e. vein *M* forks distally of the *r–m* cross-vein; the costa extends more than halfway and less than two-thirds along the front margin of the wing

and the radius forms two small, more or less equal cells (the radial cells) (Figs 7 and 8), one or both of which may be obliterated. Distinct humeral pits are located anterolaterally on the scutum, and the claws of both sexes are small, equal and simple.

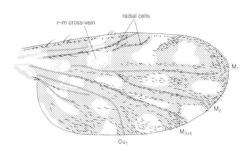

Fig. 7. Wing of female *Culicoides marmoratus* (× 88).

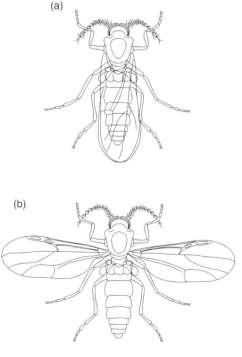

Fig. 8. Female *Culicoides* at rest (a) with wings folded scissor-like and (b) with wings spread laterally.

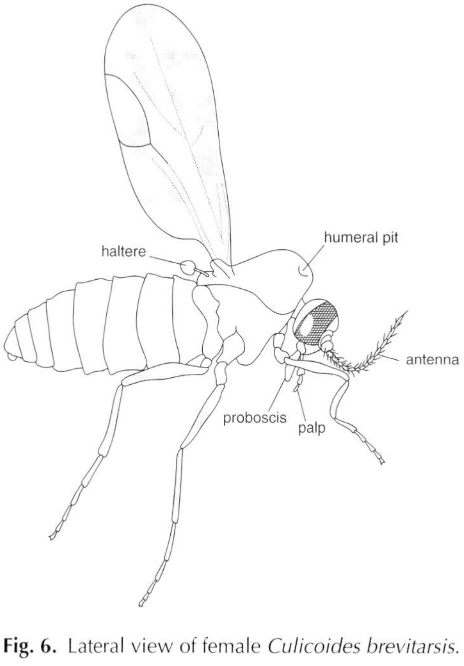

Fig. 6. Lateral view of female *Culicoides brevitarsis.*

3. MORPHOLOGY

Ceratopogonids are small, compact flies with short legs, a short vertical proboscis and pendulous palps. The wing venation is reduced, with only two veins in the posterior half of the wing, and they are both two-branched. They can be distinguished from the closely related non-biting midges (family Chironomidae – see entry for 'Non-biting Midges') by the following characters: ceratopogonids are generally smaller, with blood-sucking species rarely having a wing length greater than 2 mm (many tropical species have wing lengths less than 1 mm); at rest, they fold their wings scissor-like over the abdomen (Fig. 8); and their swarms are small and inconspicuous. Conversely, chironomids are larger; at rest, they hold their wings roof-like over the abdomen; and they often form large obvious swarms near water. Morphologically, ceratopogonids have a forked media vein (M_{1s} M_2) and piercing mouthparts, the front pair of legs is not lengthened and the postnotum is gently rounded and without a longitudinal groove. Chironomids have an unbranched media vein and reduced mouthparts, the front pair of legs is lengthened and the postnotum is more prominent and usually with a median longitudinal groove.

4. LIFE CYCLE

The typical life cycle of *Culicoides* follows, while behavioural differences shown by *Leptoconops* and *Lasiohelea* are discussed further below.

4.1 Egg

The eggs of *Culicoides* are laid in batches, which vary from 30 to 40 for *Culicoides brevitarsis* and up to 450 (with a mean of 250) for *Culicoides circumscriptus*. The eggs are small, dark in colour and slender, measuring 350–500 μm in length and 65–80 μm in breadth. They are covered with small projections (ansulae), which are particularly evident on the concave side and probably function as a plastron by retaining a film of air

in contact with the egg, facilitating diffusion of oxygen for respiration when the egg is covered with water. In most species, the eggs hatch in a few days at favourable temperatures, but those of the northern hemisphere species, *Culicoides grisescens*, do not hatch for 7–8 months in the laboratory and this species probably overwinters in the egg stage. Another Palaearctic species, *Culicoides vexans*, breeds in temporary, open pools and has a single generation in the spring. Its eggs lie dormant over the summer and hatch in the autumn, when the breeding site is unlikely to dry up.

4.2 Larva

Emerging from the egg is a vermiform larva, with a well-sclerotized head, 11 body segments and no appendages (Fig. 9). The three thoracic segments are similar to the eight abdominal segments, not being fused together and no broader than the abdominal segments. There are paired tracheae but the spiracles are closed and respiration is cutaneous. Two pairs of

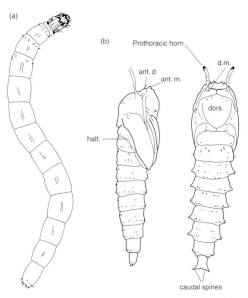

Fig. 9. (a): larva of *Culicoides impunctatus* (length 5.0 mm). (b): lateral and dorsal views of pupa of *Culicoides nubeculosus*. ant. d., ant. m., d.m. and dors. indicate various setae on the pupa; halt. = haltere. *Sources*: larva from Hill (1947) and pupa from Lawson (1951).

narrow, bifid anal papillae can be extruded from the anus or retracted into the rectum. Presumably, as in the mosquitoes, they function in the absorption of salts from the surrounding medium.

The larva swims by rapid sinuous flexions of the body, reaching speeds of 16 mm/s in water at a temperature of 20–28°C. When a larva encounters a more viscous medium (e.g. mud or sand), the movement changes from rapid to slow sinuous flexion and the rate of movement through the medium is slower than in water. Unlike mosquito larvae with their highly developed chaetotaxy, *Culicoides* larvae have only scanty, inconspicuous unbranched setae. A few species, notably those that breed in tree holes (e.g. the Australian *Culicoides angularis* and the Neotropical *Culicoides hoffmani*), have four pairs of long perianal setae. Tree-hole species probably spend more time in free water than do the larvae of other species, and the long perianal setae may serve to amplify the body oscillations and increase the larva's speed of movement, enabling it to catch prey or to avoid predators.

The most prominent internal structures in the head are the epipharynges, which occur in two main forms, heavy or light. With heavy epipharynges, the lateral arms, body and combs are strongly sclerotized; together with the hypopharynx, they act as a crushing structure, functioning like a pestle and mortar moving in one plane. Heavy epipharynges have, so far, been reported in only a small number of species belonging to the subgenus *Monoculicoides*, including the Palaearctic *Culicoides nubeculosus*, the closely related Nearctic *Culicoides variipennis* and the Afrotropical *Culicoides cornutus*. Larvae of these species occur in muddy substrates and larvae of *C. nubeculosus* browse on the surface bacterial film and on algal and fungal growth. *C. variipennis* has been mass colonized for many years, with the larvae being fed on microorganisms. With light epipharynges, the arms and medium body are only moderately sclerotized and the combs, of which there are usually two to four, are finely and delicately toothed. In spite of the lack of opposable mandibles, larvae with light pharynges are predators. Several species, for example

Culicoides austropalpalis and *Culicoides furens*, have been reared from egg to adult on free-living nematodes. In the laboratory, the activity of *C. furens* larvae increased in darkness and well-fed larvae were more active than starved or underfed larvae. *C. furens* larvae are negatively phototactic and generalist feeders, feeding below the surface on nematodes during daylight hours and feeding at the surface on algae during the night.

4.3 Pupa

There are four larval instars and the fourth ecdysis gives rise to the pupa (Fig. 9), which is culicid in appearance. The head and thorax are fused and bear a pair of moderately long, tubular prothoracic horns for respiration. These open to the atmosphere through a number of terminal and a few lateral openings (i.e. the pupa is propneustic with open spiracles). The segmented abdomen ends in a pair of caudal spines by which the pupa moves over the substrate. The pupa is a short-lived, non-feeding stage which gives rise to the winged adult. Before emergence, the pupa, which is usually buried in the substrate with only the prothoracic horns reaching the surface, moves upwards to facilitate eclosion.

4.4 Development time

The development of *Culicoides* is generally much slower than in the Culicidae (mosquitoes) under comparable conditions, although the egg stage and pupa are, with few exceptions, of short duration. In *C. variipennis*, these stages each take 12% of the development time and the intervening larval stages 76%. Development is temperature dependent, ranging from 7 weeks at 17°C to 2 weeks at 30°C in *C. variipennis*, and somewhat longer in *C. nubeculosus*, averaging 6 weeks at 25°C. In the field, the life cycle may be as short as 2–3 weeks in dung-breeding *C. brevitarsis*, 2–3 months in the subtropical salt marsh *Culicoides subimmaculatus*, 1 year in the temperate region bogland species *Culicoides impunctatus* and nearly 2 years in some Arctic species. In temperate regions, species are commonly univoltine, with one generation a year, and even in the tropics there may be only three or four generations per year.

5. BEHAVIOUR AND BIONOMICS

5.1 Reproductive behaviour

The males commonly emerge before the females and there is no permanent rotation of the terminalia to delay mating, although during mating they are inverted through 180 degrees. Male *Culicoides melleus* are thus competent to mate within minutes of emergence, with male potency reaching a peak 4–8 h after emergence, and they mate without swarming. In *C. nubeculosus*, mating can occur with or without swarming, but in most species, for example *C. brevitarsis* and *C. impunctatus*, swarming occurs around sunset, forming columnar or ovoid groups of males over well-defined markers. Swarm size varies from 10 to 1000 individuals: most commonly 50 in *C. brevitarsis* and 200 in *C. impunctatus*. A pheromone is secreted by virgin females when they are unfed and again when they are gravid. This pheromone attracts males and stimulates mating, but its effect on female behaviour has not been studied. A contact mating pheromone has been found in *C. melleus*.

Both males and females feed on nectar. Most females require a protein meal for maturation of the ovaries; some midges obtain this by feeding on pollen; in only four genera is the protein meal obtained by feeding on warm-blooded animals. However, autogeny has been shown to occur in some *Culicoides* species, including anthropophilic species such as *C. furens* and *C. subimmaculatus*, but they all require a blood meal to mature subsequent egg batches and many of them (e.g. *C. brevitarsis* and *Culicoides algecirensis*) require a blood meal to mature the first egg batch. Female *Culicoides variipennis variipennis* are anautogenous and develop the first batch of eggs to the resting stage in less than 2 days at 22°C. Following a blood meal, egg development is completed in a further 3 days. In *L. becquaerti*, there are two forms, a short-winged, autogenous form and a long-winged, anautogenous form (the latter being responsible for most of the nuisance caused by this species to humans). When *Culicoides arakawae* feeds on poultry, the size of its blood meal has been estimated at 0.36 mg (about 12% of that taken by *Culex pipiens* when feeding on a bird). Oogenesis is temperature dependent and in *Culicoides variipennis sonorensis* takes 2–10 days at 30°C and 13°C, respectively. Fecundity of female *C. variipennis* is correlated positively with size, which in turn is correlated with temperature; females emerging in late summer have a wing length of 1.48 mm, markedly shorter than the 2.02 mm wing length of females emerging in late winter.

5.2 Distribution and abundance

Bloodsucking midges are exophilic and exophagic, although in areas of high density, biting midges will enter houses (but in much smaller numbers than are found outside). *Culicoides* adults rest mainly among the herbage. *Leptoconops lucidus* has been found in cracks on the trunks of trees and in the upper layer of sand, and *Leptoconops mediterraneus* on the roots of *Limonium caspium*, a desert plant.

There is no record of hibernation or aestivation in adult *Culicoides*; however, in temperate regions, many species have only one generation a year, with the adults usually emerging in the summer. Other species may have several generations. *Culicoides obsoletus* has been reported as having two generations in the north of England and three generations in southern England, with a generation time of 7 weeks. In the warm summers of north-eastern Colorado, the generation time of *C. variipennis* may be as short as 2 weeks and seven generations may be completed in a year. In the tropics and subtropics, generations are likely to overlap and adults are present all year round. In subtropical Florida, adult *Culicoides mississippiensis* were present throughout the year and *C. furens* and *Culicoides barbosai* were absent in winter. In subtropical eastern Australia, *C. brevitarsis*, *Culicoides marmoratus* and *Culicoides victoriae* were taken in truck traps throughout the year, while *Culicoides longior* was absent in winter. In Kenya, some of these species were found all year round and although the populations fluctuated, no clear correlation with rainfall was noticed.

Activity at all times of the day is affected negatively by wind and positively by temperature. Truck trap catches of female *C. brevitarsis* were related inversely and linearly to wind speed and curvilinearly to temperature. Many nocturnal species are more active when the moon is shining, for example *C. variipennis*, *Culicoides nipponensis*, *C. subimmaculatus* and *C. mississippiensis*. It is commonplace for coastal species to show a lunar periodicity, determined by the tides. In Oceania, *Culicoides peleliouensis* has two peaks of emergence each month, corresponding to the neap tides. Similar tidal dependence has been shown for *C. subimmaculatus*, *Culicoides ornatus*, *Culicoides austeni*, *L. spinosifrons*, *C. furens* and *C. barbosai*. Although *L. becquaerti* is coastal in distribution, the numbers of this species biting were more dependent on rainfall than tidal movements.

5.3 Blood feeding

Blood-feeding species feed on a range of hosts: some, for example *C. furens, Leptoconops albiventris* and *Forcipomyia (Lasiohelea) taiwana* are anthropophilic; others, such as *C. brevitarsis*, feed mainly on cattle; yet others are ornithophilic, including the chicken-biting species *C. arakawae* and *Culicoides odibilis*, and *Leptoconops specialis* feeds on lizards and *Forcipomyia (Lasiohelea) phototropia* on Amphibia. *Culicoides anophelis* has been reported to feed on mosquitoes, but this may be a case of phoresy. *C. furens* is attracted to traps emitting CO_2 or octenol (a mammalian emanation) and, in combination, the two compounds have a synergistic effect that increases catches greatly.

The biting cycle follows a circadian rhythm. Species of *Leptoconops* and *Lasiohelea* are diurnal, as are a small number of *Culicoides* (e.g. *C. nubeculosus* and *Culicoides heliophilus*), but most *Culicoides* are crepuscular

and/or nocturnal. Diurnal species commonly show two peaks of activity, in the morning and in the afternoon. The morning peak occurs 2–3 h after sunrise and is commonly the larger of the two peaks, for example *L. becquaerti* and *Culicoides phlebotomus*. The afternoon peak occurs close to sunset and in the case of *F. (L.) sibirica*, it may occur as late as 21:00 h in the long summer days of high latitudes. The diurnal biting cycle of *F. (L.) taiwana* is unimodal, with a peak between 13:00 and 15:00 h.

Environmental conditions may modify basic activity patterns. In Jamaica, the pest species *C. furens* and *C. barbosai* showed peaks of activity at sunrise and sunset, presumably being initiated by rapidly changing light intensity; between these peaks, biting continued at a reduced rate throughout the night, with a smaller peak around midnight, and both species continued to bite during the morning until such activity was terminated by adverse meteorological conditions, especially increasing wind speed and temperature. *C. impunctatus*, mainly a crepuscular and nocturnal species, has been observed biting throughout the 24 h; even at times when females were not actively seeking a host, the arrival of a host induced activity in nearby resting, hungry females.

Female *C. variipennis* become active at sunset, reaching a peak just after sunset and having minor peaks in the middle of the night and at sunrise. This pattern, however, is not followed by all females. Nulliparous females predominate in catches between sunrise and sunset, which is probably the time when mating occurs. The catch of gravid females reached a maximum at sunset, which may be the peak time for oviposition, and parous females dominated catches after the end of twilight. Parous females were more active at new moon and gravid females at full moon.

The frequency of feeding also varies with species and situation. In Israel, *Culicoides imicola* has a 2-day cycle with daily survival of 0.65, which is considerably lower than that obtained for the same species in East Africa (0.80). In Israel also, a longer ovarian cycle (4 days) was found and, in late August, the survival rate rose to 0.75 and was followed by an increase in bluetongue infections in

September. Using a similar analysis, a 3-day gonotrophic cycle was obtained for *C. variipennis* in California, with 0.62 daily survival. In Trinidad, the daily survival rate of *C. phlebotomus* was found to be about 0.90 in the first 3 days of life, declining to 0.69 at 6 days; however, these observations were made on midges infected with *Mansonella ozzardi*, which may have affected their survival. The discovery of a burgundy-red pigment that is laid down on the ventral surface of the abdomen during the first ovarian cycle has proven to be largely beneficial to the study on the age structure of populations of *Culicoides*, and this has now been confirmed for many species of *Culicoides*.

5.4 Breeding sites

The breeding sites of bloodsucking ceratopogonids are commonly found in wet soil in the ecotone between aquatic and terrestrial habitats, or in moist, decaying vegetable material. *Culicoides* larvae burrow into the surface of the substrate and only rarely swim freely in the overlying water. Open, muddy sites, often contaminated with animal excreta, are the breeding grounds of *C. nubeculosus* and *C. variipennis*.

Female *Leptoconops* burrow into the substrate in the laboratory, and presumably do so in the field to rest and oviposit. Gravid female *C. nubeculosus* oviposit more readily when in groups than when held singly in tubes, and this may indicate the release of a pheromone.

Several coastal species of *Culicoides* breed in sand, for example *C. melleus*, *Culicoides hollensis* and *Culicoides molestus*. In more sandy habitats of coastal eastern Australia, *C. subimmaculatus* breeds in association with the surface-tunnelling soldier crab, *Mictyris livingstonei*. In the same area, larvae of *C. longior* occurred in peaty muds under mangroves, where they were associated with larvae of *C. marmoratus*, and on their own in fluid muds subject to heavy water movement. *C. marmoratus* larvae were also found in algal mats on the soil surface. In the Caribbean, *C. furens* and *C. barbosai* breed in association with mangroves, but only *C. barbosai* is

dependent on them. In Europe, salt mud flats breed *Culicoides halophilus* and *C. circumscriptus*, vegetated salt marshes breed *Culicoides maritimus* and there is an even greater range of freshwater breeding sites. Open, muddy sites, often contaminated with animal excreta, are the breeding grounds of the subgenus *Monoculicoides*, and temporary pools in pasture in Europe produce *C. vexans*. Vegetated swamps, where the water table is above the soil surface, are the breeding grounds of *Culicoides pulicaris* and *C. odibilis* in Europe, while in Japan the latter species occurs in the same type of habitat and also in rice fields, along with *C. arakawae*, which appears to be restricted to that habitat. Marshland areas, where the water in winter is below the soil level, breed *Culicoides pallidicornis* in Europe and *Culicoides marksi* in Australia. Edges of lakes breed *C. austropalpalis* in Australia and *Culicoides fascipennis*, *Culicoides achrayi* and *Culicoides duddingstoni* in Europe.

In Australia, larvae of *Culicoides bundyensis* and *Culicoides bunroensis* occur in sandy creek beds. In Canada, *Culicoides denningi* breeds in the Saskatchewan River and hibernates as larvae under ice in winter. In oligotrophic peaty areas, characterized by the mosses *Sphagnum* and *Polytrichum*, the *Culicoides* fauna is endemic and includes, in Europe, *C. impunctatus*, *Culicoides truncorum* and *Culicoides albicans*, and in North America, *Culicoides sphagnumensis*. The faunas of saltwater, eutrophic freshwater and oligotrophic bogland habitats are largely exclusive. The immature stages of many species of *Culicoides* are found only in small, specialized habitats, usually of vegetable origin. These include tree holes, producing *Culicoides fagineus*, *C. angularis*, *C. hoffmani* and *Culicoides guttipennis*. A comparable habitat to tree holes is water collecting in dugout canoes, from which, in West Africa, two species of *Culicoides* were bred. Species of the subgenus *Avaritia*, such as *C. brevitarsis*, *C. imicola*, *Culicoides dewulfi* and *Culicoides chiopterus*, breed in dung, especially that of cattle.

The New World *Culicoides coptosus* group of species breed in rotting cacti, and *Culicoides loughnani* was introduced into Australia when the moth *Cactoblastis cactorum* was introduced to control prickly pear. In West Africa, *Culicoides* have been reared from the rotting stems of bananas; in Trinidad, ten species were recorded, including the anthropophilic *Culicoides paraensis*, breeding in decaying cocoa pods, while *Culicoides scoticus* were bred from large fungi. *Culicoides heliconiae* breeds in the axils of the epiphytic bromeliad, *Heliconia*. Woodland leaf litter, which in other countries has proved an unrewarding habitat to examine for *Culicoides*, breeds the anthropophilic *Culicoides sanguisuga* in the eastern USA.

In French Polynesia, larvae of *L. albiventris* occur in a narrow strip a few metres wide above the high tide level and near creeping vegetation (*Ipomoea pes-caprae*). They are found to a depth of 24 cm, although most (78%) usually localize within the top 6 cm. The breeding sites are composed of fine, well-sorted sand, particle size 180–190 μm, with low conductivity and humidity. Larvae of *Holoconops* and *Styloconops* occur in coastal areas in almost pure sand at the high spring tide level, or even above it in sites inundated by exceptionally high tides. Larvae of the subgenus *Leptoconops* occur inland in clay-silt soils, where they may occur at considerable depths, e.g. *L. torrens* in the Sacramento Valley in California. When the soil cracks during the dry season, the larvae of *L. torrens* pupate and the adults emerge to feed and oviposit deep within the soil.

5.5 Flight range and dispersal

In general, most species disperse only short distances from their breeding sites and control of all breeding sites within 500 m is enough to reduce substantially the nuisance caused by species such as *C. molestus* and *C. subimmaculatus*. However, there is a difference between dispersal by active flight and passive wind carriage. In woodland, *C. impunctatus* disperses only a short distance from its breeding site, the density decreasing rapidly to one-tenth of the initial value of 70 m from the breeding site. However, in the open air, *C. impunctatus* disperses downwind for over 1000 m. Marked specimens of *C. mis-*

sissippiensis dispersed for an average of 2 km from the point of release and wind did not appear to be involved in this dispersal. Wind plays an essential part in the dispersal of *C. brevitarsis*, a species that feeds on cattle and breeds in their dung. By day the midges rest in the ground herbage and at sunset become airborne, with substantial numbers of males and females being captured 4 and 6 m above the ground. This behaviour allows the species to track the movements of its host. In Australia, wind carriage has dispersed *C. brevitarsis* 130–200 km from the Hunter Valley to cause outbreaks of Akabane disease in drought-stricken areas of inland New South Wales. In the UK, the wind-assisted dispersal of *Culicoides* infected with bluetongue virus from continental Europe was responsible for a major outbreak of this exotic disease in 2007.

Culicoides are mainly troublesome under calm conditions and numbers decline rapidly with increasing wind speed until few are encountered at wind speed exceeding 2.5 m/s. Female *F. (L.) sibirica* cease to bite at wind speeds in excess of 1.3 m/s, while those of *C. phlebotomus*, another day-biting species, remain active at wind speeds of 3.3 m/s. Female *L. becquaerti* continue biting in wind speeds up to 5.0 m/s and have been collected in wind traps at greater wind speeds. Such small creatures are unlikely to be able to orientate to a host in other than winds of low speed. There was some evidence of a dispersal flight by *C. furens*, when large numbers were trapped in a truck trap, but only negligible numbers were biting.

The limits of the distribution of a species within a region are constantly changing. In south-eastern Australia, *C. brevitarsis* is established on the coastal plains from which, in wet years, it extends further south and inland, where its continuing survival is dependent on the winter temperature. Some species, notably *C. imicola*, an important vector of bluetongue virus and African horse sickness virus, has recently been found in parts of Europe and appears to be expanding its range in response to climate warming. Modelling studies suggest the distribution of *C. imicola* in Spain, Greece and Italy could be extended further, and the vector potentially could invade parts of Albania, Yugoslavia, Bosnia and Croatia. Given warming of up to 2°C, the potential spread of *C. imicola* in Europe would be even more extensive.

6. MEDICAL AND VETERINARY IMPORTANCE

To most people, biting midges are synonymous with acute discomfort and irritation on calm, humid summer days. Ceratopogonids infest coastal mangrove, salt marsh and beach areas of the eastern USA, Caribbean, South America, Africa, Australia and islands of the Pacific and Indian Oceans. The nuisance impact of midges is greatest on newcomers to an infested area and hence the greater sensitivity of tourists to local pest species. In spite of their small size, they often cause severe local skin reactions. Not just *Culicoides* species are a problem – species of *Lasiohelea* and *Leptoconops* produce particularly persistent reactions, which may blister and weep serum from the site of the bite in sensitive people.

In the USA, salt marsh and mangrove *Culicoides* such as *C. furens*, *C. barbosai* and *C. mississippiensis* are problems on the Atlantic coast, southern Florida and Gulf coasts, respectively, while *Leptoconops* species such as *Leptoconops linleyi* and *L. becquaerti* cause problems in coastal beach areas of the southern USA and the Caribbean. In Australia and the Pacific islands, various species, including *C. ornatus*, *C. sub-immaculatus* and *L. albiventris*, can be problematic. Of the freshwater pest species, there are tree-hole species such as *C. paraensis* in the eastern USA and forest and bushland species such as *Forcipomyia (Lasiohelea) townsvillensis* in eastern and northern Australia and *Leptoconops stygius* in eastern Australia (the latter, which tends to bite on the head, can cause a painless swollen 'bung eye' that may last for a few days). There are even some semi-arid zone species, for example *Leptoconops kerteszi* group in the south-western USA, that can be problematic. Biting midges can reach pest proportions, even in northern temperate regions at high latitudes, such as *C. impunctatus* in the highlands of Scotland and *F. (L.) sibirica* in

Siberia, both of which can make life miserable for residents and visitors.

In all areas of the world where horses are subject to intense attack by *Culicoides* species, they suffer from a hypersensitivity condition known as 'sweet itch'. Various causative species are involved; for example *C. brevitarsis* in eastern Australia, *C. obsoletus* in western Canada and *Culicoides insignis* in Florida.

Additionally, as discussed below, various species of ceratopogonids are important for humans and/or other animals as vectors of arboviruses, blood-dwelling protozoa and filarial worms.

6.1 Arboviruses

6.1.1 The Bunyaviridae
Although primarily mosquito-borne (e.g. La Crosse and Californian encephalitis viruses), the Bunyaviruses also include the Culicoides-borne Akabane and Oropouche viruses.

AKABANE VIRUS (AKAV) AND AINO VIRUS (AINOV)
These viruses are members of the teratogenic Simbu group (along with Douglas, Peaton and Tinaroo) and are credited with causing disease in cattle in Australia and Japan. Akabane occurs also in Israel, much of Asia and most of Africa. The distribution of AINOV is not well understood and it may have a wide distribution similar to AKAV. Infection of a pregnant cow with Akabane virus 3–4 months into pregnancy results in a calf with limb deformities (arthrogryposis) and at 5–6 months the results can be brain deformities (hydranencephaly). The offspring of sheep and goats are also affected with malformation of the central nervous system, especially the brain in sheep. AINOV appears to be less important, but the pathological effects are similar to those for AKAV. The vector of AKAV is *C. brevitarsis*, which is well adapted to transmitting pathogens among cattle because its mode of life is dependent on them: feeding readily on cattle and ovipositing only in naturally lying cattle dung where the larvae and pupae complete their development.

A previously unknown but apparently related Simbu group virus, Schmallenberg virus (SBV), emerged in western Europe in 2011, associated with a transient illness in adult ruminants and late abortion or birth defects in newborn cattle, sheep and goats. Circumstantial evidence suggested that it was being transmitted by *Culicoides* midges. There was no evidence that the virus affected humans.

OROPOUCHE VIRUS (OROV)

This was first isolated in 1955 in Trinidad and since then has caused several human epidemics in Brazil. Between 1961 and 1979, eight outbreaks were recorded in Pará State, Brazil, in both small and large urban communities. After an incubation period of 4–8 days, infection with Oropouche virus causes an acute febrile illness with general aches and pains, lasting for 2–5 days. No deaths have been reported, although a proportion of patients become severely ill. In the 1967 outbreak in Bragança and in 1975 in Santarém, over 30,000 people became infected.

Oropouche virus has been isolated from *C. paraensis* and the mosquito *Culex quinquefasciatus*, and the former has proven to be the more efficient vector in the laboratory, with transmission rates varying from 25% to 83% for *C. paraensis* and less than 5% for *Cx. quinquefasciatus* under the same conditions. The maximum duration of the urban cycle is apparently only 6 months, and a sylvatic cycle is also likely to exist. Isolations of virus were made from the three-toed sloth (*Bradypus tridactylus*), and antibodies against Oropouche virus were found in several genera of monkeys. *C. paraensis* is active during the daytime, reaching peak activity just before sunset, feeding on humans both inside and outside houses.

6.1.2 The Rhabdoviridae

BOVINE EPHEMERAL FEVER VIRUS (BEFV)

This is a disease of cattle, which is enzootic in Africa, Asia, Australia and the Middle East. Commonly known as the 'three-day sickness', it is characterized by an initial generalized inflammation and toxaemia, followed by a short-term paralysis that may resolve spontaneously or result in death; fatality rates are typically low at about 3%, but there are reports of up to 30% mortality in fat cows. Recovered animals are considered to have a life-long sterile immunity. The outcome of the infection consists of a sharp drop in milk production, deaths of dairy and beef animals and abnormally delayed calf conception, resulting in major economic losses. The virus has been recovered from *Culicoides* species in Africa and from *C. brevitarsis*, *Anopheles bancroftii* and a mixed pool of culicine mosquitoes in Australia. The virus needs to be injected into the circulatory system for disease transmission. Intradermal, subcutaneous and intramuscular inoculations of BEFV do not infect cattle and these observations favour transmission by capillary-feeding mosquitoes rather than pool-feeding ceratopogonids.

6.1.3 The Reoviridae

BLUETONGUE VIRUS (BTV)

There are 24 known distinct serotypes of the bluetongue virus (BTV) and multiple strains exist within each serotype. BTVs cause severe disease in sheep, involving fever, inflammation of the mucous membranes of the oral cavity and nasal passages, enteritis and lameness. In cattle, African wild ruminants and North American cervids, BTV causes little clinical disease, although in highly susceptible cattle a few animals may become severely affected. BTV has little clinical effect on goats, which are not infected in nature. The virus was originally enzootic in Africa, but in the past 50 years has become widely distributed throughout the world and now occurs from 40°N to 35°S. To date, outbreaks have occurred in Africa, southern Europe, the Middle East, Pakistan, India, Japan and the USA. In Australia, Brazil, Canada and the West Indies, BTV is present but no clinical disease has been reported.

In susceptible sheep, BTV can cause morbidities of 50–75% and mortalities of 20–50%. Losses are both direct (i.e. mortality) and indirect through abortion of pregnant ewes, reduction in quality and quantity of the fleece and the prolonged period of convalescence required for full recovery. However, pathogenicity varies between virus strains.

Previously, it was considered that only a small number of species of *Culicoides*, perhaps as few as six, were able to transmit the virus naturally in the field. Four of the six, *Culicoides actoni, Culicoides fulvus, C. imicola* and *Culicoides wadai*, are in the subgenus *Avaritia*, and *C. variipennis* and *C. nubeculosus* are in the subgenus *Monoculicoides*. There are two other common vectors, *C. (Avaritia) brevitarsis* and *C. (Hoffmania) insignis* (with the latter being closely associated with BTV transmission in the Caribbean and Central America). However, more recently, as changes in virus distribution have brought it into contact with other *Culicoides* species, it has been found that others, such as the *C. obsoletus* and *C. pulicaris* complexes, may also be competent vectors in Europe.

The main vectors are *C. imicola* in Africa, the Middle East and southern Europe and *C. variipennis* in North America. Within the latter, three subspecies are recognized; of these, *C. v. sonorensis* is the best vector of BTV, being superior to *C. v. variipennis* and *Culicoides variipennis occidentalis*, and there is justification for recognizing the three entities as separate species. In Central and South America, it is *C. insignis* (as mentioned above). In Australia, *C. actoni* and *C. fulvus* are restricted to areas where the summer rainfall exceeds 1000 mm, which would exclude them from the drier sheep-rearing areas. However, *C. wadai*, similarly restricted initially, has extended its range and is now verging on some of the major sheep-rearing areas.

Control of BTV faces a number of practical problems. Strict quarantine measures can prevent the introduction of infected material into a country but are powerless to prevent the wind carriage of infected insects. This explanation was suggested for outbreaks in Portugal in 1956 and Cyprus and Turkey in 1977; it has also been shown that the introduction of BTV serotype 2 into Florida in 1982 could have been by wind carriage of infected *Culicoides* from Cuba. The outbreak in the UK in 2007 was attributed to the windborne movement of infected *Culicoides* from continental Europe.

Many BTV serotypes, affecting both domestic and wild ruminants, have now entered northern European countries. In particular, a new serotype (BTV8/net06), identified in August 2006, caused the largest BTV outbreak ever recorded. In 2008, mass vaccination with inactivated vaccines against serotype 8 was carried out in many European countries, effecting control of this serotype. However, other European BTV serotypes are entering north-west Europe, despite the availability of vaccination against serotype 8.

The development of protective vaccines is impaired by the existence of 24 different serotypes of BTV, most of which were isolated originally from South Africa (17) and Australia (4). Once the disease has been introduced, vaccination against the serotype involved is the only satisfactory control procedure. Control is further complicated by the existence of inapparent infections in cattle and wild ruminants. Infection rates of up to 48% have been recorded in cattle, with infection lasting for up to 81 days.

Little is known concerning the maintenance of the virus during periods when there is no active transmission. A number of wild ruminants are susceptible to infection with BTV, but present information suggests that their viraemias are relatively short-lived, so they are unlikely to act as reservoirs. It is possible that the virus may survive in apparently recovered animals. Sheep that recover from infection normally develop a solid immunity to the strain with which they were infected. However, virus has been isolated from sheep 4 months after an outbreak, and in some cases after longer periods, and latent virus in cattle has been demonstrated by recovering it from *C. variipennis* that had fed on cattle and then been maintained for a period to allow viral multiplication. It is also possible that transmission may occur through low-level cycling through midges as vectors. If the winters are short and mild, some transovarial transmission may also occur through the vectors, or there may be transplacental infection of lambs/calves from infected dams. The precise mechanisms, however, have not yet been proven.

AFRICAN HORSE SICKNESS VIRUS (AHSV)

This virus causes a highly fatal disease among susceptible equines. The disease is enzootic in Africa, from where, in the early years of the

1900s, it made occasional excursions across the Red Sea and along the Nile to Palestine and Syria. In 1959, AHSV spread eastwards into Iraq, Iran, Afghanistan, India and Pakistan, and in the same year westwards to Cyprus and Turkey. In 1965–1966, the disease appeared in Spain.

A serological survey in southern Egypt found antibodies to AHSV in sheep, goats, buffalo, dogs and camels, ranging from one in three sheep to one in thirty camels being positive. The virus is moderately resistant to drying and heating, and can survive for 2 years in putrid blood. Dogs can become infected by eating infected meat, and they develop a mild disease. The zebra is highly resistant to infection.

African horse sickness is a disease of the vascular endothelium, with three clinical expressions, all with fever. An acute or pulmonary form found in susceptible equines has an incubation period of 5–7 days. In enzootic areas, the virus produces a more slowly developing and persistent cardiac or subacute disease and a milder fever, which can be overlooked. The mortality rate in susceptible horses is about 90%, while mules suffer a lower mortality (50%) and donkeys are even less susceptible. Nevertheless, the disease is a crippling one to mules and donkeys, causing gross debilitation. It has been conservatively estimated that 300,000 equines died during the first phase of the 1960 epizootic in the Near East and South Asia. The spread of AHSV has been attributed to the introduction of infected equines into an area, but there is evidence that the disease can be spread by the wind carriage of infected vectors from enzootic areas into previously disease-free areas.

Evidence for AHSV being transmitted by nocturnal biting flies was provided by the fact that horses, accommodated in mosquito-proof stables during the hours of darkness, were protected from infection. C. variipennis was incriminated in the transmission of AHSV in South Africa. In southern Egypt, wild-caught Hyalomma dromedarii ticks were shown to harbour the virus and to transmit it to camels and horses. The brown dog tick, Rhipicephalus sanguineus, has also transmitted the virus experimentally from sick dogs to healthy dogs and to horses. In both ticks, there was trans-stadial but not trans-ovarial transmission of the virus.

Nine different antigenic serotypes have been recognized which exhibit no cross-immunity, and there are some 42 strains within the serotypes that have antigenic differences. The vaccine used in the 1960 outbreak in the Middle East and India contained seven strains of attenuated virus and proved to be effective, producing solid immunity for 1 year. Foals born to immune mares possess passive immunity for 5–6 months. Typically, the vectors are exophagic and, in South Africa, the risk of African horse sickness has been reduced greatly by simply stabling horses at night.

EPIZOOTIC HAEMORRHAGIC DISEASE VIRUS (EHDV)

Epizootic haemorrhagic disease has caused epizootics in the white-tailed deer (Odocoileus virginianus) in the USA and among cattle in Japan. In Australia, there are five serotypes of EHDV that infect cattle, buffalo and deer without causing clinical disease. The major importance of EHDV relies on its similarity to BTV and AHSV, which are all vectored by Culicoides species. EHDV was recovered from C. variipennis during an outbreak of the disease in Kentucky, USA, in 1971, and two strains of the virus have been transmitted from infected deer to uninfected deer by the bite of the same species. The virus has been shown to multiply in C. variipennis, both after oral ingestion and after intrathoracic inoculation, but in the closely related C. nubeculosus, multiplication of the virus only occurred after intrathoracic inoculation, thus indicating the existence of a mesenteron (midgut) barrier.

6.2 Apicomplexa

The Apicomplexa is a large group of unicellular, spore-forming protist parasites of both vertebrates and invertebrates. Motile structures such as flagella or pseudopods are present only in certain gamete stages.

6.2.1 Hepatocystis (Plasmodiidae)

With Hepatocystis, the pre-erythrocytic cycle occurs in the liver of mammals and gametocytes occur in the erythrocytes. Thirteen species of *Hepatocystis* have been described. They are mostly parasites of arboreal tropical mammals, namely lower monkeys, bats and squirrels. There is also one species which occurs in mouse deer (*Tragulus* spp.) and another in the hippopotamus. Most work has been done on *Hepatocystis kochi*, a parasite of monkeys in Africa, for which the vectors are species of *Culicoides*, including *Culicoides adersi* on the East African coast and *Culicoides fulvithorax*, and probably other species of *Culicoides*, in both inland and coastal areas.

The sporogonic cycle in *Culicoides* follows the usual pattern, with rapid exflagellation of the microgametocyte and the formation of eight microgametes. In *H. kochi*, the ookinete penetrates the basement membrane and enters the haemocoele. Oocysts are free in the haemocoele and accumulate anteriorly in the head, particularly near the eyes and supraoesophageal ganglia, where they mature in 5 days at 27°C and produce hundreds of sporozoites, which are transmitted via the mouthparts but not with the saliva.

In monkeys, the sporozoites invade the hepatic parenchyma cells and develop into slowly growing schizonts (merocysts). Most of the released merozoites invade erythrocytes to form gametocytes, but some re-invade the liver and repeat schizogony. In the erythrocyte, the merozoite produces haemozoin pigment. The presence of free oocysts in *H. kochi* may not be typical of the genus, because oocysts of *Hepatocystis brayi* occur in the usual location on the midgut of *C. nubeculosus* and *C. variipennis*. In some areas, infections of *H. kochi* in monkeys can be 100%, but its pathogenicity is doubtful. An unusual feature is that parasitaemia increases with the age of the host and there appears to be little or no immunity to *H. kochi*.

6.2.2 *Parahaemoproteus* (Haemoproteidae)

Parahaemoproteus nettionis, a parasite of ducks, is transmitted by *Culicoides downesi* and other species of *Culicoides*. In the cycle of *P. nettionis* in *Culicoides*, the oocyst grows very little, relatively few sporozoites are produced and they escape gradually from the oocyst. The minimum time for the sporogonic cycle is 7–10 days.

6.2.3 *Leucocytozoon* (Leucocytozoidae)

About 70 species of *Leucocytozoon* have been named and they are all parasites of birds. Schizogony occurs in the tissues and only the gametocytes appear in the peripheral circulation. Gametocytes occur in both leucocytes and erythrocytes, but no pigment is produced in the latter. Economically important species include *Leucocytozoon caulleryi*, which parasitizes chickens in South-east Asia and Africa.

L. caulleryi is placed in the subgenus *Akiba*, which is sometimes raised to generic rank. Species of *Akiba* are characterized by the disappearance of the nucleus of the parasitized cell as the gametocyte matures and by being transmitted by biting midges of the genus *Culicoides* (the vectors of most species of *Leucocytozoon* are species of Simuliidae of various genera).

With the *L. caulleryi* cycle in *C. arakawae*, sporozoites are produced in 3 days at 25°C and 6 days at 15°C and remain infective for 3–5 weeks. There is no evidence of a specific relationship between vector and parasite; *L. caulleryi* develops equally well in its major vector, *C. arakawae*, as well as in *C. odibilis* (a minor vector which feeds on chickens) and also in *Culicoides schultzei* (which feeds on cattle).

Chickens infected with *L. caulleryi* show some or all of the following symptoms: lethargy, loss of appetite, diarrhoea, convulsions and anaemia. The condition may be fatal. The anaemia cannot be accounted for by simple parasitization of the erythrocytes, but involves intravascular haemolysis. In general, *Leucocytozoon* infections are more severe in domestic than wild species, being fatal in young domestic birds.

6.3 *Leishmania* infections

Leishmania is a genus of Trypanosomatid protozoan parasites which are generally characterized by being transmitted to humans and other animals by Phlebotomine sand flies (see entry for 'Sand Flies'). However, there has been a recent discovery of an undescribed species of ceratopogonid in the *Lasiohelea* subgenus of *Forcipomyia* as the apparent vector of a novel *Leishmania* species causing cutaneous lesions in species of kangaroo in the Northern Territory, Australia.

6.4 Filarioid infections

6.4.1 *Mansonella*

Three species of *Mansonella* transmitted by Ceratopogonidae infect humans, for whom they are usually no more than mildly pathogenic. Their unsheathed microfilariae are found in the circulating blood (*Mansonella perstans*), in the dermis (*Mansonella streptocerca*), or in both the circulating blood and the subcutaneous tissues (*M. ozzardi*). *Mansonella* microfilariae in the skin need to be differentiated from the sheathed microfilariae of *Onchocerca volvulus*, which occur in the skin of humans in the same parts of the world. The three species will be considered separately.

MANSONELLA OZZARDI

This is found only in the Neotropical region, where it occurs in Central America, the north coast of South America, Brazil, Colombia, the northern province of Argentina and certain Caribbean islands. Human infection rates can be high (e.g. 37.5% in the Caribbean island of St Vincent). When *C. furens* fed on a carrier, microfilariae reached the thorax in 24 h and infective larvae were present in the head after 7–8 days. In Haiti, *C. barbosai* is considered to be as important a vector as *C. furens*. In northern Argentina, the vector is *C. paraensis* and in Trinidad *C. phlebotomus*. No particular symptoms are associated with *M. ozzardi*; in Trinidad, a chronic arthritis appears to be associated with infection, but a detailed relationship is unknown. Treatment with diethylcarbamazine is ineffective against *M. ozzardi*.

MANSONELLA PERSTANS

Previously known as *Dipetalonema* or *Acanthocheilonema perstans*, this is the most widespread of the three *Mansonella* species, occurring in the tropical rainforests of West and Central Africa and extending south to sylvatic foci in Zimbabwe. It is also found in limited foci among rainforest-dwelling

Amerindian communities in Central and South America, where the infection rate may exceed 50%. When microfilariae of *M. perstans* were ingested by *C. austeni*, they escaped from the midgut into the haemocoel in 6 h and by 20–30 h had reached the thorax. After 7 days, infective larvae were present in the head and emerged from the membranous end of the labium 8–10 days after the infected blood meal. The role of *Culicoides grahamii* is less clear, but with *Culicoides inornatipennis*, it appears to be a competent vector in some situations. Infections are usually asymptomatic, but where it is suspected that the infection is responsible for clinical symptoms, treatment with diethylcarbamazine is effective.

MANSONELLA STREPTOCERCA

Previously known as *Dipetalonema* or *Acanthocheilonema streptocerca*, this is limited to the rainforest areas of West and Central Africa. In limited areas of Zaire, the infection rate in the human population may be up to 90%. The adult worms are found in the dermis of the upper trunk and shoulder girdle. *M. streptocerca* develops well in 1.2% of wild-caught *C. grahamii*, with infective larvae being produced in 7–8 days. Differently, *Culicoides milnei* (synonymous with *C. austeni*) takes in very few microfilariae, less than one-tenth of those assumed by *C. grahamii*, thus being considered a poor vector of *M. streptocerca*. Infection results in a chronic itching dermatitis, and axillary and inguinal lymphadenopathy is common. Infection responds to treatment with diethylcarbamazine, which kills both adult worms and microfilariae.

6.4.2 *Onchocerca*

Seven species of *Onchocerca* are considered to be of veterinary importance – five in cattle (*Onchocerca gibsoni*, *Onchocerca gutturosa*, *Onchocerca lienalis*, *Onchocerca ochengi* and *Onchocerca armillata*), two in horses (*Onchocerca cervicalis* and *Onchocerca reticulata*) and one in dogs, cats and wild carnivores (*Onchocerca lupi*); however, the species are not necessarily host specific.

ONCHOCERCA IN BOVINES

Onchocerciasis of cattle and water buffaloes occurs widely throughout the world wherever these domestic animals have been introduced and there are suitable vectors. *O. lienalis* has been thought of as an unobtrusive parasite which produces no marked pathological changes or evidence of clinical disease. Losses in the beef industry arise because 'free worms cause aesthetically displeasing blemishes' to the carcass and encapsulated adult worms in nodules have to be trimmed from the carcass. Microfilariae occur in the skin and subcutaneous lymph, with the exception of those of *O. armillata*, which occur in the blood. Microfilariae are ingested by bloodsucking flies and develop in various species of *Culicoides* and *Lasiohelea* (but also species of *Simulium*).

Adult *O. lienalis* occur in the gastrosplenic ligament and its microfilariae are concentrated in the region of the umbilicus, while adult *O. gutturosa* are found in the *ligamentum nuchae* and its microfilariae in the skin of the head, neck and back. Nodule-forming onchocercas are classified in terms of their geographical distribution and the location of the adults in the bovine host. In Africa, adults of *Onchocerca dukei* are subcutaneous and those of *O. ochengi* are dermal parasites. In Asia, adults of *O. gibsoni*, and possibly those of *Onchocerca indica*, occur in the subcutaneous tissues, *Onchocerca cebei* and possibly *Onchocerca sweetae* in the dermis and *O. armillata* in the wall of the thoracic aorta. *O. ochengi* causes dermatitis on the scrotum and udder, resembling mange or pox. Nodules of *O. gibsoni* are most prevalent on the brisket and also occur on the stifle and thigh.

In Malaysia, *Culicoides pungens* ingest microfilariae of *O. gibsoni* when feeding on infected cattle. Microfilariae of *O. gibsoni* have their maximum concentration at a depth of 50–200 μm from the surface of the skin. The infection rate in *C. pungens* was less than 1%, but this was compensated for by the very large numbers biting cattle. The microfilariae of *O. gibsoni* complete their development in another ceratopogonid, *F. (L.) townsvillensis*, in 6 days at 30°C and 85% RH (relative humidity).

Ingested microfilariae of *O. gutturosa* develop to infective third stage when ingested by *C. nubeculosus* but, unlike *O. lienalis*, they do not develop in *Simulium ornatum*. Although microfilariae of *O. gutturosa* occur

mainly on dorsal surfaces of cattle, particularly in the withers, *C. brevitarsis*, which preferentially attacks the dorsal surface of cattle, does not ingest any microfilariae.

ONCHOCERCA IN EQUINES

New infections with *O. reticulata* may cause swelling of the suspensory ligament in horses, making the affected animals temporarily lame. After the swelling subsides, the ligament remains thickened. *O. cervicalis* causes fibrotic, calcified lesions in the *ligamentum nuchae* without visible clinical signs. Hypersensitivity to the microfilariae of *O. cervicalis* can cause alopecia, scaliness and pruritus along the ventral abdomen, which may become more extensive.

After feeding on an infected host, infective larvae of *O. cervicalis* developed in *C. nubeculosus*, *C. obsoletus*, *Culicoides parroti* and *C. variipennis*, but not in *C. pulicaris*, and infective forms were produced in 14–15 days at 23°C. When *C. nubeculosus* fed on an infected horse, 17% of the flies ingested microfilariae, with an average intake of 1.9 microfilariae per fly, while the ingestion of microfilariae of *O. cervicalis* by *C. v. sonorensis* was independent of the time spent feeding and the amount of blood ingested.

Microfilariae of *O. cervicalis* escaped from the midgut of *C. nubeculosus* within 5 min of the fly finishing feeding and 60% of the microfilariae reached the haemocoele within 1 h. About 40% of the microfilariae failed to escape from the midgut. Most of those that entered the haemocoele reached the thorax in 16–36 h. *C. variipennis* (subspecies unknown) is a less efficient vector, with about two-thirds of microfilariae being retained within the midgut. Early death of the midge can occur when large numbers of microfilariae penetrate the gut wall.

Microfilariae of *O. cervicalis* are predominantly (95%) present in the skin along the abdominal midline of the host, which brings them into close contact with *C. nubeculosus* since 85% of them feed on the ventral midline of the horse (i.e. from the front legs to the mammae or sheath). While the numbers of microfilariae in the whole skin remain unchanged over the year, during the active season of *C. nubeculosus* (June–September) microfilariae are most abundant just under the epidermis, favouring their ingestion by bloodsucking insects. During the cooler months of the year, October–February, the microfilariae are deeper (1–2 mm) in the skin. A similar seasonal movement of microfilariae of *O. gutturosa* in the skin of cattle has been shown to coincide with the period of activity of the vector *S. ornatum*.

7. PREVENTION AND CONTROL

Control of biting midges has been essential in many areas for the development of an expanding tourist industry. In many parts of the world, the ceratopogonid midge problem is concentrated on the coastal area and hence the popular but misleading name of 'sand flies'; as mentioned elsewhere, the term 'sand flies' should be reserved for species of phlebotomids.

From the foregoing, it is clear that species of *Culicoides* are able to exploit a wide range of moist habitats, but individual species utilize only a very limited range of breeding sites. It should be noted that although there is an association between mangroves and many tropical anthropophilic *Culicoides*, the breeding sites of the midges form only a minor part of the mangrove forest. Larvicides, therefore, have rarely been an effective approach to managing midge populations, because the specific habitats are often difficult to define and, in many situations, such chemical approaches are inadvisable because of the non-target impacts that can be associated with their use in estuarine environments. Further, physical approaches such as felling mangroves and filling the habitat with soil to control *Culicoides* are economically wasteful and ecologically damaging. This is of great practical importance because, where it is necessary to carry out larval control, whatever measures employed should be restricted to the defined breeding sites of the target species, minimizing costs and environmental damage.

It is essential to know the range of species in a locality before attempting control by habitat modification. For a tourist resort in Jamaica, removing mangroves from a swamp and filling it with sand dredged from offshore had the desired effect of eliminating the pest problem posed by *C. furens*. However, the sand-filled swamp created ideal conditions for the breeding of *L. becquaerti*, which had previously been a rarity in the area, and *L. becquaerti* was a far greater pest nuisance because its biting activity coincided with the periods of maximum tourist relaxation on the beach, making the new situation worse than the original. Effective control required the swamp to be filled to a depth greater than that to which subsoil water could ascend by capillarity action (750–1100 mm) and/or covering the surface with marl and establishing and binding the surface with grass.

Where knowledge of larval habitat and natural movement of adults of particular species is available, buffer zones that are cleared of vegetation between human habitation and midge habitats (e.g. mangrove areas) can reduce dispersal of pestiferous midges into residential communities, but these may need to be up to 1 km wide to be effective against some species such as *C. ornatus* in northern Australia.

Adulticiding with insecticidal fogs can be an effective approach for some situations, but the insecticides have to be applied strategically as fogs or mists during the evening periods of midge activity, and they provide only short-term relief, as new emergences and dispersal will replenish the adult pest populations. Barrier spraying, however, where residual insecticides are applied to peripheral vegetation, fences, walls and screens, can provide longer-term protection for individual houses on the edge of residential communities.

Personal protection for residents or visitors in outdoor situations can be afforded with protective clothing (full-length shirts and trousers of light colour) and topical insect repellents containing diethyl toluamide (DEET) or picaridin. Many midges will not enter buildings, but for those that do, the mesh-size of ordinary window-screening is too large to prevent their access and screens for tents should be approximately half the mesh size used against mosquitoes.

For stock animals at risk of attack by midges and possible infection with pathogens, studies in recent years have shown that protection of various degrees and durations can be provided by the use of sheltered or internal housing, pyrethroid-treated barriers surrounding external pens, pyrethroid-impregnated ear tags and direct spray-on formulations of residual pyrethroids.

SELECTED BIBLIOGRAPHY

Anderson, G.S., Belton, P. and Kleider, N. (1991) *Culicoides obsoletus* (Diptera: Ceratopogonidae) as a causal agent of *Culicoides* hypersensitivity (Sweet Itch) in British Columbia. *Journal of Medical Entomology* 28, 685–693.

Aussel, J.P. (1993a) Ecology of the biting midge *Leptoconops albiventris* in French Polynesia. I. Biting cycle and influence of climatic factors. *Medical and Veterinary Entomology* 7, 73–79.

Aussel, J.P. (1993b) Ecology of the biting midge *Leptoconops albiventris* in French Polynesia. II. Location of breeding sites and larval microdistribution. *Medical and Veterinary Entomology* 7, 80–86.

Aussel, J.P. (1993c) Ecology of the biting midge *Leptoconops albiventris* in French Polynesia. III. Influence of abiotic factors on breeding sites. Towards ecological control? *Medical and Veterinary Entomology* 7, 87–93.

Baylis, M., Parkin, H., Kreppel, K., Carpenter, S., Mellor, P.S. and Mcintyre, K.M. (2010) Evaluation of housing as a means to protect cattle from *Culicoides* biting midges, the vectors of bluetongue virus. *Medical and Veterinary Entomology* 24, 38–45.

Boorman, J. (1993). Biting midges (Ceratopogonidae). In: Lane, R.P. and Crosskey, R.W. (eds) *Medical Insects and Arachnids*. Chapman and Hall, London, pp. 288–301.

Borkent, A. and Wirth, W.W. (1997) World species of biting midges (Diptera: Ceratopogonidae). *Bulletin of the American Museum of Natural History* 233, 1–257.

Calvete, C., Estrada, R., Miranda, M.A., Borras, D., Calvo, J.H. and Lucientes, J. (2009) Ecological correlates of bluetongue virus in Spain: predicted spatial ocurrence and its relationship with the observed abundance of the potential *Culicoides* spp. vector. *Veterinary Journal* 182, 235–243.

Calvete, C., Estrada, R., Miranda, M.A., Del Rio, R., Borras, D., Beldron, F.J., *et al.* (2010) Protection of livestock against bluetongue virus vector *Culicoides imicola* using insecticide-treated netting in open areas. *Medical and Veterinary Entomology* 24, 169–175.

Carpenter, S., Mellor, P.S. and Torr, S.J. (2008) Control techniques for *Culicoides* biting midges and their application in the UK and northwestern Palaearctic. *Medical and Veterinary Entomology* 22, 175–187.

Carpenter, S., Wilson, A. and Mellor, P.S. (2009) *Culicoides* and the emergence of bluetongue virus in northern Europe. *Trends in Microbiology* 17, 172–178.

Dardiri, A.H. and Salama, S.A. (1988) African horse sickness: an overview. *Equine Veterinary Science* 8, 46–49.

Davies, J.B. (1969) Effect of felling mangroves on emergence of *Culicoides* spp. in Jamaica. *Mosquito News* 29, 566–571.

De Liberato, C., Farina, F., Magliano, A., Rombolà, P., Scholl, F., Spallucci, V., *et al.* (2010) Biotic and abiotic factors influencing distribution and abundance of *Culicoides obsoletus* group (Diptera: Ceratopogonidae) in Central Italy. *Journal of Medical Entomology* 47, 313–318.

Dougall, A.M., Alexander, B., Holt, D.C., Harris, T., Sultan, A.H., Bates, P.A., *et al.* (2011) Evidence incriminating midges (Diptera: Ceratopogonidae) as potential vectors of *Leishmania* in Australia. *International Journal for Parasitology* 41, 571–579.

Gibbs, E.P.J. and Greiner, E.C. (1988) Bluetongue and epizootic hemorrhagic disease. In: Monath, T.P. (ed.) *The Arboviruses: Epidemiology and Ecology*, Vol II. CRC Press, Boca Raton, Florida, pp. 39–70.

Gorman, B.M. (1990) The bluetongue viruses. *Current Topics in Microbiology and Immunology* 162, 1–19.

Greiner, E.C., Mo, C.L., Homan, E.J., Gonzalez, J., Oviedo, M.T., Thompson, L.H., *et al.* (1993) Epidemiology of bluetongue in Central America and the Caribbean: initial entomological findings. *Medical and Veterinary Entomology* 7, 309–315.

Hagan, C.E. and Kettle, D.S. (1990) Habitats of *Culicoides* spp. in an intertidal zone of southeast Queensland, Australia. *Medical and Veterinary Entomology* 4, 105–115.

Hendry, G. (2003) *Midges in Scotland*, 4th edn. Mercat Press, Edinburgh, UK.

Hess, W.R. (1988) African horse sickness. In: Monath, T.P. (ed.) *The Arboviruses: Epidemiology and Ecology*, Vol II. CRC Press, Boca Raton, Florida, pp. 1–18.

Holbrook, F.R. (1985) An overview of *Culicoides* control. In: Barber, T.L. and Jochim, M.M. (eds) *Bluetongue and Related Orbiviruses*. Alan R. Liss, New York, pp. 607–608.

Jennings, M. and Mellor, P.S. (1989) *Culicoides*: biological vectors of Akabane virus. *Veterinary Microbiology* 21, 125–131.

Kettle, D.S. (1977) Biology and bionomics of bloodsucking ceratopogonids. *Annual Review of Entomology* 22, 33–51.

Kettle, D.S. and Elson, M.M. (1976) The immature stages of some Australian *Culicoides* Latreille (Diptera: Ceratopogonidae). *Journal of the Australian Entomological Society* 15, 303–332.

Linley, J.R. (1985) Biting midges (Diptera: Ceratopogonidae) as vectors of non-viral animal pathogens. *Journal of Medical Entomology* 22, 589–599.

Linley, J.R. and Davies, J.B. (1971) Sandflies and tourism in Florida and the Bahamas and Caribbean area. *Journal of Economic Entomology* 64, 264–278.

Linley, J.R., Hoch, A.L. and Pinheiro, F.P. (1983) Biting midges (Diptera: Ceratopogonidae) and human health. *Journal of Medical Entomology* 20, 347–364.

Mellor, P.S., Boorman, J. and Baylis, M. (2000) Culicoides biting midges: their role as arbovirus vectors. *Annual Review of Entomology* 45, 307–340.

Monath, T.P. (ed.) (1988) *The Arboviruses: Ecology and Epidemiology*, Vol II. CRC Press, Boca Raton, Florida.

Mullen, G.R. and Hribar, L.J. (1988) Biology and feeding behavior of ceratopogonid larvae (Diptera: Ceratopogonidae) in North America. *Bulletin of the Society for Vector Ecology* 13, 60–81.

Murray, M.D. (1987) Akabane epizootics in New South Wales: evidence for long-distance dispersal of the biting midge *Culicoides brevitarsis*. *Australian Veterinary Journal* 64, 305–308.

Murray, M.D. (1995) Influences of vector biology on transmission of arboviruses and outbreaks of disease: the *Culicoides brevitarsis* model. *Veterinary Microbiology* 46, 91–99.

Nathan, M.B. (1981) Transmission of the human filarial parasite *Mansonella ozzardi* by *Culicoides phlebotomus* (Williston) (Diptera: Ceratopogonidae) in coastal north Trinidad. *Bulletin of Entomological Research* 71, 97–105.

Nolan, D.V., Dallas, J.F., Piertney, S.B. and Mordue (Luntz), A.J. (2008) Incursion and range expansion in the bluetongue vector *Culicoides imicola* in the Mediterranean basin: a phylogeographic analysis. *Medical and Veterinary Entomology* 22, 340–351.

Ottley, M.L., Dallemagne, C. and Moorhouse, D.E. (1983) Equine onchocerciasis in Queensland and the Northern Territory of Australia. *Australian Veterinary Journal* 60, 200–203.

Papadoulos, E., Bartram, D., Carpenter, S., Mellor, P. and Wall, R. (2009) Efficacy of alphacypermethrin applied to cattle and sheep against the biting midge *Culicoides nubeculosus*. *Veterinary Parasitology* 163, 110–114.

Parsonson, I.M. (1990) Pathology and pathogenesis of bluetongue infections. *Current Topics in Microbiology and Immunology* 162, 119–141.

Reeves, W.K., Lloyd, J.E., Stobart, R., Stith, C., Miller, M.M., Bennett, K.E., *et al.* (2010) Control of *Culicoides sonorensis* (Diptera: Ceratopogonidae) blood feeding on sheep with long-lasting repellent pesticides. *Journal of the American Mosquito Control Association* 26, 302–305.

Reynolds, D.G. and Vidot, A. (1978) Chemical control of *Leptoconops spinosofrons* in the Seychelles. *Pest Articles and News Summaries* 24, 19–26.

Rogers, D.J. and Randolph, S.E. (2006) Climate change and vector-borne diseases. *Advances in Parasitology* 62, 345–381.

Saegerman, C., Mellor, P., Uyttenhoef, A., Hanon, J.B., Kirschvink, N., Haubruge, E., *et al.* (2010) The most likely time and place of introduction of BTV8 into Belgian ruminants. *PLoS One* 5(2), e9405.

Savini, G., Afonso, A., Mellor, P., Aradaib, I., Yadin, H., Sanaa, M., *et al.* (2011) Epizootic haemorrhagic disease. *Research in Veterinary Science* 91, 1–17.

Service, M.W. (ed.) (2001) *Encyclopedia of Arthropod-transmitted Infections*. CAB International, Wallingford, UK.

Simonsen, P.E., Onapa, A.W. and Asio, S.M. (2011) *Mansonella perstans* filariasis in Africa. *Acta Tropica* 120 (Supplement 1), S109–120.

Tabachnick, W.J. (1992) Genetic differentiation among populations of *Culicoides variipennis* (Diptera: Ceratopogonidae), the North American vector of bluetongue virus. *Annals of the Entomological Society of America* 85, 140–147.

Tabachnick, W.J. (1996) *Culicoides variipennis* and bluetongue virus epidemiology in the United States. *Annual Review of Entomology* 41, 23–43.

Venail, R., Mathieu, B., Setier-Rio, M.L., Borba, C., Alexandre, M., Viudes, G., *et al.* (2011) Laboratory and field-based tests of deltamethrin insecticides against adult *Culicoides* biting midges. *Journal of Medical Entomology* 48, 351–357.

Venter, G.J., Wright, I.M., Van Der Linde, T.C. and Paweska, J.T. (2009) The oral susceptibility of South African field populations of *Culicoides* to African horse sickness virus. *Medical and Veterinary Entomology* 23, 367–378.

Venter, G.J., Wright, I.M., Del Rio, R., Lucientes, J. and Miranda, M.A. (2011) The susceptibility of *Culicoides imicola* and other South African livestock associated *Culicoides* species to infection with bluetongue virus serotype 8. *Medical and Veterinary Entomology* 25, 320–326.

Walker, A.R. (1977) Seasonal fluctuations of *Culicoides* species (Diptera: Ceratopogonidae) in Kenya. *Bulletin of Entomological Research* 67, 217–233.

Ward, M.P. (1994) The epidemiology of bluetongue virus in Australia – a review. *Australian Veterinary Journal* 71, 3–7.

Wilson, A.J. and Mellor, P.S. (2009) Bluetongue in Europe: past, present and future. *Philosophical Transactions of the Royal Society B-Biological Sciences* 364, 2669–2681.

Wirth, W.W., Ratanaworabhan, N.C. and Blanton, F.S. (1974) Synopsis of the genera of Ceratopogonidae (Diptera). *Annales de Parasitologic Humaine et Comparee* 49, 595–613.

Wittmann, E.J. and Baylis, M. (2000) Climate change: effects on *Culicoides*-transmitted viruses and implications for the UK. *Veterinary Journal* 160, 107–117.

Black Flies (Diptera: Simuliidae)

1. INTRODUCTION

Simuliids are generally known as black flies. However, in some places (e.g. southern New Zealand and parts of inland eastern Australia) they are called 'sand flies', but they should not be confused with the true sand flies (family Psychodidae, subfamily Phlebotominae – see entry for 'Sand Flies') or the biting midges (family Ceratopogonidae – see entry for 'Biting Midges') that are also incorrectly called 'sand flies' in some places. Black flies are widely distributed in both hemispheres and can be important biting pests for humans and some animals (particularly livestock); they are also vectors of filarial nematode worms to various animals (including *Onchocerca volvulus*, the agent of 'river blindness' in humans) and *Leucocytozoon* blood parasites to birds.

2. TAXONOMY

The Simuliidae comprises two subfamilies, the Parasimuliinae, which includes *Parasimulium* genus, and the Simuliinae. The latter includes the tribes Simuliini (with the genera *Astrosimulium* and *Simulium*) and Prosimuliini, which includes all other genera. About 2120 species and 26 genera have been described, of which four (*Simulium*, *Prosimulium*, *Cnephia* and *Austrosimulium*) are of medical or veterinary importance.

The largest genus is *Simulium*, with about 1720 species arranged in 37 subgenera occurring in all zoogeographical regions, with the greatest number being found in the Palaearctic area.

Prosimulium (75 species) is largely confined to the Holarctic region, whereas *Cnephia* and *Austrosimulium* are restricted to Australia and New Zealand. The other large genus, *Gigantodax* (including 65 species), is diffused in the Neotropical, while *Metacnephia* (53 species) is found in the Holarctic region.

In *Simulium* and *Austrosimulium*, the radial sector on the wing is unbranched; the costa bears spiniform setae and hairs. The hindleg has a rounded lobe (calcipala) at the inner apex of the first tarsal segment and a dorsal groove (pedisulcus) near the base of the second tarsal segment. In *Cnephia*, the calcipala is lacking and pedisulcus is less evident or absent. In *Prosimulium*, the radial sector is branched (sometimes only slightly), the costa bears only hairs and there is neither the calcipala nor pedisulcus on the hindleg. Species identification is difficult and uses, among other characters, the structure of the male and female terminalia, the pupal respiratory organ and the larval head.

3. MORPHOLOGY

Simuliids are small, dark, stout-bodied, hump-backed flies (Fig. 10). They are larger than the bloodsucking ceratopogonids and have wing lengths of 1.5–6.0 mm. They are largely diurnal and vision plays an important role in their behaviour. In the female, the individual ommatidia of which the eyes are composed are small (10–15 μm) and the eyes are well separated above the antennae (i.e. dichoptic, Fig. 11b). In the male, the eyes are larger and are broadly contiguous above the antennae (i.e. holoptic, Fig. 11a), and the lower ommatidia are similar to those of the female but the upper ones are greatly enlarged, measuring 25–40 μm. The antennae are the same in both sexes and consist of small, globular segments, compacted together to give a beaded appearance (Fig. 11). The commonest number of antennal segments is 11;

© R.C. Russell 2013. *The Encyclopedia of Medical and Veterinary Entomology* (R.C. Russell *et al.*)

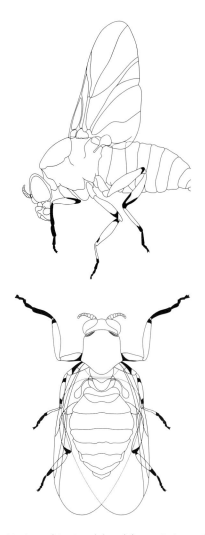

Fig. 10. Lateral (top) and dorsal (bottom) views of a female *Simulium*.

occasionally, there are 10 (*Austrosimulium*) and, rarely, there are 9 in some North American prosimuliines. The five-segmented, pendulous palps are considerably longer than the short proboscis. In males, and in a few species in which the females do not bite, the mandibles and maxillae are not toothed.

The wings are short and broad, with a large anal lobe (Fig. 12). The venation is characteristic, with well-developed radial veins along the anterior margin of the wing and weaker median and cubital veins posteriorly. In spite of its weak appearance, the wing is highly efficient and, in still air, simuliids are capable of flying in excess of 100 km. The radial sector may be unbranched or have two branches. Between the median (M_2) and the cubital (Cu) veins there is a forked submedian fold. The male terminalia are compact and relatively inconspicuous, particularly when compared to the prominent terminalia of male phlebotomines. The female has a single subspherical spermatheca.

4. LIFE CYCLE

In their early stages, simuliids are limited to fluvial ecosystems, breeding in running (often swiftly) water. The eggs are commonly laid in batches of 200–300 (range of 30–800) on objects in or near running water or directly into water. Several females deposit collective egg masses in close proximity, either dropped directly into the water, where they sink to the bottom, or on emergent objects close to the waterline, where they are either directly wetted by water or are in the splash zone. Females of several species crawl up to 15 cm below the water surface to oviposit on submerged substrates.

4.1 Egg

Eggs are 100–400 μm long and ovoid – subtriangular in shape. Their surface is comparatively smooth, lacking the patterned chorion found in the eggs of *Culicoides* and culicids. The gelatinous substance in which the eggs are embedded is formed by adherent outer membranes of the individual eggs (i.e. exochorions). The apparent egg 'shell' is the inner egg membrane, or endochorion, and the chorionic plastron is poorly developed. A well-developed plastron would not be necessary in eggs laid near the surface of running water, where the oxygen tension would be high. Simuliid eggs are sensitive to desiccation; even those of *Astrosimulium pestilens*, which survive for many months in wet river deposits, desiccate rapidly when exposed to relative humidities of 96% or less. Eggs laid near the

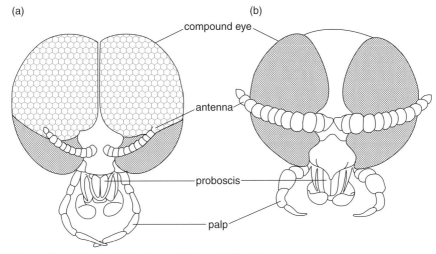

Fig. 11. Front view of heads of (a) male and (b) female *Simulium*.

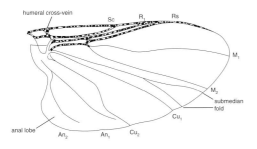

Fig. 12. Wing of female *Simulium* (× 50). Rs = radial sector.

surface hatch when the embryo has completed development, a matter of days under favourable conditions. Other species produce dormant eggs in which the adverse conditions of summer and/or winter are passed.

4.2 Larva

The egg hatches to produce a larva, much of whose behaviour revolves around the secretion of silk by the long salivary glands. The larva spins a web of silk on the substrate, which is continued into a silken thread on which the larva drifts downstream with the current in search of a suitable object on which to settle, using its posterior circlet of hooks (Fig. 13).

Larvae also produce copious amounts of protein glue, which is used to attach them firmly to the substrate. In four species of *Prosimulium*, production of the glue is dependent on the larva being in running water. Larvae remain near the surface of the water and are usually found at depths of less than 300 mm. The larva can change its location by drifting downstream on its thread or by looping over the surface using the posterior circlet and the hooks on the anterior proleg to retain a hold on secreted silk. Some species disperse further from the oviposition site than others; larvae of *Simulium ornatipes* are more sessile than those of *Astrosimulium bancrofti*, which move from the quieter waters of the oviposition site to rapids. In very large rivers, with fast-flowing water, larvae have been found at depths of several metres.

The larva has a distinct, sclerotized head with paired, simple eyes (stemmata) and an elongated body, in which the thorax and posterior part of the abdomen are broader than the anterior segments. There is a single anterior proleg, surmounted by a circlet of hooks and the abdomen ends in a posterior circlet. The anus opens dorsally of the posterior circlet and from it may be extruded the rectal organ, which probably, by analogy with the anal papillae of culicid larvae, is concerned with osmoregulation.

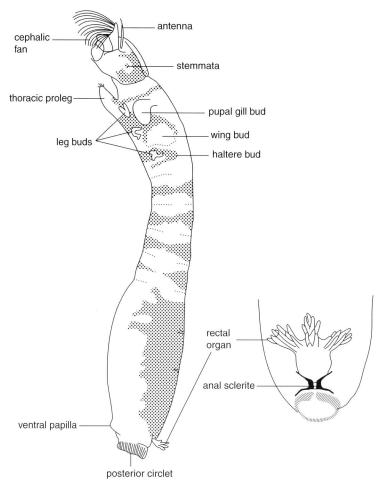

cephalic fan
antenna
stemmata
thoracic proleg
pupal gill bud
leg buds
wing bud
haltere bud
rectal organ
anal sclerite
ventral papilla
posterior circlet

Fig. 13. Lateral view of a *Simulium* larva and dorsal view of posterior end. Redrawn from Crosskey (1990).

The head bears a pair of cephalic (labral) fans, homologous structures to the lateral, palatal brushes of the Culicidae. They do not create a current but filter water passing over the larva. Larvae are anchored posteriorly and extended in the direction of the current with the head leading. The body is twisted through 90–180 degrees so that the fans and mouthparts face towards the surface of the water. The water current is divided by the proleg and directed towards the fans. A sticky secretion produced by the cibarial glands enables the fans to capture fine particles, which are transferred to the cibarium by the mandibular brushes. Larvae of *Simulium*

piperi defend their territory and are aggressive to their upstream neighbours, who would be competing for the incoming food. Territorial defence declines dramatically when food is abundant.

Simuliid larvae are capable of ingesting particles of colloidal size (0.091 µm) and up to 350 µm, but the most commonly ingested particles are 10–100 µm. Algae pass apparently unchanged through the black fly gut, but diatoms may form as much as 50% of the gut contents and are digested. Larvae of *Simulium ornatum* and *Simulium equinum* ingested particles with a maximum diameter of 25–30 µm, with diatoms forming the main food early in the year and small particles of detritus predominating later in the year. Some filter-feeding larvae also browse, and others (e.g. *Twinnia* and *Gymnopais*) do not filter feed at all but graze on the substrate.

Simuliid larvae are particularly abundant where the water current accelerates, as at rapids, and where larvae presumably will strain a greater volume of water per unit time. Heavy larval concentrations are to be found at the outflows of large lakes, where the water will be rich in phytoplankton for larval food. Movement of water over the body surface provides the larva with adequate dissolved oxygen for respiration. In deoxygenated water, larvae detach and drift downstream.

Larvae pass through six to nine instars and the number is not constant even within a species (e.g. *Prosimulium mixtum* may have six or seven larval instars).

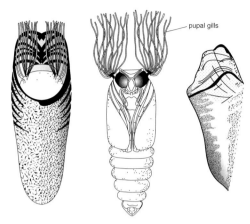

Fig. 14. Pupa and cocoon of *Simulium simile*. From left to right: dorsal view of cocoon containing a pupa; ventral view of pupa, removed from cocoon; lateral view of cocoon. The cocoon is arranged with its closed end directed into the current. *Source*: Reproduced by permission of the Minister, Supply and Services, Canada. From Cameron, A.E. (1922). *Agriculture Canada Entomological Bulletin No 20. The Morphology and Biology of a Canadian Cattle-infesting Blackfly,* Simulium simile *Mall.*

4.3 Pupa

The mature larva is actually a pharate pupa within the larval skin and may move to a different site before pupating (Fig. 14). In most species, the pharate pupa spins a cocoon, often slipper-shaped with the closed end directed upstream and the open end downstream. This alignment prevents the cocoon being torn off the substrate by the current. Construction of the cocoon takes about 1 h and then the larval skin is shed.

The head and thorax of the pupa are combined into a single cephalothorax and there is a segmented abdomen. The latter bears spines and hooks, which engage with the threads of the cocoon and retain the pupa in place. The cephalothorax bears a pair of elongate, branched pupal gills, which trail downstream of the cocoon. They are homologous, with the respiratory horns of the Culicidae and Ceratopogonidae, but they do not have open spiracles. The tubular branches of the gills bear vertical struts, which support a very thin, outer, minutely perforated, trilaminate epicuticle and an inner fine meshwork. The enclosed air-filled space around the struts functions as a plastron and the water–air interface is about 50% of the

total plastron area. The shapes of the cocoon and gills are important characters in the identification of species.

The pupa, which does not feed, becomes progressively darker as the adult develops within, but the mature pupa takes on a silvery appearance as a film of air is secreted between the pharate adult and the pupal cuticle. When the pupal exuviae split, the adult floats up to the surface in a bubble of air and immediately takes flight. Alternatively, the newly emerged adult crawls up some emergent object to reach the air.

4.4 Development time

The length of the life cycle varies with the species and environmental conditions. In temperate regions, species may have one generation a year, while continuous breeding occurs with tropical species. The larval stage of *Simulium damnosum* can be completed in as little as 8 days and the life cycle from egg to adult can be completed in less than 2 weeks.

5. BEHAVIOUR AND BIONOMICS

Adult emergence occurs predominantly in the daytime, depending on light and temperature. In *S. damnosum*, 60–90% of the day's emergence has occurred by midday and there is no emergence at night. At 24–28°C, peak emergence of *S. damnosum* occurs at 06:00–09:00 h and, when the water temperature is lower (20–24°C), the peak is reached later in the morning between 09:00 and 12:00 h.

5.1 Mating

Mating occurs in close association with the breeding site and in a few species occurs on the ground, but in the large majority of species it occurs on the wing, when males form small swarms in association with visual markers. In *A. pestilens*, male swarms orientate to *Callistemon viminalis* bushes along the banks of the semi-permanent stream in which they breed. Male simuliids recognize the female up to a distance of 50 cm and pursue the female and attempt to couple. There appears to be no contact pheromone, because males will attempt to mate with other males and with individuals of other species. In some species, male swarms and mating occur in close proximity to the female feeding sites or on the host. *S. ornatum* and *Simulium erythrocephalum* are usually on the navel and ears of cattle, respectively. During mating, a double-chambered spermatophore is transferred to the female.

5.2 Feeding

When feeding, the female simuliid anchors its proboscis to the host by small hooks on the labrum and hypopharynx. The maxillae are protruded alternately, penetrating downwards and anchoring the proboscis more firmly. The mandibles cut into the skin with rapid, scissor-like movements and penetrate to a depth of about 400 μm. Blood is ingested, using the cibarial and pharyngeal pumps, in about 4–5 min.

Ornithophilic and mammalophilic species are distinguished morphologically by the shape of the claws. In mammal feeders, the claws are simple, but in bird feeders they are toothed. This must be an adaptation to holding and penetrating feathers, but this host preference is not absolute and many species feed indiscriminately on both avian and mammalian hosts.

5.3 Biting habits

Some female simuliids show primiparous autogeny, that is females mature the first egg batch without a blood meal but need blood for each subsequent ovarian cycle, while others show obligate anautogeny (i.e. females need blood to mature every egg batch). Overall, most simuliids require a blood meal to develop eggs. *A. pestilens* feeds on many mammals, including humans, and is probably an opportunistic feeder, which may be related to the fact that it remains close to the breeding sites. However, the closely related *A. bancrofti* appeared more selective and did not bite any of six host species offered for feeding – four mammals, including a human, and two birds. In Scotland, it was found that *Simulium tuberosum* fed on a wide range of hosts, including humans, other mammals and birds; *Simulium latipes* fed largely on birds; and *Simulium reptans* and *Simulium monticola* took over 90% of their feeds on bovines.

The gonotrophic cycle may be remarkably short, being completed in 24 h in *A. pestilens* and in 2 days in *Simulium metallicum*. In *S. damnosum*, the first cycle from blood meal to oviposition takes 3–4 days and thereafter eggs are laid at intervals of 4–5 days, the additional time being required for nectar feeding before the blood meal.

Simuliids are exophilic, exophagic and largely diurnal, and in open sunny situations *S. damnosum* tends to have a bimodal pattern of activity, with peaks around 09:00 h in the morning and 17:00 h in the afternoon; but in shaded areas, biting is more evenly distributed throughout the day. The circadian rhythm of biting activity varies with the age of the flies, with parous females feeding earlier in the day than nullipars. Some species will enter the natural openings of the body, nose, ear and eye, which behaviour is particularly worrying to livestock. As in many of the other small bloodsucking flies, both males and females feed on nectar, which is stored in the crop, and only the females are haematophagous, with blood passing directly to the midgut.

Although mainly diurnal, ten species of simuliids have been taken in light traps in Scotland, three different species have been trapped in Norway and large numbers (3600/ night) of ovipositing *Simulium squamosum* (a member of the *S. damnosum* complex) were collected in light traps in Ghana. Activity of simuliids is also influenced by barometric pressure, and in Ghana the numbers of simuliids biting increased with rising humidity and falling barometric pressure.

5.4 Host finding

Three phases in host location by female black flies have been recognized. The first involves nectar feeding, mating and dispersal to bring the insect into the host's habitat; the second phase involves host searching, non-orientated flight driven by endogenous activity rhythms and hunger; finally, during the third phase, simuliids are orientated by external stimuli to find a proper host. Within the third phase, there are three stages: long-range attraction (initiated by host odour leading to an upwind response by the fly), nearer the host (orientation to carbon dioxide emitted by the host) and, within 1.8 m, visual orientation. Simuliids can discriminate between colours of the same reflectance, with *Simulium venustum* being attracted to blue, *P. mixtum* to black and *Simulium vittatum* to black, red and blue, but not to yellow.

The visual component in host finding is supported by the fact that *S. erythrocephalum*, which feeds on the ears of cattle, will attack protruding parts of a dummy and *S. ornatum*, which feeds near the navel or belly, attacks the flat underparts of a model. Simuliid visual attraction is exploited successfully in designing silhouette traps. Higher numbers of *A. bancrofti*, another cattle feeder, may be taken

on a horizontally elongated trap than on a similar trap arranged vertically or on a square trap of the same surface area; the horizontal trap being the one most closely resembling cattle. Most *S. damnosum* (93%) land on the ankles of humans, but then ascend the leg and feed on the calves. Females of the anthropophilic *Simulium ochraceum* complex feed mainly on the back and shoulders and very little on the legs below knee level. After landing, other stimuli, such as odour, sweat and other chemicals, are likely to be involved in stimulating probing. The North American cattle-feeding *Simulium arcticum* is attracted by the emission of carbon dioxide and less so by cattle urine on its own, but the combination of both is highly effective, increasing catches by 3–20 times. The ornithophilic species, *Simulium annulus* (= *Simulium euryadminiculum*), is attracted strongly to extracts of the uropygial gland of the common loon (*Gavia immer*).

5.5 Breeding sites

The gonotrophic cycle of *S. damnosum* s.s. and *Simulium sirbanum* is normally 3–4 days, made up of a period of less than 24 h between oviposition and feeding, ovarian development of 48 h and a variable time (less than 12 h) between the eggs becoming fully mature and oviposition ensuing.

Simuliids breed in running water, ranging from torrential mountain streams to slow-moving lowland rivers, and a few species are adapted to streams in which there is little perceptible current. In Newfoundland, the most significant factors affecting the distribution of simuliid larvae were current speed, substrate type and water depth. In western Africa, the distribution of sites of larvae of the *S. damnosum* complex could be classified on the pH and conductivity of the water.

S. damnosum females oviposit communally in the short period between tropical sunset and darkness. Dense swarms of females lay their eggs on vegetation trailing in the water, achieving densities of 2000–3000 eggs/cm². *A. pestilens* forms swarms of ovipositing females, which scatter their eggs over the surface of the water, where they become incorporated in the sandy river bed and can survive 2.5 years, providing they are kept permanently damp. In Guatemala, *S. ochraceum* drops its eggs directly into the water; *Simulium callidum* lays its eggs one at a time on the inclined surfaces of rocks and *S. metallicum* lays its eggs on leaves without landing on fast-flowing water, but it does land on slower-flowing water. Eggs of *Simulium argyreatum* can withstand dryness during autumn and winter when the temperatures are low, while eggs of *Simulium pictipes* resist frost and ice to survive the winter and hatch in the spring.

When eggs are deposited in dense masses, it is essential that the first instar larvae disperse. Larvae drift downstream attached to a silken thread, or they can break the thread and drift with the current. Larvae drift throughout the 24 h and early instars of *A. bancrofti* show a diurnal tendency, with a greater proportion of older instars drifting at night.

A small number of species have evolved a phoretic association with decapod Crustacea (crabs, prawns) or Ephemeroptera (mayflies) in Africa and the Himalayan region. Larvae and pupae of *Simulium nyasalandicum* and *Simulium woodi* occur on the sides, the chelipeds, and the basal segments of the walking legs of the crab *Potamonautes pseudoperlatus*. They also occur on other species of crab. Eggs are not laid on the crab and the young larva must find its own phoretic partner. The most important of these phoretic simuliid species is the *Simulium neavei* group, vectors for *O. volvulus* in eastern Africa. In Africa, mayfly phoretics are found mainly in heavily shaded forest streams and crab phoretics are found in small forest streams and larger and more open rivers.

5.6 Flight range and dispersal

Adult females of many species of simuliids disperse far from their breeding sites. In the laboratory, it has been observed that *S. venustum* is capable of flying 116 km in still air, following a sugar meal. In the field, this species disperses on average 9–13 km, but a few individuals have been found to cover 35 km in 2 days. Females of *S. arcticum* dispersed for distances of at least 150 km from the Athabasca River in western Canada in sufficient numbers to be a pest. Different categories of dispersal have been distinguished in *S. damnosum*. There is linear dispersal along river courses in the gallery forest of the West African savannah, radial dispersal in the savannah during the rainy season and in the forest region throughout the year and differential dispersal (when nullipars disperse further than parous females, which tend to remain near the breeding sites). There is also long distance (e.g. 200 km in West Africa) windborne dispersal of black flies, and this is often of parous flies, indicating that, under certain conditions, parous females can disperse as widely, if not more widely, than nullipars.

5.7 Longevity, seasonality and response to varying conditions

Adult *S. damnosum* are regarded as having a maximum lifespan of 3–4 weeks, but this does not explain the observation that in some cases adults may appear before the rivers begin to flow again after the dry season. It is not clear whether these adults arrive from unidentified breeding sites, which have persisted during the dry season, or whether the flies have been aestivating.

Many species of *Simulium* have several generations a year. When such species overwinter as growing larvae, as in *S. monticola*, there are likely to be large size variations between adults produced in the different seasons. Winter larvae produce larger adults than summer larvae. Changes in the total biovolume of the adult are found to be related inversely to mean water temperature. In addition, size may also be influenced by the quantity of food available and the photoperiod.

Simuliids have to survive periods when the temperature is too low to sustain normal activities and when rivers cease to flow in the dry season, which may be of indeterminate duration. Larvae of *P. mixtum* and *Prosimulium fuscum* grow actively during the winter at temperatures near freezing point, but in certain populations of *P. fuscum* larvae are dormant below 4°C, thus reducing their respiratory rates substantially and replacing trehalose in the haemolymph by high polyhydric alcohols (and when the temperature rises above 4°C, the reverse process is completed rapidly and growth resumes). *Simulium vernum* has been taken in light traps in Scotland in every month of the year, but it may be present only from March or April to November. Perhaps this species has overwintering adults which can be active or dormant, depending on the temperature.

In the severe climate of high latitudes, simuliids show various adaptations to survival, involving reducing the time spent in the adult stage, which is the most vulnerable to low temperatures and high winds. Eight of the nine species restricted to the Canadian tundra, for example *Simulium baffinense*, are autogenous and have reduced mouthparts; their females not only do not need to seek a blood meal, but they are unable to feed. The risk to the species' survival, if the males had to swarm, is avoided by mating occurring on the ground, where adults cluster near the breeding site. As an adaptation to that behaviour, the eyes of the male are sometimes dichoptic, as in the female, for example with *Gymnopais dichopticus*. As a further adaptation, a species may be both autogenous and parthenogenetic, with the adult female becoming gravid in the pupa, for example with *Prosimulium ursinum*, and there is virtually no free adult life.

Aridity poses other concerns. *A. pestilens* survives a dry season of uncertain length as viable eggs deep in the moist, sandy beds of transient rivers. Eggs of *Cnephia pecuarum* laid in April remain dormant until November before developing and hatching in December, so carrying the species over the summer months when many breeding places cease to flow.

6. MEDICAL AND VETERINARY IMPORTANCE

Simuliids are important as serious nuisance pests in their own right, as well as being vectors of pathogens.

6.1 Black flies as pest species

Simuliids have a well-deserved reputation as pests of humans, but mostly of livestock, in many areas of the world, particularly in North America, where *S. venustum* and *P. mixtum* are troublesome, but also in Europe and Australia, where *Simulium posticatum* and *A. pestilens*, respectively, are notorious for the severe reactions to their bites. In Iceland, *S. vittatum* is a renowned pest for humans and livestock and is said to occur at densities among the highest ever recorded for black flies in the world (around 1 million/m^2). Overall, 29 species of Simuliidae have been listed as pests of humans or domestic animals, of which 24 are species of *Simulium*, three of *Austrosimulium* and one each of *Prosimulium* and *Cnephia*.

Along the lower reaches of the Mississippi River, *C. pecuarum* caused the death of large numbers of livestock before flood control eliminated them as pests. In Serbia and Romania, the production of vast numbers of the Golubatz fly, *Simulium columbaschense*, associated with the Danube River, has caused thousands of deaths of livestock over a very wide area in past times. Even when the numbers of simuliids biting do not cause deaths, livestock can be less thrifty, with lower weight gains in beef cattle and decreased milk production in dairy cattle (e.g. following the floods of 1974, *A. pestilens* reduced milk yields by up to 15% in parts of Queensland, Australia).

6.2 Vectors of disease

The most serious human disease associated with simuliids is onchocerciasis (known as river blindness), due to infection with the filarial worm, *O. volvulus*. Thirteen species plus four unnamed forms of *Simulium* in Africa and eight species of *Simulium* in Latin America are involved in the transmission of human onchocerciasis. Further, simuliids are vectors of other *Onchocerca* spp. to livestock, they transmit blood-dwelling protozoa of the genus *Leucocytozoon*, which causes a malaria-like infection in birds (including domestic poultry), and they have been shown to be able to transmit vesicular stomatitis virus to cattle and pigs.

6.2.1 Human onchocerciasis (*Onchocerca volvulus*)

Human onchocerciasis is an economically important disease which is not directly fatal but causes untold misery in certain areas of Africa, Latin America and Yemen. It is caused by infection with the nematode, *O. volvulus*, and is transmitted by black flies of the genus *Simulium*. It places an intolerable burden on whole communities and denies vast fertile areas to human settlement and agricultural development.

Onchocerciasis by *O. volvulus* is not a zoonosis, although natural infections have been found in a spider monkey (*Ageles geoffroyi*) in Guatemala and a gorilla in the Congo, and chimpanzees can be infected in the laboratory. The helminth is passed from person to person by the bites of various *Simulium* species and can only rarely be passed before the third blood meal (providing the simuliid becomes infected at its first blood meal).

It has been estimated that there are 120 million people worldwide at risk of infection and about 18 million people infected, of which 270,000 are blind, 500,000 have some sort of visual impairment and 6.5 million suffer from itching and dermatitis. The greatest incidence is in tropical Africa, where the disease occurs throughout the northern Sudano-Guinean savannah of West Africa, east to the Ethiopian highlands and Uganda; it extends south through the rainforest of West Africa and the Democratic Republic of Congo to Angola and eastwards to Tanzania and Malawi. In Yemen, it is present in a few permanent wadis. In Latin America, onchocerciasis is present in adjoining areas of Mexico/Guatemala, Colombia/Ecuador, Brazil/Venezuela and in northern Venezuela.

THE DISEASE AND THE PARASITE

Onchocerciasis has three manifestations: an unsightly and irritating dermatitis, subcutaneous nodules and eye lesions which result in blindness. The long, slender, long-lived adult worms are to be found free in the subcutaneous tissue, or more commonly encapsulated in nodules, which may be in clusters 10 cm across. The larger female worm measures 23–50 cm in length and the males are 16–42 cm long. Apart from the unsightly nodules, the adult worms are innocuous. It is the migrating microfilariae, measuring 220–360 μm, which cause damage and provoke inflammatory lesions when they degenerate. The density of microfilariae is greatest in the vicinity of adult worms.

The distribution of nodules and microfilariae differs with the strain of parasite, being most abundant on the lower part of the body in West Africa, mainly around the buttocks and upper thigh in East Africa and on the torso in Central America. This distribution is correlated with, but not necessarily determined by, the biting habits of the vectors. In Central America, *S. ochraceum* attacks the upper part of the body and in West Africa, *S. damnosum* bites on the lower parts of the body.

Further, the local disease endemicity is related to the local vector species; in parts of West Africa, the disease can be hyperendemic (77–94%) in areas where *Simulium yahense* is the vector, but hypoendemic (26–50%) in areas where *Simulium soubrense* is the vector, because *S. yahense* is highly anthropophilic and highly susceptible to *O. volvulus*, while *S. soubrense* is predominantly zoophilic and has low capacity for transmission of *O. volvulus* to humans.

The presence of large numbers of active microfilariae in the skin causes intense itching and scratching, leading to loss of pigment in patches in the affected area. These contrast strikingly with the normal dark skin and are an obvious sign of infection. In the later stages, there is thickening of the skin and a loss of elasticity, giving the sufferer a prematurely aged appearance. Scarring in the lymph node can cause regional lymphoedema and result in a condition known as 'hanging groin'.

Onchocerciasis develops slowly, with blindness being rare in people under 20 years of age and rising to more than 50% in those more than 50 years old. Living microfilariae may be found in many parts of the eye and cause little damage, but when they die they induce lesions. Snowflake ocular opacities in the cornea are temporary but sclerosing keratitis of the cornea is progressive, eventually leading to blindness.

DEVELOPMENT OF *ONCHOCERCA VOLVULUS* IN *SIMULIUM*

It was shown in the 1920s that when *O. volvulus* was ingested by *S. damnosum* feeding on a host with dermal microfilariae, the nematodes developed in the thoracic muscles, giving rise to infective (third-stage) larvae, which escaped from the labium of the fly during feeding. The worm undergoes a very similar cycle of development to that of *Wuchereria bancrofti* in its mosquito vector. In the human host, infective larvae moult to the fourth stage in 3–7 days and to juvenile worms several weeks later. There follows a premature period of 9–12 months, after which mated females begin to produce microfilariae. Female worms are sessile in nodules but males regularly leave nodules.

A feeding female simuliid takes up microfilariae as it scrapes its way through the skin, and hence the uptake of microfilariae is more likely to be related to the distribution of microfilariae in the skin and the time taken to penetrate it than to the size of the blood meal. Indeed, there is a great deal of variation in the uptake of microfilariae by flies feeding on the same host, varying the intake of microfilariae with their concentration in the skin.

Microfilariae must avoid being trapped within the peritrophic membrane, which is secreted around the blood meal. In *S. ochraceum* and *S. metallicum*, most microfilariae have escaped before the membrane hardens and becomes impenetrable. In *S. neavei*, the thin, delicate membrane is a less formidable barrier, but in *S. damnosum*, the membrane is a major source of worm mortality. Once microfilariae have escaped from the midgut, a high proportion (~90%) may complete development to become infective larvae (third stage). However, the success rate for ingested microfilariae developing into infective forms can be around 35–40%, taking at least 8 days at 20°C and 4 days at 30°C.

Development of *O. volvulus* may or may not damage the simuliid host. Black flies with a high infection rate arrive in West Africa from localities several hundred kilometres away, but in Latin America, female *S. metallicum* and *Simulium exiguum* that have fed on heavily infected carriers of *O. volvulus* suffer high mortality as a result of physical damage. *S. ochraceum* does not suffer this damage because a high proportion of the ingested microfilariae are damaged by its buccopharyngeal armature, similar to the destruction wrought on microfilariae of *W. bancrofti* by the comparable armature of *Anopheles gambiae*.

STRAINS OF *ONCHOCERCA VOLVULUS*

The strains of *O. volvulus* in Central America and Africa behave quite differently, partly in adaptation to the vectors. In Central America, microfilariae are seven to ten times as abundant in the face, neck and arms as in the legs, making them more accessible to the vector, *S. ochraceum*, which feeds preferentially on the upper parts of the body. In contrast, in Africa, microfilariae are more abundant in the legs and lower parts of the body, where *S. damnosum* does 98% of its biting.

When fed on carriers with comparable densities of microfilariae in the skin, *S. ochraceum* ingests 20–25 times more of the Guatemalan strain of *O. volvulus* than of either the forest or Sudan savannah strains from West Africa, since in the latter cases there is no concentration of microfilariae. The success rate of microfilariae development of the forest strain *O. volvulus* in *S. damnosum* was 47% compared with 2% for microfilariae of the same strain in *S. ochraceum*, although slightly more microfilariae were ingested by *S. ochraceum*.

Two parasite–vector complexes, one existing in the forest and Guinea savannah and the other in the Sudan savannah, have been recognized. Microfilariae of *O. volvulus* from the forest area of Cameroon develop well in *S. damnosum* of the forest and Guinea savannah zones, but not in *S. damnosum* from the Sudan savannah, and vice versa for microfilariae from the Sudan savannah. This difference is considered to reside in the peritrophic membrane because when microfilariae of the savannah and forest strains are injected into the thorax, they develop equally well in both savannah (*S. damnosum*, *S. sirbanum*) and forest (*S. squamosum*, *Simulium mengense*) vectors.

The differences between vectors can be related to sibling species of the *S. damnosum* complex. No comparable separation is available for the recognition of forms of *O. volvulus*, but there is evidence for the existence of two strains of *O. volvulus* in West Africa differing in their pathogenicity, biochemical structure and vectors. The savannah strain is associated with *S. damnosum* s.s. and *S. sirbanum* and causes severe disease involving serious eye lesions and blindness, urinary excretion of microfilariae and depressed immunity in the host. The forest strain of *O. volvulus* has low human pathogenicity, is poorly transmitted by *S. damnosum* and *S. sirbanum*, but is well adapted to local vectors and shows certain characteristic biochemical features.

SIMULIID VECTORS OF *ONCHOCERCA VOLVULUS* IN AFRICA AND YEMEN

S. damnosum had been considered to be a single species distributed widely throughout the Afrotropical region before examination of the polytene chromosomes of different populations led to the recognition of four different forms in 1966, which by 1987 had increased to 42 and, in 2009, the *S. damnosum* complex was considered to comprise 57 cytoforms grouped into six subcomplexes.

At least 13 taxa (nine named species and four siblings) in the *S. damnosum* complex are thought to be vectors of *O. volvulus*. They include the widespread *S. damnosum* s.s., which occurs from West Africa eastwards to the southern Sudan and Uganda, and *S. sirbanum*, which occurs from Senegal to the southern Sudan and northwards along the Nile. They are long-range migrants which regularly recolonize large, open seasonal rivers. Other vectors in West Africa are *Simulium sanctipauli* and *S. soubrense* in forest areas, *S. squamosum* and *S. mengense* in the forest–savannah mosaic and *S. yahense* in upland forest. *S. sanctipauli* and *S. squamosum* breed in smaller, permanently flowing, shaded rivers, from which they disperse little. *Simulium kilibanum* is a vector in some foci in eastern Africa and *Simulium rasyani*, which breeds in sun-warmed, west-flowing wadis of variable water flow, is the vector in Yemen. Of the four siblings, two are vectors in Tanzania, one in eastern Zaire and the other in Ethiopia.

The distributions of these species can be related to ecological zones. With increasing distance from the coast, zones of increasing aridity can be recognized. Onchocerciasis extends into the Sudan savannah, but not further inland. *S. sirbanum* is predominant in the Sudan savannah, spreads into the Guinea savannah and is almost absent from the forest zone. *S. damnosum* s.s. is abundant in the Guinea savannah, well represented in the Sudan savannah and present in smaller numbers in the forest. *S. squamosum* is mainly a forest species, being present in heavily shaded or forested areas in the Guinea savannah but almost absent from the Sudan savannah. The other three species are found mainly in the forest zone, although *S. soubrense* also occurs in the Guinea savannah.

Three species of the *S. neavei* group are vectors of *O. volvulus* in eastern Africa. *S. neavei* is a localized vector in Uganda and Zaire and was the vector in Kenya before it was eradicated from that country. *S. woodi* is the vector in the Usambara and Uluguru mountains of eastern Tanzania and *Simulium ethiopiense* is the vector in south-western Ethiopia. The larvae and pupae of species of the *S. neavei* group have a phoretic association with crabs. Those of *S. neavei* are found on the carapace of *Potamonautes* (*Potamon*) *niloticus*, which lives in the rockier parts of rivers near cascades, and the eggs of *S. neavei* are deposited on vegetation in clusters near cascades. *S. woodi* has a bimodal cycle of biting activity, with morning and afternoon peaks in which nulliparous flies are commoner in the morning and parous flies in the afternoon. When *S. woodi* feeds on humans, it attacks, almost exclusively, the legs – where the greatest concentration of microfilariae is to be found. *Simulium albivirgulatum*, which breeds in slow to swift but smooth-flowing rivers lined by gallery forest, is a vector in the Zaire River basin.

SIMULIID VECTORS OF *ONCHOCERCA VOLVULUS* IN LATIN AMERICA

The vectors of onchocerciasis in Latin America are designated as primary or secondary vectors, depending on whether or not they are considered capable of sustaining transmission on their own, respectively. The position is complicated by the existence of cytoforms within species. *S. ochraceum* s.l. is a predominantly anthropophilic species occurring in high rainfall areas between 500 and 1500 m above sea level (a.s.l.), where it breeds in trickles and streams less than 0.5 m in width under deep vegetation and can disperse 10 km from the breeding site. It is diurnal, with peak activity occurring mid-morning, when it attacks selectively the upper region of the body where the microfilariae of *O. volvulus* are most abundant. In northern Venezuela, *S. metallicum* s.l. maintains foci of onchocerciasis up to 1000 m a.s.l.. It is largely an anthropophilic species, feeding below the waist where microfilarial densities are high. In Amazonia, *Simulium guianense* maintains hyperendemic onchocerciasis 250 m a.s.l., where it is the most common human biter. In lowland Amazonia, *Simulium oyapockense* is considered to be a poor vector, maintaining onchocerciasis by its high biting rate supplemented by the movement of infected humans from the highlands. In Colombia/Ecuador, a single endemic focus of onchocerciasis is present in lowland coastal forest up to a height of 200 m a.s.l., where the vector *S. exiguum* breeds in rivers more than 10 m wide. Transmission occurs at the end of the rainy season, when low vector infectivity rates are countered by high human-biting rates.

CONTROL OF ONCHOCERCIASIS

The problems of controlling onchocerciasis are particularly formidable because of the longevity of the parasite and the habits of the vector, which is both exophilic and exophagic. This pattern of behaviour renders ineffective most methods of personal protection, but some protection can be achieved with repellents and the wearing of suitable clothing. Adult simuliids disperse widely and the population is most concentrated in the immature stages, which occur in running water. Thus, control measures have been directed against the immature stages, with some success.

Control requires either the eradication of the vector or for the vector population to be kept at a level at which transmission does not occur for 15 years, the maximum life of the adult worm. Four stages in the development of control measures have been recognized: the

pre-DDT era, 1932–1944, the chlorinated hydrocarbon period, 1944–1972, the temephos period, 1972–1992, and the recent period of integrated control involving the use of the microfilaricide ivermectin and the use of a range of larvicides (including three organophosphorus compounds (temephos, phoxim, pyraclofos), permethrin, carbosulfan and *Bacillus thuringiensis israelensis* (Bti)) to minimize the development of insecticide resistance.

Foci of onchocerciasis sustained by *S. neavei* tend to be compact and isolated, favouring eradication – as in the inland hilly region of western Kenya, where an area of 12,000 km^2 was cleared from onchocerciasis by eradicating *S. neavei* from 323 rivers by treating them with DDT at a dosage which killed the larvae but appeared to have no effect on fish or on the crabs with which *S. neavei* had a phoretic association. The two early successes against *S. damnosum* s.l. were in Uganda. One involved its eradication along the Nile from Lake Victoria to Lake Kyoga by taking advantage of the controlled water outlet from the Owen Falls hydroelectric station to introduce DDT into the released water and control breeding of *S. damnosum* for 80 km below the dam. This measure allowed its eradication over 4000 km^2 and made the area available for more intensive settlement and agriculture without the fear of onchocerciasis. In the Ruwenzori second focus, the anthropophilic sibling *S. damnosum* was replaced by the zoophilic *Simulium adersi* and *Simulium griseicolle*, possibly due to the larvae having different feeding habits. With careful applications, chemicals such as temephos could be effective for some 20–40 km downstream from its application point, but the problem of treating all watercourses was a major one and the capacity of the flies to disperse widely and re-colonize rivers and streams after treatments was also problematic.

DIAGNOSIS AND TREATMENT

Onchocerciasis is commonly diagnosed by a skin snip, in which a small piece of skin is removed and examined for the presence of living microfilariae. In Guatemala and Mexico, where nodules on the head are common, their surgical removal is popular and this may have reduced the incidence of ocular onchocerciasis. In Ecuador, the more drastic removal of all palpable nodules improves the clinical condition of the patient, but nodulectomy is not practised widely in Africa.

For human onchocerciasis, chemotherapy usually involves a microfilaricide such as ivermectin that is less toxic than diethylcarbamazine and safe enough for large-scale use. It is a microfilarial suppressant with no lethal action on the adults (although it reduces their fecundity and longevity). It has been extremely useful in reducing community blindness and, in some areas, it has eliminated onchocerciasis transmission with at least 15 years of annual treatment. However, this chemoprophylactic protocol is not always successful, since there have been reports of possible ivermectin resistance arising in parts of Africa.

Moxidectin is a drug closely related to ivermectin, but with higher potency in some animal models and an ability to kill adult worms; it has been undergoing field trials, but its use may be limited by its close relationship to ivermectin, where resistance to that drug occurs.

In the continuing search for a safe and effective macrofilaricidal drug, a recent novel approach has been the investigation of the antibiotic doxycycline as an anti-*Wolbachia* drug to exploit the symbiosis between the onchocercal nematode and its *Wolbachia* bacteria.

THE ONCHOCERCIASIS CONTROL PROGRAMME (OCP) IN WEST AFRICA

This ambitious programme was launched in 1974 (covering 30 million people in 11 countries at its peak) with the aim of controlling onchocerciasis in a selected area of Sudan and Guinea savannah covering 654,000 km^2 extending across seven countries – Benin, Burkina Faso, Ghana, Ivory Coast, Mali, Niger, Togo – where a million people suffered from onchocerciasis and 70,000 were blind or had seriously impaired vision. The control area was based on the Volta River system and involved the weekly treatment of 14,500 km of rivers. In 1978, the area was increased by the addition of another 110,000 km^2 in the Ivory Coast, bringing the area covered to 764,000

km^2 and involving treatment in the wet season of 18,000 km of rivers.

In 1988, the area was extended to include another 8 million people in Guinea, Guinea Bissau, Senegal and Sierra Leone, bringing the area covered to 1,300,000 km^2 and involving treatment of 46,000 km of rivers. The scheme was outstandingly successful. Transmission was reduced to zero in 90% of the original central area and the prevalence of onchocerciasis fell from 70% to 3%. After 7–8 years of control, the *O. volvulus* population in the OCP was ageing and dying, with the productivity index of the parasite in ten villages being reduced by 97% to 55%. In the seven countries included in the scheme from its onset, 16.5 million people are no longer at risk of the disease.

The OCP was very successful in reducing river blindness in West Africa and it was closed in 2002, when most countries in the programme area had a prevalence of onchocerciasis of less than 5%. Building on the knowledge and experience gained in the OCP, a second programme, named the African Programme for Onchocerciasis Control (APOC), was established to combat the rest of Africa's river blindness. This was a bigger partnership programme than the OCP and included 19 participating countries with active involvement of their governments and their affected communities, international and local Non-Government Development Organizations, private sector, donor countries and United Nations agencies, including the World Health Organization.

6.2.2 Onchocerciasis of veterinary concern

Seven species of *Onchocerca* are considered to be of veterinary importance – five in cattle (*Onchocerca gibsoni*, *Onchocerca gutturosa*, *Onchocerca lienalis*, *Onchocerca ochengi* and *Onchocerca armillata*), two (*Onchocerca cervicalis* and *Onchocerca reticulata*) in horses and one in dogs, cats and wild carnivores (*Onchocerca lupi*); however, the species are not necessarily host specific. Onchocerciasis of cattle, water buffaloes and horses occurs widely throughout the world, wherever these domestic animals have been introduced and there are suitable vectors. In the past, *O. lienalis* has been thought an unobtrusive parasite which produces no marked patho-logical changes or evidence of clinical disease. However, losses in the beef industry arise because free worms cause aesthetically displeasing blemishes to the carcass and encapsulated adult worms in nodules have to be trimmed from the carcass. Microfilariae occur in the skin and subcutaneous lymph (with the exception of those of *O. armillata*, which occur in the blood). Microfilariae are ingested by bloodsucking flies and develop in various species of *Simulium* (but also in species of *Culicoides* and *Lasiohelea* cera-topogonids; see entry for 'Biting Midges').

ONCHOCERCA SPECIES IN DOMESTIC BOVINES
Adult *O. lienalis* occur in the gastrosplenic ligament and its microfilariae are concentrated in the region of the umbilicus, while adult *O. gutturosa* are found in the ligamentum nuchae and its microfilariae in the skin of the head, neck and back. Nodule-forming onchocercas are classified in terms of their geographical distribution and the location of the adults in the bovine host. In Africa, adults of *Onchocerca dukei* are subcutaneous and those of *O. ochengi* are dermal parasites. In Asia, adults of *O. gibsoni* and possibly those of *Onchocerca indica* occur in the subcutaneous tissues, *Onchocerca cebei* and possibly *Onchocerca sweetae* in the dermis and *O. armillata* in the wall of the thoracic aorta. *O. ochengi* causes dermatitis on the scrotum and udder resembling mange or pox. Nodules of *O. gibsoni* are most prevalent on the brisket and also occur on the stifle and thigh.

Microfilariae of *O. lienalis* and *O. gutturosa* occur at a depth of about 1 mm from the skin surface and are ingested by *S. ornatum* when feeding. Development is relatively slow, with the 'sausage' stage being reached in 10 days and infective forms being present in the head 19 days after the infective feed. The number of microfilariae ingested is a function of the period of time spent feeding and not of the volume of blood (usually about 3 mg) ingested.

Microfilariae of *O. ochengi* are located in the umbilical area and legs of cattle. They developed normally in females of a species of the *S. damnosum* complex, probably *S. sanctipauli*, with infective larvae of *O. ochengi* being present 6 days after an infective feed.

Although microfilariae of *O. gutturosa* and *O. dukei* were also ingested by *S. damnosum* s.l., they did not develop. In North Cameroon, *Simulium bovis* is an efficient vector of *O. dukei* but not of *O. ochengi*.

ONCHOCERCA SPECIES OF ZOONOTIC CONCERN
Cases of zoonotic onchocerciasis have been reported worldwide and attributed to four nematode species: *O. gutturosa* and *O. cervicalis* from cattle and horses, respectively, *Onchocerca jakutensis* from the European deer in Austria and *Onchocerca dewittei japonica* from wild boar in Japan and *O. lupi* from dogs in Turkey and Tunisia. The above zoonotic *Onchocerca* spp. were found mostly in the subcutaneous tissues and only *O. gutturosa*, *O. cervicalis* and *O. lupi* presented an ocular localization.

6.2.3 *Leucocytozoon* (Leucocytozoidae)
About 70 species of *Leucocytozoon* have been named and they are all parasites of birds. With species of *Leucocytozoon*, schizogony occurs in the tissues and only the gametocytes appear in the peripheral circulation. Gam-etocytes occur in both leucocytes and erythrocytes, but no pigment is produced in the latter.

Economically important species include *Leucocytozoon simondi*, a parasite of domestic and wild ducks and geese in Europe, North America and South-east Asia, *Leucocytozoon smithi*, a parasite of turkeys in Europe and North America, and probably the most important, *Leucocytozoon caulleryi*, which parasitizes chickens in South-east Asia and Africa. Although the vectors of most species of *Leucocytozoon* are species of Simuliidae of various genera, including *Simulium*, *Prosimulium*, *Eusimulium* and *Cnephia*, *L. caulleryi* is transmitted by biting midges of the genus *Culicoides* and is not further mentioned here (see entry for 'Biting Midges').

The life cycle of *L. simondi* has been studied in considerable detail. An infective simuliid injects sporozoites when feeding and in a susceptible host they develop in the parenchyma of the liver. These hepatic schizonts contain several thousand merozoites, which invade erythrocytes and erythroblasts to become rounded gametocytes in 48 h. Maximum parasitaemia is reached after 10–12 days, after which the infection can remain chronic for 2 or more years in ducks. The host cell in which a megaloschizont develops becomes hypertrophied. The density of gametocytes in the peripheral circulation shows a diurnal periodicity, with peak numbers being reached in the daytime when simuliids are active, and hence favours gametocytes being taken up by these vectors. The gametocytes undergo exflagellation and when a microgamete penetrates a macrogamete, the two nuclei fuse and the zygote becomes a motile ookinete. The ookinete penetrates between the cells of the midgut and forms an oocyst beneath the basement membrane. A mature oocyst does not rupture, but sporozoites escape gradually from the cyst and move to the salivary glands, which they penetrate. Sporogony occurs at a variable rate even under identical conditions. In the same vector species, sporogony can take 6–18 days at 18–20°C.

The cycle of *L. smithi* in turkeys is slightly different. The primary cycle occurs in the liver and merozoites released from that cycle either form rounded gametocytes or invade the liver or kidney, producing a second cycle of schizogony; however, the schizont and its host cell are not hypertrophied and no megaloschizont is formed.

During winter, the parasitaemia in the avian host is low and a small increase occurs in early spring in response to the host's developing reproductive cycle. This ensures that gametocytes will be available in the peripheral blood when the vectors appear in the spring, and when susceptible nestlings will be available for infection.

Species, such as *L. simondi*, which include a megaloschizont stage in the reticuloendothelial system in their cycle are more pathogenic than those, such as *L. smithi*, in which schizogony occurs in the liver and kidneys. Ducks infected with *L. simondi* show some or all of the following symptoms: lethargy, loss of appetite, diarrhoea, convulsions and anaemia, and the condition may be fatal. The anaemia cannot be accounted for by simple parasitization of the erythrocytes but involves intravascular haemolysis. There is conflicting evidence as to the pathogenicity of *L. smithi* in turkeys, and it may be more important in the presence of other diseases. In general, *Leucocytozoon* infections are more severe in domestic than wild species, and deaths are commoner in young birds.

7. PREVENTION AND CONTROL

Typically, it is the larvae of black flies that are targeted for control because their habitats are usually obvious. It is much more difficult to attack the adults, which are diurnally active and often disperse considerable distances, but fogging and misting may give some temporary local relief. Pour-on insecticides have been useful with livestock in some circumstances, and topical repellents containing diethyl toluamide (DEET) can protect humans.

For many years, organochlorine insecticides were used effectively against black fly larvae in riverine habitats, but they were replaced gradually by organophosphates, principally temephos, and then by the biocides containing *B. t. israelensis* (known as Bti), which has been used increasingly to control pest and vector species with little adverse impacts on non-target animals in the habitats.

SELECTED BIBLIOGRAPHY

Adler, P.H. and Crosskey, R.W. (2012) World blackflies (Diptera: Simuliidae): a comprehensive revision of the taxonomic and geographical inventory. 119 pp (http://www.nhm.ac.uk/research-curation/research/projects/blackflies/Intro.pdf, accessed 12 June 2012).

Adler, P.H., Currie, D.C. and Wood, D.M. (2004) *The Black Flies (Simuliidae) of North America*. Cornell University Press, Ithaca, New York, 941 pp.

Adler, P.H., Cheke, R.A. and Post, R.J. (2010) Evolution, epidemiology, and population genetics of black flies (Diptera: Simuliidae). *Infection Genetics and Evolution* 10, 846–865.

Basáñez, M.G., Churcher, T.S. and Grillet, M.E. (2009) Simulium interactions and the population and evolutionary biology of *Onchocerca volvulus*. *Advances in Parasitology* 68, 236–313.

Bass, J.A.B. (1998) *Last-instar Larvae and Pupae of the Simuliidae of Britain and Ireland. A Key with Brief Ecological Notes*. Freshwater Biological Association, Cumbria, UK, Scientific Publication 55, 102 pp.

Bennett, G.P., Peirce, M.A. and Ashford, R.W. (1993) Avian haematozoa: mortality and pathogenicity. *Journal of Natural History* 27, 993–1001.

Boatin, B.A. and Richards, F.O. Jr (2006) Control of Onchocerciasis. *Advances in Parasitology* 61, 349–394.

Boatin, B., Molyneux, D.H., Hougard, J.M., Christensen, O.W., Alley, E.S., Yameogo, L., *et al.* (1997) Patterns of epidemiology and control of onchocerciasis in West Africa. *Journal of Helminthology* 71, 91–101.

Collins, R.C. (1979) Development of *Onchocerca volvulus* in *Simulium ochraceum* and *Simulium metallicum*. *American Journal of Tropical Medicine and Hygiene* 28, 491–495.

Crosskey, R.W. (1981) Geographical distribution of Simuliidae. In: Laird, M. (ed.) *Blackflies*. Academic Press, New York, pp. 57–68.

Crosskey, R.W. (1990) *The Natural History of Blackflies*. John Wiley, Chichester, UK, 711 pp.

Crosskey, RW. (1993) Blackflies (Simuliidae). In: Lane, R.P. and Crosskey, R.W. (eds) *Medical Insects and Arachnids*. Chapman and Hall, London, pp. 241–287.

Crosskey, R.W. and Howard, T.M. (1997) *A New Taxonomic and Geographical Inventory of World Blackflies (Diptera: Simuliidae)*. The Natural History Museum, London, 144 pp.

Cupp, E.W. (1986) The epizootiology of livestock and poultry diseases associated with black flies. In: Kim, K.C. and Merritt, R.W. (eds) *Black Flies*. Pennsylvania State University, University Park, Pennsylvania, pp. 387–395.

Dang, P.T. and Peterson, B.V. (1980) Pictorial keys to the main species and species groups within the *Simulium Damnosum* Theobald complex occurring in West Africa (Diptera: Simuliidae). *Tropenmedizin und Parasitologie* 31, 117–120.

Davies, J.B. (1994) Sixty years of onchocerciasis vector control: a chronological summary with comments on eradication, reinvasion, and insecticide resistance. *Annual Review of Entomology* 39, 23–45.

Diawara, L., Traore, M.O., Badji, A., Bissan, Y., Doumbia, K., Goita, S.F., *et al.* (2009) Feasibility of onchocerciasis elimination with ivermectin treatment in endemic foci in Africa: first evidence from studies in Mali and Senegal. *PLoS Neglected Tropical Diseases* 3, e497.

dos Santos, R.B., Lopes, J. and dos Santos, K.B. (2010) Spatial distribution and temporal variation in composition of black fly species (Diptera: Simuliidae) in a small watershed located in the northern of Parana State, Brazil. *Neotropical Entomology* 39, 289–298.

Duerr, H.P., Raddatz, G. and Eichner, M. (2011) Control of onchocerciasis in Africa: threshold shifts, breakpoints and rules for elimination. *International Journal for Parasitology* 41, 581–589.

Duke, B.O.L. (1962) Studies on factors influencing the transmission of onchocerciasis. II. The intake of *Onchocerca volvulus* microfilariae by *Simulium damnosum* and the survival of the parasite in the fly under laboratory conditions. *Annals of Tropical Medicine and Parasitology* 56, 255–263.

Duke, B.O.L. and Taylor, H.R. (1991) Onchocerciasis. In: Strickland, G.T. (ed.) *Hunter's Tropical Medicine*. Saunders, Philadelphia.

Dunbar, R.W. and Vajime, C.G. (1981) Cytotaxonomy of the *Simulium damnosum* complex. In: Laird, M. (ed.) *Blackflies*. Academic Press, London, pp. 31–43.

Eichler, D.A. (1971) Studies of *Onchocerca gutturosa* (Neumann, 1910) and its development in *Simulium ornatum* (Meigen, 1818). II. Behaviour of *S. ornatum* in relation to the transmission of *O. gutturosa*. *Journal of Helminthology* 45, 259–270.

Eichler, D.A. (1973) Studies on *Onchocerca gutturosa* (Neumann, 1910) and its development in *Simulium* (Meigen, 1818). 3. Factors affecting the development of the parasite in its vector. *Journal of Helminthology* 47, 73–88.

Fallis, A.M. and Desser, S.S. (1977) On species of *Leucocytozoon*, *Haemoproteus* and *Hepatocystis*. In: Kreier, J.P. (ed.) *Parasitic Protozoa*. Academic Press, New York, Vol 3, pp. 239–266.

Fallis, A.M., Desser, S.S. and Khan, R.A. (1974) On species of *Leucocytozoon*. *Advances in Parasitology* 12, 1–67.

Figueiredo, R. and Gil-Azevedo, L.H. (2010) The role of *Neotropical blackflies* (Diptera: Simuliidae) as vector of the onchocercosis: a short review of the ecology behind the disease. *Oecologia Australis* 14, 745–755.

Garms, R., Lakwo, T.L., Ndyomugyenyi, R., Ipp, W., Rubaale, T., Tukesiga, E., *et al.* (2009) The elimination of the vector *Simulium neavei* from the *Itwara onchocerciasis* focus in Uganda by ground larviciding. *Acta Tropica* 111, 203–211.

Gebre-Michael, T. and Gemetchu, T. (1996) Anthropophilic blackflies (Diptera: Simuliidae) and onchocerciasis transmission in southwest Ethiopia. *Medical and Veterinary Entomology* 10, 44–52.

Gray, E.W., Fusco, R.A., Noblet, R. and Wyatt, R.D. (2011) Comparison of morning and evening larvicide applications on black fly (Diptera: Simuliidae) mortality. *Journal of the American Mosquito Control Association* 27, 170–172.

Hougard, J.M., Yaméogo, L., Sékétéli, A., Boatin, B. and Dadzie, K.Y. (1997) Twenty-two years of blackfly control in the onchorcerciasis control programme in West Africa. *Parasitology Today* 13, 425–431.

Hunter, D.M. and Moorhouse, D.E. (1976a) Comparative bionomics of adult *Austrosimulium pestilens* Mackerras and Mackerras and *A. bancrofti* (Taylor) (Diptera: Simuliidae). *Bulletin of Entomological Research* 66, 453–467.

Hunter, D.M. and Moorhouse, D.E. (1976b) The effects of *Austrosimulium pestilens* on the milk production of dairy cattle. *Australian Veterinary Journal* 52, 97–99.

Katabarwa, M.N., Eyamba, A., Nwane, P., Enyong, P., Yaya, S., Baldiagai, J., *et al.* (2011) Seventeen years of annual distribution of ivermectin has not interrupted onchocerciasis transmission in north region, Cameroon. *American Journal of Tropical Medicine and Hygiene* 85, 1041–1049.

Kim, K.C. and Merritt, R.W. (eds) (1987) *Black Flies. Ecology, Population Management and Annotated World List*. The Pennsylvania State University, University Park, Pennsylvania.

Krueger, A. (2006) Guide to blackflies of the *Simulium damnosum* complex in eastern and southern Africa. *Medical and Veterinary Entomology* 20, 60–75.

Lewis, D.J. and Bennett, G.F. (1974) The blackflies (Diptera: Simuliidae) of insular Newfoundland. II. Seasonal succession and abundance in a complex of small streams on the Avalon Peninsula. *Canadian Journal of Zoology* 52, 1107–1113.

Lewis, D.J. and Bennett, G.F. (1975) The blackflies (Diptera: Simuliidae) of insular Newfoundland. III. Factors affecting the distribution and migration of larval simuliids in small streams on the Avalon Peninsula. *Canadian Journal of Zoology* 53, 114–123.

Mackenzie, C.D., Homeida, M.M., Hopkins, A.D. and Lawrence, J.C. (2012) Elimination of onchocerciasis from Africa: possible? *Trends in Parasitology* 28, 16–22.

Mead, D.G., Howerth, E.W., Murphy, M.D., Gray, E.W., Noblet, R. and Stallnecht, D.E. (2004) Black fly involvement in the epidemic transmission of vesicular stomatitis New Jersey virus (Rhabdoviridae: Vesiculovirus). *Vector-Borne and Zoonotic Diseases* 4, 351–359.

Molyneux, D.H. (1995) Onchocerciasis control in West Africa: current status and future of the onchocerciasis control programme. *Parasitology Today* 11, 399–402.

Molyneux, D.H. and Davies, J.B. (1997) Onchocerciasis control: moving towards the Millennium. *Parasitology Today* 13, 418–425.

Morales-Hojas, R. and Krueger, A. (2009) The species delimitation problem in the *Simulium damnosum* complex, blackfly vectors of onchocerciasis. *Medical and Veterinary Entomology* 23, 257–268.

Otranto, D., Sakru, N., Testini, G., Gürlü, V.P., Yakar, K., Lia, R.P., *et al.* (2011) Case report: first evidence of human zoonotic infection by *Onchocerca lupi* (Spirurida, Onchocercidae). *American Journal of Tropical Medicine and Hygiene* 84, 55–58.

Otranto, D., Dantas-Torres, F., Cebeci, Z., Yeniad, B., Buyukbabani, N., Boral, O.B., *et al.* (2012) Human ocular onchocerciasis: further evidence on the zoonotic role of *Onchocerca lupi*. *Parasites and Vectors* 5, 84.

Post, R.J., Onyenwe, E., Somiari, S.A.E., Mafuyai, H.B., Crainey, J.L. and Ubachukwu, P. (2011) A guide to the *Simulium damnosum* complex (Diptera: Simuliidae) in Nigeria, with a cytotaxonomic key for the identification of the sibling species. *Annals of Tropical Medicine and Parasitology* 105, 277–297.

Remme, J.H.F. (1995) The African programme for onchocerciasis control: preparing to launch. *Parasitology Today* 11, 403–406.

Rodriguez-Perez, M.A., Unnasch, T.R., Dominguez-Vazquez, A., Morales-Castro, A.L., Pena-Flores, G.P., Orozco-Algarra, M.E., *et al.* (2010) Interruption of transmission of *Onchocerca volvulus* in the Oaxaca Focus, Mexico. *American Journal of Tropical Medicine and Hygiene* 83, 21–27.

Ross, D.H. and Merritt, R.W. (1987) Factors affecting larval black fly distributions and population dynamics. In: Kim, K.C. and Merritt, R.W. (eds) *Black Flies. Ecology, Population Management and Annotated World List*. The Pennsylvania State University, University Park, Pennsylvania, pp. 90–108.

Service, M.W. (2001) *Encyclopedia of Arthropod-transmitted Infections*. CAB International, Wallingford, UK.

Shelley, A.J. (1988) Vector aspects of the epidemiology of onchocerciasis in Latin America. *Annual Review of Entomology* 30, 337–366.

Shelley, A.J., Maia-Herzog, M., Dias, A.P.A.L., Camargo, M., Costa, E.G., Garritano, P., *et al.* (2001) Biting behaviour and potential vector status of anthropophilic blackflies in a new focus of human onchocerciasis at Minacu, central Brazil. *Medical and Veterinary Entomology* 15, 28–39.

Smith, P.F., Howerth, E.W., Carter, D., Gray, E.W., Noblet, R. and Mead, D.G. (2009) Mechanical transmission of vesicular stomatitis New Jersey Virus by *Simulium vittatum* (Diptera: Simuliidae) to domestic swine (*Sus scrofa*). *Journal of Medical Entomology* 46, 1537–1540.

Sutcliffe, J.F. (1986) Black fly host location: a review. *Canadian Journal of Zoology* 64, 1041–1053.

Taylor, M.J., Bandi, C. and Hoerauf, A. (2005) *Wolbachia* bacterial endosymbionts of filarial nematodes. *Advances in Parasitology* 60, 245–284.

Wetten, S., Collins, R.C., Vieira, J.C., Marshall, C., Shelley, A.J. and Basáñez, M.G. (2007) Vector competence for *Onchocerca volvulus* in the *Simulium* (*Notolepria*) *exiguum* complex: cytoforms or density-dependence? *Acta Tropica* 103, 58–68.

Wilson, M.D., Cheke, R.A., Flasse, S.P.J., Grist, S., Osei-Ateweneboana, M.Y., Tetteh-Kumah, A., *et al.* (2002) Deforestation and the spatio-temporal distribution of savannah and forest members of the *Simulium damnosum* complex in southern Ghana and south-western Togo. *Transactions of the Royal Society of Tropical Medicine and Hygiene* 96, 632–639.

Blow Flies and Screw-worm Flies (Diptera: Calliphoridae)

1. INTRODUCTION

The blow flies (Calliphoridae) belong to a large family of over 1000 species. Most blow fly species breed primarily in carrion. However, a small number have evolved the ability to act as parasites in their larval stage, feeding on living vertebrate hosts, including humans. This is known as myiasis. Some species, such as the Old World screw-worm, *Chrysomya bezziana*, and the New World screw-worm, *Cochliomyia hominivorax*, are obligatory agents of myiasis, meaning that they can only complete their larval development on a living host. Others are facultative myiasis agents, meaning that they can breed either on living host or in carrion. However, among the facultative species, the degree of adaptation to the living hosts varies widely and reflects the ability of these flies to initiate myiases. Primary flies, such as the greenbottle (or sheep blow fly), *Lucilia cuprina*, will readily initiate myiasis and, while it can breed in carrion, it survives relatively poorly there as a result of intense competition. Secondary flies, such as *Chrysomya rufifacies*, are largely unable to initiate myiasis and are more commonly found in carrion, but they will readily participate in myiasis once an animal has been infested. Tertiary flies, which are predominantly carrion breeding, become involved in myiasis at a late stage, when the host animal is almost dead.

The various species of blow fly produce different clinical types of myiasis, which can be classified according to the tissue and part of the body affected. Most commonly, blow flies cause dermal and subdermal myiases, which include wound (traumatic) myiasis. Occasionally, some species such as *Cordylobia anthropophaga* cause a furuncular myiasis, in which a boil-like condition is produced. Sanguinivorous myiasis is very rare and is caused by the bloodsucking larvae of a small number of species, for example species of the genera *Auchmeromyia* and *Protocalliphora*; the larvae of *Auchmeromyia* species (floor maggots) suck the blood of sleeping humans, while the larvae of *Protocalliphora* species (bird blow flies) live in bird nests and suck the blood of nestlings.

The ability of blow fly larvae to feed on decaying organic matter of animal origin may be exploited beneficially in human medicine. Maggot debridement therapy (also known as larvatherapy or biosurgery) uses the careful placement of disinfected or 'sterile' maggots (often of *Lucilia sericata*) to remove necrotic tissue in the treatment of chronically infected wounds where antibiotics are ineffective and surgery is impracticable.

2. TAXONOMY

All the flies that act as important agents of myiasis are calypterate Diptera in the superfamily Oestroidea. Within this superfamily there are three major families of myiasis-producing flies: Calliphoridae, Oestridae (see entries for 'Nostril Flies' and 'Stomach Bot Flies') and Sarcophagidae (see entry for 'Flesh Flies'). The Calliphoridae is a large family, composed of over 1000 species divided between 150 genera. At least 80 species have been recorded as causing traumatic, cutaneous myiasis. These species are found largely in four important genera: *Cochliomyia*, *Chrysomya*, *Lucilia* and *Calliphora*. The genera *Protophormia* and *Phormia* are of minor importance. Most of these species are either primary or secondary facultative invaders. Only two species of Calliphoridae, the screw-worm flies *Ch. bezziana* and *C. hominivorax*, are obligate agents of myiasis.

3. MORPHOLOGY

The Calliphoridae are medium to large calypterate Diptera, most of which have a

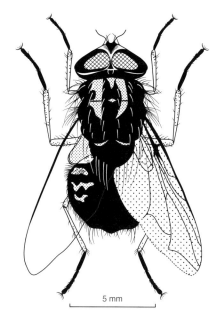

Fig. 15. Male *Calliphora stygia*. *Source*: Colless and McAlpine (1991).

metallic-blue or green sheen as adults (Fig. 15), while a few are testaceous (red-brown). The Calliphoridae are characterized by the presence of a strong row of hypopleural bristles on the thorax, with a weak or absent post-scutellum. There is a sharp bend in the wing discal vein (M_{1+2}), towards vein R_{4+5}, but the two do not meet before the margin (Fig. 16).

The genera *Cochliomyia* and *Chrysomya* are found in the warmer parts of the world and the adults may be distinguished by the fact that the stem vein (stem of veins R_1, R_{2+3} and R_{4+5}) bears a row of short, fine hairs posteriorly, the bristles on the scutum are poorly developed and there is no external posthumeral bristle. *Cochliomyia* are green to violet-green blow flies with tiny palps, three prominent black, longitudinal stripes on the scutum and the upper surface of the lower squama mostly bare. *Chrysomya* are green to bluish-black blow flies with normal palps, no bold black stripes on the scutum and the upper surface of the lower squama hairy (Fig. 17). *Cochliomyia* is restricted to the New World and *Chrysomya* to the Old World, although occasional inadvertent introductions may occur.

In the genera *Chrysomya*, *Lucilia*, *Protophormia* and *Phormia*, the radial stem vein is bare, the bristles on the scutum are well developed and there are either two notopleural bristles and two anterior plus one posterior sternopleural bristles, or three notopleurals and one plus one sternopleurals. The external posthumeral is located lateral to the level of the presutural.

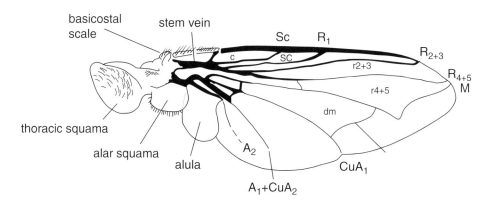

Fig. 16. The veins and cells of the wing of a typical calliphorid, *Calliphora vicina*.

The genera *Lucilia* and *Calliphora* are widely known (particularly in the northern hemisphere), respectively, as greenbottles and bluebottles, but as green blow flies and brown blow flies in Australia and sheep and golden blow flies in New Zealand. Adult *Lucilia* have glossy green or coppery green thorax and abdomen, a bare lower squama and measure 6–9 mm. *Calliphora* are larger flies (10–14 mm), often with black thorax, a steely blue to blue-black, slightly metallic abdomen (e.g. *Calliphora vicina*), but others are more grey-brown coloured (e.g. *Calliphora stygia*) or distinctively golden brown (e.g. *Calliphora ochracea*) in appearance and the lower squama has a hairy upper surface.

The Phormiinae, containing the monotypic genera *Protophormia* (*Protophormia terraenovae*) and *Phormia* (*Phormia regina*) are dark blue to black holarctic calliphorids with a ciliated stem vein and a bare lower squama. The testaceous calliphorid, *Auchmeromyia*, is found in tropical Africa; it is reddish-yellow or reddish-brown, with reddish-yellow legs.

The larvae of *Ch. bezziana* and *C. hominivorax*, which are obligatory agents of myiasis, are armed with broad, encircling bands of spines, which give them an undulating outline and the common name of screw-worms. The posterior spiracles consist of three straight slits surrounded by an incomplete peritreme with an indistinct button in the unsclerotized zone (Fig. 18). The larvae of some other *Chrysomya* species (*Ch. rufifacies* in the Australian and Oriental regions and *Chrysomya albiceps* in the Palaearctic and Afrotropical regions) are known as hairy maggots and have lateral rows of fleshy tubercles; they feed on the vertebrate host and are also predatory on primary blow fly larvae. The larvae of *Lucilia* and *Calliphora* are smooth-bodied maggots in which the posterior spiracles have three slits surrounded by a continuous peritreme ring which includes the button (Fig. 19); larvae of *Calliphora* have an additional accessory sclerite between the mouth-hooks, but this is lacking in *Lucilia*.

4. LIFE CYCLE

Most blow flies are oviparous, laying 200–300 eggs directly on to the host. The egg stage usually lasts about 24 h and is followed by three larval stadia, during which feeding occurs. When feeding is complete, the third-stage larva enters a wandering phase, wherein it leaves the host and locates a suitable site for pupation, usually burrowing into the ground. After pupation, the newly emerged adult breaks out of the puparium and works its way to the surface. The adults of most species of blow fly require a protein meal to initiate egg production.

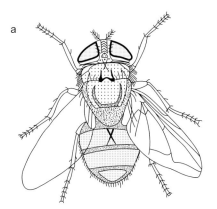
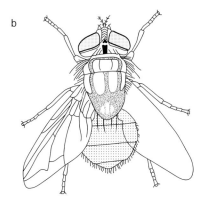

Fig. 17. (a) Female *Chrysomya bezziana* and (b) *Cochliomyia hominivorax. Source*: (a) redrawn from James (1947); (b) redrawn from Spradbery (1991).

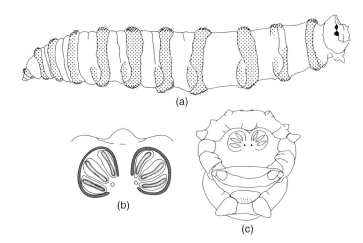

Fig. 18. Third-stage larva of *Chrysomya bezziana*. (a) Lateral view of larva; (b) posterior spiracles; (c) posterior view of larva.

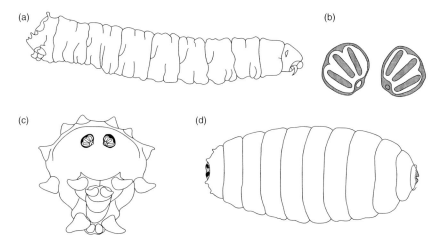

Fig. 19. (a) Third-stage larva of *Calliphora* (lateral view); (b) posterior spiracles of larva; (c) posterior view of larva; (d) puparium of *Calliphora*.

5. BEHAVIOUR AND BIONOMICS

5.1 Screw-worm myiasis

The primary screw-worm flies *C. hominivorax* and *Ch. bezziana* cause wound myiasis in a wide range of vertebrate hosts, including livestock and humans, and untreated infestations can be fatal.

The historic distribution of *C. hominivorax* extended from the southern USA to most of Latin America. As a result of intensive control efforts, however, it is currently limited to Central and South America. The eggs of *C. hominivorax* are laid in batches on the dry surfaces at the edge of 2- to 10-day-old wounds. They hatch in 11–21 h and the larvae aggregate to feed with their posterior spiracles exposed. They are fully developed in 4–8 days, leave the host in the morning and pupate in the surface layers of the soil. If they are exposed to temperatures below 9.5°C for 3 months, they die. They also require soil moisture of less than 16%. The adults emerge around dawn. Females mate only once and, after a preoviposition period of 5–10 days, deposit batches of about 300 eggs. Their activity is reduced by hot, dry conditions, strong winds and during rain, but increases after rain. They feed on wounds, dung and fresh meat, to obtain protein for ovarian development.

C. hominivorax is active all year round in areas where the temperature is above 16°C and, during the summer, disperses widely from its overwintering areas.

Ch. bezziana is widely distributed in the Afrotropical and Oriental regions, extending as far south as New Guinea, but not to Australia. It attacks a wide range of hosts, but there are few records from wild animals. Cases of human myiasis are common in the Oriental region, but rare in Africa. Females are attracted to wounds for oviposition and eggs are laid on the upper, dry side of wounds, usually in the late afternoon, when 100–250 eggs are deposited in a single batch. The eggs hatch 12–16 h later but, if exposed to sunlight for more than 2 h, suffer a high mortality and all are dead after 6 h. Larvae feed initially on blood and serum and later lacerate tissue with their mouth-hooks. They aggregate and tunnel deeply (15 cm) into the host's tissue, causing considerable destruction. Several females may oviposit at the same site, probably attracted by pheromones emitted by the first ovipositing female. As a result, several thousand larvae may occur in a wound. In 6–8 days, the larvae are fully developed, leave the host as wandering larvae and pupate in the ground, with the pupal period lasting 8–10 days. In the field, mean life expectancy has been estimated to be about 9 days, which is not long enough to complete two ovarian cycles. The population of Ch. bezziana is usually relatively low, of the order of 1–200 in every 25 ha. They range widely and labelled females have deposited egg masses 100 km from their point of release.

5.2 Sheep myiasis

Species of Lucilia and Calliphora are the primary blow fly agents of cutaneous myiasis in sheep, although they will also infest a variety of vertebrate hosts, including humans, if appropriate conditions for oviposition exist. They are attracted to sheep which have areas of wet, soiled fleece, are suffering from bacterial decomposition of the fleece (fleece rot) or have heavily wrinkled skins (as found particularly in some sheep breeds such as Merinos). In a severe outbreak, 30% of a sheep flock may die. Sheep myiasis is particularly important in Australia, New Zealand, South Africa and north-western Europe. Oviposition on the host is referred to as 'blow' and the establishment of larvae as 'strike'. In Australia and South Africa, the main species involved is L. cuprina, which initiates over 90% of strikes, although species of Calliphora such as C. stygia may also be involved. In north-western Europe, the primary species responsible is the closely related L. sericata.

The eggs are deposited in a cluster of 100–300, depending on the size of the female. In the fleece of sheep, the eggs hatch in 8–12 h. However, the eggs and first-stage larvae are highly susceptible to desiccation. Newly hatched larvae move down to the skin to feed on the protein-rich exudates produced in response to the saliva and ammoniacal excreta of the larvae. Second- and third-stage larvae also abrade the skin with mouth-hooks, causing and extending lesions. The mature third-stage larvae drop from the host at night and burrow into the soil to a depth of 1–2 cm, where they pupate, providing the temperature is above 10°C and the humidity low. Larvae may diapause over winter but breeding all year round may occur where seasonal conditions permit. The pupal stage lasts approximately 6 days at 30°C and 20–25 days at 15°C, with a survival of 75–95%. Development is highly temperature dependent, but the time required to complete a generation may be as little as 3–4 weeks. In parts of Australia, where conditions are suitable for year-round development, there are about eight generations of L. cuprina a year, whereas in north-western

Europe, there may be only 3–4 generations of *L. sericata* per year, followed by winter diapause. Adult females need sources of carbohydrate, such as nectar, and protein, which can be obtained from carrion, wounds and, to a lesser extent, dung. A protein meal is necessary for ovarian development.

Although *L. cuprina* and *L. sericata* are often early colonizers of fresh carrion, competition from other blow flies is believed to restrict their survival in carrion.

5.3 Other Calliphoridae

Ph. regina, the black blow fly, and *Pr. terraenovae* breed mainly in carrion but can be facultative agents of myiasis. *Ph. regina*, a common sheep blow fly in the south-west USA, has been associated with wound myiasis in humans. *Pr. terraenovae* is a more northern species capable of causing fatal wound myiasis in cattle, sheep and reindeer.

Three species of testaceous calliphorids, *Auchmeromyia senegalensis* (formerly *Auchmeromyia luteola*), the Congo floor maggot, and *C. anthropophaga*, the Tumbu or mango fly (and the closely related *Cordylobia rodhaini*) (see also the separate entry for 'Tumbu Fly'), cause myiasis in humans in Africa. The adults may occur in huts or houses and can be distinguished by the second visible abdominal segment being obviously longer than the third segment in *A. senegalensis*, and the two segments being of similar length in *C. anthropophaga*. The eyes of male *A. senegalensis* are widely separated and those of *C. anthropophaga* almost contiguous (Fig. 20).

A. senegalensis, the Congo floor maggot, is one of a few species of the genus that are found in sub-Saharan Africa and causes an unusual form of sanguiniverous myiasis in humans. Adult *A. senegalensis* have lapping mouthparts and feed on faeces, particularly human faeces, and fermenting fruit, but the larvae are haematophagous. A female adult lays batches of about 50 eggs, and in her lifetime may lay a total of 300 eggs. They are laid in dry, dusty soil or sand in the earth floor of huts and hatch in 36–60 h at 26–28°C and 50–60% RH (relative humidity). There are three larval stages (Fig. 21) which feed at night on the inhabitants of the hut sleeping on the ground. Larvae take at least two feeds during each larval stage and 6–20 feeds in the course of development. Larvae of *A. senegalensis* are associated with humans, warthogs and aardvarks. There is some evidence that *Auchmeromyia* larvae may act as mechanical vectors of *Trypanosoma brucei*. There are four other species of *Auchmeromyia*, of which one, *Auchmeromyia bequaerti*, may be locally important in parts of southern Africa.

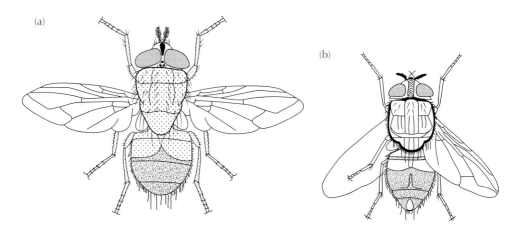

(a)

(b)

Fig. 20. (a) Female *Cordylobia anthropophaga*; (b) female *Auchmeromyia senegalensis*. *Source*: *Cordylobia anthropophaga* redrawn from James (1947) and *Auchmeromyia senegalensis* from Crosskey and Lane (1993).

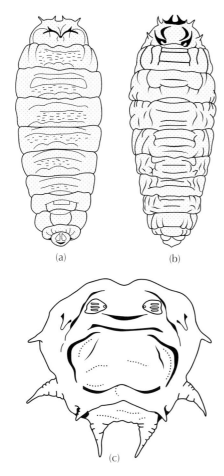

(a)　　　　　　　　(b)

(c)

Fig. 21. Third-stage larva of *Auchmeromyia senegalensis*. (a) Dorsal view; (b) ventral view; (c) posterior view. *Source*: Zumpt (1965).

Larval populations die out if the ground is damp or wet, but areas are repopulated rapidly when they dry out.

C. *anthropophaga*, the Tumbu (or mango) fly (and C. *rodhaini*, Lund's fly), causes a furuncular (or cutaneous) myiasis in humans in Africa (see also the separate entry for 'Tumbu Fly' for further details).

6. MEDICAL AND VETERINARY IMPORTANCE

In humans, calliphorid myiasis is rarely fatal, unless it occurs in an infant, elderly or otherwise debilitated patient. However, it can be alarming and cause discomfort.

The economic importance of screw-worm flies stems largely from the fact that they cause myiasis in cattle. In domestic stock, the areas most susceptible to attack are the navels of newborn animals, surgical wounds produced during castration, docking and dehorning, and tick bites; the condition is complicated by secondary infections. Preventive measures include delaying surgery to the cooler season of the year, when screw-worm flies are less active, dressing wounds and twice weekly inspection of livestock.

Blow flies are primarily of economic importance because of their effects on sheep. The common sites of blow fly infestation are the perineum, anal and tail (breech and crutch) areas, particularly where the wool has been contaminated with faeces and is moist. Less common sites of infestation are the prepuce of rams and poll strike on the dorsum of the head following fight-induced wounds. In older and heavily fleeced animals, body strike may be important. Strikes can occur in wounds such as caused during shearing, and by other conditions such as foot rot. Sheep struck by *Lucilia* show a rapid increase in body temperature and respiratory rate. The animals become anaemic and suffer severe toxaemia, with both kidney and heart tissues affected. The feeding activity of the larvae may cause extensive tissue damage which, in combination with the larval proteases produced, results in the development of inflamed, abraded or undermined areas of skin. Infested animals become lethargic, appear depressed and cease feeding, resulting in loss of weight and anorexia. If untreated, multiple infestation leads rapidly to death from toxaemia, usually within about 2 weeks of the initial infestation, although the precise time required will depend on the intensity of infestation.

7. PREVENTION AND CONTROL

Myiasis prevention is based largely on the prophylactic treatment of livestock with insect-icides, particularly organophosphates and insect growth regulators, such as dicyclanil and cyromazine. Organophosphates may be applied by hand spraying, plunge dipping, in a spray race or by jetting. Insect growth regulators are applied as pour-on formulations. The problems associated with strike management are the relatively short period spent by the larvae on the host, the repeated infestation challenge (which persists for many months or year-round in some areas) and the rapidity with which severe damage occurs. Any insecticide used must therefore not only kill the larvae, but also persist on the host. Commercial traps have been developed to attract and kill *Lucilia* using synthetic baits or offal.

The control of screw-worm flies is complicated by the fact that they cause myiasis not only in cattle and humans but also in wild animals, which act as a reservoir and maintain the fly population even when good levels of management are practised in domestic livestock. Both *C. hominivorax* and *Ch. bezziana* may be trapped using a chemical attractant called 'Swormlure', or modifications of this mixture of 11 organic compounds, of which dimethyl disulfide is the most important. The attracted flies may be trapped on adhesive surfaces or non-return traps.

C. hominivorax was the first insect to be eradicated from an area by the use of the sterile insect technique, which required the release of sterilized flies in numbers adequate to swamp the wild population. Initial successful trials on Sanibel Island in Florida and on Curacao in the Netherlands Antilles were followed by the eradication of *C. hominivorax* from Florida in 1959. In 1962, a programme then began to deal with the overwintering areas along the Mexico–US border, with California, Arizona and New Mexico becoming screw-worm free in 1979, and Texas likewise in 1982. This programme involved the production of 100 million sterile male flies per week and their release along a barrier 400 km wide and more than 3000 km in length. The programme was subsequently extended south and Mexico was declared screw-worm free in

1991. The sterile insect technique is now in use to eliminate the fly in other areas of central and southern America.

Elsewhere, *C. hominivorax* became established in the Old World in Libya in 1988, leading to FAO and the Libyan Government undertaking a massive eradication programme. It again used the sterile insect technique which had proved so effective in North America. An area of 25,000 km² around Tripoli on the North African Mediterranean coast was put under quarantine. Livestock within the area were inspected and, if necessary, treated and movement of infested livestock out of the area prohibited. The first release of sterile flies from the Mexico–American Commission was made in December 1990 at a rate of 3.5 million flies/week, increasing to 40 million/week by May 1991. Release of sterile flies ceased in October 1991, 6 months after the last detected case of screw-worm myiasis in Libya.

In humans, no prophylactic treatments are practised routinely and treatment for infestation is almost always reactive. For wound myiasis, with, for example, *C. hominivorax*, *Ch. bezziana* or *Lucilia*, *Calliphora* or *Phormia* species treatment usually requires the physical removal (by forceps and irrigation) of all maggots, surgical debridement of necrotic tissues, followed by appropriate wound dressing and antibiotic therapy. For cutaneous/furuncular myiasis, with, for example, *Cordylobia* species, surgical removal of the maggot is usually unnecessary and occluding the breathing hole with a thick substance (such as petroleum jelly) will usually force the maggot to emerge far enough to be grasped with forceps. Application of macrocyclic lactone endectocides, such as ivermectin, to kill the larva can be effective (see also entry for 'Tumbu Fly'). For cavity myiasis involving the eye, nose, ears, oral region and the urogenital areas, specialist medical assistance should be sought immediately.

SELECTED BIBLIOGRAPHY

Amendt, J., Richards, C.S., Campobasso, C.P., Zehner, R. and Hall, M.J.R. (2011) Forensic entomology: applications and limitations. *Forensic Science, Medicine, and Pathology* 7, 379–392.

Ashworth, J.R. and Wall, R. (1994) Responses of the sheep blowflies *Lucilia sericata* and *L. cuprina* to odour and the development of semiochemical baits. *Medical and Veterinary Entomology* 8, 303–309.

Bisdorff, B. and Wall, R. (2008) Sheep blowfly strike risk and management in Great Britain: a survey of current practice. *Medical and Veterinary Entomology* 22, 303–308.

Blueman, D. and Bousfield, C. (2012) The use of larval therapy to reduce the bacterial load in chronic wounds. *Journal of Wound Care* 21, 244–253.

Cruickshank, I. and Wall, R. (2002) Population dynamics of the sheep blowfly *Lucilia sericata*: seasonal patterns and implications for control. *Journal of Applied Ecology* 39, 493–501.

Francesconi, F. and Lupi, O. (2012) Myiasis. *Clinical Microbiology Reviews* 25, 79–105.

Geary, M.J., Hudson, B.J., Russell, R.C. and Hardy, A. (1999) Exotic myiasis with Lund's fly (*Cordylobia rodhaini*). *Medical Journal of Australia* 171, 654–655.

Geary, M.J., Smith, A. and Russell, R.C. (2009) Maggots Down Under. *Wound Practice and Research* 17, 36–42.

Hakeem, M.J.M.L. and Bhattacharyya, D.N. (2009) Exotic human myiasis. *Travel Medicine and Infectious Disease* 7, 198–202.

Hall, M.J.R. and Wall, R. (1995) Myiasis in humans and domestic animals. In: Baker, J.R., Muller, R. and Rollinson, D. (eds) *Advances in Parasitology*, Vol 35. Academic Press, London, pp. 258–334.

Harvey, B., Bakewell, M., Felton, T., Stafford, K., Coles, G.C. and Wall, R. (2010) Comparison of traps for the control of sheep blowfly in the U.K. *Medical and Veterinary Entomology* 24, 210–213.

Hendrix, C.M., King-Jackson, D.A., Wilson, M., Blagburn, B.L. and Lindsay, D.S. (1995) Furunculoid myiasis in a dog caused by *Cordylobia anthrophaga*. *Journal of the American Veterinary Medicine Association* 207, 1187–1189.

Krafsur, E.S., Whitten, C.J. and Novy, J.E. (1987) Screwworm eradication in north and central America. *Parasitology Today* 3, 131–137.

Lindquist, D.A., Abusowa, M. and Hall, M.J.R. (1992) The New-World screwworm fly in Libya – a review of its introduction and eradication. *Medical and Veterinary Entomology* 6, 2–8.

McDonagh, L.M. and Stevens, J.R. (2011) The molecular systematics of blowflies and screwworm flies (Diptera: Calliphoridae) using 28S rRNA, COX1 and EF-1alpha: insights into the evolution of dipteran parasitism. *Parasitology* 138, 1760–1777.

Myers, J.H., Savoie, A. and van Randen, E. (1998) Eradication and pest management. *Annual Review of Entomology* 43, 471–491.

Norris, K.R. (1965) The bionomics of blow flies. *Annual Review of Entomology* 10, 47–68.

Sherman, R., Hall, M. and Thomas, S. (2000) Medicinal maggots: an ancient remedy for some contemporary afflictions. *Annual Review of Entomology* 45, 55–81.

Siddig, A., Al Jowary, S., Al Izzi, M., Hopkins, J. and Hall, M.J.R. (2005) Seasonality of Old World screwworm myiasis in the Mesopotamia valley in Iraq. *Medical and Veterinary Entomology* 19, 140–150.

Stevens, J.R., Wallman, J.F., Otranto, D., Wall, R. and Pape, T. (2006) The evolution of myiasis in humans and other animals in the Old and New Worlds (part II): biological and life-history studies. *Trends in Parasitology* 22, 181–188.

Urech, R., Bright, R.L., Green, P.E., Brown, G.W., Hogsette, J.A., Skerman, A.G., *et al.* (2012) Temporal and spatial trends in adult nuisance fly populations at Australian cattle feedlots. *Australian Journal of Entomology* 51, 88–96.

Wall, R. and Ellse, L. (2011) Climate change and livestock disease: integrated management of blowfly strike in a warmer environment. *Global Change Biology* 17, 1770–1777.

Wall, R., Cruickshank, I., Smith, K.E., French, N.P. and Holme, A.S. (2002) Development and validation of a simulation model for blowfly strike of sheep. *Medical and Veterinary Entomology* 16, 335–346.

Wardhaugh, K.G. and Morton, R. (1990) The incidence of flystrike in sheep in relation to weather conditions, sheep husbandry, and the abundance of the Australian sheep blowfly, *Lucilia cuprina* (Weidemann) (Diptera: Calliphoridae). *Australian Journal of Agricultural Research* 41, 1155–1167.

Wardhaugh, K.G., Morton, R., Bedo, D., Horton, B.J. and Mahon, R.J. (2007) Estimating the incidence of fly myiases in Australian sheep flocks: development of a weather-driven regression model. *Medical and Veterinary Entomology* 21, 153–167.

Zumpt, F. (1965) *Myiasis in Man and Animals in the Old World*. Butterworth's, London.

Butterflies and Moths (Lepidoptera: Particularly the Erebidae, Geometridae, Limacodidae, Lymantridae, Megalopygidae, Notodontidae, Pyralidae, Saturnidae, Sphingidae, Thaumetopoeidae and Thyatiridae but also the Anthelidae, Arctiidae, Lasiocampidae, Morphoidae, Noctuidae, Nolidae and Nymphalidae)

1. INTRODUCTION

Butterflies and moths belong to the order Lepidoptera and are distributed globally. With a few cryptic exceptions, the adults are of little medical or veterinary importance, except for the potential of their body hairs/scales to cause inhalant allergies; however, the larval stages of some species have a general covering or grouping of hairs, with or without associated toxin glands, which can cause mild to serious urtication on human skin. Additionally, ingestion of forage infested by such caterpillars can cause serious gastrointestinal problems for livestock.

2. TAXONOMY

There is no longer any readily acceptable division of the Lepidoptera into simply moths and butterflies. However, of the more than 100 families, 12 families of moths and 2 of butterflies (representing more than 60 genera and 100 species worldwide) have larvae that cause health-related problems, and 6 families include species in which the adults feed from animals.

3. MORPHOLOGY

Lepidoptera are endopterygote insects whose adult body, legs and wings are covered with detachable scales that are morphologically flattened hairs and who have a coiled proboscis for feeding on nectar (Fig. 22). The larval stage is a caterpillar with chewing mandibulate mouthparts, three pairs of thoracic legs and five pairs of abdominal prolegs. The caterpillars

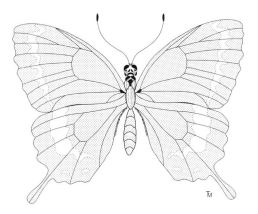

Fig. 22. Dorsal view of a lepidopteran, *Papilio canopus.*

that have urticating hairs generally have one or other or both of 'spicule hairs'. Detachable setae are readily shed, can be incorporated into pupal cases and can become windborne to cause remote dermatitis; mostly, these cause mechanical damage only, but some have associated toxins. The caterpillars may also have 'spine hairs', which cause stinging only when there is direct contact between skin and the caterpillar, due to toxin glands associated with nerve cells.

4. LIFE CYCLE

Development is holometabolous, with eggs being laid on, sometimes specific, host plants that provide the larval food. The caterpillars go through up to ten moults, depending on species, before pupation, which may occur on the plant or in a protected site on the ground below. There often is only one generation per year, with overwintering in temperate species taking place in the egg or as a pupa.

5. BEHAVIOUR

Larval and adult behaviours are relatively diverse, and relevant behaviours associated with the particular species of significance in the present context are addressed below.

6. MEDICAL AND VETERINARY IMPORTANCE

Cutaneous reactions from contact with adults (butterflies, moths), larvae (caterpillars) or pupae or their cases/cocoons are often called erucism, while systemic symptoms can be termed lepidopterism, and there are various other conditions associated with problems following contact with the insects (e.g. ophthalmia nodosa – an ocular inflammation; dendrolimiasis and pararamose – dermatitis with arthritis/chondritis). As mentioned above, adult Lepidoptera are of little relevance in this context, although in parts of Central and South America, female moths of a number of species of *Hylesia* (Saturnidae), such as

Hylesia metabus, *Hylesia alinda* and *Hylesia lineata*, have abdomens covered in barbed setae that become dispersed aerially and cause urticarial dermatitis and conjunctivitis. Otherwise, adult moths and butterflies can induce allergic responses (dermatological and/or respiratory) in sensitive individuals with exposure to their shed scales and hairs, and this has been reported as an occupational hazard for some entomologists.

In South and South-east Asia, dozens of species of Geometridae (e.g. *Hypochrosis* and *Chiasmia* syn. *Semiothisa* spp.), Pyralidae (e.g. *Filodes* and *Microstega* spp.), Notodontidae (e.g. *Tarsolepis* spp.), Erebidae (e.g. *Lobocraspis griseifusa* and *Arcyophora* spp.), Thyatiridae (e.g. *Chaeopsestis ludovicae*) and Sphingidae (*Rhagastis olivacea*) suck tears from the eyes of cattle and other large domestic and wild ungulates, elephants and occasionally humans. Except the Erebidae, they also feed on animal wounds and various body secretions. They do not pierce the skin, as their proboscides are not so adapted. In Africa, mainly *Arcyophora* spp. are involved. Overall, however, there is no evidence of transmission of disease agents.

While the 17 species of erebid moths of the largely Asian genus *Calyptra* are obligatory fruit piercing, males of eight of the species actually pierce the skin of mammals and feed on blood; the feeding is nocturnal and facultative. The hosts are essentially the same as those of the lachryphagous moths, although they have evolved on separate lines. Five *Calyptra* species (*Calyptra bicolor*, *Calyptra fasciata*, *Calyptra ophideroides*, *Calyptra parva* and *Calyptra pseudobicolor*) have been documented to attack and pierce human skin in the wild. Again, there is no evidence of transmission of disease-causing agents to such animals or humans, possibly because most *Calyptra* are rare.

The larval stages of Lepidoptera, however, are more relevant, since many caterpillars are covered with long hairs (setae) and spines, each of which may have an associated venom gland. A number of families are of concern in this context, the most important being the Limacodidae, Lymantridae, Megalopygidae, Saturnidae and Thaumetopoeidae (which is sometimes considered a subfamily of Noto-dontidae), while the Anthelidae, Arctiidae, Lasiocampidae, Morphoidae, Noctuidae, Nolidae and Nymphalidae also contain species that have hairs that can cause urticaria and other effects. The toxins involved include histamine or histamine-like compounds, esterases and other enzymes, peptides and globulins that overall are responsible for a range of effects including wheal formation and oedema, bleeding, blood-cell destruction and local necrosis. In general, most contacts result in mild to moderately severe urticarial dermatitis, with vesicle formation, radiating pain and lymphadenopathy recorded in some severe cases. A few caterpillars of the family Notodontidae secrete defensive noxious fluids from a ventral prothoracic gland, for example formic acid by species of the genera *Cerura* and ketones by species of *Heterocampa*; however, these are not known to cause problems for humans or larger animals.

Veterinary concerns relate mostly to the ingestion by livestock (e.g. horses and cattle) of forage or stored hay or similar containing caterpillars with urticating or toxic hairs (such as some of the above-mentioned), resulting in gastrointestinal problems that range from the mild to the severe and life threatening.

6.1 Limacodidae (e.g. *Doratifera, Natada, Parasa, Sibine* spp.)

A widely distributed group, with most species being tropical or subtropical and variously known as 'slug moths', 'cup moths', 'nettle grubs', 'saddlebacks' and 'chinese junks'; as a group, these are found on a wide range of native and ornamental plants, including fruit trees. They are often brightly coloured (perhaps as a warning sign for predators) and have venomous spines, often in the form of clusters on protrusions front and rear, or all along the body, with tips that break easily on contact and release a toxin (with a high concentration of histamine in some species) that generally produces a sharp stinging sensation, erythema and some localized swelling.

6.2 Lymantridae (e.g. *Euproctis, Dasychira, Lymantria, Orgyia* spp.)

Generally called 'tussock moths' and, as a group, distributed globally and infesting a wide range of trees; relatively few species are urticating, but those caterpillars that are have numerous tubercles bearing long barbed setae with shorter barbed stinging setae. In general, the long setae are not stinging hairs but do cause urticaria, as they readily break off and can be widely distributed. Serious outbreaks associated with direct contact with cast skins or pupal cocoons or from windblown hairs have been caused by the yellow-tail moth (*Euproctis similis*) in Europe, the brown-tail moth (*Euproctis chrysorrhoea*) and the gypsy moth (*Lymantria dispar*) in Europe and North America, the Douglas-fir tussock moth (*Orgyia pseudotsugata*) in North America, *Euproctis subflava* in Japan, Korea, Siberia and China, *Euproctis flavociliata* in Malaysia and *Euproctis edwardsi* in Australia. The last larval instar of *E. similis* has been estimated to have 2 million urticating setae per caterpillar. Reactions to tussock moth hairs vary from mild to severe pruritus, sometimes with papular responses, and individuals can develop hypersensitivity. Toxins associated with hairs of some *Euproctis* species have been shown to contain histamine and esterase- and phospholipase-containing fractions.

6.3 Megalopygidae (e.g. *Lagoa, Megalopyge, Norape* spp.)

This group contains the so-called 'puss caterpillar' (*Megalopyge opercularis*), which causes the most severe reactions of caterpillars in North America. It is found on a range of trees and shrubs, and its very hairy appearance hides the short toxic spines, arranged as clusters along the back and sides, whose tips break off with skin contact and release toxin from a basal cavity. Reactions include an initial burning sensation with numbness and swelling, through to nausea and vomiting, and sometimes inflammation of the lymphatics persisting for many hours.

6.4 Saturnidae (e.g. *Automeris*, *Dirphia*, *Hemileuca*, *Hylesia*, *Lonomia* spp.)

Automeris io feeds on a wide range of host plants in North America. It has stinging spines on fleshy tubercles along the back and sides, and the tips of the spines readily break off and allow toxin to be released, resulting in a stinging sensation and swelling that may produce large wheals. *Hemileuca maia*, *Hemileuca oliviae* and other *Hemileuca* spp. closely related to *Automeris* are also capable of significant stinging, producing itching and prolonged swelling.

In the northern areas of South America, problems occur with caterpillars of *Hylesia* spp., which have poisonous spines, and some species of *Dirphia* in Brazil and Peru have caterpillars with long and strong barbed venomous spines. Caterpillars of *Lonomia archilous* in Venezuela and *Lonomia obliqua* in Brazil have a toxin that causes an immediate sharp burning sensation and various systemic effects, but which also contains a potent fibrinolytic enzyme that has an anticoagulant effect and can bring on severe bleeding from various orifices; there have been deaths associated with the condition.

6.5 Thaumetopoeidae (*Ochrogaster*, *Thaumetopoea*, *Trichiocercus* spp.)

Known as 'processionary caterpillars', this group contains a few species of significance. In Europe, where there are oak- and pine-feeding species, some *Thaumetopoea* have been recorded as causing severe dermatitis, conjunctivitis, laryngitis and pharyngitis. In Australia, *Ochrogaster lunifer* (the bag-shelter moth) has gregarious hairy larvae that live in silken bags in the host tree (often *Acacia* spp.), with masses of larval skins and broken-off hairs that are dispersed with wind action from holes left as the caterpillars move out to feed and as the bag eventually breaks down. The long barbed hairs readily penetrate human skin, causing rash and severe irritation, and inflammation of the eyes (ophthalmia) has been reported. Generally, these caterpillars have irritating hairs rather than envenomating spines, but hairs of some species (e.g. *Thaumetopoea pityocampa*) act as histamine liberators and hairs of other species (e.g. *Thaumetopoea wilkinsoni*) have been shown to contain toxins, with another (*Trichiocercus* sp.) in Australia having been shown to have directly envenomating spine-like setae.

7. PREVENTION AND CONTROL

Avoidance of environmental contact through recognition of dangerous species in casual circumstances and protection from contamination through the use of protective clothing, masks and goggles in occupational situations is usually advised. Careful manual removal of larval nests and cocoons may be appropriate in some circumstances and insecticidal control of threatening populations can be a pre-emptive approach in others, but not necessarily in those where the setae and spines from dead bodies are able to perpetuate the problem.

Hairs can often be removed effectively from skin using repeated stripping with adhesive tape/plaster, and application of cold (ice) packs, iodine tincture, oral antihistamines and topical corticosteroids are reported to relieve the burning, itching and inflammatory reactions. If there is ocular or respiratory involvement, specialist medical treatment should be sought immediately. Hypersensitivity reactions can be treated with antihistamines and adrenaline (epinephrine) if necessary, and may require medical assistance if anaphylaxis is a risk. If envenomation with a *Lonomia* species is suspected and haemorrhagic symptoms are apparent, consultation with a haematologist is indicated for appropriate treatment, but the species involved must be known as different treatments can be required.

SELECTED BIBLIOGRAPHY

Alexander, J.O'D. (1984) *Arthropods and Human Skin*. Springer-Verlag, Berlin, pp. 177–197.

Arlian, L.G. (2002) Arthropod allergens and human health. *Annual Review of Entomology* 47, 395–433.

Balit, C.R., Ptolemy, H.C., Geary, M.J., Russell, R.C. and Isbister, G.K. (2001) Outbreak of caterpillar dermatitis caused by airborne hairs of the mistletoe browntail moth (*Euproctis edwardsi*). *Medical Journal of Australia* 175, 641–643.

Balit, C.R., Geary, M.J., Russell, R.C. and Isbister, G.K. (2003) Prospective study of definite caterpillar exposures. *Toxicon* 42, 657–662.

Balit, C.R., Geary, M.J., Russell, R.C. and Isbister, G.K. (2004) Clinical effects of exposure to the white-stemmed gum moth (*Chelepteryx collesi*). *Emerging Medicine Australasia* 16, 74–81.

Bänziger, H. (1980) Skin-piercing blood-sucking moths III: feeding act and piercing mechanism of *Calyptra eustrigata* (Hmps.) (Lep., Noctuidae). *Mitteilungen der Schweizerischen Entomologischen Gesellschaft* 53, 127–142.

Bänziger, H. (1989) Skin-piercing blood-sucking moths V: attacks on man by 5 *Calyptra* spp. (Lepidoptera, Noctuidae) in S and SE Asia. *Mitteilungen der Schweizerischen Entomologischen Gesellschaft* 62, 215–233.

Bänziger, H. (1992) Remarkable new cases of moths drinking human tears in Thailand (Lepidoptera: Thyatiridae, Sphingidae, Notodontidae). *Natural History Bulletin of the Siam Society* 40, 91–102.

Bänziger, H. (1995) *Microstega homoculorum* sp. n. – the most frequently observed lachryphagous moth of man (Lepidoptera, Pyralidae: Pyraustinae). *Revue Suisse de Zoologie* 102, 265–276.

Bänziger, H. (2007) Skin-piercing blood-sucking moths VI: fruit-piercing habits in *Calyptra* (Noctuidae) and notes on the feeding strategies of zoophilous and frugivorous adult Lepidoptera. *Mitteilungen der Schweizerischen Entomologischen Gesellschaft* 80, 271–288.

Battisti, A., Holm, G., Bengt, F. and Larsson, S. (2011) Urticating hairs in arthropods: their nature and medical significance. *Annual Review of Entomology* 56, 203–220.

Bettini, S. (ed.) (1978) *Arthropod Venoms*. Springer-Verlag, Berlin.

Bruchim, Y., Ranen, E., Saragusty, J. and Aroch, I. (2005) Severe tongue necrosis associated with pine processionary moth (*Thaometopoea wilkinsoni*) ingestion in three dogs. *Toxicon* 45, 443–447.

Carrijo-Carvalho, L.C. and Chudzinski-Tavassi, A.M. (2007) The venom of the *Lonomia* caterpillar: an overview. *Toxicon* 49, 741–757.

Delgado, A. (1978) Venoms of Lepidoptera. In: Bettini, S. (ed.) *Arthropod Venoms*. Springer-Verlag, Berlin, pp. 555–611.

Derraik, J.G.B. (2007) Three students exposed to *Uraba lugens* (gum leaf skeletoniser) caterpillars in a West Auckland school. *New Zealand Medical Journal* 120, No.1259.

Diaz, J.H. (2005) The evolving global epidemiology, syndrome classification, management, and prevention of caterpillar envenoming. *American Journal of Tropical Medicine and Hygiene* 72, 347–357.

Everson, G.W., Chapin, J.B. and Normann, S.A. (1990) Caterpillar envenomations: a prospective study of 112 cases. *Veterinary and Human Toxicology* 32, 114–119.

Gottschling, S., Meyer, S., Dill-Mueler, D., Wurm, D. and Gortner, L. (2007) Outbreak report of airborne caterpillar dermatitis in a kindergarten. *Dermatology* 215, 5–9.

Holland, D.L. and Adams, D.R. (1998) 'Puss caterpillar' envenomation: a report from North Carolina. *Wilderness and Environmental Medicine* 9, 213–216.

Horng, C.T., Chou, P.I. and Liang, J.B. (2000) Caterpillar setae in the deep cornea and anterior chamber. *American Journal of Ophthalmology* 129, 384–385.

Hossler, E.W. (2009) Caterpillars and moths. *Dermatology Therapy* 22, 353–366.

Hossler, E.W. (2010a) Caterpillars and moths: Part I. Dermatologic manifestations of encounters with Lepidoptera. *Journal of the American Academy of Dermatology* 62, 1–10.

Hossler, E.W. (2010b) Caterpillars and moths: Part II. Dermatologic manifestations of encounters with Lepidoptera. *Journal of the American Academy of Dermatology* 62, 13–28.

Huang, D.Z. (1991) Dendrolimiasis: an analysis of 58 cases. *Journal of Tropical Medicine and Hygiene* 94, 79–87.

Kawamoto, F. and Kumada, N. (1984) Biology and venoms of Lepidoptera. In: Tu, A.T. (ed.) *Handbook of Natural Toxins*, Vol 2. Dekker, New York, pp. 291–330.

Kozer, E., Lahat, E. and Berkovitch, M. (1999) Hypertension and abdominal pain: uncommon presentation after exposure to a pine caterpillar. *Toxicon* 37, 1797–1801.

Kuspis, D.A., Rawlins, J.E. and Krenzelok, E.P. (2001) Human exposures to stinging caterpillar: *Lophocampa caryae* exposures. *American Journal of Emergency Medicine* 19, 396–398.

Maier, H., Spiegel, W., Kinaciyan, T., Krehan, H., Cabaj, A., Schopf, A., *et al.* (2003) The oak processionary caterpillar as the cause of an epidemic airborne disease: survey and analysis. *British Journal of Dermatology* 149, 990–997.

Mulvaney, J.K., Gatenby, P.A. and Brookes, J.G. (1998) Lepidopterism: two cases of systemic reactions to the cocoon of a common moth, *Chelypteryx collessi. Medical Journal of Australia* 168, 610–611.

Natsuakai, M. (2002) Immediate and delayed-type reactions in caterpillar dermatitis. *Journal of Dermatology* 29, 471–476.

Pinson, R.T. and Morgan, J.A. (1991) Envenomation by the puss caterpillar (*Megalopyge opercularis*). *Annals of Emergency Medicine* 20, 562–564.

Redd, J.T., Voorhees, R.E. and Torok, T.J. (2007) Outbreak of lepidopterism at a Boy Scout camp. *Journal of the American Academy of Dermatology* 56, 952–955.

Rothschild, M., Reichstein, T., von Euw, J., Aplin, R. and Harman, R.R.M. (1970) Toxic Lepidoptera. *Toxicon* 8, 293–296.

Sood, P., Tuli, R., Puri, R. and Sharma, R. (2004) Seasonal epidemic of ocular caterpillar hair injuries in the Kangra district of India. *Ophthalmic Epidemiology* 11, 3–8.

Southcott, R.V. (1978) Lepidopterism in the Australian Region. *Records of the Adelaide Children's Hospital* 2, 87–173.

Southcott, R.V. (1983) Lepidoptera and skin infestations. In: Parish, L.C., Nutting, W.B. and Schwartzman, R.M. (eds) *Cutaneous Infestations of Man and Animal.* Praeger, New York, pp. 304–343.

Watson, P.G. and Sevel, D. (1966) Ophthalmia nodosa. *British Journal of Ophthalmology* 50, 209–217.

Wirtz, R.A. (1984) Allergic and toxic reactions to non-stinging arthropods. *Annual Review of Entomology* 29, 47–69.

Centipedes (Scolopendromorpha: Scolopendridae, Scolopocryptopidae and Cryptopidae)

1. INTRODUCTION

Centipedes (class Chilopoda) are long, soft-bodied, dorsoventrally flattened terrestrial arthropods (Fig. 23) of medical significance because they can inflict venomous bites that may be serious or even fatal for humans.

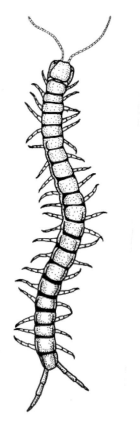

Fig. 23. Dorsal view of a chilopod, *Scolopendron morsitans*. Redrawn from Lewis (1981).

2. TAXONOMY

About 3000 species of centipede have been described (although many more are estimated to exist) within 5 orders and 23 families; the most significant order (from a medical perspective) is the Scolopendromorpha (which includes some of the larger species) and arguably contains 3 families (Scolopendridae, Scolopocryptopidae and Cryptopidae).

3. MORPHOLOGY

Centipedes have a distinct head, long antennae and three pairs of appendages associated with the mouth. The first pair of legs acts as ventrally located powerful jaws, which are pierced by a duct through which the secretion of the venom glands is injected into prey. Behind the head, the body is metamerically segmented and composed of at least 15 segments, each of which, with the exception of the last segment, bears a pair of legs. The soil-dwelling (worm-like) forms have up to 181 pairs of legs, while the surface-dwelling types have 15–21 pairs.

4. LIFE CYCLE

Reproduction does not involve copulation; males deposit a spermatophore, which the female collects. Eggs (10–50/batch) are laid in moist conditions in soil or its surface humus. In tropical areas, centipedes seem to breed year-round, but in temperate areas, egg laying occurs in spring and summer. Hatch may not occur for some weeks to months, and in some species the females appear to tend the egg batch. The immatures may undergo up to 10 moults over some months to years before reaching adulthood, and the adults can live for up to 5 years.

5. BEHAVIOUR AND BIONOMICS

Centipedes are essentially predators, usually sheltering by day and emerging at night to hunt for soil annelids and arthropods (and even small vertebrates for the larger centipedes), killing them with their jaws and injecting venom. The Venezuelan giant centipede, *Scolopendra gigantea*, reaches 30 cm in length and will catch and kill mice and bats. Centipedes are relatively agile and can move quickly.

6. MEDICAL AND VETERINARY IMPORTANCE

Some species (e.g. of *Otostigmus* in Southeast Asia) have glands that produce (presumably defensive) body fluids that can cause blistering on human skin, but it is bites from centipedes that are of most importance.

The so-called house (or 'feather') centipede, *Scutigera* spp., is relatively harmless, but there are a few records of humans being bitten. It is the larger centipedes of the genus *Scolopendra* (often 15 and up to 45 cm long) which are most likely to inflict serious bites. However, most contacts are incidental, with attacks occurring when the centipedes are disturbed as they shelter in dark, humid places or when picked up and handled. The bite is characterized by the puncture wounds being intensely painful and lasting some hours, and there may be associated inflammation, oedema, anxiety, dizziness, vomiting, headaches, lymphadenitis and lymphangitis. Because of the centipede's ground-dwelling and feeding habits, there is a risk of secondary infections with bites, and superficial necrosis at the bite site (that may take some days to heal) has been reported. Only very rarely is the bite of a chilopod fatal to humans, but there is one authentic case of a child from the Philippines who had been bitten on the head by *Scolopendra subspinipes*; the venom is reported to be a cytolysin-based compound, but the bites are usually uncomplicated and resolve within a day or so.

7. PREVENTION AND CONTROL

In cases of serious domestic infestations, the administration of residual insecticides may be required. Removal of harbourage sites (accumulated rubbish and trash, rocks and stones, stacks of wood and timber, grass piles and compost heaps) around homes will help in reducing the populations of centipedes.

Handling centipedes should be avoided, and poking them with a stick can be dangerous as they can run rapidly up the stick and attack the handler.

Treatment for bites includes site washing, while cold packs and analgesics will relieve pain. If secondary infection is an issue, antibiotics may be required.

SELECTED BIBLIOGRAPHY

Babak, V., Hassan, R.A. and Abbas, M.S. (2007) Two cases of Chilopoda (centipede) biting in human from Ahwaz, Iran. *Pakistan Journal of Medical Sciences* 23, 956–958.

Balit, C.R., Harvey, M.S., Waldock, J.M. and Isbister, G.K. (2004) Prospective study of centipede bites in Australia. *Journal of Toxicology and Clinical Toxicology* 42, 41–48.

Bettini, S. (ed.) (1978) *Arthropod Venoms*. Springer-Verlag, Berlin.

Bush, S.P., King, B.O., Norris, R.L. and Stockwell, S.A. (2001) Centipede envenomation. *Wilderness and Environmental Medicine* 12, 93–99.

Forrester, M.B. (2006) Epidemiology of centipede exposures reported to Texas poison control centers, 1998–2004. *Toxicological and Environmental Chemistry* 88, 213–218.

Lewis, J.G.E. (1981) *The Biology of Centipedes*. Cambridge University Press, Cambridge, UK.

Lin, T.J., Yang, G.Y., Ger, J., Tsai, W.L. and Deng, J.F. (1995) Features of centipede bites in Taiwan. *Tropical and Geographic Medicine* 47, 300–302.

Southcott, R.V. (1976) Arachnidism and allied syndromes in the Australian regions. *Records of the Adelaide Children's Hospital* 1, 97–187.

Undheim, E.A.B. and King, G.F. (2011) On the venom system of centipedes (Chilopoda), a neglected group of venomous animals. *Toxicon* 57, 512–524.

Cockroaches (Blattodea: Blattidae, Blatellidae and Blaberidae)

1. INTRODUCTION

A number of cockroach species are considered to be nuisance pests, closely associated with human communities and of concern for human health. Cockroaches are active, nocturnal insects. Apart from their presence being considered unacceptable, cockroaches can produce a characteristic offensive odour. They are scavengers which are attracted to any organic material that can serve as food. They usually aggregate in the sewers and in kitchens of urban developments. The sewers of tropical cities often support very large populations of cockroaches, which on warm nights emerge and are attracted to lights around and inside houses.

As synanthropic insects, i.e. adapted to living closely with people, they have been carried around the world with commercial trade and the goods and chattels of humans. The main pest species are the large *Periplaneta americana* and the smaller *Blatella germanica* and *Blatta orientalis*. Cockroaches feed on human and animal food, excreta and sputa and various waste materials. This wide-ranging feeding habit makes cockroaches potential mechanical vectors of pathogens; however, their precise vectorial role needs to be assessed individually and not assumed generally.

When cockroaches feed on pathogen-infected material, their legs and mouthparts become contaminated and they can introduce the infection into the human environment. In addition, they defecate while feeding, and some pathogens can remain fully viable after passage through the cockroach gut. In some situations, cockroaches may be more important than house flies as mechanical vectors of human disease pathogens. Furthermore, they also represent a source of respiratory allergens, particularly in situations where populations are highly abundant and in close association with humans.

2. TAXONOMY

There are more than 4000 species of cockroach, allocated to five families, of which three (Blattidae, Blatellidae and Blaberidae) contain most of the so-called 'domestic pest species'. *B. orientalis* is in the Blattidae, while the species of *Blatella*, *Periplaneta* and *Supella* are in the Blatellidae.

3. MORPHOLOGY

Cockroaches are dorsoventrally flattened, exopterygote, terrestrial insects with long antennae and wings (when present) folded flat over the body (Fig. 24). The adult mouthparts are mandibulate for chewing. The legs are cursorial, well developed, often with heavily spined femurs and tibiae, and adapted for running. In adults, there are generally two pairs of wings; the front pair is hardened (tegmina), overlaps and covers much of the dorsal surface of the body, protecting the more delicate membranous hindwings. In most species, the hindwings are effective flying organs, although some species have shortened or only vestigial wing buds and cannot fly. There are large compound eyes situated above the antennal sockets. The mandibles are strongly toothed for effective biting and chewing, and the maxillary and labial palps are well developed.

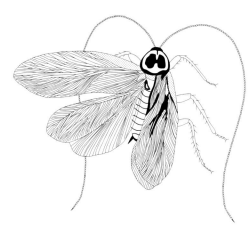

Fig. 24. Dorsal view of a cockroach, *Periplaneta australasiae.*

The abdomen ends in paired, jointed cerci. The terminal cerci are accompanied by styli in some nymphs and males and, with the external genitalia, can be used to distinguish males from females.

4. LIFE CYCLE

Most species are oviparous, including the domestic pests, and the eggs (6–40 depending on species) are contained within a hard case (ootheca) and embryonic development occurs, perhaps for up to 8 weeks, depending on the temperature and the species. The ootheca may be deposited by the female soon after formation or carried until hatching is imminent. The nymphs are miniature versions of the adults, but without wings. The number of instars varies (e.g. 5–13) with species and environmental conditions and may last from a few months to more than a year.

5. BEHAVIOUR AND BIONOMICS

The need for shelter, food and water determines the behaviour of cockroaches, 'domestic', 'peridomestic' or 'feral' species alike. However, the ecological features of the domestic species that are of greatest medical or public health concern vary.

5.1 American cockroach, *Periplaneta americana*

This is a large (34–53 mm), reddish-brown species with winged adults that can fly. This probably is the most cosmopolitan pest species and is established in most tropical and many temperate regions of the world, where they are found in a range of habitats such as refuse dumps and drainage and sewage systems, as well as the various spaces within domestic, industrial and commercial buildings. There are usually 12–16 embryos per ootheca, up to 13 nymphal instars over 168–700 days and the adults can generally survive for 1–2 years. Other *Periplaneta* species (such as *Periplaneta australasiae*, *Periplaneta brunnea* and *Periplaneta fuliginosa*) can be similarly important in different regions and situations.

5.2 Oriental cockroach, *Blatta orientalis*

This species is black, 25–33 mm long, with males that are winged but do not fly and females with only shortened wings. They are usually found in damp or wet conditions in cooler temperate regions of Europe, Asia, Australia and the Americas. Their ootheca generally contains 16 eggs, they have 7–10 nymphal instars, full development is completed within 200–800 days and the adults may survive for many months.

5.3 German cockroach, *Blatella germanica*

Arguably, this species is the most important domestic pest species throughout the developed world. The species is generally found in warm, moist conditions, such as in kitchens. This cockroach is relatively small (16 mm) and the ootheca contains approximately 30 embryos. There are 5–7 nymphal instars and development is generally completed within 50–100 days.

5.4 Brownbanded cockroach, *Supella longipalpa*

This cockroach is usually found in warm, indoor environments in temperate regions.

The adult is relatively small (13–14.5 mm), their oothecae are often affixed to furniture or house fittings and contain 16–18 eggs, there are 6–8 nymphal instars and their development completes within 90–276 days.

6. MEDICAL AND VETERINARY IMPORTANCE

Cockroaches appear as a threat to human and animal health because of their potential to contaminate food with their excreta, because of the pathogens they may carry externally or internally and because of their role as intermediate hosts for parasites. In addition, they may cause allergic reactions and bite (there are rare reports of cockroaches biting and feeding on human tissues).

Although there is a substantial literature on the potential for cockroaches to be involved in the mechanical transmission of pathogens, their vectorial role has been controversial. The range of pathogens isolated from cockroaches includes many species of bacteria (e.g. *Escherichia coli* and various *Salmonella*, *Shigella*, *Staphylococcus* and *Streptococcus* species), as well as various protozoa, fungi, viruses and eggs of helminths. There is circumstantial evidence of cockroach involvement in bacterial disease outbreaks and their cessation following cockroach control. Overall, however, their public health significance relates to the actual possibility of contaminating material and human foodstuffs. This situation is less likely in modern developed societies than in developing communities with less effective sanitation. While cockroaches have been found contaminated with eggs of helminths of human concern, such as hookworms, roundworms, whipworm, pinworm and tapeworms, their development in cockroaches has never been demonstrated; cockroaches do serve as intermediate hosts for a number of parasitic helminths of other animals (e.g. dogs, cats, rats, cattle and poultry), but most of these are nematodes and none are of particular significance.

The importance of cockroach allergies has been recognized increasingly during recent decades, mostly with inner-city residents and with respiratory symptoms (perhaps leading to anaphylaxis) being most often reported. Patients with atopic rhinitis, atopic dermatitis and asthma are often reactive to cockroach allergens, especially of *B. germanica*, *B. orientalis* and *P. americana*; the potent antigens have been associated with the exoskeletons, saliva and body secretions, faecal material and dead bodies.

7. PREVENTION AND CONTROL

Cockroach control in human habitation traditionally has employed a range of inorganic and organic chemicals, including boric acid powder and silica dust, organochlorine, organophosphate, carbamate and pyrethroid insecticide residual sprays and, in more recent years, gel and paste bait formulations (with some of the above-mentioned active ingredients). Insect growth regulators are also now used as part of an integrated pest management approach to cockroach control, using environmental as well as chemical, physical and biological 'tools'. Mitigation of cockroach allergen problems typically relies on a combination of pesticide control of the cockroaches and thorough cleaning of the living quarters, and it has been shown that the cockroach populations must be eliminated (and not just reduced) for full resolution of allergic reactions.

SELECTED BIBLIOGRAPHY

Alexander, J.B., Newton, J. and Crowe, G.A. (1991) Distribution of Oriental and German cockroaches, *Blatta orientalis* and *Blattella germanica* (Dictyoptera) in the United Kingdom. *Medical and Veterinary Entomology* 5, 395–402.

Arlian, L.G. (2002) Arthropod allergens and human health. *Annual Review of Entomology* 47, 395–433.

Bettini, S. (ed.) (1978) *Arthropod Venoms*. Springer-Verlag, Berlin.

Blazar, J.M., Lienau, E.K. and Allard, M.W. (2011) Insects as vectors of foodborne pathogenic bacteria. *Terrestrial Arthropod Reviews* 4, 5–16.

Brenner, R.J., Koehler, P.G. and Patterson, R.S. (1987) *Infections in Medicine.* 4, 349–355, 358–359, 393.

Burgess, N.R.H. and Chetwyn, K.N. (1981) Association of cockroaches with an outbreak of dysentery. *Transactions of the Royal Society of Tropical Medicine and Hygiene* 75, 332–333.

Cohn, R.D. (2011) Cockroach allergens: exposure risk and health effects. In: Nriagu, J.O. (ed.) *Encyclopedia of Environmental Health.* Elsevier, Amsterdam, pp. 732–739.

Cornwell, P.B. (1968) *The Cockroach, Vol 1. A Laboratory Insect and an Industrial Pest.* Hutchinson, London, 391 pp.

Cornwall, P.B. (1976) *The Cockroach, Vol 2. Insecticides and Cockroach Control.* Associated Business Programmes, London, 557 pp.

Crissman, J.R., Booth, W., Santangelo, R.G., Mukha, D.V., Vargo, E.L. and Schal, C. (2010) Population genetic structure of the German cockroach (Blattodea: Blattellidae) in apartment buildings. *Journal of Medical Entomology* 47, 553–564.

Devi, S.J.N. and Murray, C.J. (1991) Cockroaches (*Blatta* and *Periplaneta* species) as reservoirs of drug-resistant *Salmonellas. Epidemiology and Infection* 107, 357–361.

Eggleston, P.A. (2003) Cockroach allergen abatement in inner-city homes. *Annals of Allergy, Asthma and Immunology* 91, 512–514.

Eggleston, P.A. and Arruda, L.K. (2001) Ecology and elimination of cockroaches and allergens in the home. *Journal of Allergy and Clinical Immunology* 107, S422–429.

Fakoorziba, M.R., Eghbal, F., Hassanzadeh, J. and Moemenbellah-Fard, M.D. (2010) Cockroaches (*Periplaneta americana* and *Blattella germanica*) as potential vectors of the pathogenic bacteria found in nosocomial infections. *Annals of Tropical Medicine and Parasitology* 104, 521–528.

Fathpour, H., Emtiazi, G. and Ghasemi, E. (2003) Cockroaches as reservoirs and vectors of drug resistant *Salmonella* spp. *Iranian Biomedical Journal* 7, 35–38.4

Gore, J.C. and Scha, L.C. (2007) Cockroach allergen biology and mitigation in the indoor environment. *Annual Review of Entomology* 52, 439–463.

Katial, R.K. (2003) Cockroach allergy. *Immunology and Allergy Clinics of North America* 23, 483–499.

Kinfu, A. and Erko, B. (2008) Cockroaches as carriers of human intestinal parasites in two localities in Ethiopia. *Transactions of the Royal Society of Tropical Medicine and Hygiene* 102, 1143–1147.

Kopanic, R.J., Sheldon, B.W. and Wright, C.G. (1994) Cockroaches as vectors of *Salmonella*: laboratory and field trials. *Journal of Food Protection* 57, 125–132.

LeGuyader, A., Rivault, C. and Chaperon, J. (1989) Microbial organisms carried by brown-banded cockroaches in relation to their spatial distribution in a hospital. *Epidemiology and Infection* 102, 485–492.

Nalyanya, G., Gore, J.C., Linker, H.M. and Schal, C. (2009) German cockroach allergen levels in North Carolina schools: comparison of integrated pest management and conventional cockroach control. *Journal of Medical Entomology* 46, 420–427.

Oothuman, P., Jeffery, J., Aziz, A.H.A., Bakar, E.A. and Jegathesan, M. (1989) Bacterial pathogens isolated from cockroaches trapped from paediatric wards in peninsular Malaysia. *Transactions of the Royal Society of Tropical Medicine and Hygiene* 83, 133–135.

Rabito, F.A., Carlson, J., Holt, E.W., Iqbal, S. and James, M.A. (2011) Cockroach exposure independent of sensitization status and association with hospitalizations for asthma in inner-city children. *Annals of Allergy, Asthma and Immunology* 106, 103–109.

Roth, L.M. and Willis, E.R. (1957) The medical and veterinary importance of cockroaches. *Smithsonian Miscellaneous Collections* 134, 1–147.

Roth, L.M. and Willis, E.R. (1960) The biotic associations of cockroaches. *Smithsonian Miscellaneous Collections* 141, 1–470.

Schal, C. and Hamilton, R.L. (1990) Integrated suppression of synanthropic cockroaches. *Annual Review of Entomology* 35, 521–551.

Sherron, D.A., Wright, C.G., Ross, M.H. and Farrier, M.H. (1982) Density, fecundity, homogeneity, and embryonic development of German cockroach (*Blatella germanica* (L.)) populations in kitchens of varying degrees of sanitation (Dictyoptera: Blattellidae). *Proceedings of the Entomological Society of Washington* 84, 376–390.

Short, J.E. and Edwards, J.P. (1991) Reproductive and developmental biology of the oriental cockroach *Blatta orientalis* (Dictyoptera). *Medical and Veterinary Entomology* 5, 385–394.

Tee, H.S., Saad, A.R. and Lee, C.Y. (2011) Population ecology and movement of the American cockroach (Dictyoptera: Blattidae) in sewers. *Journal of Medical Entomology* 48, 797–805.

Eye Flies (Diptera: Chloropidae and Drosophilidae)

1. INTRODUCTION

Eye flies include chloropids, generally known as 'grass flies', and selected drosophilids of the Steganinae subfamily. These species are known as 'eye flies' because of their attraction to human and animal eyes and mucous membranes. In addition, chloropids also feed on body secretions from wounds, lacerations and scratches. By moving from one individual to another, chloropids can act as mechanical vectors of pathogens, causing illnesses such as yaws and conjunctivitis. Steganinae are considered of medical and veterinary importance due to their role as vectors of the eyeworm spirurid, *Thelazia callipaeda*, which infects domestic and wild carnivores, and humans.

2. TAXONOMY

The Chloropidae include ~2000 described species in 160 genera. The medically important genera in the Nearctic and Neotropical regions are *Hippelates* and *Liohippelates*, with the *Liohippelates pusio* group the most significant; in Eastern and South-eastern Asia, the genus *Siphunculina* is important, with *Siphunculina funicola* the dominant pest and vector species.

Knowledge of species ranked within the Steganinae (family Drosophilidae) is limited. While Steganinae larvae often exhibit zoophilic feeding habits, in the adult stages this behaviour is confined to three genera: *Amiota*, *Phortica* and the very rare extra-European genus *Apsiphortica*. The species of medical and veterinary importance are *Phortica variegata*, the vector of *T. callipaeda* in Europe, and *Phortica okadai*, *Phortica magna* and *Amiota nagatai*, which act as vectors of *Thelazia* spp. in Asian countries.

3. MORPHOLOGY

Chloropidae are small, shiny flies <5 mm (and often only 1–2 mm) long (Fig. 25). Their antennae are short, with the third segment being nearly globular and the arista either bare or with very short branches. The ocellar triangle is very large. The thorax has no distinct transverse suture and the squamae are small. The wings have no markings, the subcosta is rudimentary and the discal cross-vein and vein vi are absent.

P. variegata is about 3.5–5 mm in length. Males are generally darker and smaller than the females, with a short and plump scutum and a scutellum presenting numerous greyish spots and eight irregular rows of acrostichal

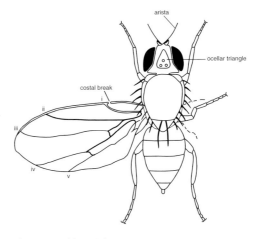

Fig. 25. A chloropid.

© R.C. Russell 2013. *The Encyclopedia of Medical and Veterinary Entomology*
(R.C. Russell *et al.*)

setae. The legs are characterized by a dark coxa, a dark femur with a yellow base and apex and with three conspicuous dark bands around the tibia. The wing is hyaline, but the two cross-veins are shaded and thus clearly visible. The costal vein bears two interruptions. A pale ring is present around the red eyes and the antennae are golden yellow.

4. LIFE CYCLE

Many species breed and feed in grass, other plants or vegetable debris, although others (e.g. the important *L. pusio* group) oviposit in freshly cultivated or otherwise disturbed soil where there is incorporated decaying plant or animal material. *Liohippelates* eggs are deposited in batches of up to 50 and the larvae feed on organic matter and develop within 1–2 weeks, after which they pupate. The adult will emerge after about 1 week, although in colder months larval and pupal stages may endure for many weeks. Development from egg to egg, under favourable conditions, takes about 4 weeks.

S. *funicola* breeds in various terrestrial habitats, particularly those characterized by organic pollution. In these habitats, the life cycle is much shorter, i.e. 10 days from egg to egg.

The life cycle of *P. variegata* is completed within a minimum of 9 days (in July and August) to a maximum of 18 days (in June). Wild *P. variegata* have been observed to lay eggs about 2 days after collection, with hatching from 2 to 12 days following oviposition, depending on the conditions of temperature and humidity.

The ecology and feeding habits of the pre-imaginal stages of most species remain unexplored.

5. BEHAVIOUR AND BIONOMICS

The biology of *Liohippelates collusor* has been well studied in California and that of S. *funicola* in Thailand. In the south-west USA, summer rains bring out heavy populations of *L. collusor* in California, Arizona and New Mexico, and irrigation of soils in some areas can have a similar effect. *Liohippelates* are generally strong fliers and some species are known to disperse as far as 7.5 km. For feeding, the adult females of some species (e.g. *Liohippelates* spp.) and both sexes in others (e.g. S. *funicola*) are attracted during daylight hours to the secretions of body orifices; they feed on sweat, sores with blood and pus and they are particularly attracted to eyes, especially those with a copious discharge. They scrape the conjunctival surface of the eye with the spiny tips of the pseudotracheal rings on the labella and this irritation increases the flow of secretion, which, in turn, attracts more eye flies. In Thailand, S. *funicola* are diurnally active, hovering around and feeding on hosts when there is little wind and particularly when temperatures are 32–35°C. Large masses of the flies often congregate on hanging objects in outdoor shaded structures or in dwellings, bringing them in close contact with humans and other animals (particularly dogs). Males are often more common at human hosts than females (in contrast to *L. collusor*, with which only females are attracted to humans).

Phortica spp. feed on lachrymal secretions of animals and humans, as well as on decaying fruit and slime fluxes. *P. variegata* displays a characteristic flight pattern around the bait (slow, small, vertical circles) before landing; this has been described as 'searching' behaviour. It has been shown that specimens of adult *P. variegata* collected in flight from around the eyes of humans and animals are mostly male, while flies collected on baits (i.e. fermented fruit) are characterized by a relatively balanced sex ratio. A variation of the sex ratio throughout the year has been recorded. The abundance of males around human eyes in late summer/early autumn might be attributed to their need for dietary (e.g. protein) supplementation. In southern Italy, the largest number of *Phortica* flies has been collected during summer (July–August) in hilly to mountainous wooded areas with undergrowth constituted mainly of holly and characterized by high RH (relative humidity). The biological activity of adult *P. variegata* is highest when temperature and RH are between 20–25°C and 50–75%, respectively. The use of GIS predictive modelling (integrating climatic and environmental data) has also suggested that this species should be limited to hilly areas with relatively high precipitation and continental temperatures, and this is supported by identification of infestations by *T. callipaeda* in dogs, cats and foxes from France, Portugal, Spain and Switzerland.

6. MEDICAL AND VETERINARY IMPORTANCE

Both *Liohippelates* and *Siphunculina* flies cause considerable nuisance to humans and animals. By moving from one individual to another, visiting contaminated tissues and scarifying other tissues with the spines on the labella, eye flies can be seen to be predisposed to mechanical transmission. They have been implicated or suspected, mostly via circumstantial evidence (e.g. isolation of pathogens from the flies), to be associated with transmission of various organisms/diseases, including anaplasmosis, mastitis and vesicular stomatitis of livestock, and pinkeye, *Streptococcus*, *Staphylococcus*, *Haemophilus* and yaws (*Treponema*) in humans, but there is little definitive evidence.

In the USA, the significant species are *L. collusor* in the west and *L. pusio* and *Liohippelates bishoppi* in the east. In Central America and the Caribbean, *Liohippelates currani*, *Liohippelates flavipes*, *Liohippelates pallipes* and *Liohippelates peruanus* have been indicated as being potentially significant. In the Orient, South and South-east Asian regions, *S. funicola* (known as the Oriental eye fly) is the main species visiting humans, and it can be an intolerable nuisance in parts of India. In Sri Lanka, *S. funicola* appears to have been replaced by the closely related *Siphunculina ceylonica*.

P. variegata acts as vector of *T. callipaeda* when feeding on the conjunctiva of vertebrates. While feeding, flies ingest the first-stage larvae released in the conjunctival sac by adult female nematodes of an infested animal. Within the body cavities of the arthropod vector, these larvae develop to second-stage larvae and, subsequently, to infective third-stage larvae, which reach the proboscis of the fly and are deposited into the conjunctival sac of a new,

receptive host after about 4 weeks. While both adult female and male *P. variegata* have been shown to harbour larvae of *T. callipaeda* under experimental settings, only males have been demonstrated to act as vectors of this parasitic nematode under natural conditions. The biological bases of this phenomenon are still unclear.

T. callipaeda nematodes live in the orbital cavities and associated host tissues, causing ocular disease for a range of carnivores (e.g. dogs, cats and foxes and other wild carnivores). Due to their zoonotic potential, these nematodes represent a potential public health concern. Human thelaziosis is considered to be an underestimated parasitic disease, which is prevalent in poor socio-economic settings in many Asian countries, mostly in China. The disease can be subclinical or symptomatic, being associated with epiphora, conjunctivitis, keratitis, excessive lachrymation, corneal opacity and/or ulcers.

7. PREVENTION AND CONTROL

Sanitation to reduce or eliminate potential breeding sites of eye flies, such as reducing manure and organic matter in soil and reducing loose soil matter, may assist in reducing population sizes. Incorporation of insecticides into the soil has been used for control, but this usually is not economically or environmentally viable or acceptable. Attractant baits have been used effectively for area control in some cases. Adulticiding can provide temporary relief from adult annoyance but is not an effective solution for ongoing problems. Physical barriers, such as finely screened porches and windows on houses and fly masks or fly sheets (a fine mesh cloth) on livestock, can provide some relief from annoyance and protection from possible pathogen transmission. The use of residual insecticides and repellents applied to surfaces where flies aggregate has shown some promise for protecting nearby houses. Personal repellents containing diethyl toluamide (DEET) also can provide temporary protection.

While the prevention of human thelaziosis could include control of the fly vector populations, this approach is presently not feasible, because the biology and feeding habits of *Phortica* flies are poorly known. Therefore, prevention of infestation through the use of bed nets to protect children while sleeping, keeping their faces and eyes clean and treating all infected domestic animals (which may act as reservoirs for human infection) with effective drugs is recommended. For the treatment of canine thelaziosis, topical instillation of antihelminthics is effective. Also, the adults and larvae of *T. callipaeda* can be removed mechanically from the eyes of hosts by rinsing of the conjunctival sac with sterile physiological saline.

SELECTED BIBLIOGRAPHY

Bächli, G., Vilela, C.R., Andersson Esher, S. and Saura, A. (2004) *The Drosophilidae (Diptera) of Fennoscandia and Denmark, Vol 39, Fauna Entomologica Scandinavica*. Brill, Leiden, Netherlands.

Bassett, D.C.J. (1967) *Hippelates* flies and acute nephritis. *Lancet* 4, 503.

Bassett, D.C.J. (1970) *Hippelates* flies and streptococcal skin infection in Trinidad. *Transactions of the Royal Society of Tropical Medicine and Hygiene* 64, 138–147.

Burgess, R.W. (1951) The life history and breeding habits of the eye-gnat *Hippelates pusio* Loew in the Coachella Valley, Riverside County, California. *American Journal of Hygiene* 53, 164–177.

Chansang, U. and Mulla, M.S. (2008a) Field evaluation of repellents and insecticidal aerosol compositions for repelling and control of *Siphunculina funicola* (Diptera: Chloropidae) on aggregation sites in Thailand. *Journal of the American Mosquito Control Association* 24, 299–307.

Chansang, U. and Mulla, M.S. (2008b) Control of aggregated populations of the eye fly *Siphunculina funicola* (Diptera: Chloropidae) using pyrethroid aerosols. *Southeast Asian Journal of Tropical Medicine and Public Health* 39, 246–251.

Chansang, U., Mulla, M.S., Chantaroj, S. and Sawanpanyalert, P. (2010) The eye fly *Siphunculina funicola* (Diptera: Chloropidae) as a carrier of pathogenic bacteria in Thailand. *Southeast Asian Journal of Tropical Medicine and Public Health* 41, 61–71.

Chansang, U., Mulla, M.S. and Sawanpanyalert, P. (2011) Temporal and spatial distribution, sex ratio and fecundity of the eye fly *Siphunculina funicola* (Diptera: Chlorpoidae) at aggregation sites during diurnal and nocturnal periods. *Southeast Asian Journal of Tropical Medicine and Public Health* 42, 274–288.

Dawson, C.R. (1960) Epidemic Koch–Weeks conjunctivitis and trachoma in the Coachella Valley of California. *American Journal of Ophthmalogy* 49, 801–808.

Floore, T.G. and Ruff, J.P. (1982) A six-year population survey of eye gnats (*Hippelates pusio* Loew, Diptera: Chloropidae) at a large soybean farm in Northern Bay County, Florida. *Journal of the Florida Anti-Mosquito Association* 53, 27–30.

Francy, D.B., Moore, C.G., Smith, G.C., Jakob, W.L., Taylor, S.A. and Calisher, C.H. (1988) Epizootic vesicular stomatitis in Colorado 1982: isolation of virus from insects collected in the northern Colorado Rocky Mountain Front Range. *Journal of Medical Entomology* 25, 343–37.

Kanmiya, K. (1989) Study on the eyeflies, *Siphunculina* Rondani from the Oriental region and Far East (Diptera: Chloropidae). *Japanese Journal of Sanitary Zoology* 40 (Supplement), 65–86.

Kumm, H.W. (1935) The natural infection of *Hippelates pallipes* with the spirochaetes of yaws. *Transactions of the Royal Society of Tropical Medicine and Hygiene* 29, 265–272.

Legner, E.F. and Bay, E.C. (1970) Dynamics of hippelates eye gnat breeding in the southwest non-cultivation and cover. *California Agriculture* 24(5), 4–6.

Máca, J. (1977) Revision of Palearctic species of *Amiota* subg. *Phortica* (Diptera, Drosophilidae). *Acta Entomologica Bohemoslovaca* 74, 115–130.

Mulla, M.S. (1962) The breeding niches of *Hippelates* gnats. *Annals of the Entomological Society of America* 55, 389–393.

Mulla, M.S. and Chansang, U. (2007) Pestiferous nature, resting sites, aggregation, and host-seeking behaviour of the eye fly *Siphunculina funicola* (Diptera: Chloropidae) in Thailand. *Journal of Vector Ecology* 32, 292–301.

Mulla, M.S. and Stains, G.S. (1977) The eye gnats, pests and plague of mankind, the friendly Coachella Valley salute. *Proceedings and Papers of the California Mosquito and Vector Control Association* 45, 205–209.

Mulla, M.S., Axelrod, H. and Ikeshoji, T. (1974) Attractants for synanthropic flies: area-wide control of *Hippelates collusor* with attractive baits. *Journal of Economic Entomology* 67, 631–668.

Otranto, D., Lia, R.P., Cantacessi, C., Testini, G., Troccoli, A., Shen, J.L., *et al.* (2005) Nematode biology and larval development of *Thelazia callipaeda* (Spirurida, Thelaziidae) in the drosophilid intermediate host in Europe and China. *Parasitology* 131, 847–855.

Otranto, D., Brianti, E., Cantacessi, C., Lia, R.P. and Máca, J. (2006a) The zoophilic fruitfly *Phortica variegata*: morphology, ecology and biological niche. *Medical annd Veterinary Entomology* 20, 358–364.

Otranto, D., Cantacessi, C., Testini, G. and Lia, R.P. (2006b) *Phortica variegata* as an intermediate host of *Thelazia callipaeda* under natural conditions: evidence for pathogen transmission by a male arthropod vector. *International Journal for Parasitology* 36, 1167–1173.

Otranto, D., Stevens, J.R., Cantacessi, C. and Gasser, R.B. (2008) Parasite transmission by insect: a female affair? *Trends in Parasitology* 24, 116–120.

Otranto, D., Cantacessi, C., Lia, R.P., Grunwald Kadow, I.C., Purayil, S.K., Dantas-Torres, F., *et al.* (2012) First *in vitro* culture of *Phortica variegata* (Diptera, Steganinae), the 'oriental eyeworm' vector'. *Journal of Vector Ecology* (In press).

Paganelli, C.H. and Sabrosky, C.W. (1993) Hippelates flies (Diptera, Chloropidae) possibly associated with Brazilian purpuric fever. *Proceedings of the Entomological Society of Washington* 95, 165–174.

Payne, W.J., Cole, J.R., Snoddy, E.L. and Seibold, H.R. (1977) Eye gnat *Hippelates pusio* as a vector of bacterial conjunctivitis using rabbits as an animal model. *Journal of Medical Entomology* 13, 599–603.

Taplin, D., Zaias, N. and Rebell, G. (1967) Infection by *Hippelates* flies. *Lancet* 2, 472.

Fleas (Siphonaptera)

1. INTRODUCTION

Fleas (order Siphonaptera) are wingless ectoparasites that occur on a wide range of terrestrial mammals and birds. They are particularly associated with hosts that spend part of their life in nests, dens, holes or caves. Fleas are therefore common on rodents, carnivores, bats, rabbits and nest-making birds and virtually absent from free-ranging ungulates and primates.

Both sexes are blood feeders and cause painful and irritating bites. Fleas are generally non-host specific and will parasitize a range of hosts. It is this ability to transfer from one host species to another that makes them of particular medical importance, allowing them to act as vectors of major diseases, such as bubonic plague and murine typhus, transmitted largely from rodents to humans. In veterinary medicine, fleas are highly problematic because of the pronounced allergic responses and dermatitis that can occur in hosts such as cats and dogs as a result of their feeding activity.

2. TAXONOMY

There are more than 2500 described taxa (including subspecies) of fleas, of which approximately 90% occur on mammals and only 10% on birds. Fleas are grouped into 15 (or 17) families, of which the most important are the Ctenophthalmidae, Leptopsyllidae, Pulicidae and Tungidae. About 25% occur on only a single mammalian species or genus. Others have a wide range of hosts, notably *Tunga penetrans*, *Xenopsylla cheopis*, *Echidnophaga gallinacea*, *Ctenocephalides felis* and *Pulex irritans*.

The so-called human flea, *P. irritans*, is a normal parasite of pigs. The genus *Pulex*, which infests diurnal mammals, especially porcines, originated in the New World, to which most species in the genus are restricted, with the exception of *Pulex simulans*, which also occurs in Hawaii, and *P. irritans*, which has become cosmopolitan. It has been suggested that *P. irritans* reached Western Europe about 14,000 years ago, coming from North America via Asia.

A small number of species, mainly of the genus *Ceratophyllus*, have become secondarily adapted to birds (mostly Passeriformes and sea birds). There are about 60 species and subspecies of *Ceratophyllus* in the Holarctic region, parasitizing many families of birds but especially the Hirundinidae (swallows and martins).

3. MORPHOLOGY

Adult fleas are readily recognized by the jumping behaviour seen when they are disturbed, their brown colour and their size (1–6 mm long). Females are larger than males of the same species (Fig. 26).

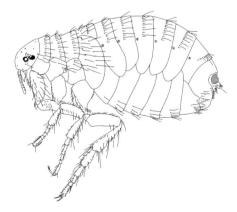

Fig. 26. Lateral view of female *Pulex irritans*.

Fleas show many adaptations to an ectoparasitic mode of life. They are laterally compressed, allowing them to move easily and quickly between the hairs of the host's body and to minimize damage from the host's reaction to their presence. Each antenna is recessed in a deep antennal fossa at the side of the head. The antennae are three-segmented, with the third segment being elaborately developed (Fig. 27). In many species, the

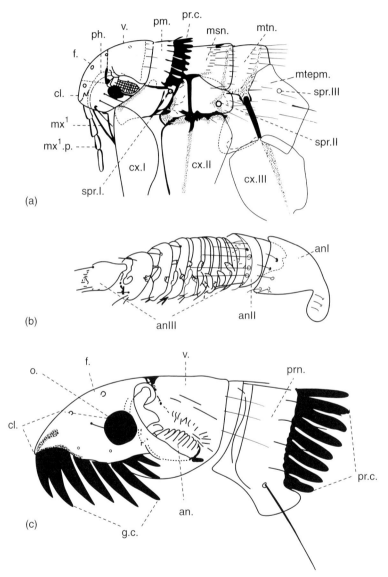

Fig. 27. (a) Head and thorax of *Nosopsyllus fasciatus* female. (b) Antenna of male *N. fasciatus*. (c) Head and pronotum of *Ctenacephalides felis*: an = antenna; anI, anII, anIII = first, second and third segments of antenna; cl = clypeus; cx.I, cx.II, cx.III = coxae of legs; f. = frons; g.c. = genal comb; msn. = mesonotum; mtepm. = metepimeron; mtn. = metanotum; mx¹ = maxilla (stipes); mx¹.p. = maxillary palp; o. = ocellus; ph. = pharynx; pre. = pronotal comb; prn. = pronotum; spr.I, spr.II, spr.III = thoracic and first two abdominal spiracles; v. = vertex. *Source*: Patton, W.S. and Evans, A.M. (1929) *Insects, Ticks, Mites and Venomous Animals. Part I: Medical.* H.R. Grubb, Croydon, UK.

antennae of the males have adhesive disks on the inner surface that are used to hold the female during mating. The head is sessile on the prothorax and the body is covered with backwardly directed setae and, in many cases, with combs (also known as ctenidia). Setae and combs may help the flea with holding on to its host, resisting the host's grooming activities, and may also protect delicate structures such as the mouthparts. The maxillary palps are well developed, with four obvious segments. There are no compound eyes, but lateral ocelli are present on both sides of the head. The ocelli are particularly large in species of *Xenopsylla* and reduced in many other species or even absent, as in the house-mouse flea, *Leptopsylla segnis*.

The thorax bears three pairs of legs, the third of which is particularly well developed for jumping, and consequently the metathorax supporting these legs is also well developed. In *Xenopsylla* and some other genera, the mesopleuron above the coxa of the second pair of legs is divided by the pleural rod into an anterior mesepisternum and a posterior mesepimeron (Fig. 28). The pleural rod is absent in *Pulex*, enabling these two combless genera to be distinguished.

The shape of the abdomen may be used to distinguish between sexes. In female fleas, both the ventral and dorsal surfaces are convex, whereas the dorsal surface of males is more or less flattened while the ventral surface is greatly curved. In addition, male fleas may be distinguished by the presence of a complex copulatory apparatus, located posteroventrally in the abdomen. A sensilium (= pygidium) is found posteriorly on the dorsal surface of both sexes, with the antesensilial seta immediately anterior to the sensilium (Figs 29 and 30).

The abdomen has ten segments, eight of which are easily recognizable externally, each of them bearing a pair of spiracles. In addition, there are two pairs of spiracles on the thorax. The ninth abdominal segment is much modified in the male, with tergum IX forming paired manubria and articulating claspers and sternum IX forming an L-shaped clasping organ, the apical arm of which is a useful character in the identification of species of *Xenopsylla*.

There is a single spermatheca in the female (Fig. 30). The spermathecal duct opens into the head of the spermatheca, which is separated from the tail by a small constriction. The relative sizes of the head and tail are useful

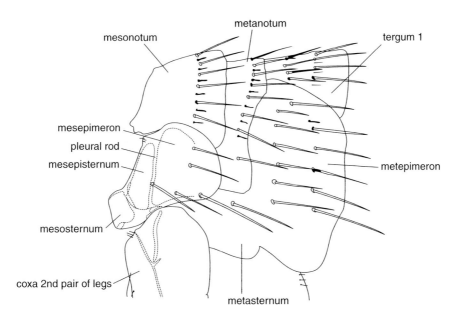

Fig. 28. Lateral view of mesothorax, metathorax and tergum 1 of *Xenopsylla cheopis*.

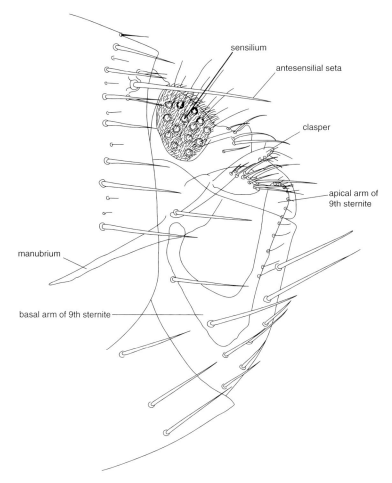

Fig. 29. Terminal segments of the abdomen of a male *Xenopsylla cheopis.*

characters for distinguishing species of *Xenopsylla*: they are of similar size in *X. cheopis*, while the head is considerably larger than the tail in *Xenopsylla brasiliensis* and the reverse occurs in *Xenopsylla astia*.

Many fleas possess combs (ctenidia), although the Pulicidae are relatively combless and there are no combs in *Xenopsylla, Pulex, Echidnophaga* and *Tunga*. In Ctenocephalides, there are both genal and pronotal combs with backwardly directed teeth (Fig. 27), and combs are particularly well developed in fleas that parasitize bats, where they may have a metathoracic to abdominal localization. In the ceratophyllid genera, *Ceratophyllus* and *Nosopsyllus*, only the pronotal comb is

present. The number of spines in the pronotal comb varies according to the type of host. In bird fleas (e.g. *Ceratophyllus* spp.), the spines are narrower and more numerous, usually exceeding 24, while in parasites of mammals, such as *Nosopsyllus* spp., the spines are broader and less numerous, usually less than 24. *Stephanocircus* spp. is characterized by a 'crest' consisting of a peripheral comb attached to the frons.

4. LIFE CYCLE

In the laboratory, the life cycle of *C. felis* (Fig. 31) takes 14 days at 32°C and 140 days at

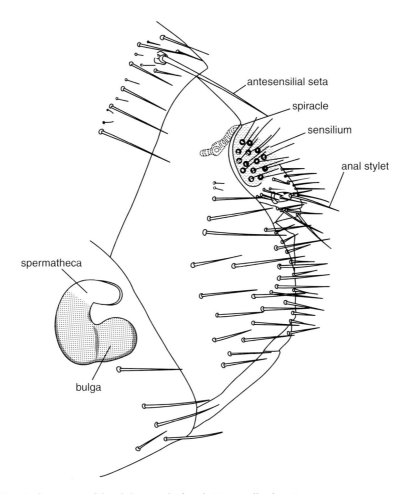

Fig. 30. Terminal segments of the abdomen of a female *Xenopsylla cheopis*.

13°C, providing a humidity above 50% is maintained. Larvae are most susceptible to desiccation and require a minimum of 50% RH (relative humidity) for 50% survival. The eggs are more tolerant, requiring 33% RH, and adults will emerge from 80% of pupae at 2% RH. At 30°C and 78% RH, pupation usually occurs about 7 days after egg hatch.

4.1 Egg

The female adult flea produces relatively large (0.3–0.5 mm), whitish, oval eggs, which are sticky in *X. cheopis* but dry in *T. penetrans* and *E. gallinacea*. Eggs are deposited within a nest or on the host (from where they fall to the ground). Female *C. felis* have six ovarioles in each ovary, half of which contain mature oocytes at any time. Egg production begins 2 days after the first blood meal and peaks in 6–7 days. Single females confined with five males in microcells on a cat were shown to produce 14 eggs/day on average and 158 in a whole lifetime. Unconfined fleas on a cat lay up to 24 eggs/day. The majority of the eggs of *C. felis* is laid during the last 8 h of the scotophase.

Eggs hatch within a few days, provided that the humidity is above 70%. At 80% RH, eggs of *X. brasiliensis* hatch in 6 days at 24°C and 4 days at 35°C, and those of *E. gallinacea*

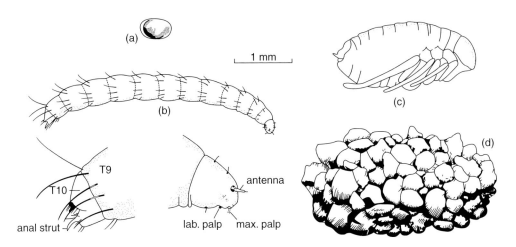

Fig. 31. Life cycle of *Ctenocephalides felis*. (a) Egg; (b) larva with caudal and head ends enlarged; (c) pupa; (d) sand-encrusted cocoon. *Source*: Dunnet and Mardon (1991).

need 3–4 days at 26°C and 85% RH. Eggs that lose water when exposed to RH of less than 70% are unable to recover when placed in a higher humidity. The threshold temperature for egg development is 12°C for the tropical rat flea, *X. cheopis*, and 5°C for the temperate region European rat flea, *Nosopsyllus fasciatus*.

4.2 Larva

The larva uses a hatching spine to emerge from the egg (Fig. 31). It has a distinct head and 13 body segments, with no distinction between thoracic and abdominal segments, and no appendages. The larva is negatively phototactic, burrowing into the material of the nest or substrate. The whitish, vermiform larva measures 4–10 mm in length when fully grown and its body segments bear a circlet of backwardly directed bristles that, together with the anal struts on the last segment, enable the larva to move. A range of external morphological features can be used to group flea larvae into families, species and subspecies. There are three larval stages in most species but only two in *T. penetrans*.

Larvae feed in their environment on organic debris supplemented, in many species, by undigested blood in the faeces of adult fleas. The faeces of adult *C. felis* contain, for instance, 7–11% of protein. In *N. fasciatus*, the interaction of larva and adult flea has become much closer; larvae actively pursue and seize adult fleas with their mandibles in the region of the sensilium, the adults respond by defecating and the larva releases its hold on the adult and eats the excreted faecal blood. The pharynx of the larva is muscular and larvae of *N. fasciatus* on the whole can imbibe blood, water and rat urine. These larvae are semi-predatory and will attack damaged adults and kill them. Adults of the rabbit flea, *Spilopsyllus cuniculi*, normally defecate every 20 min, but the frequency of defecation increases greatly shortly before oviposition, presumably to provide a more favourable environment for larval development in the rabbit burrow.

Larvae of *X. cheopis* lack the ability to close their spiracles and, consequently, they require high humidity for development. Exposure to 0% RH and 22°C for 24 h is lethal, whereas at 90% RH the lethal temperature is increased to 36°C. Larvae of *X. cheopis* will move to areas of high humidity.

The duration of the larval stage is therefore dependent on temperature and humidity. At 24°C, its duration in *X. brasiliensis* increases from 12 to 25 days as the relative humidity decreases from 93 to 70%, respectively. At 25°C and 85% RH, there are similar rates of development in four species of Pulicidae, with the larval stages being completed in 1–2 weeks. Considerable individual variation exists, however, in the rate of development of fleas, even when larvae from the same batch of eggs are reared under identical conditions.

4.3 Pupa

The mature third-stage larva empties its gut and constructs a thin, loosely woven cocoon, within which pupation occurs (Fig. 31). The cocoon is typically ovoid, about 3 mm long and 1 mm wide. It is made of a sticky material so that debris from the environment can adhere to it easily, disguising its real appearance. The main threat to adult emergence is desiccation, which prevents ecdysis of the pupal cuticle. The cocoon itself offers no protection against desiccation. Larvae of *X. brasiliensis*, for example, readily lose water in an atmosphere where the humidity is below 45%.

Temperature is the other factor affecting pupal development. Successful pupation and emergence of *X. cheopis* occurs at 18–35°C and 60% or higher RH. At lower temperatures, pupation is more variable, with no emergence of adults occurring below 14°C. As for the threshold temperature for egg development, the optimum temperature for pupal development is about 10°C higher for *X. cheopis* than for *N. fasciatus*.

The duration of the period spent in the cocoon between pupation and the emergence of the adult varies from 1 week to 6–12 months. This is a major factor in the survival of flea populations during the absence of a host or in the presence of adverse climatic conditions.

4.4 Adult

Adults moult from the pupal cuticle but may remain within the cocoon as pharate adults for long periods. When emerging from the cocoon, adults of *Ceratophyllus gallinae*, for example, use the frontal tubercle on the head to weaken the fibres of the cocoon. The tubercle may be lost later in adult life. Emergence of the adult may also be triggered by environmental cues.

Females of *X. cheopis*, *X. brasiliensis* and *E. gallinacea* usually emerge 3–4 days before the males. Given the importance of water conservation to the survival of adult fleas, it is of interest that during the first day of adult life *C. gallinae* appears to be able to take up water from air with humidity above 82% RH. Adult fleas may be long-lived; for example, unfed *S. cuniculi* can survive for 9 months at –1°C. Conversely, *C. felis* must feed every 12 h to survive and reproduce. On a cat, relatively few female *C. felis* live longer than 6 weeks, with most living for an average of about 2 weeks.

Fleas that infest diurnal hosts have well-developed eyes and those that remain in the host's nest have reduced eyes, thorax and legs as an adjustment of their more sedentary lifestyle. The geographical distribution of each flea species is limited by the availability of its preferential host(s), the presence of habitats suitable for larval development and its evolutionary history.

In sticktight (also known as stickfast) fleas (*Echidnophaga* spp.), the mouthparts are relatively much longer than they are in more mobile fleas (Fig. 32). In *E. gallinacea*, the mouthparts are one-third the length of the body, while in *P. irritans* and *X. cheopis* their length is only 10–20% of the body length. In addition, the laciniae in *Echidnophaga* spp. are strongly toothed, being used to anchor the flea at its attachment site for up to 6 weeks.

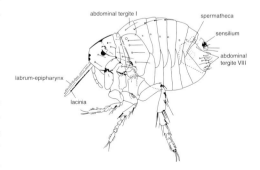

Fig. 32. Female *Echidnophaga gallinacea*. *Source*: Patton, W.S. and Evans, A.M. (1929) *Insects, Ticks, Mites and Venomous Animals. Part I: Medical.* H.R. Grubb, Croydon, UK.

5. BEHAVIOUR AND BIONOMICS

5.1 Host finding

As adult fleas feed only on blood, newly emerged individuals must find a host to survive. Some fleas, such as *C. gallinae*, actively disperse in search of a host, whereas others, such as *N. fasciatus* or the rabbit flea, *S. cuniculi*, wait in the nest or burrow for the return of their host. Both groups of fleas need to be able to detect a host when it is nearby, to orientate to it and achieve contact. Various attributes of the hosts provide stimuli to which the flea may respond. These include vibration, warmth, exhalation of carbon dioxide, characteristic odours and, in lit situations, the casting of a shadow. Female *C. felis* respond to visual and thermal stimuli. Fleas are also able to detect air currents by receptors on the sensilium. Warm airwaves generated by a living being in motion can attract fleas even in the absence of visual stimuli. Adult *X. cheopis*, for example, can detect the odour from a white rat at a distance of 30 cm.

The flea *C. gallinae* parasitizes more than 75 species of bird. It overwinters as a pharate adult in the cocoon and emerges in the spring in response to a sharp rise in temperature and/or tactile stimuli. The flea emerges in an old nest. At first, the newly emerged adult is negatively phototactic, but after 3 or 4 days becomes positively phototactic, which, associated with negative geotaxis, ensures that the adults crawl up trees and bushes. In search of a host, the flea stops periodically and faces towards the brightest source of light and jumps when the light is suddenly obscured. Readiness to jump rises to a peak 4 days after emergence, coinciding with the positive response to light, but falls off later due in part to water loss. Humans and their pets are frequently bitten by dispersing *C. gallinae* as they search for a suitable host.

5.2 Feeding

Adult fleas are generally capillary feeders (solenophages). In *X. cheopis* and *P. irritans*, the maxillae are used to penetrate the host's skin, while the tip of the labrum epipharynx serves for entering capillary vessels from which the flea imbibes blood. The flea's saliva is passed into the host by the salivary pump and appears as clear drops of fluid outside the capillary. The saliva of *X. cheopis* contains an anticoagulant substance and a material of low molecular weight with allergenic properties.

In *Ctenocephalides canis*, there are three pumps (cibarial, precerebral and postcerebral) used to convey blood to the midgut. *S. cuniculi* is probably a pool feeder and, at times, *X. cheopis* may also feed in the same manner. *C. felis* feeds to repletion in 10 min, imbibing 7 µl of blood and doubling in weight. Feeding is more frequent at higher temperatures as a result of accelerated physiological activity and increased rate of water loss. The volume of blood imbibed may vary with its composition.

5.3 Role of the proventriculus in feeding and digestion

Ingested blood passes from the foregut to the midgut (stomach). The latter has the dual functions of initial storage and digestion. The proventriculus of the posterior foregut is particularly well developed and possesses needle-like spines. In *X. cheopis*, the spines are arranged in a regular series; in females there are 15 rows of 30 spines each and in males 12 rows of 22 spines each. They seem to facilitate the passage of blood from foregut to midgut, enabling rapid feeding. In addition, the spines play a role in the fragmentation of red blood cells. Three to five peristaltic waves are followed by one antiperistaltic wave, which thrusts the blood forwards against the spines. The spines do not penetrate into the midgut but only into the posterior spineless zone of the proventriculus. Other actions contributing to the crushing of erythrocytes include the to and fro shifting movements of the proventriculus and the contractile motion of its posterior region. The anterior half of the spined region of the proventriculus serves as an effective barrier to regurgitation.

5.4 Mating and reproduction

With the exception of some fleas (e.g. bird fleas) that mate soon after emergence before even taking a blood meal, most fleas, especially females, require a considerable period of feeding before mating. In *C. gallinae*, sex recognition is achieved by a contact pheromone detected by receptors on the male palps. The pheromone is species specific and is present in both sexes. The male erects his antennae and moves under the female, grasping her with the adhesive organs on their inner surfaces. Correct alignment of the pair, and successful coupling, is probably assisted by receptors on the sensilium and the antesensilial seta. Movement of the female is inhibited by pressure of the ninth sternum of the male on hairs of the female's sensilium. The aedeagus or intromittent organ of the male is used to dilate the female's genital chamber. One penis rod is then inserted into the spermathecal duct and may reach to the spermatheca. Penetration of the rod is slow and copulation may last for up to 9 h, with the average duration being of 3 h in *C. gallinae*.

In species where the female is sessile or semi-sessile, for example *E. gallinacea* and *S. cuniculi*, the antennae of the male lack adhesive disks: copulation occurs while the female is feeding. After the male *E. gallinacea* is coupled with the female, it may lose all contact with the host's skin, with its legs free in the air. After mating, however, the male flea returns to the skin surface and may take a blood meal. In *T. penetrans*, mating occurs when the female is endoparasitic within the skin of its host. The duration of mating is thought to be about 10 min for *X. cheopis*, 15 min for *E. gallinacea* and 20 min for *T. penetrans*.

Control of mating in the rabbit flea *S. cuniculi* is a complex process and is thought to be dependent on the physiological state of the host. In this species, maturation of the female and maximum maturation of the male take place only when the fleas feed on a pregnant doe rabbit or its newborn young, 1–10 days old. The reproductive cycles of both flea and host are therefore closely coordinated.

In the tropics and subtropics, it is likely that fleas breed continuously throughout the year, although this may be moderated by very hot or very dry conditions. In more seasonal environments, *C. felis*, for example, is more abundant in the warmer months than in the cooler part of the year.

5.5 Jumping

One of the most distinctive features in the behaviour of fleas is undoubtedly their ability to jump. Jumping is carried out through the movement of the third pair of legs, with the other two pairs acting mainly as supports.

Fleas do not use direct muscular contraction to deliver the required jump. Energy for jumping is stored within the thorax, largely in the pleural arches, which are laterally placed pads made of resilin protein. These are kept compressed by muscular contraction and then held in place by cuticular catches. Once the plural arches are compressed and the catches engaged, the muscle which compressed the resilin can be relaxed. As this occurs, the femur of the third pair of legs is rotated to a vertical position and connected to the substrate by the trochanter and tibia. The jump is initiated by the release of the catches, which allows the plural arches and coxal walls to spring back into shape, releasing the stored energy. While jumping takes place, the femur rotates downwards, transmitting its thrust via the tibia to the substrate, and the flea jumps. The tarsi and claws do not seem to play a role in jumping, wherein the flea may turn over with the legs extended. In *N. fasciatus*, the second pair of legs may extend above the dorsal surface of the body. This will increase the probability of the flea holding on to a host should it encounter one during its jump.

There is considerable variation in the ability of different flea species to jump. The jump of the rat flea, *X. cheopis*, averages 18 cm but can reach up to 30 cm. The hen flea, *C. gallinae*, can jump up to 24 cm horizontally and 11 cm vertically. The nest-dwelling, semi-sessile rabbit flea, *S. cuniculi*, jumps a mere 3.5 cm vertically. The human flea, *P. irritans*, has been observed to jump a vertical height of 13 cm and is considered to be able to reach 20 cm.

6. MEDICAL AND VETERINARY IMPORTANCE

The main fleas of medical importance are the tropical rat flea, *X. cheopis*, the main vector of the diseases known as plague and murine (or endemic) typhus to humans, the sand flea (or chigoe or jigger flea), *T. penetrans*, the female of which develops as an endoparasite under the skin of people, particularly on the feet and ankles, and the human flea, *P. irritans*, which breeds in human dwellings.

Fleas of veterinary importance include the cat flea, *C. felis*, and the dog flea, *C. canis*, which are widely distributed globally, and also the sticktight (stickfast) flea of poultry, *E. gallinacea*, which is widely distributed in the warmer countries of the world. This latter flea attaches to the heads of poultry and may occur in clusters of 100 or more on the comb, wattles, back of the head and round the eyes and beak; it is a serious pest, as large numbers cause a progressive anaemia and emaciation, leading to lowered egg production in laying hens and death in young birds.

The bird flea, *C. gallinae*, may be a locally important pest of poultry, as well as many other species of birds. The European rabbit flea, *S. cuniculi*, is an important mechanical vector of myxoma virus (the agent of myxomatosis disease of rabbits) in Europe (although mosquitoes have been the principal vectors of that virus in Australia).

As nuisance pests for humans, a number of species can cause annoyance, with their bites causing mild to intense irritation and leading to severe allergic reactions (papular urticaria) in sensitive individuals. In this respect, the most common species is probably the cat flea, *C. felis*, although the human flea, *P. irritans*, and the related *P. simulans* and various rodent and bird fleas can be locally significant in different parts of the world. For hypersensitive individuals, treatment with corticosteroids or desensitizing antigens can be helpful.

Other agents of disease have been associated with fleas and their bites but they are generally considered to be of less great importance. Among others, the two most significant pathogens may be *Bartonella henselae* (the agent of the cat scratch disease) and *Francisella tularensis* (the agent of tularaemia).

6.1 Flea species of particular significance

6.1.1 *Ctenocephalides felis*

The cat flea has become a major domestic pest worldwide. It is the most common species of flea found on domestic cats and dogs throughout North America and northern Europe and in other countries such as Australia. Significantly more cats are infested with fleas than dogs. There are four distinct subspecies of *C. felis*: *Ctenocephalides felis felis* is widespread, *Ctenocephalides felis strongylus* occurs in Africa, *Ctenocephalides felis damarensis* in south-western Africa and *Ctenocephalides felis orientalis* in India, Sri Lanka and South-east Asia.

Fleas may be found on pets throughout the year, but in the northern hemisphere numbers tend to increase around late spring and early autumn, when environmental conditions are favourable for larval development. Since *C. felis* is able to survive for long periods off the host, it therefore does not require direct contact for its survival. The response to a flea bite is a raised, slightly inflamed wheal on the skin, associated with mild pruritus. As a consequence, the animal will scratch intermittently, with little distress. Since each female *C. felis* can ingest as much as 14 µl of blood/day, severe infestations may lead to iron-deficiency anaemia. Anaemia caused by *C. felis* is particularly prevalent in young animals and has been reported in cats and dogs and, very rarely, in cattle, goats and sheep. However, more importantly, after repeated flea bites over a period of several months, a proportion of dogs and cats develop flea-bite allergy, which is often associated with profound clinical signs. Flea-bite allergy is a hypersensitive reaction to components of the flea saliva released into the skin during feeding. The allergy shows seasonality in temperate areas, appearing in summer when flea activity is highest, though in centrally heated homes and in warmer regions, exposure may be continuous. The most commonly affected areas in both dogs and cats are the preferential biting sites of the fleas, which are the back, the ventral abdomen and the inner thighs. Flea allergy dermatitis is one of the most common causes of dermatological disease of dogs and cats and is characterized by intense pruritus and reddening of the skin, with itching persisting up to 5 days after the bite. The resultant licking, chewing and scratching can lead to hair loss, self-induced trauma and secondary infection. Other symptoms include restlessness, irritability and weight loss, though the intensity of irritation varies greatly with the individual attacked.

In some parts of the world, severe infestations of calves, lambs and kids by *C. felis* may be found among young ruminants maintained on straw bedding in barns, which provides ideal breeding sites for fleas introduced by farm dogs and cats.

Cat fleas also act as intermediate hosts for the common tapeworm of dogs and cats, *Dipylidium caninum*, and for the subcutaneous filaioird, *Acanthocheilonema reconditum*, infesting dogs worldwide. Both helminths may occasionally be found in humans, particularly young children who, when playing with pets, may ingest infected fleas inadvertently.

6.1.2 *Tunga penetrans*

Known variously as the sand flea, or jigger or chigoe, this is an important subcutaneous parasite of humans in the Neotropical and Afrotropical regions, where people often walk barefooted. Although it primarily infests humans, *T. penetrans* is thought to be naturally associated with pigs as hosts in South America; it was introduced into West Africa in the middle of the 19th century, probably with the slave trade, and nowadays it is widespread in tropical Africa. Besides humans and pigs, *T. penetrans* infests various other animal hosts, including dogs, cows and horses.

The male *T. penetrans* is very small and free-living. The female burrows into the skin of the feet and ankles of humans. It begins as one of the smallest of the fleas, but then undergoes considerable hypertrophy of the abdomen, particularly the second and third abdominal

segments, becoming the shape and size of a pea (Fig. 33). The female feeds head down, and consequently the spiracles on abdominal segments 5–8 are very large, while the other abdominal spiracles are not developed. The female deposits about 200 eggs that are passed out to the exterior. They hatch to produce larvae that follow the normal cycle of development. The presence of a number of adult *T. penetrans* in the foot can be crippling and the damage to the skin can facilitate the entry of other pathogens, leading to secondary infection and ulceration.

Overall, there are 11 species of *Tunga* known to date, mostly infecting anteaters, armadillos, pigs, rodents and various domestic animals. Among these others, there is one species, *Tunga trimamillata*, which has been reported infesting humans (as well as cows, sheep, goats and pigs) in Ecuador and Peru.

In endemic areas, the wearing of closed footwear is the best method of preventing infestations and, in uncomplicated cases, infestations can be readily dealt with by extracting the flea with a sterile needle and applying a topical antibiotic to prevent secondary infections. Various therapies based on poultices of natural products are also used in endemic countries. Occlusive medications based on salicylate vaseline have been shown to be effective for diffused infestations and more sophisticated approaches with oral or topical thiabendazole and ivermectin have also given good results.

6.2 Flea-borne disease

6.2.1 Plague

Plague is a disease of rodents caused by the bacterium *Yersinia pestis* (formerly *Pasteurella pestis*), a Gram-negative, facultatively anaerobic, non-motile microorganism to which humans are susceptible. It occurs worldwide, although most cases are reported from countries in Africa (e.g. Congo, Madagascar, Uganda), Asia (e.g. Vietnam, India, Myanmar) and the Americas (e.g. USA, Brazil, Peru), and more than 200 species of rodents are known to be able to harbour the bacillus. *Y. pestis* can produce a coagulant enzyme that has the effect of causing ingested blood to clot in the

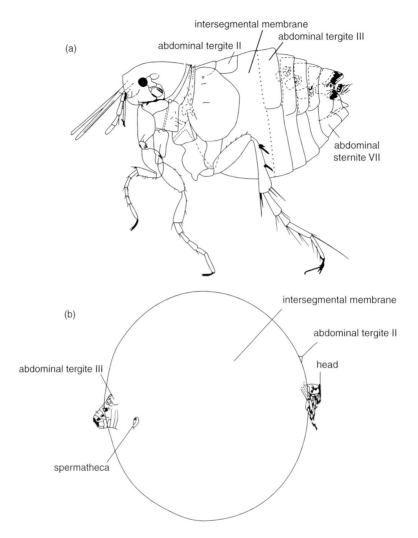

Fig. 33. Female *Tunga penetrans*. (a) Recently fertilized female with the intersegmental membrane between segments II and III beginning to be stretched; (b) gravid female, fully expanded. *Source*: Patton, W.S. and Evans, A.M. (1929) *Insects, Ticks, Mites and Venomous Animals. Part I. Medical*. H.R. Grubb, Croydon, UK.

proventriculus of the flea, blocking the gut; this is a prerequisite for transmission.

The most common clinical form of plague is an acute lymphadenitis, called bubonic plague, which can lead to septicaemic and pneumonic manifestations. Less commonly, plague can cause meningitis and pharyngitis. Following an infected flea bite, infection by *Y. pestis* undergoes an incubation period, proliferating in the regional lymph nodes for 2–8 days. Infected persons then suffer from fever, headache and painful swellings (buboes) in the lymph nodes of groin, axilla and/or neck. This condition can proceed to septicaemic plague when there is a substantial increase in *Y. pestis* in the circulating blood (where it is readily available to bloodsucking fleas). Bloodstream spread of the bacilli to the lungs can result in pneumonic plague, which is highly contagious by airborne transmission and in which the fleas play no vectorial role. While a mild form of disease (pestis minor) and asymptomatic cases may occur during outbreaks, serious progress of this illness can be swift and mortality in untreated cases is estimated to exceed 50%.

Plague is considered to have originated in Central Asia, from which it has spread virtually throughout the world. A first pandemic spread out from Arabia into the Byzantine Empire. A second pandemic, the so-called 'Black Death' of the Middle Ages, ravaged Europe from the 14th to the 17th centuries. The third pandemic is currently ongoing. In the 40 years from the mid-1950s, almost 40 countries reported cases to the WHO (~80,000 cases and 6500 deaths), with more than 50% of the cases recorded in Asia. More recently, most cases have occurred in Africa (>75% of cases and >80% of deaths) and there have been limited outbreaks in Asia and South America.

PLAGUE TRANSMISSION AND EPIDEMIOLOGY
Plague is spread among rats by rat fleas. In the absence of fleas, no epizootics are recorded among rats. When a flea feeds, blood is propelled along the oesophagus to the midgut by the pharyngeal pump, with the proventriculus acting as a valve to prevent regurgitation. If the blood is infected, Y. pestis multiplies in the midgut and in the interstices of the spines of the proventriculus. The culture of Y. pestis in the flea's midgut forms a cohesive, gelatinous body, which gradually fills up both the midgut and proventriculus, effectively 'blocking' the gut.

When a 'blocked flea' attempts to feed, the pharyngeal pump is unable to force blood through the proventriculus into the midgut and its pumping distends the oesophagus. When the pharyngeal pump ceases to work, the stretched oesophagus recoils and forces blood, contaminated with fragments of the gelatinous bacterial culture, into the host on which the flea has been attempting to feed. Being unable to feed successfully, a blocked flea will attempt to ingest blood repeatedly and, in so doing, enhances the chances to infect many hosts. However, the process of blockage in the flea is a complicated one and may be affected by the species and strain of flea, the strain of Y. pestis, the environmental temperature, the

frequency and duration of feeding and the host species.

Being unable to take in liquid food, a blocked flea is susceptible to desiccation and survives only a short time under hot, dry conditions. Blockage of the gut is, however, not necessarily a permanent or fatal condition in the flea. A passage may be re-established through the plug, giving rise to a partially blocked flea. Such a flea is even more dangerous than a fully blocked flea, because not only is it able to feed and therefore live longer but also its proventriculus is unable to function as an effective valve and infective material is regurgitated from the midgut into the flea's host.

Transmission of Y. pestis via fleas can also occur through the faeces (where the bacilli can remain infective for at least 3 years) being scratched into the skin or through contacting mucous membranes. Furthermore, infections can also result from skin lesion contamination following direct contact with infected sick or dead vertebrate hosts.

Plague can spread from rats living in relatively close association with humans to wild rodents in the field; this is called 'sylvatic plague' and is now established in all continents except Australia. This adds greatly to the complexity of the epidemiology of plague, due to the numbers of small mammals and their flea parasites that may be infected. Excluding the cosmopolitan species of rats and commensal mice, more than 200 taxa (species and subspecies) of Rodentia and almost 20 of Lagomorpha have been found infected with Y. pestis. At least 100 taxa of fleas, associated with these small mammals, have been found infected with Y. pestis in nature.

Plague can survive the hot dry season in India by persisting in aestivating Indian gerbils. It can also overwinter in hibernating rodents and latent infections in rodents can relapse, later to become active and initiate an epizootic. It is possible for flea larvae to ingest Y. pestis when feeding on the faeces of infected adult fleas, but there is little evidence that this results in infected emerging adults.

RAT FLEA VECTORS OF PLAGUE
X. cheopis has a worldwide distribution between 35°S and 35°N, where it is a common

ectoparasite of rats in cities, ports and rural areas. It is undoubtedly the most important vector of human plague throughout the world. In a comparison of ten species of fleas, *X. cheopis* showed the highest rate of blockage (58%), the highest ratio of transmissions (35 transmissions from 53 fleas) and the lowest rate of eliminating *Y. pestis* from its body to become plague free.

X. brasiliensis is an ectoparasite of *Rattus* spp. in rural settings in Africa, South America and India. It is as efficient as *X. cheopis* as a vector of plague and is regarded as the major vector of plague to humans in rural situations in Africa and in the hilly, woody tracts of Bombay State in India.

X. astia is an ectoparasite of *Rattus* spp. in South-east Asia, where it occurs in fields, villages and ports. It is regarded as a mediocre vector of plague. Feeding more readily on rats than on humans, it may play a greater role in maintaining plague among the rodent population than in transmitting it to the human host.

N. fasciatus is widely distributed throughout the world in cool, temperate areas and can maintain plague among *Rattus* spp. It is, however, reluctant to feed on people, and human cases are rare in epizootics when *N. fasciatus* is the vector.

L. segnis is a cosmopolitan ectoparasite of mice and it also occurs on rats. It is infected with *Y. pestis* less readily than *N. fasciatus* but may play a minor role in plague epizootics; however, it feeds only reluctantly on humans and is considered to play a negligible role in the transmission of plague to humans.

PREVENTION, CONTROL AND TREATMENT OF PLAGUE

Control of sylvatic activity is generally not practical, so surveillance of mortality in urban rodent populations and human cases is essential. In endemic areas, management of urban rodent and flea populations should be the focus of prevention. Rat-proofing buildings and use of rodenticides (the anticoagulants are usually recommended) and insecticides such as organophosphates, carbamates or pyrethroids as dusts/powders in rat runways are usually advised. However, there is widespread flea

resistance to some chemicals. During outbreaks, the first priority should be the killing of flea vectors – not the destruction of rodent reservoir populations, because that will result in infected fleas searching for new hosts and increase the transmission risk to humans.

While untreated plague has a high (>50%) mortality rate, prompt treatment with streptomycin is very effective and can reduce mortality to less than 5%. Gentamicin, tetracycline, doxycycline and chloramphenicol can also be used effectively. Variably effective vaccines have been made, not appropriate for immediate community protection in outbreak situations but useful for protection of persons at high risk (e.g. field, hospital and laboratory personnel). There are new plague vaccines undergoing development but it is not expected that these will be commercially available for some time.

6.2.2 Murine typhus (endemic typhus)

Rickettsia typhi (formerly known as *Rickettsia mooseri*) is a flea-borne organism placed in the 'typhus group' of the genus *Rickettsia* together with *Rickettsia prowazekii* (which is carried by lice). Murine typhus is distributed worldwide and, being a disease associated with commensal rodents (particularly *Rattus norvegicus* and *Rattus rattus*) and their fleas (especially *X. cheopis* and *L. segnis*), has been disseminated by shipping. It therefore tends to have a coastal distribution, clustering along portside buildings or houses inhabited by rats and their fleas in urban environments, although it may also occur inland in rural areas in India, Burma, Thailand, Pakistan and the southern USA, as well as in many other countries (e.g. Australia, China, Greece and Israel). Murine typhus is possibly the most prevalent rickettsial infection reported in humans. However, due to the difficulty of its diagnosis, this infection is under-reported, with perhaps only one-fifth of the actual number of infections being recorded.

R. typhi causes a milder disease than *R. prowazekii* but it is still a serious debilitating illness, with high fever, headaches, myalgias and (usually) a maculopapular rash. There is an incubation period of 1–2 weeks, followed by a similar period of clinical disease, with a mortality of less than 5% in untreated cases and zero using appropriate anti-rickettsial drugs.

R. typhi harms neither the rat nor the flea. In X. cheopis, it multiplies in the cells of the midgut, from which it tends to escape after 3–5 days without damaging the cells and spreads to the entire midgut lining within 7–9 days. Fleas begin to pass infected faeces 10 days after an infective feed and continue to pass R. typhi for 40 days. Once infected for more than 3 weeks, the flea is capable of transmitting R. typhi while feeding, although self-inoculation by scratching of crushed infected fleas or their contaminated faeces into the skin remains the main source of infection. Fleas remain infected for the rest of their lifetime and the rickettsias are transmitted transovarially. Although X. cheopis is primarily an ectoparasite of rats, it readily feeds on humans in the absence of its main host. In X. cheopis, feeding and defecation are closely associated, and infected fleas feeding on people would deposit R. typhi in the human environment.

A similarly infecting rickettsia (Rickettsia felis) is also found widely in various vertebrates (including humans) and the cat flea, C. felis, although this rickettsia belongs to the spotted fever group rather than the typhus group, as with R. typhi. R. felis has a similar clinical profile to R. typhi.

In hyperendemic foci, R. typhi infection rates of 3–10% have been found in X. cheopis and 9–16% in L. segnis; although the latter is a semi-sessile flea, usually disregarded as an important vector of murine typhus, it may prove to be a significant vector in parts of the world where X. cheopis is absent or very rare. R. typhi (and the similar R. felis) has been found also in domestic cats and opossums in areas without obvious infected rats and their fleas. These hosts appear responsible for some human infections in the USA, and in Australia there is evidence of a high prevalence of R. felis in dogs that may indicate a role of the canine host as reservoir. It appears that some bloodsucking mites (e.g. Ornithonyssus bacoti) and lice (e.g. Polyplax spinulosa) can be involved in maintaining infections in rodent populations. Once R. typhi has become established in a human population, it could be transmitted by P. irritans, acquired by Pediculus humanus and become epidemic. This, however, does not appear to occur in nature, likely because R. typhi localizes intracellularly in P. humanus, causing the death of the louse, perhaps even more rapidly than does R. prowazekii.

During outbreaks, rodent and flea control as reported above for plague can be advised. For treatment of infected humans, the usual drug of choice is doxycycline, although tetracycline and chloramphenicol can also be effective.

7. PREVENTION AND CONTROL

For optimal control, adult fleas already infesting the host should be killed and reinfestation from the environment should be prevented. This can be achieved using a number of different strategies.

Direct treatment of infested animals usually involves the use of insecticides, mainly in the form of powders, sprays, shampoos, collars or spot-on preparations. These are generally organophosphorus compounds, pyrethrum and its pyrethroid derivatives or carbamates. Fipronil, given by either spray or spot-on, is highly effective against fleas of dogs and cats and achieves a protection lasting for 2–3 months. Similarly, the macrocyclic lactone, selamectin, provides effective flea control. Imidacloprid is also an effective, relatively new systemic neurotoxic insecticide, chemically related to the tobacco toxin, nicotine. There are also oral and in-feed formulations of insect growth regulators (IGRs) for use against fleas in dogs such as the benzoylurea derivative, lufenuron, which is ingested by fleas during feeding and is transferred to the eggs. Within the egg, it blocks the formation of chitin, thereby inhibiting the development of flea larvae. This not only gives ovicidal and/or larvicidal activity but also delivers it effectively to the sleeping areas most likely to be infested. Since in-contact animals may also harbour fleas without developing allergy, these should also be treated.

A second aspect of effective flea control is treatment of the environment where eggs, larvae and pupae are present. This may be achieved with those neurotoxic insecticides which have appropriate levels of residual activity; pressurized sprays containing organophosphorus or pyrethroid insecticide may give good control for up to 2 months. IGR formulations of methoprene, pyriproxyfen and fenoxycarb may also be used. Pyriproxyfen and fenoxycarb are stable in UV light and thus can be applied both indoors and outside. However, it must be noted that IGRs do not kill adult fleas and are not usually considered to be suitable by themselves for effective control of established flea infestations. Methoprene kills larvae in the third stage, while another IGR, diflubenzuron, which inhibits chitin synthesis, kills in all stages but must have been acquired by early second stage. The key issue in environmental control is the fact that, because larvae and pupae are usually hidden within the floor covering or bedding, they are often difficult to reach directly. The compound applied must therefore have sufficient residual activity to be picked up by the adult fleas as they emerge at the end of pupation. Alternatively, treatment must be repeated at appropriate intervals. Frequent vacuuming can also help to reduce environmental infestation and pet bedding should be washed at high temperatures.

Removal of rodent reservoirs is another method of control, particularly for the rat fleas that may transmit disease to humans. This can be achieved through dusting rodent burrows with insecticides such as pyrethroids, or trapping and killing the rodents using baited stations, where rodents are exposed to either rodenticides or flea control agents. Monitoring and surveillance of flea populations and their vectors can indeed help in preventing human infections.

SELECTED BIBLIOGRAPHY

Abramowicz, K.F., Rood, M.P., Krueger, L. and Eremeeva, M.E. (2011) Urban focus of *Rickettsia typhi* and *Rickettsia felis* in Los Angeles, California. *Vector-Borne and Zoonotic Diseases* 11, 979–984.

Azad, A.F. (1990) Epidemiology of murine typhus. *Annual Review of Entomology* 35, 553–569.

Azad, A.F. and Beard, C.B. (1998) Interactions of rickettsial pathogens with arthropod vectors. *Emerging Infectious Diseases* 4, 179–186.

Azad, A.F., Radulovic, S., Higgins, J.A., Noden, B.H. and Troyer, M.J. (1997) Flea borne rickettsioses: some ecological considerations. *Emerging Infectious Diseases* 3, 319–328.

Bacot, A.W. and Martin, C.J. (1914) Observations on the mechanism of the transmission of plague by fleas. *Journal of Hygiene* 13, 423–439.

Baldo, B.A. (1993) Allergenicity of the cat flea. *Clinical and Experimental Allergy* 23, 347–349.

Barnes, A.M. (1982) Surveillance and control of bubonic plague in the United States. *Symposia of the Zoological Society of London* 50, 237–270.

Bennet-Clark, H.C. and Lucey, E.C.A. (1967) The jump of the flea: a study of the energetics and a model of the mechanism. *Journal of Experimental Biology* 47, 59–76.

Bibikova, V.A. (1977) Contemporary views on the interrelationships between fleas and the pathogens of human and animal diseases. *Annual Review of Entomology* 22, 23–32.

Boisier, P., Rahalison, L., Rasolomaharo, M., Ratsitorahina, M., Mahafaly, M., Razafimahefa, M., *et al.* (2002) Epidemiologic features of four successive annual outbreaks of bubonic plague in Mahajanga, Madagascar. *Emerging Infectious Diseases* 8, 311–316.

Brianti, E., Gaglio, G., Napoli, E., Giannetto, S., Dantas-Torres, F., Bain, O., *et al.* (2012) New insights into the ecology and biology of *Acanthocheilonema reconditum* (Grassi, 1889) causing canine subcutaneous filariosis. *Parasitology* 139, 530–536.

Buckland, P.C. and Sadler, J.P. (1989) A biogeography of the human flea, *Pulex irritans* L. (Siphonaptera: Pulicidae). *Journal of Biogeography* 16, 115–120.

Butler, T. (1994) *Yersinia* infections: centennial of the discovery of the plague bacillus. *Clinical Infectious Diseases* 19, 655–663.

Craven, R.B., Maupin, G.O., Beard, M.L., Quan, T.J. and Barnes, A.M. (1993) Reported cases of human plague infections in the United States, 1970–1991. *Journal of Medical Entomology* 30, 758–761.

Debboun, M. and Strickman, D. (2012) Insect repellents and associated personal protection for a reduction in human disease. *Medical and Veterinary Entomology*, doi: 10.1111/j.1365-2915. 2012.01020.x

Drancourt, M. (2011) Finally, plague is plague. *Clinical Microbiology and Infection* 18, 105–106.

Dryden, M.W. and Rust, M.K. (1994) The cat flea – biology, ecology and control. *Veterinary Parasitology* 52, 1–19.

Dryden, M.W., Payne, P.A., Smith, V., Riggs, B., Davenport, J. and Kobuszewski, D. (2011) Efficacy of dinotefuran-pyriproxyfen, dinotefuran-pyriproxyfen permethrinand fipronil-(S)-methoprene topical spot-on formulations to control flea populations in naturally infested pets and private residences in Tampa, FL. *Veterinary Parasitology* 182, 281–286.

Eisele, M.J., Heukelbach, J., Van Marck, E., Melhorn, H., Meckes, O., Franck, S., *et al.* (2003) Investigations on the biology, epidemiology, pathology and control of *Tunga penetrans* in Brazil. I. Natural history of tungiasis in man. *Parasitology Research* 90, 87–99.

Eisen, R.J., Enscore, R.E., Biggerstaff, B.J., Reynolds, P.J., Ettestad, P., Brown, T., *et al.* (2007a) Human plague in the southwestern United States, 1957–2004: spatial models of elevated risk of human exposure to *Yersinia pestis*. *Journal of Medical Entomology* 44, 530–537.

Eisen, R.J., Wilder, A.P., Bearden, S.W., Montenieri, J.A. and Gage, K.L. (2007b) Early-phase transmission of *Yersinia pestis* by unblocked *Xenopsylla cheopis* (Siphonaptera: Pulicidae) is as efficient as transmission by blocked fleas. *Journal of Medical Entomology* 44, 678–682.

Foongladda, S., Inthawong, D., Kositanont, U. and Gaywee, J. (2011) *Rickettsia, Ehrlichia, Anaplasma,* and *Bartonella* in ticks and fleas from dogs and cats in Bangkok. *Vector-Borne and Zoonotic Diseases* 11, 1335–1341.

Gage, K.L. and Kosoy, M.Y. (2005) Natural history of plague: perspectives from more than a century of research. *Annual Review of Entomology* 50, 505–528.

Goddard, J. (1998) Fleas and murine typhus. *Infections in Medicine* 15, 438–440.

Goddard, J. (1999) Fleas and plague. *Infections in Medicine* 16, 21–23.

Hii, S.F., Kopp, S.R., Abdad, M.Y., Thompson, M.F., O'Leary, C.A., Rees, R.L.. *et al.* (2011) Molecular evidence supports the role of dogs as potential reservoirs for *Rickettsia felis*. *Vector-Borne and Zoonotic Diseases* 11, 1007–1012.

Hinkle, N.C., Rust, M.K. and Reierson, D.A. (1997) Biorational approaches to flea (Siphonaptera: Pulicidae) suppression – present and future. *Journal of Agricultural Entomology* 14, 309–321.

Holland, G.P. (1964) Evolution, classification, and host relationships of Siphonaptera. *Annual Review of Entomology* 9, 123–146.

Humphries, D.A. (1967) The mating behaviour of the hen flea *Ceratophyllus gallinae* (Schrank) (Siphonaptera: Insecta). *Animal Behaviour* 15, 82–90.

Humphries, D.A. (1968) The host-finding behaviour of the hen flea *Ceratophyllus gallinae* (Schrank) (Siphonaptera). *Parasitology* 58, 403–414.

Hutchinson, M.J., Jacobs, D.E., Fox, M.T., Jeannin, P.H. and Postal, J.M. (1998) Evaluation of flea control strategies using fipronil on cats in a controlled simulated home environment. *Veterinary Record* 142, 356–357.

Krämer, F. and Mencke, N. (2001) *Flea Biology and Control*. Springer, Berlin, Heidelberg, New York.

Keeling, M.J. and Gilligan, C.A. (2000) Bubonic plague: a metapopulation model of a zoonosis. *Proceedings of the Royal Society of London, Series B* 267, 2219–2230.

Lewis, R.E. (1993) Fleas (Siphonaptera). In: Lane, R.P. and Crosskey, R.W. (eds) *Medical Insects and Arachnids*. Chapman and Hall, London, pp. 529–575.

Lewis, R.E. (1998) Resumé of the Siphonaptera of the world. *Journal of Medical Entomology* 35, 377–389.

Mosbacher, M.E., Klotz, S., Klotz, J. and Pinnas, J.L. (2011) *Bartonella henselae* and the potential for arthropod vector-borne transmission. *Vector-Borne and Zoonotic Diseases* 11, 471–477.

Osbrink, W.L.A. and Rust, M.K. (1984) Fecundity and longevity of the adult cat flea, *Ctenocephalides felis felis* (Siphonaptera: Pulicidae). *Journal of Medical Entomology* 21, 727–731.

Osbrink, W.L.A. and Rust, M.K. (1985a) Seasonal abundance of adult cat fleas, *Ctenocephalides felis* (Siphonaptera: Pulicidae), on domestic cats in southern California. *Bulletin of the Society for Vector Ecology* 10, 30–35.

Osbrink, W.L.A. and Rust, M.K. (1985b) Cat flea (Siphonaptera: Pulicidae): factors influencing host-finding behaviour in the laboratory. *Annals of the Entomological Society of America* 78, 29–34.

Osbrink, W.L.A., Rust, M.K. and Reierson, D.A. (1986) Distribution and control of cat fleas in homes in southern California (Siphonaptera: Pulicidae). *Journal of Economic Entomology* 79, 135–140.

Pampiglione, S., Fioravanti, M.L., Gustinelli, A., Onore, G., Mantovani, B., Luchetti, A., *et al.* (2009) Sand flea (*Tunga* spp.) infections in humans and domestic animals: state of the art. *Medical and Veterinary Entomology* 23, 172–186.

Parola, P. (2011) *Rickettsia felis*: from a rare disease in the USA to a common cause of fever in sub Saharan Africa. *Clinical Microbiology and Infection* 17, 996–1000.

Perry, R.D. and Fetherston, J.D. (1997) *Yersinia pestis*: etiologic agent of plague. *Clinical Microbiological Reviews* 10, 35–66.

Pilgrim, R.L.C. (1991) External morphology of flea larvae (Siphonaptera) and its significance in taxonomy. *Florida Entomologist* 74, 386–395.

Pollitzer, R. (1954) *Plague*. WHO, Geneva, Switzerland.

Pollitzer, R. (1960) Review of recent literature on plague. *Bulletin of the World Health Organization* 23, 313–400.

Poulin, R., Krasnov, B.R., Mouillot, D. and Thieltges, D.W. (2011) The comparative ecology and biogeography of parasites. *Philosophical Transactions of the Royal Society, B-Biological Sciences* 366, 2379–2390.

Ratsitorahinha, M., Chanteau, S., Rahalison, L., Ratsifasomanana, L. and Boisier, P. (2000) Epidemiological and diagnostic aspects of the outbreak of pneumonic plague in Madagascar. *The Lancet* 355, 111–113.

Reif, K.E. and Macaluso, K.R. (2009) Ecology of *Rickettsia felis*: a review. *Journal of Medical Entomology* 46, 723–736.

Rothschild, M. (1975) Recent advances in our knowledge of the order Siphonaptera. *Annual Review of Entomology* 20, 241–259.

Rothschild, M., Schlein, Y., Parker, K. and Sternberg, S. (1972) Jump of the oriental rat flea *Xenopsylla cheopis* (Roths.). *Nature* 239, 45–48.

Rust, M.K. and Dryden, M.W. (1997) The biology, ecology and management of the cat flea. *Annual Review of Entomology* 42, 451–473.

Rust, M.K. and Reierson, D.A. (1989) Activity of insecticides against the preemerged adult cat flea in the cocoon (Siphonaptera: Pulicidae). *Journal of Medical Entomology* 26, 301–305.

Samia, N.I., Kausrud, K.L., Heesterbeek, H., Ageyev, V., Begon, M., Chan, K.S., *et al.* (2011) Dynamics of the plague–wildlife–human system in Central Asia are controlled by two epidemiological thresholds. *Proceedings of the National Academy of Sciences of the United States of America* 108, 14527–14532.

Simon, N.G., Cremer, P.D. and Graves, S.R. (2011) Murine typhus returns to New South Wales: a case of isolated meningoencephalitis with raised intracranial pressure. *Medical Journal of Australia* 194, 652–654.

Silverman, J., Rust, M.K. and Reierson, D.A. (1981) Influence of temperature and humidity on survival and development of the cat flea, *Ctenocephalides felis* (Siphonaptera: Pulicidae). *Journal of Medical Entomology* 18, 78–83.

Slapeta, J., King, J., McDonell, D., Malik, R., Homer, D., Hannan, P., *et al.* (2011) The cat flea (*Ctenocephalides f. felis*) is the dominant flea on domestic dogs and cats in Australian veterinary practices. *Veterinary Parasitology* 180, 383–388.

Sleeman, D.P., Smiddy, P. and Moore, P. (1996) The fleas of Irish terrestrial mammals: a review. *Irish Naturalists Journal* 25, 237–248.

Traub, R., Wisseman, C.L. Jr and Azad, A.F. (1978) The ecology of murine typhus – a critical review. *Tropical Diseases Bulletin* 75, 237–317.

Twigg, G.L. (1978) The role of rodents in plague dissemination: a worldwide review. *Mammal Review* 8, 77–110.

Velimirovic, B. (1972) Plague in South-East Asia. *Transactions of the Royal Society of Tropical Medicine and Hygiene* 66, 479–504.

Vogler, A.J., Chan, F., Wagner, D.M., Roumagnac, P., Lee, J., Nera, R., *et al.* (2011) Phylogeography and molecular epidemiology of *Yersinia pestis* in Madagascar. *PLOS Neglected Tropical Diseases* 5, e1319.

Wade, S. and Georgi, J.R. (1988) Survival and reproduction of artificially fed cat fleas *Ctenocephalides felis* Bouche (Siphonaptera: Pulicidae). *Journal of Medical Entomology* 25, 186–190.

Wheeler, C.M. and Douglas, J.R. (1945) Sylvatic plague studies. V. The determination of vector efficiency. *Journal of Infectious Diseases* 77, 1–12.

Whiting, M.F., Whiting, A.S., Hastriter, M.W. and Dittmar, K. (2008) A molecular phylogeny of fleas (Insecta: Siphonaptera): origins and host associations. *Cladistics* 24, 1–31.

Williams, M., Izzard, L., Graves, S.R., Stenos, J. and Kelly, J.J. (2011) First probable Australian cases of human infection with *Rickettsia felis* (cat-flea typhus). *Medical Journal of Australia* 194, 41–43.

Wisseman, C.L. (1991) *Rickettsial infections*. In: Strickland, G.T. (ed.) *Hunter's Tropical Medicine*. W.B. Saunders, Philadelphia, pp. 256–286.

Witt, L.H., Linardi, P.M., Meckes, O., Schwalfenberg, S., Ribeiro, R.A., Feldmeier, H., *et al.* (2004) Blood-feeding of *Tunga penetrans* males. *Medical and Veterinary Entomology* 18, 439–441.

World Health Organization (1999) *Plague Manual. Epidemiology, Distribution, Surveillance and Control.* WHO/CDS/CSR/EDC/99.2. WHO, Geneva, Switzerland, 172 pp.

Yeruham, I., Rosen, S. and Hadani, A. (1989) Mortality in calves, lambs and kids caused by severe infestations with the cat flea, *Ctenocephalides felis felis* (Bouche, 1835) in Israel. *Veterinary Parasitology* 30, 351–356.

Zimba, M., Pfukenyi, D., Loveridge, J. and Mukaratirwa, S. (2011) Seasonal abundance of plague vector *Xenopsylla brasiliensis* from rodents captured in three habitat types of periurban suburbs of Harare, Zimbabwe. *Vector-Borne and Zoonotic Diseases* 11, 1187–1192.

Flesh Flies (Diptera: Sarcophagidae)

1. INTRODUCTION

The flesh flies (Sarcophagidae) are distributed throughout the world, with most species of the family breeding as larvae on carrion, dung or decaying material, some species on other insects and a few species are agents of myiasis, since they infest living tissues of vertebrate animals.

The only genus containing species which act as important agents of myiasis is *Wohlfahrtia*, and the species of flesh fly of greatest medical and veterinary importance is *Wohlfahrtia magnifica* (Fig. 34). This is an obligate agent of traumatic myiasis, which can only complete its larval development by feeding on a vertebrate host. It causes traumatic cutaneous myiasis of warm-blooded vertebrates throughout the Mediterranean basin, eastern and central Europe, the Middle East and Asia. *Wohlfahrtia*

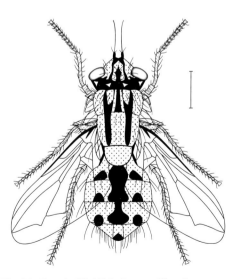

Fig. 34. Female *Wohlfahrtia magnifica*. *Source*: James (1947).

vigil is an obligate agent of myiasis in North America, and the Holarctic species *Wohlfahrtia opaca* (formerly *Wohlfahrtia meigeni*) behaves in North America in a similar manner to *W. vigil*, causing a furuncular, boil-like myiasis in smaller animals. Another important species in the same genus is *Wohlfahrtia nubia*, which is a facultative species that breeds in carrion or living hosts in North Africa and the Middle East, where it can be locally important, particularly in camels.

2. TAXONOMY

All the flies that act as important agents of myiasis are calypterate Diptera in the superfamily Oestroidea. Within this superfamily, there are three major families of myiasis-producing flies: Calliphoridae (see entry for 'Blow Flies'), Oestridae (see entries for 'Nasal Bot Flies' and 'Stomach Bot Flies') and Sarcophagidae. The family Sarcophagidae contains over 2000 species in 400 genera.

3. MORPHOLOGY

The Sarcophagidae are generally dull grey in appearance, with three black stripes on the scutum. The abdomen usually has a chequered or spotted pattern. In the genus *Wohlfahrtia*, the arista of the antenna is bare, the notopleuron and katepisternum bear two setae each and the abdomen is grey, with a pattern of black spots which is unaffected by the angle at which they are viewed. In *W. magnifica*, the adults are large, 8–14 mm in length and bristly with long, black legs. The body is elongated and grey coloured, with three dark stripes on the thorax and a number of distinct, separate, rounded, dark patches on the abdomen.

The mature sarcophagid larva has its posterior spiracles recessed in a posterior depression and hidden from view. The spiracles have three slits, which are orientated more or less dorsoventrally and surrounded by an incomplete peritreme.

4. LIFE CYCLE

W. magnifica adults occur in fields and orchards, where they feed on the nectar of flowers. They are active during the bright, hot period of the day from 10:00 h to 16:00 h. Females are attracted by exudates of natural orifices (e.g. anus, vulva and prepuce) and other body openings, such as the ear, nose and eyes or wounds of the hosts, where they deposit first-stage larvae. A female may contain 120–170 larvae. Once deposited, the larvae grow rapidly, burrow into the tissues and cause extensive damage, which may prove fatal. The larvae are fully grown in about 6–7 days, when they leave the host and pupate in the ground. They require soil temperatures of above 12–14°C to emerge as adults. In Kazakhstan, *W. magnifica* overwinters as diapausing pupae, a stage which lasts 6–10 months. The larvae are extremely resilient, pupating and producing normal adults even after exposure to 95% alcohol for 1 h (and they can even survive for a considerable time in concentrated hydrochloric acid).

5. BEHAVIOUR AND BIONOMICS

The Sarcophagidae are viviparous or ovoviviparous, depositing either live first-stage larvae or eggs, which hatch immediately on deposition. This gives members of this family considerable advantage in competing for carrion with the Calliphoridae, which are oviparous.

6. MEDICAL AND VETERINARY IMPORTANCE

In general, the larvae are deposited in groups and can only penetrate thin skin (or broken skin and wounds); for that reason, human and animal cases are usually restricted to the young, in whom they cause a furuncular myiasis.

W. magnifica can cause rapid and severe myiasis (wohlfahrtiosis) in most livestock, particularly sheep and camels, and also geese, but cattle, horses, pigs and dogs may also be infested. Levels of infestation appear to be high, particularly in sheep in Eastern Europe. Infestation will lead rapidly to death unless treated. In humans, infection of the ears by *W. magnifica* can lead to deafness, loss of an eye and cause severe damage to the tissue of the nasal region. Faecal soiling in sheep has been recorded as an important predisposing factor for breech myiasis by *W. magnifica*. In a 4-year period, cases of myiasis by *W. magnifica* were recorded in 45 out of 195 sheep flocks in Bulgaria, affecting between 23 and 41% of sheep each year. Only 0.5–1.0% of cows and goats were affected over the same period. In one study in Romania, 80–95% of sheep were infested, with 20% fatalities of newborn lambs.

This myiasis causes severe economic losses to animal productions, as the rapid growth of large numbers of larvae in multiple infested sites results in failure of animal condition, reproduction problems, lameness, blindness and even death of infested sheep.

In North America, *W. opaca* (formerly *W. meigeni*) and *W. vigil* are also obligate agents of myiasis. Although Palaearctic in origin, *W. opaca* has not been recorded as a myiasis agent in that part of the world; however, in North America, along with *W. vigil*, it can cause substantial mortality to young mink and foxes in fur farms, and rabbits, dogs and cats are also attacked occasionally. The adult females deposit active maggots on the host and the larvae can penetrate intact skin if it is thin and tender, and hence young animals tend to be most affected. The myiasis caused by these two species is furuncular rather than cutaneous and the swellings produced may contain up to five larvae. *W. nuba* is not strictly parasitic but it infests wounds of livestock in North Africa and the Middle East, feeding only on dead or diseased tissues; this species has also been used in maggot therapy.

7. PREVENTION AND CONTROL

Prevention and control of flesh fly infestation in areas of livestock and wildlife involves the combination of a range of different strategies. The odour of urine and faeces attract flesh flies and thus good sanitation and hygiene will reduce sarcophagid numbers. Activities that may cause injury, broken skin and open wounds, such as dehorning, castrating and branding of animals, should be limited to the non-fly seasons and harmful materials (such as barbed wire) removed. This will prevent injury, reduce the amount of broken skin and open wounds exposed and thus reduce the sites available in which females can larviposit. Similarly, frequent inspection of animals and the early treatment and covering of wounds will reduce the number of flies attracted to an individual and reduce the incidence of myiasis. Insecticides are commonly used. Larvicides can be applied directly to the animal to prevent fly strike or to kill larvae that already exist. Dipping is another method of insecticide application, where the animal is required to move through a bath of insecticide. Other insecticides can be applied via pour-on, injection or as boluses (which are placed in the stomach of the animal and act systemically).

To reduce the incidence of human infection by flesh flies, unnecessary exposure to the outdoors during the fly season should be avoided. Preventing infestation by *W. vigil* and *W. meigeni* can be achieved by fly screening prams containing sleeping infants left outdoors.

SELECTED BIBLIOGRAPHY

Farkas, R., Hall, M.J.R. and Kelemen, F. (1997) Wound myiasis of sheep in Hungary. *Veterinary Parasitology* 69, 133–144.

Giangaspero, A., Traversa, D., Trentini, R., Scala, A. and Otranto, D. (2011) Traumatic myiasis by *Wohlfahrtia magnifica* in Italy. *Veterinary Parasitology* 175, 109–112.

Hall, M.J.R. (1997) Traumatic myiasis of sheep in Europe: a review. *Parassitologia* 39, 409–413.

Hall, M.J.R., Adams, Z.J.O., Wyatt, N.P., Testa, J.M., Edge, W., Nikolausz, M., *et al.* (2009) Morphological and mitochondrial DNA characters for identification and phylogenetic analysis of the myiasis-causing flesh fly *Wohlfahrtia magnifica* and its relatives, with a description of *Wohlfahrtia monegrosensis* sp n. *Medical and Veterinary Entomology* 23, 59–71.

Lehrer, Z., Lehrer, M. and Verstraeten, C. (1988) Myiasis in sheep provoked by *Wohlfahrtia magnifica* (Schiner). *Annales de Medecine Veterinaire* 132, 475–481.

Sotiraki, S., Farkas, R. and Hall, M.J.R. (2009a) Fleshflies in the flesh: epidemiology, population genetics and control of outbreaks of traumatic myiasis in the Mediterranean Basin. *Veterinary Parasitology* 174, 12–18.

Sotiraki, S., Farkas, R. and Hall, M.J.R. (2009b) Traumatic myiasis in dogs caused by *Wohlfahrtia magnifica* and its importance in the epidemiology of wohlfahrtiosis of livestock. *Medical and Veterinary Entomology* 23, 80–85.

Zumpt, F. (1965) *Myiasis in Man and Animals in the Old World*. Butterworths, London.

Horn Flies (Diptera: Muscidae, *Haematobia*)

1. INTRODUCTION

Adult horn flies (*Haematobia*) are parasitic blood-feeding true flies (Diptera). They are generally not important in a medical context; however, they can have a huge economic impact on the health and productivity of cattle and are one of the most widespread and economically relevant pests of cattle.

There are two common species of *Haematobia* in temperate habitats: *Haematobia irritans* (sometimes referred to as *Lyperosia irritans*), known as the horn fly, which is found in Europe, North and South America and Australia, and *Haematobia stimulans* (also referred to as *Haematobosca stimulans*), which is found only in Europe. A third species, *Haematobia titillans* (also referred to as *Lyperosia titillans*), is found in southern Europe and in Asia, often in association with *H. irritans*. Of these, *H. irritans* is the most economically important. It primarily attacks cattle, but horses in adjacent pastures may be also attacked. A fourth species, *Haematobia exigua* (also referred to as the subspecies *Haematobia irritans exigua*), known as the buffalo fly, is of importance throughout Asia to Australasia.

2. TAXONOMY

Within the large dipteran family Muscidae, the subfamily Stomoxyinae contains ten genera, including the blood-feeding *Haematobia* (the horn flies) and *Stomoxys* (see the entry for 'Stable Flies' for details). Species of *Haematobia* may be readily recognized by the possession of an elongate, sclerotized proboscis, with palps that are as long as the proboscis and which are grooved internally. The palps extend in front of the head and are distinctly dilated towards their tips. Except for *H. stimulans*, whose arista carries hairs on the dorsal and ventral side, in both *Stomoxys* and other species of *Haematobia*, the arista carries hairs only on the dorsal side.

3. MORPHOLOGY

Adult *Haematobia* are small, brownish flies (Fig. 35). Adult *H. irritans* are about 3–4 mm in length, which is about half the size of a typical house fly (*Musca domestica*) and smaller than *H. stimulans* (6 mm in length). The palps of the adults are long relative to the proboscis and are club-shaped apically. The eggs are bright yellowish-brown in colour and lack a terminal horn. From the eggs, small slender maggots emerge. These larvae have an anterior spiracle with approximately five lobes.

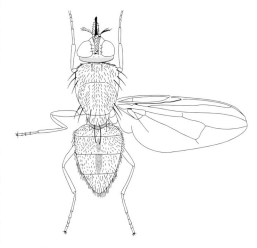

Fig. 35. *Haematobia exigua*. Drawn from specimens and Greenberg (1971).

Their posterior spiracle has three very sinuous slits. Puparia are dull reddish-brown and 3–4 mm in length.

4. LIFE CYCLE

Adult *H. irritans* generally remain on the host animal day and night, usually congregating on the back, withers, belly and around the head, resting around the horn region when not feeding. They only leave the animal briefly to mate and lay eggs in fresh dung. Each female is capable of producing 300–400 eggs in batches of 20–30 eggs. These hatch within 18–24 h of being deposited. There are three larval stadia, lasting about 1–2 weeks, and pupation occurs in the dung pat or in the nearby soil. The egg-to-adult life cycle may be completed in as little as 10–14 days and 3–4 generations may occur in one summer. Horn flies overwinter as diapausing pupae in the soil below cowpats, emerging as adults the following spring.

5. BEHAVIOUR AND BIONOMICS

Both sexes are obligatory blood feeders, largely biting cattle and buffalo, but occasionally other animals, including humans, closely associated with bovines. They take frequent small blood meals, with each fly consuming on average 10 μl of blood/day. They do not walk over the surface of the bovid but fly to change position. Females usually only mate once and this occurs on the host. The female oviposits only into fresh dung, leaving the host when dung is dropped to deposit small batches of eggs in the dung pat, before returning to the host after a few minutes.

H. exigua feeds primarily on buffalo and cattle and, like *H. irritans*, rarely leaves the host unless disturbed. Emergence of adult *H. exigua* occurs in the afternoon and early evening (12:00–20:00 h). Unfed flies live less than 24 h. Its distribution is determined by an annual rainfall of 500 mm and a temperature of 22°C; in addition, the immature stages require dung with a moisture content of greater than 68%.

H. stimulans is a less strictly obligatory blood-feeding species; it blood feeds on cattle but it can often be found feeding on flowers.

6. MEDICAL AND VETERINARY IMPORTANCE

In some parts of its range, very large populations of *H. irritans* may occur. It has been estimated that a cow being fed on by 3000 flies will lose on average approximately 30 ml of blood/day. The overall blood volume of an adult cow is 25 l and thus overall this daily loss of blood is not substantial. However, the bites cause pain, irritation to the host, cutaneous lesions and lead to an increase in heart rate and respiration and a general increasing in metabolic activity, with marked negative effects on production. Heavy attack by horn fly in the USA has been shown to be able to reduce milk production by up to 25–50%. A meta-analysis of published studies has suggested that, for *H. exigua*, the threshold number of flies below which no adverse effects would be observed is 30. At a moderate level of infestation of about 200 flies/cow, daily milk yield loss would be about 520 ml and the daily loss of live weight gain would be about 28 g. The horn fly is also believed to cause damage to cattle hides, while horn fly wounds may attract other flies.

The sedentary nature of these species would operate against their being important mechanical vectors of pathogens. However, the horn fly is an intermediate host or vector for the spirurid nematode *Stephanofilaria stilesi*, which causes stephanofilariasis, a type of dermatitis in cattle. This is a granular form of skin inflammation that occurs mainly on the udder, belly, prepuce and scrotum of cattle, located mainly in the western USA and Canada. *Haematobia* spp. may also cause a periorbital and ventral ulcerative dermatitis in horses, the lesions of which may become infected by *Habronema* nematodes. In southern-eastern European countries, *H. irritans* is vector of *Habronema microstoma*, a nematode responsible for equid habronemosis.

7. PREVENTION AND CONTROL

The benefits of prevention and control of horn flies have been demonstrated by studies in the USA and Canada that have shown the prevention of horn fly attack can lead to an increase in the average daily growth rate and milk production. The advantages of horn fly control will vary depending on the density of the attendant fly population and the degree of control implemented, and the existence of other parasites (both internally and externally). The prevailing weather will also influence the effectiveness of control methods.

Before control and management programmes are implemented, their economic viability must be assessed. Thresholds should be used to judge whether or not control is economically justified. Thresholds are usually a measurement of the average number of horn flies per animal side and, when densities exceed these thresholds, the increase in growth rates of the cattle in response to fly control strategies will compensate economically for the cost of implementing the control strategies.

The most effective approach to controlling horn fly populations is to use a combination of different control methods in an integrated pest management programme. For example, surveillance of adult abundance can be used to determine the amount of sanitation and adulticides required to prevent populations from exceeding the tolerable thresholds. In some circumstances, larvicides can be sprayed directly on to breeding populations or added to the feed of cattle. Insecticide residues are then passed through the animal's digestive system into the faeces or soil bedding, killing any breeding populations within. This latter approach needs considerable care, however, because the residues will also kill the beneficial insects in dung that are responsible for its breakdown and recycling.

To exterminate adult populations, adulticides can be used. These are often formulated into topical insecticides, which can be applied directly to animals. Topical insecticides are particularly effective for adult horn flies because of their habit of remaining on their host continuously. Active compounds can be applied in the form of sprays, pour-ons or wipe-ons. Self-applicators such as plastic ear tags, oilers, back rubbers and dust bags can be used to dispense insecticides regularly on to animals.

To provide temporary relief from horn flies, chemical repellents can be applied by hand directly to individuals. These disrupt the host-finding mechanisms of the horn flies, but most repellents have limited effectiveness.

Although insecticides have been widely used for fly control, success has been limited because of the development of insecticide resistance in all countries where the horn fly is present. This problem has driven increasing investigation into the use of alternative control methods (resistant cattle selection, semio-chemicals and biological control) and the development of vaccines.

SELECTED BIBLIOGRAPHY

Birkett, M.A., Agelopoulos, N., Jensen, K.M.V., Jespersen, J.B., Pickett, J.A., Prijs, H.J., *et al.* (2004) The role of volatile semiochemicals in mediating host location and selection by nuisance and disease-transmitting cattle flies. *Medical and Veterinary Entomology* 18, 313–322.

Guglielmone, A.A., Curto, E., Anziani, O.S. and Mangold, A.J. (2000) Cattle breed-variation in infestation by the horn fly *Haematobia irritans*. *Medical and Veterinary Entomology* 14, 272–276.

Jensen, K.M.V., Jespersen, J.B., Birkett, M.A., Pickett, J.A., Thomas, G., Wadhams, L.J., *et al.* (2004) Variation in the load of the horn fly, *Haematobia irritans*, in cattle herds is determined by the presence or absence of individual heifers. *Medical and Veterinary Entomology* 18, 275–280.

Jones, C.J., Patterson, R.S., Koehler, P.G., Bloomcamp, C.L., Hagenbuch, B.E. and Milne, D.E. (1988) Horn fly (Diptera: Muscidae) control on horses using pyrethroid tags and tail tapes. *The Florida Entomologist* 71, 205–207.

Jonsson, N.N. and Mayer, D.G. (1999) Estimation of the effects of buffalo fly (*Haematobia irritans exigua*) on the milk production of dairy cattle based on a meta-analysis of literature data. *Medical and Veterinary Entomology* 13, 372–376.

Kuramochi, K. (2000) Survival, ovarian development and bloodmeal size for the horn fly *Haematobia irritans irritans* reared *in vitro*. *Medical and Veterinary Entomology* 14, 201–206.

Lockwood, J.A., Byford, R.L., Story, R.N., Sparks, T.C. and Quisenberry, S.S. (1985) Behavioural resistance to the pyrethroids in the horn fly, *Haematobia irritans* (Diptera: Muscidae). *Environmental Entomology* 14, 873–880.

Lysyk, T.J., Kalischuk-Tymensen, L.D., Rochon, K. and Selinger, L.B. (2010) Activity of *Bacillus thuringiensis* isolates against immature horn fly and stable fly (Diptera: Muscidae). *Journal of Economic Entomology* 103, 1019–1029.

Maldonado-Siman, E., Martinez-Hernandez, P.A., Sumano-Lopez, H., Cruz-Vazquez, C. de, Lara, R.R. and Alonso-Diaz, M.A. (2009) Population fluctuation of horn fly (*Haematobia irritans*) in an organic dairy farm. *Journal of Animal and Veterinary Advances* 8, 1292–1297.

Oyarzun, M.P., Quiroz, A. and Birkett, M.A. (2008) Insecticide resistance in the horn fly: alternative control strategies. *Medical and Veterinary Entomology* 22, 188–202.

Oyarzun, M.P., Palma, R., Alberti, E., Hormazabal, E., Pardo, F., Birkett, M.A., *et al.* (2009) Olfactory response of *Haematobia irritans* (Diptera: Muscidae) to cattle-derived volatile compounds. *Journal of Medical Entomology* 46, 1320–1326.

Oyarzun, M.P., Li, A.Y. and Figueroa, C.C. (2011) High levels of insecticide resistance in introduced horn fly (Diptera: Muscidae) populations and implications for management. *Journal of Economic Entomology* 104, 258–265.

Sheppard, D.C. (1994) Dispersal of wild-captured, marked horn flies (Diptera: Muscidae). *Environmental Entomology* 23, 29–34.

Horse Flies (Diptera: Tabanidae)

1. INTRODUCTION

Tabanids are variously known as horse flies, march flies, deer flies and clegs. They can be serious nuisance biters and vectors of various pathogens infecting humans and animals, including the nematode causing loiasis in humans and the trypanosome causing surra in livestock.

2. TAXONOMY

More than 4000 species of tabanids have been described and ranked within four subfamilies, identifiable based on the morphology of the male and female terminalia. One of these families, the Scepsidinae, includes only eight species of non-blood-feeding tabanids; seven of these are found in Africa and the remaining one in South America. The Pangoniinae are of little economic importance, whereas the Chrysopsinae (main genus *Chrysops*) and the Tabaninae (main genera *Tabanus* and *Haematopota*) are of considerable economic significance.

3. MORPHOLOGY

Tabanids are usually large, stout-bodied flies (Fig. 36) that range in wing length from 6 to 30 mm. The head is much broader than long and the eyes are particularly well developed. Like the Simuliidae, the sexes can be differentiated based on the distance between the compound eyes, the males being holoptic and the females dichoptic. The frons, separating the eyes in the female, is wider in *Haematopota* (cleg fly) than in *Tabanus* (horse fly or march fly) species. The eyes are often brilliantly coloured (although the hues disappear shortly after death), and are spotted in *Chrysops* (deer flies), with zigzag bands in *Haematopota* and unicolorous or horizontally banded in *Tabanus*.

The palps are two-jointed, with the second segment being particularly prominent. The antennae are porrect, i.e. projecting stiffly forwards, composed of three segments, the last of which is greatly enlarged and subdivided into four to eight subsegments. The mouthparts are adapted for both blood feeding and lapping. Both sexes feed on nectar; however, females of most species are also blood feeders.

The wing venation is well developed, with all wing veins being obvious (Fig. 37). The venation is characteristic, but not diagnostic, as similar venations occur in related brachyceran families. Both the radius and the media are four-branched, with R_4 and R_5 forks to form a large Y across the apex of the wing. The discal cell is located more or less in the centre of the wing. The anal cell may be open or closed, i.e. the first anal vein may join Cu, before the wing margin. The squamae are large and obvious. The stout legs end in three pads, because the empodium is pad-like and similar to the pulvilli.

Species within the Pangoniinae subfamily have a head with functional ocelli and the proboscis is longer than the head (Fig. 36a). The antennae bear seven or eight segments in the flagellum. A pair of large spurs is present apically on the hind tibiae (all tabanids have apical spurs on the mid-tibiae) and the wings are clear or dusky. These species are largely tropical or subtropical in distribution.

Chrysops spp. are characterized by a head with functional ocelli; the proboscis is not longer than the head (Fig. 36b); the antennal flagellum has five segments; there are apical

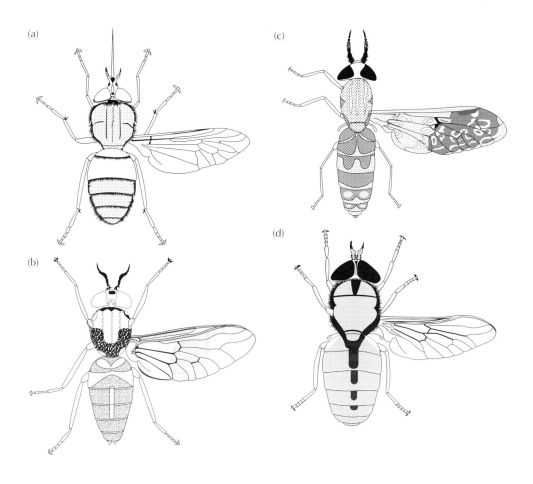

Fig. 36. Female Tabanidae. (a) *Pangonia*; (b) *Chrysops*; (c) *Haematopota*; and (d) *Tabanus parvicallosus*. Drawn from pinned specimens supplemented by colour plates in Chvala *et al.* (1972).

spurs on the hind tibiae that are small and may be hidden by hair; the wings usually have a dark costal and a single, broad, transverse, dark band (Fig. 37). They are distributed mainly through Holarctic and Oriental areas.

Tabanus spp. are distributed worldwide and usually have only vestigial ocelli on the head; the proboscis is shorter than the head; the antennal flagellum has five segments; there are no apical spurs on hind tibia; the wings are usually clear, but may be dark or banded.

Haematopota spp. bear four segments in the antennal flagellum and the wings are mottled (Fig. 36c). These species are distributed

in Palaearctic, Afrotropical and Oriental areas. This genus is absent from Australia and rare in the Americas.

4. LIFE CYCLE

4.1 Eggs

Eggs are laid in masses (of 200–1000 eggs each) on leaves or rocks or debris overhanging water. Eggs of *Chrysops* are often laid in a single layer, whereas those of *Tabanus* and *Haematopota* are usually stratified into three

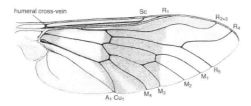

Fig. 37. Wing of *Chrysops australis*.

to four and two to three layers, respectively. The creamy white eggs darken with age and are dark grey to black at hatching, which occurs >4 days after oviposition, depending on temperature.

4.2 Larva

The larva emerges from the egg using a hatching spine and moults soon after emergence. The second-stage larva does not feed and is positively phototactic, moving over the surface of the substrate in close association with water. After 3–6 days, it moults into the third larval stage, which is negatively phototactic, and burrows into the substrate, where it spends several months. The mature larva is a greyish white, soft-bodied, cylindrical grub (Fig. 38) with a reduced head that is retractile into the thorax. It bears a pair of simple eyes and piercing mandibles. The larvae of *Tabanus* and *Haematopota* are carnivorous and cannibalistic, whereas those of *Chrysops* feed on plant remains. Consequently, *Chrysops* larvae are often found in considerable density compared with carnivorous larvae. The larval thorax and abdomen merge imperceptibly.

Ventral abdominal and smaller dorsal pads facilitate movement of the larva. The larva is metapneustic, with the spiracles opening at the end of a siphon, located dorsally on the eighth abdominal segment. The siphon is variable in shape, being short and conical in *Haematopota pluvialis* and moderately long and blunt in *Tabanus septentrionalis*.

4.3 Pupae

At pupation, the larva moves to the edge of aquatic habitats or to the surface of edaphic habitats (i.e. related to soil). The tabanid pupa is obtect, with limited movement. There is no compound cephalothorax, but head, thorax and abdomen are distinct. Respiration occurs through kidney-shaped thoracic spiracles and seven pairs of abdominal spiracles on short, lateral projections (Fig. 39). The abdominal segments are each fringed with a row of stout bristles and the terminal segment bears a spiny aster. Spines give the pupa purchase, enable it to move up and down in the substrate, away from adverse conditions at the surface, or up for emergence at the appropriate time. The head is often rugose and is used to effect escape from the soil.

Pupae of *Tabanus biguttatus* are found in temporary ponds and display a pattern of behaviour which is designed to avoid being exposed to predators and parasites when the pond dries out and the mud cracks. The larva comes near to the surface and then descends on a spiral course to a depth of 8–10 cm, isolating a central core of mud. It then moves upwards on the outside of the core and, near the surface, burrows into it, hollowing out the

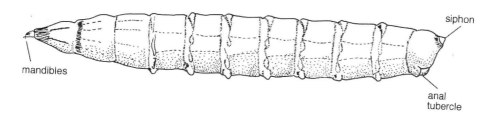

Fig. 38. Ninth larval stadium of *Haematopota pluvialis*. Redrawn from Cameron (1934).

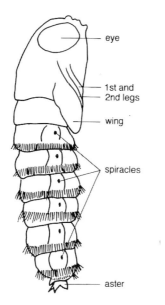

eye

1st and
2nd legs

wing

spiracles

aster

Fig. 39. Lateral view of pupa of a species of
Hybomitra. Redrawn from Chvala *et al.* (1972).

5. BEHAVIOUR AND BIONOMICS

5.1 Adult mating and feeding

The newly emerged female mates before
seeking a blood meal. Males form swarms,
usually early in the morning, and virgin females
are seized by the males, with copulation
beginning in the air and insemination being
completed on the ground within about 5 min.
In forests, male swarms occur above the
canopy. Recognition of the female by the male
is visual. Males of *C. fuliginosus* become
active earlier in the day than females. While
mating occurs at 19–20°C, females become
host seeking at 24–25°C, and the effect of this
is for mating to occur early in the morning and
blood feeding later in the day.

5.2 Feeding

The mated female seeks a blood meal, which is
required in most species for the development
of the ovaries. Some species (e.g. *Tabanus
nigrovittatus*) are autogenous for the first
oviposition, but require a blood meal from the
second oviposition onwards. Other genera
(e.g. *Pangonia* and *Scaptia*) include many
flower-feeding, non-bloodsucking species.

Tabanid adults feed largely on sugars,
although these nutrients are inadequate for
egg production in haematophagous species.
With constant access to water and either blood
or sugar solution, *Chrysops silacea* has been
found to survive considerably longer on sugar,
but lays no eggs.

Most species are diurnal, while others, for
example *Tabanus paradoxus*, are nocturnal.
Activity is higher on warm, sunny days with
low wind speeds. European species are not
active below 13–15°C and reach maximum
activity at 25°C, providing that the wind speed
is less than 4 m/s. Tabanids have evolved in
close association with the ungulates and feed
mainly on mammals and, rarely, on birds.
Visual cues are the most important to host-
seeking tabanids when the host is moving,
while a stationary host is located either visually
or olfactorily, since the two stimuli reinforce
each other. The presence of dry ice may

interior to form a pupal cell. The entrance to
the cell is blocked to deter predators. When
the mud dries, the soil cracks at the line of
weakness made by the spiral, but the mud cell
remains intact. At emergence, the pupa rasps
its way through the cap of the cell and the
adult emerges. Pupae of *Tabanus taeniola* are
able to reach the surface from a depth of 2 m
in dry sand but from only 10 cm when the
sand is wet, which is the more natural situation.

The number of larval moults is variable,
from 7–11 stadia, being variable even within a
species. The pupal stage lasts about 1–3
weeks. Although the larvae of *Chrysops
fuliginosus* are widespread through their salt
marsh habitat, adult emergences occur within
9 m of the ecotone where the influence of
fresh water is apparent. Speed of development
is temperature dependent. For example, *T.
taeniola* completes its life cycle from egg to
adult in 10–11 weeks at high temperature (i.e.
32 and 35°C) and in 42 weeks at 22°C. At 32
and 35°C there is no diapause, while at 22°C
it always occurs, with the final larval stage
being prolonged.

double the trap catch of tabanids such as *Tabanus molestus* and is considerably better than semiochemicals such as octenol. However, in combination, these two compounds may act synergistically, the catch being three times as large as that of dry ice on its own.

Species feed on different body areas of the host. *C. fuliginosus* attacks the head of humans, while *Chrysops atlanticus* the upper body regions when walking and the lower limbs. In Zimbabwe, most tabanid species were found to alight preferentially on the lower legs of cattle.

While feeding, the female will imbibe 20–200 mg of blood. However, the effective blood loss is frequently greater due to the large puncture caused by feeding tabanids, which continues to ooze blood after the mouthparts have been withdrawn. An average-sized female *Tabanus fuscicostatus* (wing length – 8.80 mm) produces 174 eggs from a bovine blood meal of 55 mg.

5.3 Oviposition

Gravid female tabanids are often highly selective in their oviposition on plant species. Thus, *Chrysops discalis* oviposits only on *Scirpus americanus* and not on other plants, while various species of *Chrysops* have been found to select *Pontederia cordata* and three other plants (out of the range available) for oviposition. In the field, *T. taeniola* lay an average of 259 eggs. The eggs of *C. discalis* hatch in 5–6 days and the young larvae disperse widely from the site of hatching, being carried by water movement as they float at the surface. Mature larvae of *C. discalis* have been found 10 m from the shore in shallow, alkaline lakes, under 60 cm of water and at a depth of 5–10 cm in the underlying mud.

5.4 Breeding sites

Three broad divisions of breeding site have been recognized, depending on the proportion of water that they contain. *Chrysops* larvae are generally hydrobionts, occurring in the wettest situations; *Tabanus* larvae are hemi-hydrobionts, occurring in soil near water; while *Haematopota* larvae are edaphic, being found in soil. *Chrysops* larvae have been found to be associated with large amounts of decaying organic matter, on which presumably they feed; *Hybomitra* in mosses, larvae of *Sylvius* in sand and silt above the margins of flowing water, while larvae of *Tabanus* are generally more widely distributed, with those of *Tabanus punctifer* occurring in every semi-aquatic habitat except tree holes. Within a habitat, larvae may be generally distributed, as are those of *C. fuliginosus* and *T. nigrovittatus* in *Spartina* salt marshes.

Interestingly, with the extension of pasture and rice production in Japan (Hokkaido) in recent years, there has been a change in the tabanid fauna from one dominated by forest/eurytopic species to an openland/eurytopic fauna. This corresponded to a decrease in *Haematopota tristis* and an increase in *Tabanus nipponicus*.

5.5 Adult activity and dispersal

The activity of tabanids is influenced greatly by meteorological conditions, especially light intensity and temperature. The flight activity of three species of salt marsh tabanids has been shown to respond differently to light intensity: *C. fuliginosus* was most active under very bright conditions (100,000 lux), *T. nigrovittatus* under less bright but warm conditions (40,000 lux; 25°C), while *C. atlanticus* was most active under overcast but hot conditions (5000 lux; 30°C).

Barometric pressure appears to be the most important meteorological factor influencing the flight of *Tabanus pallidescens* and *Tabanus fulvulus*. Both species were influenced by evaporation, but the catches of *T. fulvulus* were related more closely to changes in rate of evaporation and those of *T. pallidescens* to the actual rate of evaporation. Three species of *Tabanus* showed marked bimodal, diurnal activity in sunny weather, but activity became irregular under cloudy conditions unless the air temperature was above 25°C, when it remained bimodal. Some species, for example *C. fuliginosus* and *T. taeniola*, normally have only one period of activity during the day.

In the open, tabanids such as *T. nigrovittatus* remain close to the ground, but in the forest many species (e.g. *Chrysops langi*) occur in the canopy, while *C. silacea* feeds in the canopy and only comes to the forest floor in clearings or when attracted by wood smoke. Wood smoke increases the biting of *C. silacea* at ground level by more than tenfold and that of *Chrysops dimidiata* by nearly fivefold. When a canopy nest of *Bembix bequaertii dira* wasp was examined, it contained 26 species of tabanids, including the newly discovered males of 10 species, four new species and two previously considered 'rare' species.

Although tabanid flies have a potential flight range of over 50 km, they are not known for their wide dispersal. The salt marsh tabanids *T. nigrovittatus*, *C. fuliginosus* and *C. atlanticus* disperse less than 200 m from the marshes from which they emerge, *C. silacea* and *Tabanus iyoensis* disperse 1–3 km and *C. discalis* can be numerous up to 7 km from its breeding site. However, these flights represent only a small fraction of the potential flight range of tabanids.

5.6 Survival

In the field, adults live a maximum of 3–4 weeks and produce five to six batches of eggs. The survival rate of *T. iyoensis* has been calculated to be 0.73/day, which gives a survival of 11% after 1 week and 0.13% after 3 weeks. Making certain reasonable assumptions, the emergence of *T. iyoensis* in a linear habitat has been calculated to be 14,000/m of river.

Tabanids shelter from adverse conditions in the larval stage. Little is known of the ways in which tabanids survive drought, except for the pupal mud cells constructed by *T. biguttatus* and *Tabanus conspicuus*, but presumably larvae survive deep in soil, where the atmosphere is moist. The cold season can be passed in the larval stage, but a problem arises when the soil freezes. *Tabanus autumnalis* overwinters as a larva at a depth of 2–20 cm and 50–100 cm above the waterline of small

lakes, where it is warmed by the winter sun. Larvae of this species can survive −4°C and, as the temperature at 10 cm deep in the soil remains above that threshold, the larvae survive. After cold acclimation, 60% of *T. autumnalis* larvae can survive −6°C. *Chrysops caecutiens* larvae remain in the soil below water, where they are protected by a layer of ice.

6. MEDICAL AND VETERINARY IMPORTANCE

6.1 Tabanids as pest species

Independent of their role as disease vectors, tabanids represent a serious pest for humans and livestock, especially when particularly abundant. They cause painful nuisance biting in humans, and some species can cause severe skin ulceration. In Australia, *Mesomyia tryphera* and *Mesomyia silvester* have been reported to cause serious symptoms (e.g. hives, secondary infections, fever, wheezing and anaphylaxis), leading to hospitalization in some cases. In the veterinary context, the biting activity of many tabanid species is known to disturb livestock greatly, causing loss of production.

6.2 Vectors of disease

The mouthparts of tabanids are particularly well suited to the mechanical transmission of blood-dwelling pathogens from host to host. This ability is enhanced when the species is a determined feeder such as *T. taeniola*, which passes readily from host to host. For the mechanical transmission of *Trypanosoma evansi*, which causes surra in camels and horses (and also infects members of the kangaroo family Macropodidae), all tabanids are better vectors than biting muscids (e.g. *Stomoxys*), and species of *Tabanus* are more efficient vectors than those of *Chrysops* and *Haematopota*. *T. evansi* and *Trypanosoma vivax viennei* are only transmitted mechanically, and tabanids are the most important of the mechanical vectors. However, the importance of mechanical transmission in settings where alternative routes of infection exist is difficult to assess. Thus, the significance of tabanids in the transmission of pathogenic trypanosomes in areas where tsetse flies (*Glossina*) occur is probably low.

With regard to other diseases, there is evidence that tabanids are important mechanical vectors of *Anaplasma marginale* infecting cattle, of *Francisella tularensis* (the causative organism of tularaemia) infecting humans and livestock and of *Bacillus anthracis* (which causes anthrax) infecting humans and animals. Tabanids are also the vectors of three viral diseases, bovine leukaemia, equine infectious anaemia and hog cholera, and they may play a significant role in the transmission of the rinderpest virus.

Tabanids are biological vectors of three species of filarial worms: *Elaeophora schneideri*, the arterial worm of sheep, *Loa loa*, the cause of Calabar or fugitive swellings in humans in West Africa, and *Pelecitus roemeri*, a parasite of connective tissues in macropodid marsupials in Australia. Tabanids are also biological vectors of the blood-dwelling sporozoan, *Haematoproteus metchnikovi* of turtles, and there is evidence that tabanids are biological vectors of *Trypanosoma theileri*, a benign parasite of cattle.

6.2.1 *Anaplasma*

Species of *Anaplasma* (Rickettsiales, Anaplasmataceae) are very small (0.3–1.0 μm in diameter) parasites of the erythrocytes of ruminants, especially bovids and cervids. *Anaplasma marginale* has a wide range of hosts, including zebu cattle, water buffalo, African antelopes, American deer and camels, and is a pathogen of cattle in the tropics and subtropics, occurring in South Africa, Australia, Asia, Europe, South America, the former Soviet Union and the USA. It causes severe debility, anaemia, jaundice and abortion in adult cattle. Young animals are relatively resistant to infection but in cattle more than 3 years old a peracute condition may develop, with death occurring within 24 h. Recovered animals require prolonged convalescence and will continue to be infected for the rest of their lives. *A. marginale* is transmitted primarily by ticks, but also mechanically by tabanids (and some other bloodsucking flies) in some regions,

for example the south-eastern USA and in parts of Tanzania, where the cattle tick is absent and cases of anaplasmosis have been attributed to *T. taeniola*.

6.2.2 Tularaemia (*Francisella tularensis*)

Francisella tularensis is a Gram-negative, obligate aerobic organism, pleomorphic, non-motile coccobacillus which causes tularaemia in humans and many other warm-blooded animals, including sheep, horses, pigs, cattle and birds. Tularaemia is ubiquitous in the northern hemisphere between latitudes 30° and 71° north, occurring in Japan, Russia, Canada, Mexico and all the mainland states of the USA, but not on the Iberian Peninsula or in Great Britain. At least three subspecies of *F. tularensis* have been recognized as causes of disease in humans: *Francisella tularensis* subsp. *tularensis* (causing type A tularaemia), *Francisella tularensis* subsp. *holarctica* (causing type B tularaemia) and *Francisella tularensis* subsp. *novicida* (a relatively non-virulent strain formerly known as *Francisella novicida*). Of the two subspecies most commonly associated with human disease, *F. tularensis* subsp. *tularensis* is the more virulent agent and is found predominantly in North America, while *F. tularensis* subsp. *holarctica*, which causes a milder disease, is endemic in both the Palaearctic and Nearctic regions of the northern hemisphere (but has been reported also from the southern hemisphere, e.g. Australia). *F. novicida* is known only from a few immunocompromised cases in North America, and a fourth subspecies, *Francisella tularensis* subsp. *mediasiatica*, has been reported from central Asia, but little is known about it.

F. tularensis can be transmitted from host to host by a variety of routes, including bloodsucking arthropods, water, food and inhalation. Since it can penetrate unbroken skin, this organism can be transmitted via direct contact with infected material. It usually produces a marked reaction at the site of entry, which is an ulcer in 70–80% of cases.

Around 100 species of wild mammals, 25 species of birds and more than 50 species of arthropods have been found to be naturally infected. In North America, *F. t. tularensis* occurs particularly in lagomorphs and wild rodents and is transmitted by the bites of ticks and tabanid flies. The ability of *F. tularensis* to survive away from its vertebrate host favours mechanical transmission by bloodsucking flies, especially tabanids. *C. discalis* is recognized as being a major mechanical vector of *F. tularensis* in the USA, and in the summer of 1971, up to three-quarters of the human cases of tularaemia in Utah (USA) were attributed to the biting of *C. discalis*.

Tularaemia is most severe in sheep, in which the morbidity rate may be as high as 40%, with a mortality of 50% (in North America) especially in young animals. Horses display fever and foals are affected more seriously than older animals. In swine, tularaemia causes fever in piglets, but is latent in adult pigs.

Treatment with streptomycin was found to reduce a mortality rate of 7% in humans to virtually zero. Human tularaemia is an acute, febrile, infectious zoonotic disease with an incubation period of 3–4 days. The introduction of as few as ten organisms subcutaneously can cause disease in non-vaccinated individuals. In the former USSR, there were 10,000 cases a year before vaccination was introduced; large-scale vaccination reduced the incidence to less than 200 per annum.

6.2.3 Trypanosomiases

The trypanosomiases are diseases of humans and livestock caused by parasitic flagellate Protozoa of the order Kinetoplastida (Phylum Sarcomastigophora). At different stages of the developmental cycle, the parasite takes various forms, with a flagellum absent (amastigote) or located anteriorly (promastigote), with a long flagellar pocket (opisthomastigote), forming an undulating membrane along the body to the anterior end (epimastigote) or emerging laterally and forming an undulating membrane (trypomastigote).

The economically important trypanosomes transmitted by insects are divided into two sections, the Salivaria and the Stercoraria, which differ in their modes of transmission. Salivarian trypanosomes either undergo cyclic development in the insect before being transmitted with the saliva or are transmitted mechanically. Stercorarian trypanosomes also undergo development in the insect but, with

the exception of *Trypanosoma rangeli*, the infective forms are deposited in the faeces of the vector.

Salivarian trypanosomes are transmitted by species of *Glossina*, with two exceptions: *T. evansi* and *T. vivax viennei*, infecting cattle and other livestock and being transmitted mechanically by tabanids and *Stomoxys* in areas where *Glossina* species are not present. Of the Stercorian trypanosomes, *Trypanosoma theileri* is a large trypanosome which produces a benign infection of cattle. In the tabanid *H. pluvialis*, the trypomastigotes of *T. theileri* give rise by multiplication and transformation to epimastigotes within 24 h of ingestion. From day 5 after infection, trypanosomes may be found in the hindgut, and from day 6, small metacyclic forms, presumably infective to cattle, may be present.

6.2.4 Loiasis

Loiasis is a human filarial disease caused by infection with *Loa loa*. It occurs in the rainforests of tropical Africa, extending roughly from 10°N to 10°S and from 0° to 30°E, ranging from Nigeria to south-west Sudan and south to Zaire and north-west Angola. The disease is characterized by recurrent, temporary subcutaneous swellings, mostly in the region of the wrists and forearms. This condition, known as Calabar or fugitive swellings, is a response to antigenic material released by migrating worms. The thin, transparent adult female and male worms measure 50–70 × 0.5 mm and 30–35 × 0.3 mm, respectively. The adult worms live a nomadic life moving through loose connective tissue and at times can be seen moving under the skin, causing minimal local reaction. The most disquieting demonstration of their mobility is when they move across the eye under the conjunctiva at a speed of 1 cm/min. No permanent damage of the eye occurs; however, it becomes oedematous. The sheathed microfilariae occur in the circulating blood.

LOA LOA IN THE VERTEBRATE HOST

Two strains of *L. loa* are recognized. Microfilariae of the human strain have a diurnal periodicity in the circulating blood and develop in day-biting *Chrysops*. The microfilariae of the strain which parasitizes monkeys, especially the drill (*Mandrillus leucophaus*), have a nocturnal periodicity and develop in night-biting *Chrysops*. Hybrids derived from the two strains have a characteristic periodicity, differing markedly from that of either parent or from a 50:50 mixture of the two periodicities.

In monkeys, the worms become mature and microfilariae appear in the circulating blood 4–5 months after the introduction of infective larvae, while in humans maturation of the worm takes 6–12 months before the appearance of microfilariae in the peripheral bloodstream. In monkeys, and probably also in humans, the female passes microfilariae into the connective tissue, from which they enter the vascular system and accumulate in the pulmonary blood before entering the peripheral circulation about 3 weeks after being released. The long-lived adult worms are considered to live for 4–17 years.

DEVELOPMENT OF *LOA LOA* IN CHRYSOPS

In *C. silacea*, the worm develops in the fat body of the fly, where it undergoes two moults before reaching the infective stage. At 28–30°C and 92% RH (relative humidity), microfilariae develop to the infective stage within 7 days, increasing in length from 275 μm to more than 2 mm in the third developmental stage. Most microfilariae develop in the fat body of the abdomen, and a smaller number in the fat body of the thorax and head. In the initial stages of development, the parasite is intracellular, but it later becomes free. Infective larvae move to the head, where they accumulate in the subcibarial haemocoelic space and escape when the fly is feeding, by rupturing the delicate labiohypopharyngeal membrane.

When *C. silacea* and *C. dimidiata* feed, they ingest about twice their body weight of blood, with the heavier species, *C. silacea*, taking in about 20–25% more blood. *C. silacea* ingests only about half the number of microfilariae that would be expected from the size of the blood meal and the density of the microfilariae in the circulating blood. However, mortality of ingested microfilariae during development is minimal and field studies have shown that the numbers of *L. loa* in *C. silacea* and *C. dimidiata* are almost identical.

ECOLOGY OF THE VECTOR AND EPIDEMIOLOGY OF LOIASIS

It is considered that *C. silacea* and *C. dimidiata* are the most important vectors, but the latter is usually less numerous. *Chrysops distinctipennis* and *Chrysops zahrai* are less effective local or subsidiary vectors, whereas *C. langi* and *Chrysops centurionis* are zoophilic and responsible for the transmission of the monkey strain. Transmission of *L. loa* from human to monkey is thought to occur rarely, while the reverse transmission from monkey to human is considered most unlikely.

The biting cycle of *C. silacea* and *C. dimidiata* in the forest canopy is bimodal, with morning and afternoon peaks in which nulliparous females are more numerous in the morning and parous females are more numerous in the afternoon. *C. langi* and *C. centurionis* are crepuscular, biting from about 17:00 h to 21:00 h. It is unlikely that *C. silacea* and *C. dimidiata* would acquire infections by feeding on monkeys, which during the daytime would quickly catch tabanids attempting to feed. Even if some fed successfully, there would be virtually no microfilariae in the monkeys' peripheral blood during the daytime. At night, *C. langi* and *C. centurionis* would find it easier to feed on sleeping monkeys, at a time when microfilarial density was high in their circulating blood, favouring infection.

In a Congo rainforest where *C. silacea* was three times more abundant than *C. dimidiata*, both species were markedly anthropophilic, with 90% of their blood meals being taken from humans. In the presence of a fire, the catch of *C. silacea* increased by eightfold at ground level and by fivefold in the canopy, while that of *C. dimidiata* remained unchanged. In both species, 0.6% of the flies were infective, carrying on average ten infective third-stage larvae. The annual transmission potential was about 140 for the more numerous *C. silacea* and 95 for *C. dimidiata*.

The distribution of immature *Chrysops* follows that of the biting adults, suggesting that these species do not disperse strongly, particularly across open ground. Hence, the risk of infection with *L. loa* should decrease with distance from forest. Certainly, the biting intensity decreases with distance, with the rate of decline depending on the degree of cover available. Even in the tropical rainforest, the biting densities of *C. silacea* and *C. dimidiata* are not constant throughout the year but show seasonal cycles, with *C. silacea* being abundant from April to December and *C. dimidiata* having two peaks of abundance, from November to January and March to May.

PREVENTION AND CONTROL

Loiasis is a relatively benign disease which can be treated with the microfilaricide, diethyl-carbamazine (DEC). Care is needed in treating patients with high microfilaraemia, because DEC can cause meningoencephalitis, a serious and potentially fatal condition. Ivermectin is effective as a single dose repeated after 1 month; it will reduce microfilariae numbers and is more suited to mass treatment programmes. Although it is more tolerable, reactions still occur, particularly in infections with eye involvement. Preventive action against bites through wearing protective clothing and repellents, rather than direct vector control measures, is advisable. None the less, in areas where high transmission may occur there have been antivector efforts that have involved clearing of forest areas and applying oil to known larval habitats. This approach is not necessarily ecologically acceptable in modern times.

6.2.5 *Elaeophora schneideri* (in sheep)

The filaria *Elaeophora schneideri* occurs mainly in North America but is also recorded in some European countries in sheep which have been grazed at high altitudes (1800 m) in the summer months. It has an incidence of about 1% and is commoner in 4- to 6-year-old sheep. Sheep that survive the establishment of the worm show a severe dermatitis on the head and feet. *E. schneideri* is a benign parasite of the mule deer, *Odocoileus hemionus*, which causes clinical disease in abnormal hosts such as elk (*Cervus canadensis*) and moose (*Alces alces*). Lesions caused by the nematodes include occlusion of the cephalic and other arteries, which can result in severe disease and death. Hypersensitivity to the microfilariae of *E. schneideri* can produce a severe dermatitis in the head of domestic sheep (*Ovis aries*) and Barbary sheep (*Ammotragus lervia*). At least 16 species of *Hybomitra* and *Tabanus* are biological vectors of *E. schneideri*.

In Gila National Park, New Mexico, the vector of *E. schneideri* is *Hybomitra laticornis*, of which in one survey 16% were infected, with an average of 25 developing worms. Ingested microfilariae escape from the midgut into the haemocoel and enter the fat body for the initial stage of development. Older larvae leave the fat body and develop in the abdomen to infective larvae measuring 4.5 mm × 50 μm. Subsequently, they move to the head and mouthparts, from which they escape while the fly is feeding.

7. PREVENTION AND CONTROL

Few attempts are usually made to control tabanids. The larvae and pupae are typically inaccessible in their substrate habitat; therefore, water or land management may not be feasible and it can be counterproductive in enhancing populations of some species. Aerial applications of insecticides against adults are rarely effective for more than a few days.

In areas where population densities of adult flies are high, long-sleeved shirts, long trousers and headnets may provide the most effective protection for humans; repellents are variously effective or ineffective for humans, depending on tabanid species. Insecticide-impregnated ear tags can provide good control for livestock, but results are dependent on the fly species targeted. The flies are often readily attracted to dark colours and solid shapes, and sticky traps, box traps, Manitoba or pyramid traps can be useful in reducing numbers locally.

SELECTED BIBLIOGRAPHY

Anderson, J.F. (1985) The control of horse flies and deer flies (Diptera: Tabanidae). *Myia* 3, 547–598.

Centers for Disease Control and Prevention (CDC) (2005) Tularemia transmitted by insect bites – Wyoming, 2001–2003. *Morbidity and Mortality Weekly Report (MMWR)* 54, 170–173.

Chainey, J.E. (1993) Horse-flies, deer-flies and clegs (Tabanidae). In: Lane, R.P. and Crosskey, R.W. (eds) *Medical Insects and Arachnids*. Chapman and Hall, London, pp. 310–332.

Chippaux, J.-P., Bouchite, B., Demanov, M., Morlais, I. and LeGoff, G. (2000) Density and dispersal of the loiasis vector *Chrysops dimidiata* in southern Cameroon. *Medical and Veterinary Entomology* 14, 339–344.

Chvala, M., Lyneborg, L. and Moucha, J. (1972) *The Horse Flies of Europe (Diptera, Tabanidae)*. Entomological Society of Copenhagen, Copenhagen.

Cilek, J.E. and Olson, M.A. (2008) Effects of carbon dioxide, an octenol/phenol mixture, and their combination on Tabanidae (Diptera) collections from French 2-tier box traps. *Journal of Medical Entomology* 45, 638–642.

Desquesnes, M. and Dia, M.L. (2004) Mechanical transmission of *Trypanosoma vivax* in cattle by the African tabanid *Atylotus fuscipes*. *Veterinary Parasitology* 119, 9–19.

Desquesnes, M., Biteau-Coroller, F., Bouyer, J., Dia, M.L. and Foil, L. (2009) Development of a mathematical model for mechanical transmission of trypanosomes and other pathogens of cattle transmitted by tabanids. *International Journal for Parasitology* 39, 333–346.

Duke, B.O.L. (1955) The development of Loa in flies of the genus *Chrysops* and the probable significance of the different species in the transmission of loiasis. *Transactions of the Royal Society of Tropical Medicine and Hygiene* 49, 115–121.

Duke, B.O.L. (1991) Loiasis. In: Strickland, G.T. (ed.) *Hunter's Tropical Medicine*. Saunders, Philadelphia, pp. 727–729.

Foil, L.D. (1989) Tabanids as vectors of disease agents. *Parasitology Today* 5, 88–96.

Hawkins, J.A., Love, J.N. and Hidalgo, R.J. (1982) Mechanical transmission of anaplasmosis by tabanids (Diptera: Tabanidae). *American Journal of Veterinary Research* 43, 732–734.

Hayes, R.O., Doane, O.W. Jr, Sakolsky, K. and Berrick, S. (1993) Evaluation of attractants in traps for greenhead fly (Diptera: Tabanidae) collections on a Cape Cod, Massachusetts, salt marsh. *Journal of the American Mosquito Control Association* 9, 436–440.

Hollander, A.L. and Wright, R.E. (1980) Impact of tabanids on cattle: blood meal size and preferred feeding sites. *Journal of Economic Entomology* 73, 431–433.

Hungerford, L.L. and Smith, R.D. (1996) Spatial and temporal patterns of bovine anaplasmosis as reported by Illinois veterinarians. *Preventive Veterinary Medicine* 25, 301–313.

Kloch, L.E., Olsen, P.F. and Fukushima, T. (1973) Tularaemia epidemic associated with the deerfly. *Journal of American Medical Association* 226, 149–152.

Kocan, K.M., de la Fuente, J., Blouin, E.F. and Garcia-Garcia, J.C. (2004) *Anaplasma marginale* (Rickettsiales: Anaplasmataceae): recent advances in defining host–pathogen adaptations of a tick-borne rickettsia. *Parasitology*, 129 Suppl, S285–300.

Krinsky, W.L. (1976) Animal disease agents transmitted by horse flies and deer flies. *Journal of Medical Entomology* 13, 225–275.

Lang, J.T., Schreck, C.E. and Pamintuan, H. (1981) Permethrin for biting-fly (Diptera: Muscidae; Tabanidae) control on horses in central Luzon, Philippines. *Journal of Medical Entomology* 18, 522–529.

Manet, G., Guilbert, X., Roux, A., Vuillaume, A. and Parodi, A.L. (1989) Natural-mode of horizontal transmission of bovine leukemia-virus (BLV) – the potential role of tabanids (*Tabanus* spp.). *Veterinary Immunology and Immunopathology* 22, 255–263.

Mihok, S. and Carlson, D.A. (2007) Performance of painted plywood and cloth Nzi traps relative to Manitoba and Greenhead traps for tabanids and stable flies. *Journal of Economic Entomology* 100, 613–618.

Mihok, S. and Mulye, H. (2010) Responses of tabanids to Nzi traps baited with octenol, cow urine and phenols in Canada. *Medical and Veterinary Entomology* 24, 266–272.

Muzari, M.O., Jones, R.E., Skerratt, L.F. and Duran, T.L. (2010) Feeding success and trappability of horse flies evaluated with electrocuting nets and odour-baited traps. *Veterinary Parasitology* 171, 321–326.

Noireau, F., Nzoulani, A., Sinda, D. and Caubere, P. (1991) *Chrysops silacea* and *C. dimidiata* seasonality and loiasis prevalence in the Chaillu mountains, Congo. *Medical and Veterinary Entomology* 5, 413–419.

Pence, D.B. (1991) Elaeophorosis in wild ruminants. *Bulletin of the Society of Vector Ecology* 16, 149–160.

Perich, M.J., Wright, R.E. and Lusby, K.S. (1986) Impact of horse flies (Diptera: Tabanidae) on beef cattle. *Journal of Economic Entomology* 79, 128–131.

Reid, S.A., Husein, A., Partoutomo, S. and Copeman, D.B. (2001) The susceptibility of two species of wallaby to infection with *Trypanosoma evansi*. *Australian Veterinary Journal* 79, 285–288.

Sanford, J.P. (1991) Tularaemia. In: Strickland, G.T. (ed.) *Hunter's Tropical Medicine*. Saunders, Philadelphia, pp. 416–417.

Schultze, T.L., Hansens, E.J. and Trout, J.R. (1975) Some environmental factors affecting the daily and seasonal movements of the salt marsh greenhead *Tabanus nigrovittatus*. *Environmental Entomology* 4, 965–971.

Spratt, D.M. (1975) Further studies of *Dirofilaria roemeri* (Nematode: Filarioidea) in naturally and experimentally infected Macropodidae. *International Journal for Parasitology* 5, 561–564.

Strickland, G.T. (ed.) (2000) *Hunter's Tropical Medicine and Emerging Infectious Diseases*, 8th edn. Saunders, Philadelphia, pp. 7540–7556.

Tidwell, M.A., Dean, W.D., Tidwell, M.A., Combs, G.P., Anderson, D.W., Cowart, W.O., *et al.* (1972) Transmission of hog cholera virus by horseflies (Tabanidae: Diptera). *American Journal of Veterinary Research* 33, 615–622.

van Hennekeler, K., Jones, R.E., Skerratt, L.F., Fitzpatrick, L.A., Reid, S.A. and Bellis, G.A. (2008) A comparison of trapping methods for Tabanidae (Diptera) in North Queensland, Australia. *Medical and Veterinary Entomology* 22, 26–31.

van Hennekeler, K., Jones, R.E., Skerratt, L.F., Muzari, M.O. and Fitzpatrick, L.A. (2011) Meteorological effects on the daily activity patterns of tabanid biting flies in northern Queensland, Australia. *Medical and Veterinary Entomology* 25, 17–24.

Watson, D.W., Denning, S.S., Calibeo-Hayes, D.I., Stringham, S.M. and Mowrey, R.A. (2007) Comparison of two fly traps for the capture of horse flies (Diptera: Tabanidae). *Journal of Entomological Science* 42, 123–132.

Wilson, B.H. (1968) Reduction of tabanid populations on cattle with sticky traps baited with dry ice. *Journal of Economic Entomology* 61, 827–829.

House Flies and Other Non-biting Flies (Diptera: Muscidae and Fanniidae)

1. INTRODUCTION

The Muscidae is a family of nearly 4000 species and includes the house flies and a large number of other species of fly that are of medical and veterinary importance. The Fanniidae is a much smaller family of about 260 species, most of which are in the genus *Fannia*. The medically important members of both families are dark-coloured, medium-sized flies whose immature stages occur in fermenting organic material of vegetable origin and are often associated with the dung of herbivorous mammals. Many of these species have become cosmopolitan, spreading throughout the world with people, and flies that live closely with humans are said to be synanthropic. The Muscidae and Fanniidae contain a number of subfamilies and genera of considerable medical and veterinary importance, most notably the genera *Musca* and *Fannia*, all species of which are liquid feeders and do not have biting mouthparts but which are able to transmit mechanically various pathogens (mostly alimentary) from contaminated material to human environments.

2. TAXONOMY

The superfamily Muscoidea contains two families of medical and veterinary interest: the Muscidae and Fanniidae. In the Muscidae, the squamae are conspicuous, with the lower much larger than the upper. In the Fanniidae, the squamae are small, subcircular and of similar size.

The genus *Musca* contains about 60 species, of which the cosmopolitan house fly, *Musca domestica*, and the face fly, *Musca autumnalis*, are of particular importance. In southern Europe, *Musca larvipara* shares the same habitat and behaviour of *M. autumnalis*. In the African and Oriental regions, the bazaar fly, *Musca sorbens*, is widespread, largely replacing *M. domestica*. In Australia, the bush fly, *Musca vetustissima*, which is related very closely to *M. sorbens*, is an important nuisance pest of humans and livestock. The genus *Fannia* contains over 200 species, the most important and cosmopolitan of which is the little house fly, *Fannia canicularis*, but the latrine fly, *Fannia scalaris*, and in North America, *Fannia benjamini*, may also be locally common.

3. MORPHOLOGY

Three subspecies of *M. domestica* are recognized: *M. domestica domestica* occurs worldwide but is least abundant in Africa, where the two other subspecies occur – the endophilic *M. d. curviforceps* and the exophilic *M. d. calleva*. The *Musca* is characterized by a body with more or less distinct grey pollinosity and the presence of four dark, longitudinal stripes on the scutum (Fig. 40). The arista is plumose, with branches above and below. Medial vein (M_{1+2}) is angularly rounded at its bend and ends close to R_{4+5} (Fig. 41a). Females of *M. domestica* are 6–8 mm and males 5–6 mm in length.

The third-stage larva (Fig. 42a, b, c and d) measures $6–12 \times 1–2$ mm and has 12 visible segments (one head segment, three thoracic and eight abdominal). On the ventral side of the first segment there are two oral lobes transversed by parallel tubes, which converge on the mouth. The tubes function in a

© R.C. Russell 2013. *The Encyclopedia of Medical and Veterinary Entomology* (R.C. Russell *et al.*)

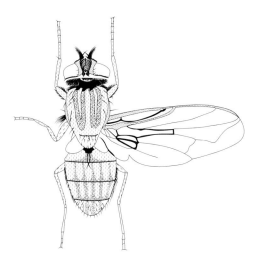

Fig. 40. *Musca domestica.* Drawn from specimens and Greenberg (1971).

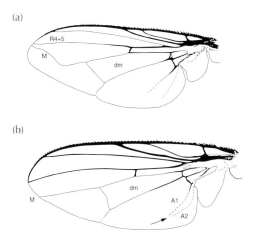

Fig. 41. The typical arrangement of veins and cells on the wing of the muscids (a) *Musca domestica* and (b) *Fannia canicularis.*

larger than the left. Respiration is amphipneustic, with fan-shaped anterior spiracles on the second segment, undeveloped in first-stage larvae and dark, flat, plate-like spiracles on the posterior surface of the body. In the third stage, the anterior spiracles have five to seven openings and the posterior spiracles are D-shaped, with three sinuous slits and a button in the middle of the straight side of the D. The button is the scar left at the moult from second to third stage. The posterior spiracle has one simple, reniform opening in the first stage and two and three nearly straight slits in the second and third stage, respectively.

Flies of the genus *Fannia* (Fig. 43) are small to medium-sized flies (usually no more than 7 mm in length) with bare arista and a characteristic venation, in which medial vein M is straight and vein A2 curved, so that if extended it would intersect with an extended vein A1 (Fig. 41b). The males are holoptic and the females dichoptic. A few species are of importance: *F. canicularis*, the lesser or little house fly, is a worldwide endophilic synanthrope; *F. scalaris*, the latrine fly, is a worldwide exophilic synanthrope. In North America, *Fannia femoralis* is an occasional pest of humans. The larvae of *Fannia* are flattened, tapering anteriorly and bearing prominent lateral processes on most segments (Fig. 44). In the mature larva, the anterior spiracles are prominent and the posterior ones elevated with three lobes. The larvae of *F. canicularis* have long lateral and dorsal processes with short basal spines; and those of *F. scalaris* have short dorsal processes and pinnate lateral processes.

4. LIFE CYCLE

The life cycles of species of *Musca* and *Fannia* are broadly similar.

The pearly-white eggs are long and narrow, measuring about 1.20×0.25 mm. Under optimal conditions (37°C), they hatch in about 8 h to give rise to legless, saprophagous larvae (maggots). There are three larval stages, of which the first two last about 24 h and the third for 3 or more days.

The fully-grown larva ceases to feed, empties its gut and moves into drier conditions

comparable manner to the pseudotracheae of the labellum of the adult fly. The head is retracted into the thorax and its dark cephalopharyngeal skeleton can be seen through the translucent body of the larva. This skeleton supports a pair of retractable mouth-hooks, which can be extended through the mouth and used for progression and tearing at the substrate; the mouth-hooks are closely apposed, with the right one being markedly

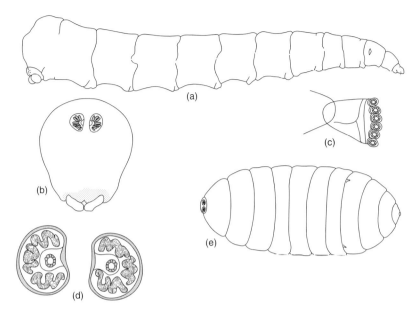

Fig. 42. (a) Third-stage larva of *Musca domestica*; (b) posterior view of larva; (c) anterior spiracle; (d) posterior spiracles; (e) puparium showing pupal respiratory horns.

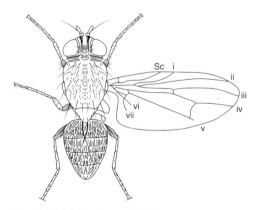

Fig. 43. Adult *Fannia canicularis*.

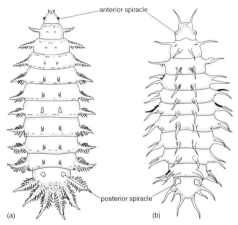

Fig. 44. (a) Third-stage larvae of *Fannia scalaris*; and (b) *Fannia canicularis*. *Source*: Hewitt, C.G. (1912) *Parasitology* 5, 164.

and buries itself into the substrate, where it pupates within the last larval skin, which forms the puparium (Fig. 42e). Immediately after moulting, the puparium is creamy white, but it steadily darkens through shades of reddish-brown until the mature puparium is almost black. The external larval structures are retained on the outside of the puparium, which measures 4–6 × 2–2.5 mm. The larval spiracles are non-functional and the pupa respires through pupal horns, which pierce the puparium between the fifth and sixth segments.

When the adult fly has completed its development, it emerges from the puparium by using its ptilinum to force off a hemispherical cap from the anterior end of the puparium and makes its way up to the surface of the soil, at which time the ptilinum is withdrawn into the head and the frontal suture closed. The body of the fly is expanded by taking air into the gut and the wings extended by pumping haemolymph through the veins. This ability of the newly emerged fly to move to the surface can render the burying of infested material ineffective as a fly control measure. Flies are able to emerge from material buried under 1.2 m of clay, loam or sand because most of the larvae move to within 30 cm of the surface before pupating, and from this distance adults can reach the surface easily. Adults emerging from dung pats with high densities of immatures are smaller than those emerging from pats of low density.

Fanniidae breed in decaying animal and vegetable matter, especially faeces (for example *F. scalaris* are often found in semi-fluid faeces of humans and pigs). The adults are more abundant in the cooler months, declining in the summer and, where necessary, the species overwinters as pupae, buried at depths of 50–80 mm in the soil. *F. canicularis* is often attracted indoors and characteristically flies in the middle of the room, does not readily settle on human food and is therefore less annoying than *M. domestica*.

5. BEHAVIOUR AND BIONOMICS

Adult *M. domestica* are diurnal and activity is favoured by high temperatures and low humidity but, as the name house fly implies, they are more active in shade than in sunlight. Mating takes place soon after adult emergence. Males will mate on the day of emergence and the mating response is highest in 3-day-old females. Two pheromones are involved in mating; one, produced by the female, attracts males and the other, produced by males, induces aggregation and receptivity in virgin females. Maturation of the eggs depends on the female having access to a diet of protein, and a batch of eggs may be laid as early as 54 h after emergence of the female.

The ovipositing female deposits her eggs in clumps in cracks and crevices of a suitable medium, and sometimes the whole batch may be deposited in a single clump. By inserting the eggs into a moist medium, the female protects them from desiccation. The ancestral breeding site was probably horse dung, but house flies now breed in the dung of a wide range of herbivores, in fowl manure, in fermenting kitchen waste and in rubbish tips. Indeed, separate populations of *M. domestica* may develop in association with stabled animals and breed in urine and dung-contaminated stable refuse, but *M. domestica* does not usually breed in cow dung, which is the main source of *M. autumnalis* and *M. larvipara*.

The duration of the larval stages is a function of temperature and the quality of the larval medium. When dung quality is not limiting, the development time of the larval stages is 145 day-degrees above a threshold of 12°C and up to an optimum of 36°C, above which development is adversely affected. This relationship implies that at 22°C, i.e. 10°C above the threshold, the larval duration would be 14.5 days. Susceptibility to high temperatures is lowest in the egg and highest in the pupa, and all stages are killed by exposure to 50°C or higher; a fact used in fly control. Tight packing of refuse dumps containing organic material will help to attain temperatures lethal to the immature stages of the house fly through the fermentation and decay of the organic material.

On average, a female will mature 120 eggs in a batch (range 100–150) and deposit four to six batches of eggs during a lifetime of 2–4 weeks in summer. With a cycle from egg to egg of 3 weeks, 10–12 generations can be produced a year in the warmer temperate regions of the world. In colder regions, breeding will be restricted to the warmer months and the winter passed as slowly growing larvae and pupae, some of which will survive to emerge when warmer conditions return.

5.1 *Musca sorbens* (bazaar fly)

The *M. sorbens* complex of species is widely distributed in the tropics and subtropics of the Afrotropical, Oriental, Australian regions and in the southern Palaearctic. *M. sorbens* s.s. occurs throughout the range, with the exception of Australia, where the only species present is *M. vetustissima*. A third member of the complex, *Musca biseta*, occurs in Africa. Flies of the *M. sorbens* complex have two broad, dark, longitudinal stripes on the scutum.

The wings are clear, with white squamae and the first abdominal segment black. The proboscis is of the normal lapping type with small prestomal teeth. The female is dichoptic and the male nearly holoptic. The primary breeding site of members of the *M. sorbens* complex is cow dung, but it may also breed in human faeces. It is common around garbage dumps, privies, carrion and various types of excrement. In Africa and Oriental regions, it abounds in bazaars, swarming on foodstuffs of all kinds. It commonly feeds on exudations from cuts and sores. This species attacks people, especially children, around the eyes.

5.2 *Musca vetustissima* (Australian bush fly)

The Australian bush fly, *M. vetustissima* (Fig. 45), has eggs that are larger than those of *M. domestica*, measuring approximately 1.7 × 0.3 mm. They have the typical muscine shape,

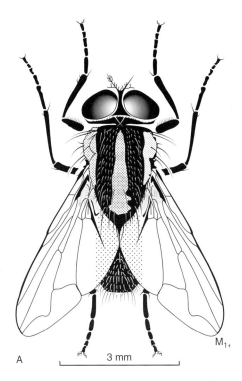

A   3 mm M_{1+}

Fig. 45. *Musca vetustissima. Source: Insects of Australia*, Melbourne University Press, Carlton, Victoria, Australia.

with a hatching strip on the concave surface which functions as a plastron facilitating respiration when the eggs are in water. Eggs of *M. vetustissima* will hatch in water but not if totally immersed in faeces, when the plastron is unable to function. The eggs are very susceptible to desiccation.

The third-stage larva is very similar to that of *M. domestica*, being of comparable size, with the right mouth-hook longer than the left, the anterior spiracle with about six openings and three sinuous slits in the posterior spiracle.

Mature larvae are sensitive to exposure to excess water and no pupation occurs in waterlogged soil. Post-feeding larvae leave the dung between midnight and dawn and bury themselves into the substrate, where they pupate at depths of 20–30 mm. The pupal stage takes about 3 days at 39°C and 18 days at 18°C. Adult emergence occurs around dawn, which is appropriate for a diurnally active adult.

Females develop faster than males and emerge first. At 27°C, mating occurs on the third day after emergence and it is believed that a second mating may occur after the second ovarian cycle. Copulation is a long process, lasting about 80 min. The female requires a protein meal to develop the ovaries. In the field, females complete two to three oviposition cycles during their lifetime. Wild flies develop 4–48 eggs in a cycle, the number of eggs maturing being reduced in each subsequent cycle. Ovarian development depends on the nutritional state of the female, and where that is inadequate, fewer eggs are matured. The female is attracted to fresh dung, to which it flies upwind and deposits its eggs in crevices in the dung. Oviposition occurs during daylight and an ovipositing female attracts other gravid females, probably through the emission of an egg-laying pheromone.

Adult survival is related inversely to temperature, being 11 days at 29°C and 7 or more weeks at 12°C, the threshold temperature for development. Survival is dependent on the adult having access to free water and an easily metabolized energy source such as sugar. They are most active at temperatures up to 35°C and wind speeds of <8 km/h. They can be displaced hundreds of kilometres in a day on hot, strong winds and are also regularly dispersed via human transport.

Several features of the bush fly biology determine its seasonal distribution. There is no diapausing stage in the life cycle. At low temperatures, all stages may survive for a long time, but temperatures below freezing are lethal. The bush fly populations are highest in the north of Australia in the autumn, following the summer rains, and decline in the winter and spring as the pasture deteriorates. They move south on warm winds and are able to exploit the spring flush of vegetation following the winter rains in the south of Australia. There, bush fly populations reach a peak in late spring and early summer, and then wane as the summer dry season advances.

5.3 *Musca crassirostris*

The Indian cattle fly occurs from Africa through the Middle East to South-east Asia. The proboscis of this fly has large prestomal teeth which can rasp away skin and enable the fly to feed on blood. It deposits small batches (40–50) of large (2 mm long) eggs into freshly dropped cow and horse dung.

5.4 *Musca autumnalis* and *Musca larvipara*

The face flies, *M. autumnalis* and *M. larvipara*, are similar to *M. domestica* in appearance; however, the abdomen in females is darker and in males is yellowish-orange with a T-like black median strip. Both species occur frequently on pastures and congregate in large numbers around the faces of cattle. While the males are often floricolous, females feed on secretions from the eyes, nose and mouth and are facultatively haematophagous, feeding from blood in wounds left by other flies, such as tabanids. *M. autumnalis* lays its eggs just beneath the surface of fresh cattle manure, within about 15 min after the dung pats are deposited. The eggs of *M. autumnalis* are about 3 mm in length and possess a short respiratory stalk. Like *M. domestica*, the larvae pass throughout three stadia before pupating to form a whitish-coloured puparium. *M. larvipara* females give birth to second

instar larvae, which they deposit on cattle manure. Summer generations require about 2 weeks to complete a life cycle. Face flies prefer bright sunshine and usually do not follow cattle into barns or heavy shade. Adults are strong fliers and can move between widely separated herds. Face flies overwinter as adults, aggregating in overwintering sites such as farm buildings. *M. autumnalis* is a Palaearctic and Afrotropical species which was introduced, probably in 1951, into North America. Both species also have well-developed prestomal teeth which can be used to reopen healing wounds but cannot penetrate unbroken skin.

5.5 *Muscina stabulans*

The false stable fly occurs throughout the USA, in southern America (Chile), New Zealand and Europe. It is a synanthropic fly of 8 mm in length, bigger and stouter than *M. domestica*; it has four characteristic dark stripes along the thorax region and a pale spot posteriorly. The scutellum is pale yellow and the legs are partially reddish-brown. The abdomen is either entirely black or grey black.

Larvae of *M. stabulans* exhibit a strong preference for decaying matter and breed in the faeces of several species of mammals, including humans, but also birds. Females lay 150–200 eggs in batches of 8–10 eggs and the life cycle lasts 20–25 days at 27°C. The eggs are similar to those of *M. domestica* and mature larvae are 15 mm in length; the posterior spiracles have three slits that are not straight and exhibit some form of curvature. The habitat of *M. stabulans* is similar to *M. domestica*. False stable flies have been spotted in animal premises, such as poultry and pigs houses and, in temperate regions, adults can often be found on the external walls of human houses, but they are rarely attracted indoors.

5.6 *Hydrotaea*

Species of the genus *Hydrotaea* closely resemble species of *Musca* and are known as the sweat flies. They feed on exudates of the eyes, nose and mouth and do not bite.

The genus contains one particularly

important species, *Hydrotaea irritans*, known as the sheep head fly, which is widespread throughout northern Europe but is not believed to be present in North America. This species is of the same size as *M. domestica*, but the abdomen is of olive-green colour and yellow-orange at the basis of the wings. Males have a characteristic femur notched or excavated on the pre-apical region of the ventral surface.

The sheep head fly worries domestic stock and people by feeding on secretions from the mouth, nose, ears, eyes and wounds. To get enough protein from these food sources to mature their eggs, they need to feed relatively frequently. Adults are most active under calm, humid, sultry conditions, particularly before and after rain, when their attacks can be particularly irritating. In Denmark, the highest densities of *Hy. irritans* occur in permanent, low-lying, fairly sheltered grassland compared to temporary, dry, wind-exposed pastures. Adults also feed on honeydew, flowers, carrion and faeces, from which they presumably acquire the necessary protein for ovarian development. The threshold temperature for activity is about 12°C, similar to that for *M. vetustissima*, and activity ceases at wind speeds above 3.6 km/h. Activity is bimodal, with peaks in the morning and evening. Although there are records of this species breeding in cow dung, the normal larval biotope of this univoltine species is in pasture soil, under long grass or on woodland edges.

Although they do not bite, they are facultative blood feeders and may ingest blood at the edges of wounds if available. While this species has the sponging mouthparts typical of most Muscidae, they also have well-developed prestomal teeth, which are used for rasping, and they can cause skin damage and enlarge existing lesions during feeding.

6. MEDICAL AND VETERINARY IMPORTANCE

The house fly, *M. domestica*, is closely associated with humans, livestock, their buildings and organic wastes. Although it may be of only minor direct annoyance to animals, its potential for transmission of viral and bacterial diseases, and protozoan and meta-

zoan parasites, is of significance. However, its pathological importance varies considerably, depending on the precise circumstances in which it occurs. Free availability of livestock or human excrement and low levels of household hygiene provides sites in which flies can breed and opportunities for flies to act as vectors to humans as they move from site to site.

The movement of house flies between faeces and food makes them efficient mechanical vectors of disease pathogens. House flies have been found to carry up to 65 human pathogens, ranging from viruses to helminths, including the viruses of poliomyelitis and infectious hepatitis, the bacteria associated with cholera, enteric infections caused by species of *Salmonella* and *Shigella*, pathogenic *Escherichia coli*, haemolytic streptococci, *Staphylococcus aureus* and agents of trachoma, bacterial conjunctivitis, anthrax, diphtheria, tuberculosis, leprosy and yaws. In addition, flies can carry the cysts of Protozoa, including those of *Entamoeba histolytica*, which causes amoebic dysentery, and the eggs of the threadworm, *Trichuris trichiura*, the hookworm, *Ancylostoma duodenale*, and of other nematodes and cestodes. It is important to note, however, that simply recovering pathogens from house flies does not necessarily mean that they are involved in the active transmission of disease.

The house fly is the biological vector and intermediate host of some cestodes of poultry (*Choanotaenia*) and of the nematodes (*Habronema* spp., *Thelazia* spp.) which cause habronemiasis in horses. House fly larvae are not usually involved in myiases (the invasion of living tissue by dipterous larvae). There are three ways in which house flies can disseminate pathogens. The body surface of the fly (particularly its legs and proboscis) can be contaminated, pathogens can be regurgitated on to food or may be deposited in its faeces. Infective material picked up on the body hairs and tarsi of house flies may survive only a short period; they will be subject to the cleaning behaviour of the fly, in which it seeks to rid itself of foreign material, and organisms exposed on the surface will be subject to desiccation, particularly in flight, and to UV sterilization in sunlight. Greater survival would be expected for organisms trapped in the mouthparts, such as between the lobes of the labellum. Dependent on the pathogen, there is a minimum number of infectious units needed to infect a human, but a lesser number of organisms deposited into a medium (e.g. milk) in which they can multiply may enable an infective dose to be ingested.

In northern Europe, the face fly, *M. autumnalis*, and in southern Europe *M. larvipara* may often be the most numerous flies worrying cattle in pasture.

M. autumnalis is also one of the most important livestock pests to invade the USA in recent times. Its introduction into North America from Europe was first detected in 1951 in Nova Scotia. From there it spread southward and, by 1959, there were many reports from cattle. It now occurs practically throughout the USA. Face flies are generally found around the eyes and nose of livestock, or on wounds where the females feed. The annoyance caused by the face flies results in cattle aggregating and bunching in the shade to escape, and contributes to reduced production rates. In the USA, *M. autumnalis* is an important vector of bovine keratoconjunctivitis caused by *Moraxella bovis*. Face flies are also intermediate hosts of *Parafilaria bovicola*, the causative agent of parafilariosis of cattle in northern Europe and elsewhere, and the irritation of the eye arising from their feeding can exacerbate the transmission of pinkeye and other conditions such as eyeworm. Adults are developmental hosts for *Thelazia* (Spirurida: Thelaziidae) nematodes, which live in the conjunctival sac and lachrymal ducts of cattle and horses, causing conjunctivitis, keratitis, photophobia and epiphora (this latter disease is an increasing problem in the USA).

M. sorbens in parts of Africa is thought to be the principal insect vector of *Chlamydia trachomatis*, the causative agent of trachoma, and *M. vetustissima* has been thought to be likewise involved in Australia.

In addition to dung feeding, adults of the bush fly, *M. vetustissima*, will attempt persistently to feed at the mouth, eyes and nose. As a result, they are of considerable significance as a nuisance pest in Australia for both livestock and humans. In the absence of native Australian dung beetles capable of disposing of the dung of introduced cattle, the dung of these herbivores is slow to decompose, allowing *M. vetustissima* ample opportunity to breed and reach large and problematic population densities. It has been estimated that 100–300 *M. vetustissima* per animal can be tolerated by cattle without adverse effect, but densities of 500–1000 and up to 5000 are found, which are considered to lower weight gains in beef cattle and milk yield in dairy herds.

Muscina flies are attracted to decaying organic material and are commonly found on corpses, urine and faeces. *M. stabulans* can carry several species of pathogens such as enterovirus and bacteria (*Proteus*, *Brucella*, staphylococci). Although larvae of *Draschia* and *Habronema*, the agents of equine habronemiasis, have been detected in *M. stabulans*, its role as vector of these nematodes has not yet been confirmed.

Hy. irritans are attracted by wounds, such as those incurred by fighting rams; swarms of flies around the head lead to intense irritation and annoyance, and can result in self-inflicted wounds. Secondary bacterial infection of wounds is common. There is circumstantial evidence for *Hy. irritans* playing a central role in transmitting summer mastitis pathogens.

7. PREVENTION AND CONTROL

Insecticide-impregnated ear tags, tail bands and halters, mainly containing pyrethroids, together with pour-on, spot-on and spray preparations are widely used to reduce fly annoyance in cattle and horses. Insecticides may also be incorporated in solid or liquid fly baits, using attractants such as various sugary syrups or hydrolysed yeast and animal proteins. Environmental preparations applied to walls and ceilings may prove beneficial but, in general, insecticides applied to the surface of manure heaps are not advised because they will serve to kill beneficial decomposers as well as the pest species. Various types of screens and electrocution grids for buildings are available to reduce fly nuisance. However, given the high rates of reproduction, high rates of dispersal and multiple generations per year, area-wide control of most of these flies is impractical. It should also be noted that overuse of insecticides has resulted in widespread resistance in many house fly populations. Hence, often the best methods of control are those aimed at improving sanitation and reducing breeding places (source reduction). For example, in stables and farms, manure should be removed or stacked in large heaps, where the heat of fermentation will kill the developing stages of flies, as well as the eggs and larvae of helminths.

Biological control agents, such as predators, parasites, competitors or pathogens, may also be used to control fly nuisance pests. For example, in poultry facilities the release of pupal parasites may reduce muscid fly populations successfully. In Australia, a range of dung beetles has been introduced as biological control agents to increase the rate of dung decomposition. One species, *Onthophagus gazella*, has proved to be highly effective, reducing bush fly populations successfully and spreading widely from centres of release. Entomopathogenic fungi and bacteria such as *Bacillus thuringiensis* could potentially become important biological agents in the control of many pests, including house flies.

Monitoring and surveillance of fly populations is also important, as it enables the right level and method of control to be tailored to the density of the fly population, promoting more rapid and efficient eradication. Computer models and forecasting systems can be used to predict the seasonal and temporal patterns of pest abundance and evaluation of the economic consequences resulting from outbreaks.

SELECTED BIBLIOGRAPHY

Ahmad, A., Nagaraja, T.G. and Zurek, L. (2007) Transmission of *Escherichia coli* O157:H7 to cattle by house flies. *Preventive Veterinary Medicine* 80, 74–81.

Allan, S.A., Day, J.F. and Edman, J.D. (1987) Visual ecology of biting flies. *Annual Review of Entomology* 32, 297–316.

Axtell, R.C. and Arends, J.J. (1990) Ecology and management of arthropod pests of poultry. *Annual Review of Entomology* 35, 101–126.

Ferrar, P. (1979) The immature stages of dung-breeding muscoid flies in Australia, with notes on the species, and keys to larvae and puparia. *Australian Journal of Zoology Supplementary Series* 73, 1–106.

Hogsette, J.A., Prichard, D.L. and Ruff, J.P. (1991) Economic effects of horn fly (Diptera: Muscidae) populations on beef cattle exposed to three pesticide treatment regimes. *Journal of Economic Entomology* 84, 1270–1274.

Howard, J.J. and Wall, R. (1996) Control of the house fly, *Musca domestica*, in livestock units: current techniques and future prospects. *Agricultural Zoology Reviews* 7, 247–265.

Krafsur, E.S. and Moon, R.D. (1997) Bionomics of the face fly, *Musca autumnalis*. *Annual Review of Entomology* 42, 503–523.

Madsen, M., Sorensen, G.H. and Nielsen, S.A. (1991) Studies on the possible role of cattle nuisance flies, especially *Hydrotaea irritans*, in the transmission of summer mastitis in Denmark. *Medical and Veterinary Entomology* 5, 421–429.

Meyer, J.A. and Mullens, B.A. (1988) Development of immature *Fannia* spp. (Diptera: Muscidae) at constant laboratory temperatures. *Journal of Medical Entomology* 25, 165–171.

Otranto, D., Tarsitano, E., Traversa, D., De Luca, F. and Giangaspero, A. (2003) Molecular epidemiological survey on the vectors of *Thelazia gulosa*, *Thelazia rhodesi* and *Thelazia skrjabini* (Spirurida: Thelaziidae). *Parasitology* 127, 365–373.

Roca, A.G., Zapater, M. and Toloza, A.C. (2009) Insecticide resistance of house fly, *Musca domestica* (L.) from Argentina. *Parasitology Research* 105, 489–493.

Traversa, D., Otranto, D., Iorio, R., Carluccio, A., Contri, A., Paoletti, B., *et al.* (2008) Identification of the intermediate hosts of *Habronema microstoma* and *Habronema muscae* under field conditions. *Medical and Veterinary Entomology* 22, 283–287.

Urech, R., Bright, R.L., Green, P.E., Brown, G.W., Hogsette, J.A., Skerman, A.G., *et al.* (2012) Temporal and spatial trends in adult nuisance fly populations at Australian cattle feedlots. *Australian Journal of Entomology* 51, 88–96.

Vriesekoop, F. and Shaw, R. (2010) The Australian bush fly (*Musca vetustissima*) as a potential vector in the transmission of foodborne pathogens at outdoor eateries. *Foodborne Pathogens and Disease* 7, 275–279.

Human Bot Fly (Torsalo)/New World Skin Bot Flies (Diptera: Oestridae: Cuterebrinae)

1. INTRODUCTION

The Cuterebrinae are generally dermal parasites of rodents and birds, with one important species, *Dermatobia hominis*, parasitizing wild and domestic animals, including cattle and humans. This is a Neotropical species, which is distributed from southern Mexico through Argentina and inhabits wooded areas along forest margins of river valleys and lowlands. It is variously known as Torsalo, the human bot fly or the American warble fly.

2. TAXONOMY

Within the calypterate Diptera, family Oestridae, the subfamily Cuterebrinae includes six genera and 83 species. The major species of medical and veterinary importance is *D. hominis*, although there are also many species of the genus *Cuterebra* that infest wild rodents and lagomorphs (and occasionally humans) in North America, and there is one species, *Alouattamyia baeri*, which parasitizes howler monkeys in Central and South America.

3. MORPHOLOGY

In cuterebrines, the postscutellum is undeveloped, the squamae large and the apical cell narrowed towards the margin. The adults of *D. hominis* (Fig. 46) are metallic in appearance, the short, broad abdomen having a bluish metallic sheen, but there are only vestigial mouthparts covered by a flap. The female measures approximately 12 mm in length. Adults have a yellow-orange head and

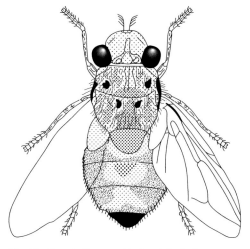

Fig. 46. Female *Dermatobia hominis*. Redrawn from James (1947).

legs and the thorax possesses a sparse covering of short setae. The arista of the antennae has setae on the outer side only.

The first-stage larva is subcylindrical, with small spines on segments 3 and 4 and stouter spines on segments 5–7, arranged in two dorsal rows and one ventral row. The second-stage larva is pyriform, with stout spines on the globular anterior portion and no spines on the narrower posterior part (Fig. 47a). The third-stage larva is elongate and ovate, measuring up to 25 mm in length, with prominent flower-like anterior spiracles, reduced spines and prominent mouth-hooks (Fig. 47b and c). The larval mouth-hooks are well developed and the posterior spiracles, which are deeply sunk, have three straight slits.

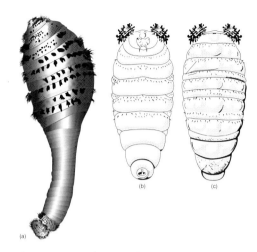

Fig. 47. Larvae of *Dermatobia hominis*. (a) Second-stage larva; (b) and (c) ventral and dorsal views of a mature third-stage larva. *Source*: James (1947).

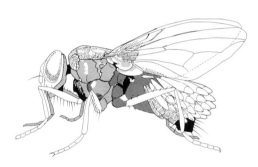

Fig. 48. *Sarcopromusca arcuata* with the eggs of *Dermatobia hominis* glued to its abdomen (length of egg 1 mm). *Source*: Catts (1982). Reproduced by permission of Annual Reviews Inc.

4. LIFE CYCLE

A female *D. hominis* produces 800–1000 eggs over its lifetime and attaches them in clusters of 5–40 to the abdomen of a zoophilic 'carrier' insect, usually a dipteran that is likely to visit a warm-blooded host (Fig. 48). Development of the egg requires at least 4–5 days but egg hatch is delayed until the stimulus of a sudden increase in temperature, which would occur when the carrier insect visits a potential warm-blooded host. The larva

transfers to the host and either enters through the feeding puncture made by the carrier, through a hair follicle or by penetrating soft, unbroken skin, which it can do in 5–10 min. Each larva penetrates individually and a boil-like swelling develops around it. The larva feeds on tissue exudate and grows slowly, requiring 4–18 weeks to complete development. The swelling has an opening through which the larva respires and secondary infections may occur. Often, the discharge from the opening has a foetid odour, which may attract other myiasis-causing flies.

The post-feeding larvae emerge from the host in the early morning and burrow into the soil, where they pupate. The pupal stage is long, taking 4–11 weeks. Females mate 24 h after emergence. The adults do not feed and are comparatively inactive, living 1–9 days.

5. BEHAVIOUR AND BIONOMICS

D. hominis is endemic to the Neotropical region, where it occurs from 18°S to 25°N, being associated with moist, cool, tropical highlands between 160 and 2000 m above sea level, especially the coffee growing areas between 600 and 1000 m above sea level. It causes cutaneous myiasis in a wide range of mammalian hosts, including humans, and is particularly important as a parasite of cattle. It has also been reported from chickens, turkeys and toucans.

D. hominis has a unique method of ensuring that its offspring reach a range of hosts. The female captures other insects to act as carriers of its eggs. These are glued carefully on to the carrier's abdomen in such a way as not to affect its flight efficiency adversely (Fig. 48). There is no evidence that eggs are laid on plants or directly on hosts in the field. The eggs hatch when the arthropod carrier feeds on the vertebrate host. Nearly 50 species of carriers have been reported, of which about half are mosquitoes and a third are muscoids (Anthomyiidae). In Costa Rica, five carrier species were reported to be used, of which the most important was the muscid *Sarcopromusca arcuata*, which carried an average load of 28 eggs, but fewer eggs (6–10) were laid on mosquitoes.

6. MEDICAL AND VETERINARY IMPORTANCE

The economic effects of *D. hominis* are particularly important in domestic cattle, with losses to livestock productions estimated at US$260 million in the early 1970s as a result of reduced calf growth and weight gains, lowered milk yields and hide damage. Infestations in susceptible animals can exceed 1000 warbles per host and result in its death. Zebu cattle appear to be relatively resistant, but Holstein and Brown Swiss breeds are highly susceptible. Sheep may become heavily infested and develop severe abscesses. Dogs are frequently attacked; cats and rabbits are less frequently attacked and equids troubled little.

Humans can also be hosts to *D. hominis* and suffer from painful, discharging, cutaneous swellings on the body. Larvae are able to penetrate clothing, and boils may be found on all parts of the body. When the larva is removed, in the absence of secondary infection, the condition clears spontaneously in about a week. Rarely, a larva infesting the scalp may penetrate the skull into the brain, with fatal results. This rare condition has been reported in a 5-month-old baby and an 18-month-old child, presumably facilitated by the incomplete ossification of the skull at that age. The long period of development of larvae in the host favours the introduction of *D. hominis* into other parts of the world. Travellers from South America have introduced *D. hominis* into various countries, including in Europe, Canada and Australia and, given its wide host range, if suitable conditions exist it is conceivable that *D. hominis* might establish itself in other parts of the tropics.

7. PREVENTION AND CONTROL

Control of adult *D. hominis* is difficult and protection from larval infestation can only be effected by prevention of bites by carrier insects such as mosquitoes. For human infestations by *D. hominis* larvae, expression should be avoided as it is painful and usually unsuccessful and surgical removal of larvae may be required, although occlusion of the breathing hole with a physical barrier (e.g. petroleum jelly) has been shown to make the larva emerge far enough to be removed with forceps. Great care should be taken during the extraction process to avoid rupturing the larva *in situ*, and an injection of lidocaine (1%) can be used to paralyse the larva and facilitate its removal. Oral treatment with the macrocyclic lactone ivermectin is not advised, as it may kill the larva *in situ* and initiate an inflammatory reaction, whereas topical ivermectin may be helpful (although there is a similar concern that the larvae may be killed trapped within the skin). Antibiotics should be prescribed if secondary infection is present or likely. Antigenic proteins of *D. hominis* have been characterized to assess their utility for a potential vaccine against dermatobiosis in cattle.

SELECTED BIBLIOGRAPHY

Baron, R.W. and Colwell, D.D. (1991) Mammalian immune responses to myiasis. *Parasitology Today* 7, 353–355.

Brant, M.P.R., Guimaraes, S., Souza-Neto, J.A., Ribolla, P.E.M. and Oliveira-Sequeira, T.C.G. (2010) Characterization of the excretory/secretory products of *Dermatobia hominis* larvae, the human bot fly. *Veterinary Parasitology* 168, 304–311.

Catts, E.P. (1982) Biology of the New World bot flies: Cuterebridae. *Annual Review of Entomology* 27, 313–338.

Colwell, D.D., Hall, M.J.R. and Scholl, P.J. (2006) *Oestrid Flies: Biology, Host–Parasite Relationships, Impact and Management*. CAB International, Wallingford, UK.

Fernandes, N.L., Zanata, S.M., Rönnau, M., Soccolm, C.R., Pandey, A. and Thomaz-Soccol, V. (2012) Production of potential vaccine against *Dermatobia hominis* for cattle. *Applied Biochemistry and Biotechnology* 167, 412–424.

Francesconi, F. and Lupi, O. (2012) Myiasis. *Clinical Microbiology Reviews* 25, 79–105.

Gomes, C.C.G., Trigo, J.R. and Eiras, A.E. (2008) Sex pheromone of the American warble fly, *Dermatobia hominis*: the role of cuticular hydrocarbons. *Journal of Chemical Ecology* 34, 636–646.

Gordon, P.M., Hepburn, N.C., Williams, A.E. and Bunney, M.H. (1995) Cutaneous myiasis due to *Dermatobia hominis*: a report of six cases. *British Journal of Dermatology* 132, 811–814.

Hall, M.J.R. and Smith, K.G.V. (1993) Diptera causing myiasis in man. In: Lane, R.P. and Crosskey, R.W. (eds) *Medical Insects and Arachnids*. Chapman and Hall, London, pp. 429–469.

Hall, M.J.R. and Wall, R. (1994) Myiasis of humans and domestic animals. *Advances in Parasitology* 35, 258–334.

Rossi, M.A. and Zucoloto, S. (1973) Fatal cerebral myiasis by the tropical warble fly, *Dermatobia hominis*. *American Journal of Tropical Medicine and Hygiene* 22, 267–269.

Sabrosky, C.W. (1986) *North American Species of Cuterebra, The Rabbit and Rodent Bot Flies (Diptera: Cuterebridae)*. Thomas Say Publication, Entomological Society of America, College Park, Maryland.

Sancho, E. (1988) *Dermatobia*, the Neotropical warble fly. *Parasitology Today* 4, 242–246.

Keds and Louse Flies (Diptera: Hippoboscidae)

1. INTRODUCTION

The Hippoboscidae are unusual bloodsucking ectoparasites of birds and mammals, probably related to the bloodsucking muscids or the tsetse flies. There are two related dipteran families (sometimes considered to be subfamilies of the Hippoboscidae), the Nycteribiidae and the Streblidae, which are bloodsucking ectoparasites of bats, but these are of little to no medical or wider veterinary significance.

The Hippoboscids are parasites of veterinary importance in many parts of the world, particularly on small ruminants and horses, where their biting and blood feeding may be extremely damaging, but they are of no medical importance. Although about 80% of hippoboscid species occur on birds, most domesticated birds, such as poultry, turkeys, ducks or geese, are free from these bloodsucking parasites. One exception is the domestic pigeon, which may be infested with *Pseudolynchia canariensis*. Keds are unusual because the female adult is viviparous; it retains the larva within a modified common oviduct until it is fully grown and deposits it as an immobile final-stage larva which pupates *in situ*.

The most important species is the sheep ked, *Melophagus ovinus*, which is originally Palaearctic in its distribution but which has spread with sheep widely throughout the world; it has established itself in temperate countries and in the cooler highlands of the tropics but is absent from the hot, humid tropics.

2. TAXONOMY

About 200 species are recognized in this family and arranged in three subfamilies – Ornithomyinae, Melophaginae (Lipopteninae)

and Hippoboscinae. The Ornithomyinae is the largest with over 150 species, mostly parasites of birds, but it also includes five species parasitic on macropods (kangaroos and wallabies) and one species on lemurs in Madagascar. The Melophaginae contains about 30 species parasitic on bovids and cervids, including the economically important sheep ked, *M. ovinus*. The Hippoboscinae contains eight species, of which six parasitize equines and ruminants, mainly bovids; one species, *Hippobosca longipennis*, parasitizes carnivores and another species is found only on ostriches.

The three main genera which occur on mammals are readily distinguished. Adult *Melophagus* are wingless (Fig. 49a), the wings being reduced to tiny, veinless, opaque knobs, and there are no halteres. In *Lipoptena*, the newly emerged fly has fully developed and functional wings, which break off close to the base after the final host is reached; they are said to be caducous. Adult *Hippobosca* are permanently winged (Fig. 49b). They are distinguished from other winged genera by the pronotum being large and clearly visible, forming an easily observable neck-like segment between the scutum and the head. In all mammal-infesting hippoboscids, the paired claws are simple, while in the majority of bird-infesting species, the claws have two separate teeth (with the basal lobe distinguished by being unevenly pigmented and sclerotized, and not pointed).

3. MORPHOLOGY

Keds appear leathery and dorsoventrally flattened, with extended mouthparts and robust legs ending in large, recurved claws. The dorsoventrally flattened body gives them a superficial resemblance to lice. Adults range in

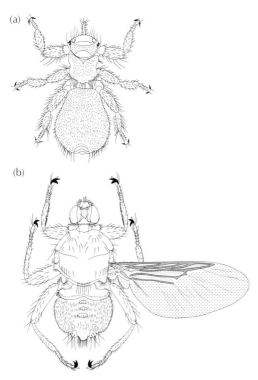

Fig. 49. (a) *Melophagus ovinus*, the sheep ked; (b) *Hippobosca equina. Source*: (b) based on Fig. 6 in Ma (1963).

size from 1.5 to 12 mm and have a soft, flexible abdominal integument. This flexibility allows for stretching and distension of the female's abdomen during feeding and when the larva is developing. The hippoboscids, in general, have relatively robust legs, consisting of an enlarged femur, a flattened tibia and short compact tarsi, with one or more basal teeth. Species that parasitize mammals can be distinguished from those that parasitize birds by their shorter and stouter legs and heavier tarsal claws.

Where present, the adult forewings are generally long and broad and the hindwings replaced by the characteristic halteres possessed by Diptera. Similar to tsetse flies, when at rest Hippoboscidae have forewings that lie flat over each other. The anterior veins of the wings are strongly developed. In the genera *Lipoptena* and *Neolipoptena*, the

adults emerge with fully developed functioning wings and are referred to as *volants*. Once the host has been located, these wings break off at their base. After the first blood meal, the flight tissues begin to break down and the leg muscles grow and strengthen. One particularly specialized genus, *Melophagus*, has both reduced wings and halteres that are non-functional. Their rudimentary, non-functioning flight muscles later atrophy. Genera that have functional wings also have well-developed compound eyes, and those that have non-functional wings or lose their wings once they have located a host have significantly reduced eyes.

The antennae are apparently immovable and placed far forwards in a deep antennal pit. The second antennal segment forms the greater part of the antenna and houses the greatly reduced third antennal segment in a ventral recess from which the arista protrudes. The Hippoboscidae are similar to *Glossina* in the structure of their mouthparts, method of reproduction and the possession of a bacteriome in the midgut. All species have mouthparts that are projected outwards, rather than downwards. They have a heavily sclerotized proboscis, with a piercing labium and labella armed with teeth; the bulb of the haustellum is withdrawn into the head and the narrow terminal portion concealed in grooves on the inner side of the palps.

4. LIFE CYCLE

The life cycle of the ked has been studied in several parts of the world with substantially the same conclusions. The newly emerged female *M. ovinus* mates within 24 h of emergence, but the ovaries have to mature before an egg is available for fertilization. This process takes 6–7 days and further development within the female takes an additional 7 days, so the first fully developed larva is deposited when the female is 13–14 days old. Thereafter, additional larvae are deposited every 7–8 days and, in a lifetime of 4–5 months, a female will produce about 15 larvae, a comparatively slow rate of increase for an insect.

The deposited larva pupates within 6 h and the duration of the pupal stage is 20–26 days.

The cycle from newly emerged adult female to the emergence of an adult of the next generation is 5 weeks. *M. ovinus* is a permanent ectoparasite on a homiothermic host and therefore living under very constant conditions, which accounts for the narrow range in the durations of the different stages.

Pupae develop over a relatively narrow range of temperature (25–34°C), with optimal development at 30°C. The puparia are glued to the fleece and carried away from the skin as the fleece grows. The temperature at the skin surface will be about 37°C and, near the surface of the fleece, it will be nearer to air temperature. Larviposition usually occurs in areas of the fleece where a suitable temperature will be found during the 3-week development period of the pupa. This is found most easily in the neck region, where the wool staple lies parallel to the skin and temperature varies slowly with increasing wool length over time. In studies of yearling sheep, over 50% of puparia were found in the neck region, while nearly 60% of the adults were found on the region of the forelegs and flanks. On lambs, puparia are concentrated on the hindlegs, neck and belly, although substantial numbers of adults may be found on the flanks and forelegs.

5. BEHAVIOUR AND BIONOMICS

5.1 *Melophagus ovinus*

Populations of *M. ovinus* show seasonal changes and, at the same time of the year, different levels of infestation on sheep of different ages. Populations of *M. ovinus* are at their highest in winter and lowest in summer.

At the start of the year in the northern hemisphere, there is an increase in keds on both young and 2- and 3-year-old ewes, but with substantially higher infestation (× 5–6) on the younger animals. Later, when the sheep are penned for lambing, there is a rapid rise in infestation on ewes due to transfer from the younger animals as a result of their close association. After lambing (late April), the ked population on both the young animals and ewes decreases sharply, coinciding with a rapid rise in infestation on lambs. This continues until shearing in late June, when puparia and keds may be removed with the fleece, and reductions of 70–98% have been reported; in one study in summer, the ked population on lambs decreased by 35–69%, but densities of keds still ranged from 36 to 66 adults/lamb.

Transfer of keds from one sheep to another occurs when they move to the surface of the fleece in response to temperature. When the air temperature was 15°C only, a small number of keds (average four) were on the surface of the fleece, but when the air temperature increased to 23°C, the number of keds on the surface increased substantially (average 98). However, keds are vulnerable when they are on the surface of the fleece. They may be dislodged and fall to the ground, where they will survive only 2–5 days, and are unlikely to find another host unless sheep are densely crowded. Additionally, keds on the surface of the fleece are subject to predation by birds such as magpies and starlings. They are also ingested by sheep biting their fleece, and this is probably the route by which sheep become infected with the benign trypanosome, *Trypanosoma melophagium*.

Several factors contribute to the fluctuations in natural populations of *M. ovinus*. It has been stated that only newly emerged keds go on lambs and that the older keds stay on the ewes and suffer considerable mortality from infection with *T. melophagium*, but the latter conclusion has been disputed. Undoubtedly, older animals have fewer keds and in a study in Wyoming, USA, numbers increased on lambs from March to May and then a natural decrease occurred in unshorn lambs until September, after which numbers increased to reach a maximum in February before declining until May.

Yearlings, which have lambs in March, have similar intensities of keds in the following summer and autumn as older ewes. Two factors contribute to this: temperature and the development of resistance in individual sheep. There are at least two factors in the development of resistance. There is a long-lasting cutaneous arteriolar vasoconstriction, which reduces the ability of keds to obtain blood, so they die from starvation, and a need for adequate amounts of vitamin A in the diet.

The fact that *M. ovinus* is more abundant in winter than summer and has not established itself in tropical areas suggests that temperature may play an important part in the decline of ked populations in summer.

5.2 Other hippoboscids of veterinary importance

- *Hippobosca longipennis* occurs in the western Oriental, southern Palaearctic and Afrotropical regions, excluding West Africa. It parasitizes carnivores, including domestic dogs.
- *Hippobosca equina* is primarily an ectoparasite of horses and cattle in the Palaearctic and western Oriental regions, but has been introduced and become more widely established in southern Europe, South-east Asia and in some island groups in the Pacific.
- *Hippobosca variegata* parasitizes equines and cattle in the Afrotropical and Oriental regions. No wild hosts of this species are known.
- *Hippobosca rufipes* occurs in the Afrotropical region, where it parasitizes wild bovids and domestic cattle and, less frequently, domestic and wild equines.
- *Hippobosca camelina* occurs where camels are present in the northern part of eastern Africa, the Mediterranean region and the southern part of the eastern Palaearctic.
- *Lipoptena capreoli* parasitizes domestic goats in the eastern Mediterranean region and eastwards through the desert countries to north-west India.
- *Lipoptena cervi*, the deer ked, affects red deer, roe deer, elk and sika deer in Europe, Siberia and northern China, white-tailed deer, elk, horses and cattle in North America and goats in Europe. *Lipoptena* spp. can also attack humans.

Species of *Hippobosca* are generally most abundant in summer and occur in low numbers during winter. Larvae are deposited off the host: those of *H. equina* in crevices in mud walls of stables; and, in keeping with this, *H. equina* is more abundant on stabled animals than free-ranging animals. The newly deposited larva is creamy in colour, with its flattened posterior end bearing dark spiracular plates. The larva becomes a pupa in 4–6 h. The puparium darkens rapidly to a dark red-black colour. It is broadly oval with posterolateral spiracular lobes. The adults are winged and fly directly to a host. The newly emerged adult does not feed for the first 24 h but thereafter feeds frequently, several times a day in the case of *H. longipennis*. The longevity of species of *Hippobosca* is about 6 weeks in summer and 8–9 weeks in winter, with females living slightly longer than males. Adult *H. equina* aggregate on the host in areas where the skin is thinner and comparatively hairless. On horses, most aggregate under the tail and around the genitalia, and on cows it is under the tail and on the udder. On buffalo, most *H. equina* are found on the genitalia and inner thighs. The average density of *H. equina* on horses has been found to be six to ten times that on cows and buffaloes. Newly emerged adults take several days (4–11) to become sexually mature.

6. MEDICAL AND VETERINARY IMPORTANCE

From a medical perspective, no hippoboscids use humans as a host, although various species, such as the sheep ked (*M. ovinus*) and those from pigeons, deer and macropods, will bite and occasionally feed on humans, reportedly causing mild pain, erythema, swelling and subsequent itching. In general, the bat flies (nycteribiids and streblids) have little contact with hosts other than bats, but there are reports that some streblids will bite humans.

With stock animals, since keds suck blood, heavy infections may lead to loss of condition and anaemia. Inflammation leads to pruritus, biting, rubbing, wool loss and a vertical ridging of the skin known as 'cockle'. Experiments with ked-free and ked-infested lambs showed that on a diet of lucerne, ked-free lambs gained 3.6 kg more in 4 months and produced 13% more wool. *M. ovinus* is also responsible for an allergic dermatitis in sheep, characterized by small nodules on the grain layer of the skin, reduced weight gain and darkened patches at the affected site. They are spread by contact, and long-wooled breeds appear to be particularly susceptible. If a sheep eats a ked, it may become infected with the metacyclic stages of a non-pathogenic trypanosoma (*T. melophagium*), which penetrates the buccal mucosa.

7. PREVENTION AND CONTROL

M. ovinus is the only species within the Hippoboscidae for which extensive control strategies are implemented. Insecticides are typically applied to sheep after shearing and before lambing. Commonly used insecticides include organophosphates and pyrethroids (applied externally) and macrocyclic lactones (applied orally). For the most effective results, the insecticide should be applied several weeks after shearing, so that the coat is of sufficient length to hold the chemical product. It is important that the insecticide used persists long enough to kill adults emerging from retained puparia.

SELECTED BIBLIOGRAPHY

Heath, A.C.G. and Bishop, D.M. (1988) Evaluation of 2 pour-on insecticides against the sheep-biting louse, *Bovicola ovis* and the sheep ked, *Melophagus ovinus*. *New Zealand Journal of Agricultural Research* 31, 9–12.

Nelson, W.A. and Slen, S.B. (1968) Weight gains and wool growth in sheep infested with sheep ked *Melophagus ovinus*. *Experimental Parasitology* 22, 223–226.

Paakkonen, T., Mustonen, A.M., Roininen, H., Niemela, P., Ruusila, V. and Nieminen, P. (2010) Parasitism of the deer ked, *Lipoptena cervi*, on the moose, *Alces alces*, in eastern Finland. *Medical and Veterinary Entomology* 24, 411–417.

Small, R.W. (2005) A review of *Melophagus ovinus* (L.), the sheep ked. *Veterinary Parasitology* 130, 141–155.

Strickman, D., Lloyd, J.E. and Kumar, R. (1984) Relocation of hosts by the sheep ked (Diptera, Hippoboscidae). *Journal of Economic Entomology* 77, 437–439.

Trout, R.T., Steelman, C.D. and Szalanski, A.L. (2010) Phylogenetics and population genetics of the louse fly, *Lipoptena mazamae*, from Arkansas, USA. *Medical and Veterinary Entomology* 24, 258–265.

Kissing Bugs (Hemiptera: Reduviidae)

1. INTRODUCTION

Most of the 25 subfamilies of Reduviidae (6800 species), known generally as assassin bugs, are predacious on other arthropods. One subfamily, the Triatominae, known generally as kissing bugs, contains species that are bloodsucking on mammals and birds and some are vectors of the protozoan *Trypanosoma cruzi*, the aetiological agent of Chagas disease.

2. TAXONOMY

There are at least 140 species of Triatominae, classified into 17 genera. The Triatominae are largely confined to the western hemisphere, although there is one tropicopolitan species (*Triatoma rubrofasciata*), a group of seven species of *Triatoma* in southern China, Southeast Asia and north-western Australia and a single genus (*Linshcosteus*) with six described species in India. In the New World, the triatomines are found from approximately 42°N to 42°S, with their greatest diversity occurring in South and Central America and only five species extending north from Mexico into the Nearctic region.

3. MORPHOLOGY

Adult triatomines are generally large insects with broad abdomens, commonly measuring 20–28 mm in length and 8–10 mm in width (although the smallest is only 5 mm and the largest is 5 cm in length). They have long, thin, four-segmented antennae (Fig. 50). The compound eyes are placed laterally and the ocelli are located dorsally behind the eyes (Fig. 51). In front of the eyes, the head is narrowed and forwardly produced (Fig. 50). The rostrum (labium) is three-segmented and straight, not arched (Fig. 51), and extends back to the prosternal stridulatory groove. The forewings are hemelytra, with a sclerotized basal area (corium and clavus) and a distal membranous portion (Fig. 52). The hindwings are entirely membranous and, in repose, are folded beneath the hemelytra. In the unfed state, the abdomen is almost flat and the hinged dorsal and ventral plates are close together and almost parallel. While feeding, the abdomen becomes greatly distended and the plates rotate on the hinge and become widely separated. In some triatomines, for example *Rhodnius* spp., abdominal expansion is further increased by the unfolding of an additional longitudinally pleated membrane.

4. LIFE CYCLE

Oviposition follows a circadian rhythm in *Triatoma infestans* and *Panstrongylus megistus*, with eggs being deposited early in the scotophase. The eggs are oval, about 2.5 mm in length, with an obvious operculum at one end. Eggs of *Rhodnius prolixus* turn a bright red within a few hours of being deposited; those of other species are pearly white when laid and become red as the embryo develops. Eggs are laid in cracks and crevices in houses and, in wild populations of *R. prolixus*, eggs may be glued on to palm fronds. A female will produce 200–300 eggs in her lifetime, laid in a number of batches. In *T. infestans*, peak fecundity is reached 11 weeks after the first oviposition.

Eggs of *R. prolixus* hatch over the range of 16–34°C, with the highest fertility (82%)

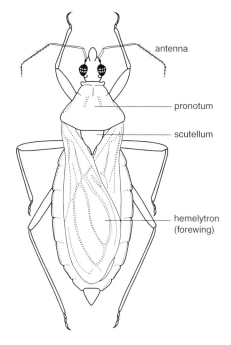

Fig. 50. Female *Panstrongylus megistus.* Redrawn from Schofield and Dolling (1993).

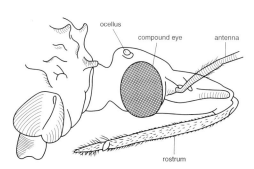

Fig. 51. Lateral view of head of *Panstrongylus megistus.* Redrawn from Schofield and Dolling (1993).

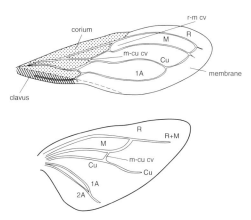

Fig. 52. Wings of a triatomine bug. Above: hemelytron; below: hindwing.

The nymph emerges by pushing off the operculum from the egg. It is a miniature version of the adult except for the absence of wings; there are five nymphal stages, all of which feed on blood (Fig. 53). If a nymph engorges fully, only one blood meal is required before moulting to the next stage. The first-stage nymph of *R. prolixus* imbibes 12 times its own body weight of blood, and the other four nymphal stages take six times their body weight, while adults take in only 1.5 times their body weight. During development, the nymph grows at each moult, from 2.7 mm in the first instar to 17 mm in the fifth instar. In mass rearing at 28°C and 60–70% RH (relative humidity) and fed every 2 weeks on hens, the life cycle of *R. prolixus* from egg to adult took 80 days, with 42% survival. About 18% of individuals in each instar either died or failed to feed. At 26–27°C and 60% RH, the life cycle of *T. infestans* takes about 23 weeks, with the fifth instar being the longest (7 weeks) and the fourth instar the shortest (11 days).

Significantly, more female *T. infestans* reach the adult stage than males, but the males are longer-lived, averaging 26 weeks compared with 16 weeks for females. Assuming a 1:1 ratio of males to females at the egg stage, then about half the female eggs give rise to adults. *R. prolixus* and *T. infestans* have the shortest life cycles among nine species of Triatominae, and *Triatoma dimidiata* and *P. megistus* take about twice as long to complete development under the same

occurring at 21–32°C. Fertility is also related to humidity. At 25–27°C all eggs hatch at humidity of 50% or higher, but at 32°C a humidity of 80% is required; eggs are resistant to desiccation; 50% of the eggs hatch at 20% humidity and 25°C. The egg stage is relatively long, taking 12 days at 32°C and nearly a month at 22°C in *R. prolixus*, and about 3 weeks in *T. infestans* at 26–27°C with 86% fertility.

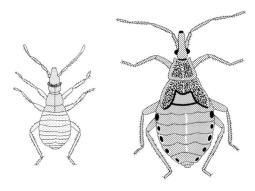

Fig. 53. Early stages of *Triatoma rubrofasciata*. First-stage nymph (left); third-stage nymph (right). Redrawn from Patton, W.S. and Evans, A.M. (1929) *Insects, Ticks, Mites and Venomous Animals. Part I: Medical.* H.R. Grubb, Croydon, UK.

conditions. A generation time of 31 weeks for *T. infestans* has been calculated, with an intrinsic rate of increase of 10% per week.

5. BEHAVIOUR AND BIONOMICS

Triatomines can be classified in five grades of increasing association with humans. Grade one includes sylvatic species with specialized hosts and habitats (e.g. *Psammolestes* spp.). Grade two species are fairly specialized but adults are occasionally found around houses (e.g. *Triatoma protracta*). Grade three species are more usually found in a sylvatic ecotope, but occasionally colonize houses (e.g. *Rhodnius ecuadoriensis*). Grade four species are commonly domestic in chicken houses and animal shelters but still retain a sylvatic ecotope (e.g. *Triatoma sordida*). This grade includes several locally important vectors of Chagas disease. Grade five species are highly adapted to the domestic environment and include species acting as the major vectors of Chagas disease (e.g. *R. prolixus*, *T. infestans* and *P. megistus*).

Most triatomines are nocturnal and feed on sleeping inhabitants in houses. For that reason, they are most abundant in sleeping areas, which may contain 50% of the total population of *R. prolixus* in a house. The bugs are

attracted to a host by warmth, carbon dioxide and odour, detected in *T. infestans* by sensory receptors on the antennae. Orientation on the host and probing involve additional receptors on the tarsi and labium.

In feeding, the mandibles pierce the skin and then cease to move. They are barbed and hold the insect in position while feeding. The flexible maxillae penetrate deeply and move until they contact a capillary. The right maxilla is longer than the left and penetrates into the capillary; the left maxilla may enter the capillary or remain closely attached on the outside because the junction between the two maxillae is the effective functional mouth.

During feeding, saliva containing an anticoagulant is injected and blood is pumped into the midgut, the first section of which acts as a storage organ. This section of the midgut contains symbionts that are important in the successful development of a triatomine. In their absence, development stops in the fourth or fifth instar. During development, the nymph acquires its symbionts by probing the contaminated surface of the egg from which it has emerged or from excreta of other members of the species.

The bites of domesticated triatomine bugs are relatively painless. On repeated exposure, some individuals may have a delayed reaction 24–48 h after being bitten and, rarely, a more severe response. It is possible that longer periods of exposure would produce an immediate response.

Triatomines feed on a wide range of hosts; domestic species feed on humans, dogs, cats, chickens and rodents infesting the house. They feed on more than one host, particularly if their feeding has been interrupted, and will often feed again before the previous meal has been completely digested. More than one-third of a population of *T. dimidiata* was shown to have fed on more than one host, and the gut contents of one adult reacted positively to antisera of six hosts – human, dog, cat, mouse, cow and opossum. In Argentina, the biting rate of *T. infestans* is controlled by temperature during the cool season and by population density in the hot season.

The time to take a full blood meal varies with the instar and size of the insect. Among five species, the time taken to engorge varied from 3 to 30 min, with *T. protracta* being the slowest feeder. *R. prolixus* and *T. infestans* feed faster and with fewer interruptions than *T. dimidiata*. Male and female *T. dimidiata* ingest over 40% and 50% of their body weight, respectively, at each feed. Fifth instar nymphs of *P. megistus* take twice as much blood as the same stage of *R. prolixus* (600 versus 300 mg).

When a heavily infested rural house in Venezuela was demolished, nearly 8000 *R. prolixus* of all stages were recovered. The information was analysed in great detail and it was calculated that in that house the feeding rate was 58 *R. prolixus* per person per day, and in 13 other houses it was nine, ranging from 0.2 to 33; the loss of blood per person per month in the 13 houses ranged from 0.7 to 40 ml; and in the heavily infested house exceeded 100 ml.

Infection with *T. cruzi* depends on the presence of infective forms in the faeces of the bug, which are deposited on the host and gain access via a wound or moist mucosa. Therefore, the earlier a bug defecates during or after feeding, the more likely is that the same species acts as an efficient vector. *R. prolixus* and *T. infestans* defecate within 10 min of finishing their blood meal (therefore the insect is still likely to be on the host while defecating) and 8% of *R. prolixus* actually defecate during the blood meal. Only two-thirds of *T. dimidiata* were found to defecate within 10 min of finishing a blood meal but, possibly because of the longer feeding time, 13% defecated during the blood meal.

The bugs are sensitive to high temperature and low humidity. Their rehydration is dependent on ingesting a blood meal; they seek refugia where they are likely to be less stressed and where blood is likely to be readily available. In South America, many species occur in the crowns of palms; however, these populations are small (5–70/tree) because blood meals depend on visiting birds and mammals, and natural predators are present. Other biotopes occupied by triatomines include bromeliads, under the bark of trees, in hollow trees and fence posts and in ground burrows. *R. prolixus* occurs in both palms and houses, and in the latter occurs equally in the walls and roof, particularly in palm thatch roofs. *T.*

infestans and *P. megistus* infest cracks in unplastered mud and cane walls and are less common in mud-brick. *T. dimidiata* occurs lower down than the other two species, rarely being more than 1 m above the ground, and in houses raised on supports it occurs on the ground or on the foundations that support the floor. Here, they feed on animals that shelter under the house and are able to enter the house at night through cracks in the floor and feed on its sleeping occupants. The bugs congregate near a food source, hence the greater concentration in the walls of bedrooms. In *T. infestans* and *R. prolixus*, this aggregation is facilitated by the production of a pheromone in the faeces which attracts the unfed nymphs of both species. The locomotory activity of fed nymphs is arrested in the presence of faecal material, which leads to their congregation in the same area.

Wild *R. prolixus* disperse mainly as an ectoparasite of birds, and this is considered to be the means by which it has spread to Mexico. *T. infestans*, the most domestic triatomine, is considered to have originated as a parasite of wild guinea pigs in the Cochabamba Valley in the Bolivian Andes and, in pre-Columbian times when guinea pigs were domesticated, *T. infestans* spread with them; it is now the most widespread domestic species occurring from the temperate areas of southern Argentina to the dry tropics of north-eastern Brazil.

Houses are invaded by bugs being introduced in palm fronds, in firewood and in household articles. Nearly one-third of *R. prolixus* released into a house found shelter in household articles. In one study, labelled *R. prolixus* dispersed less than 4 m in houses and less than 15 m outside over a period of 40 days. Adult *R. prolixus* appear to fly very little, and movement between palm trees and houses occurred only when the two habitats were close together. However, in some studies, nymphs and adults of *R. prolixus* have been found to disperse 100–500 m, while *P. megistus* moved 400 m from a natural to an artificial biotope.

Triatomines are attracted to light, which favours their establishment in houses. Movements between natural and artificial biotopes can occur. For instance, the range of blood meals identified in *T. dimidiata* in natural and artificial biotopes indicated free movement, with 22% of the bugs in natural biotopes and some of the bugs found in houses containing opossum blood.

In the laboratory, triatomines are long-lived and able to withstand long periods of starvation. The greatest resistance was shown by fourth and fifth instar nymphs, which in *T. dimidiata* survived 6 months unfed, whereas adults survived only 4–5 months. Similar conditions were observed in nymphs of *T. infestans* and *R. prolixus*. In the field, survival of *R. prolixus* was considerably lower (weeks rather than months), and predation by domestic fowl, rodents and other animals seemed to be the main cause of early mortality. Perhaps as a defence against predation, nymphs of *T. dimidiata* camouflage themselves by covering their bodies with debris.

6. MEDICAL AND VETERINARY IMPORTANCE

The saliva injected with the bites of triatomines contains proteins that can induce mild to severe allergic responses, including ana-phylactic reactions that can be fatal. However, the bugs are of greatest medical importance as vectors of Chagas disease caused by the protozoan parasite, *T. cruzi*. Natural infections with *T. cruzi* have been found in 65 species of Triatominae, belonging to eight genera and including 41 out of 68 species of *Triatoma*, 8 of 12 species of *Rhodnius* and 10 of 13 species of *Panstrongylus*. The main domestic vectors of *T. cruzi* are *T. infestans*, *T. dimidiata*, *Triatoma brasiliensis*, *P. megistus* and *R. prolixus*.

6.1 Chagas disease (*Trypanosoma* (*Schizotrypanum*) *cruzi*)

Chagas disease is most prevalent in Argentina, Brazil and Venezuela, but it occurs in 23 countries in Latin America. Towards the end of the last century, it was estimated that there were 24 million seropositive people, of which 10–30% would develop clinical disease, resulting in 70,000 deaths per annum. *T. cruzi* has a broad host range, being found in more than 100 species of mammals of 24 families, including marsupials and six orders of eutherian mammals. It is found in wild animals (and bugs) in parts of the southern USA but human cases are rare north of Mexico. Although *T. cruzi* is maintained in urban communities, disease is most common in rural areas.

In the vertebrate host, slender and stumpy trypomastigotes are found in the circulating blood. It is believed that it is the slender forms that penetrate the host's cells, where they develop into oval amastigotes and multiply, forming pseudocysts which, after 4–5 days, burst and release trypomastigotes into the bloodstream, from which they invade other cells. When a triatomine bug ingests an infected blood meal, the trypomastigotes differentiate into epimastigotes in the midgut and later adhere to the walls of the rectum, where they develop into metacyclic trypo-mastigotes that are infective to a vertebrate host. The cycle in the bug takes about 20 days, after which the bug remains infective for life and may transmit *T. cruzi* for several years. Transmission to the vertebrate is effected by metacyclic stages being deposited in excretory material on the skin of the host and entering through bite wounds or by transfer to mucous membranes.

The acute phase of Chagas disease is often asymptomatic, with clinical disease usually occurring in children. This phase lasts 4–8 weeks, followed by an indeterminate phase in which the infected individual shows no clinical signs. This may continue indefinitely; however, about 10–30% of infections develop into chronic disease, which may become noticeable up to 30 years or longer after the initial infection. In the chronic phase, patients may develop myocarditis or marked dilation of the oesophagus or colon.

The vectors of *T. cruzi* differ in their ecological requirements and geographical distributions. *T. dimidiata* is found in dry areas where the climate is not too warm and occurs in Mexico, Central America, Colombia and Ecuador. *T. infestans* occurs in warmer but equally dry habitats in Argentina, Paraguay, Bolivia, Chile and parts of Peru and Brazil; it is the most domestic of the triatomines and extends widely, being found at 3682 m in Argentina and as far south as 45°S.

P. megistus occurs in the moister areas of Brazil and Peru; it requires a high humidity (above 60% RH) for breeding, and wild peridomestic and domestic populations occur in rural and sometimes urban settings. By contrast, *T. braziliensis* flourishes under dry conditions in north-east Brazil, where *P. megistus* cannot survive. *R. prolixus* occurs north of the equator in the mainland countries bordering the Caribbean, i.e. Mexico, Central America (except Panama and Costa Rica), Colombia, Venezuela and the Guianas; it is an ancient human pest, well adapted to poor households, and in Venezuela it occurs as high as 1500 m and in Colombia up to 2600 m, but in El Salvador it is replaced by *T. dimidiata* above 340 m.

The main wild hosts of *T. cruzi* are opossums, marsupials of the genus *Didelphys*, which are widely distributed from Argentina to the USA and have a high incidence of infection. In sylvatic situations, the disease is enzootic, but when opossums become established near houses, they can infect peridomiciliary triatomines. Human infections result when these triatomines enter houses and infect the residents directly or by infecting dogs and cats, which become sources of infection for resident domiciliary triatomines, which in turn will feed on and infect the inhabitants of the house. Because *T. cruzi* is spread by faecal contamination of the host by the bug, and since bugs only visit hosts to feed, they will only be potential vectors if they defecate while feeding: some bugs, for example *T. infestans*, defecate on the host and act as vectors of Chagas disease, while others, for example *T. protracta*, defecate off the host and thus are not vectors.

6.1 Non-pathogenic trypanosomes

Trypanosoma rangeli resembles *T. cruzi* in being transmitted by triatomine bugs, particularly *R. prolixus*. It occurs in Central and South America, where it has been recovered from humans and 23 species of marsupials, carnivores, edentates, primates and rodents. *T. rangeli* is not pathogenic to vertebrates; however, in humans, it must be differentiated from *T. cruzi*. Flagellates ingested with the blood meal multiply in the

midgut of the bug and penetrate between the cells lining the posterior midgut to enter the haemocoel and multiply and form epimastigotes. These move through the haemocoel to penetrate the salivary glands, in which they produce metacyclic forms, which are passed with the saliva when the bug feeds. In the haemocoel, some flagellates parasitize haemocytes and form dividing amastigotes, which give rise to trypomastigotes. They are released when the infected cell bursts but their subsequent fate is not known. Some flagellates will proceed down the midgut to the rectum and be deposited with the faeces. Posterior station transmission is considered to occur infrequently. Six species of *Rhodnius* have been found with salivary infections in nature, and two more *Rhodnius* species and five species of *Triatoma* have produced salivary gland infections experimentally.

7. CONTROL AND PREVENTION

In the acute and early chronic stages of Chagas disease, treatment with nifurtimox and benzimidazole can be effective; they can alleviate, but not always cure, chronic Chagas disease. A new, less toxic drug, allopurinol, is currently being assessed for the treatment of asymptomatic Chagas disease. However, the main strategy for disease management should be to reduce the numbers of bugs in human domestic environments.

Bugs shelter in cracks and crevices of human and animal habitations and it is desirable that they should have a minimum of recesses in which bugs can shelter. The disease is common in poor settings, with older and rougher houses, and makeshift animal shelters offering ideal conditions for colonization by triatomines. The application of residual insecticides to the bugs' daytime resting places in bedrooms and animal shelters, previously using the organochlorine insecticide BHC (benzene hexachloride) and, more recently, synthetic pyrethroids, has proved highly effective, with the latter giving much greater residual activity on mud walls than other insecticides. Fumigating premises with pyrethroid aerosol and using insecticidal paint for walls have also been used effectively. When treatment ceases, there is the possibility that the original vector may recover or be replaced by another vector; for example, *T. infestans* can be replaced by *P. megistus* and *T. sordida*. In Mexico, *T. dimidiata* moves in and out of houses, so insecticide spraying offers little protection, while insect screens, backyard clean-up and outside insecticide spraying can be more effective.

In South America, there is a joint agreement between the governments of Argentina, Bolivia, Chile, Paraguay, Uruguay and Peru called the 'Southern Cone Initiative to Control/ Eliminate Chagas'. In the preparatory phase, the incidence of *T. cruzi* infections and domestic bug infestations are assessed. This is followed by an attack phase in which insecticides are used against the vector and repeated until less than 5% of the houses are infested. In the following vigilance phase, the community is responsible for detecting bug-infested houses. In Brazil, this has been highly successful, with the seroprevalence of schoolchildren in many endemic areas being reduced from 20–40% to 0–2%. Apart from the insecticide applications, the programme involves fixing cracks and crevices in poor-quality households, replacing adobe walls with plaster and thatch roof with metal sheeting. Perhaps more applicable to travellers than local residents, the use of insect repellents and sleeping under bed nets (preferably insecticide treated) would provide protection in endemic areas, although the latter would also be a useful protective technique for residents in areas where other vector-borne diseases, such as malaria, were an issue.

SELECTED BIBLIOGRAPHY

Araujo, C.A.C., Waniek, P.J. and Jansen, A.M. (2009) An overview of Chagas' disease and the role of triatomines on its distribution in Brazil. *Vector-Borne and Zoonotic Diseases* 9, 227–234.

Arlian, L.G. (2002) Arthropod allergens and human health. *Annual Review of Entomology* 47, 395–433.

Dias, J.C.P. (1987) Control of Chagas' disease in Brazil. *Parasitology Today* 3, 336–341.

Dias, J.C.P., Silveira, A.C. and Schofield, C.J. (2002) The impact of Chagas' disease control in Latin America – a review. *Memórias do Instituto Oswaldo Cruz* 97, 603–612.

Edwards, L. and Lynch, P. (1984) Anaphylactic reaction to kissing bugs. *Arizona Medicine* 41, 159–161.

Garcia, E.S. and Azambuja, P. (1991) Development and interactions of *Trypanosome cruzi* within the insect vector. *Parasitology Today* 7, 240–244.

Garcia-Zapata, M.T. and Marsden, P.D. (1993) Chagas' disease: control and surveillance through use of insecticides and community participation in Mambai, Goiás, Brazil. *Bulletin of the Pan American Health Organization* 27, 265–279.

Garcia-Zapata, M.T.A., McGreevey, P.B. and Marsden, P.D. (1991) American trypanosomiasis. In: Strickland, G.T. (ed.) *Hunter's Tropical Medicine*. Saunders, Philadelphia, pp. 628–637.

Heger, T.J., Guerin, P.M. and Eugster, W. (2006) Microclimatic factors influencing refugium suitability for *Rhodnius prolixus*. *Physiological Entomology* 31, 248–256.

Kjos, S.A., Snowden, K.F. and Olson, J.K. (2009) Biogeography and *Trypanosoma cruzi* infection prevalence of Chagas' disease vectors in Texas, USA. *Vector Borne and Zoonotic Diseases* 9, 41–49.

Klotz, J.H., Dorn, P.L., Logan, J.L., Stevens, L., Pinnas, J.L., Schmidt, J.O., *et al.* (2010) 'Kissing bugs': potential disease vectors and cause of anaphylaxis. *Clinical Infectious Diseases* 50, 1629–1634.

Lent, H. and Wygodzinsky, P. (1979) Revision of the Triatominae (Hemiptera: Reduviidae), and their significance as vectors of Chagas' disease. *Bulletin of the American Museum of Natural History* 163, 125–520.

Miles, M.A., Patterson, J.W., Marsden, P.D. and Minter, D.M. (1975) A comparison of *Rhodnius prolixus, Triatoma infestans* and *Panstrongylus megistus* in the xenodiagnosis of a chronic *Trypanosoma (Schizotrypanum) cruzi* infection in a rhesus monkey (*Macaca mulatta*). *Transactions of the Society of Tropical Medicine and Hygiene* 69, 377–382.

Milne, M.A., Ross, E.J., Sonenshine, D.E. and Kirsch, P. (2009) Attraction of *Triatoma dimidiata* and *Rhodnius prolixus* (Hemiptera: Reduviidae) to combinations of host cues tested at two distances. *Journal of Medical Entomology* 46, 1062–1073.

Minter, D.M. (1978) Triatomine bugs and the household ecology of Chagas's disease. In: Willmott, S. (ed.)

Medical Entomology Centenary Symposium Proceedings. Royal Society of Tropical Medicine and Hygiene, London, pp. 85–93.

Moffitt, J.E., Venarske, D., Goddard, J., Yates, A.B. and deShazo, R. (2003) Allergic reactions to *Triatoma* bites. *Annals of Allergy, Asthma, and Immunology* 91, 122–128.

Punukollu, G., Gowda, R.M., Khan, I.A., Navarro, V.S. and Vasavada, B.C. (2007) Clinical aspects of the Chagas' heart disease. *International Journal of Cardiology* 115, 279–283.

Rabinovich, J.E. (1972) Vital statistics of Triatominae (Hemiptera: Reduviidae) under laboratory conditions. *Journal of Medical Entomology* 9, 351–370.

Rabinovich, J.E., Leal, J.A. and Feliciangeli de Pinero, D. (1979) Domiciliary biting frequency and blood ingestion of the Chagas's disease vector *Rhodnius prolixus* Stahl (Hemiptera: Reduviidae), in Venezuela. *Transactions of the Royal Society of Tropical Medicine and Hygiene* 73, 272–283.

Ryckman, R.E. and Zackrison, J.L. (1987) Bibliography to Chagas' disease, the Triatominae and Triatominae-borne trypansomes of South America (Hemiptera: Reduviidae: Triatominae). *Bulletin of the Society of Vector Ecology* 12, 1–464.

Schofield, C.J. (1979) The behaviour of Triatominae (Hemiptera: Reduviidae): a review. *Bulletin of Entomological Research* 69, 363–379.

Schofield, C.J. (1988) Biosystematics of the Triatominae. In: Service, M.W. (ed.) *Biosystematics of Haematophagous Insects.* Clarendon Press, Oxford, UK, pp. 284–312.

Schofield, C.J. and Dias, J.C.P. (1999) The southern cone initiative against Chagas' disease. *Advances in Parasitology* 42, 1–27.

Schofield, C.J. and Dolling, W.R. (1993) Bedbugs and kissing-bugs (blood-sucking Hemiptera). In: Lane, R.L. and Crosskey, R.W. (eds) *Medical Insects and Arachnids.* Chapman and Hall, London, pp. 483–516.

Schofield, C.J. and Dujardin, J.P. (1997) Chagas disease vector control in Central America. *Parasitology Today* 13, 141–144.

Schofield, C.J. and Galvão, C. (2009) Classification, evolution, and species groups within the Triatominae. *Acta Tropica* 110, 88–100.

Schofield, C.J., Minter, D.M. and Tonn, R.J. (1987) XIV. The triatomine bugs – biology and control. In: *Vector Control Series: Triatominae Bugs – Training and Information Guide.* World Health Organization Vector Biology and Control Division 87.941, Paris.

Schofield, C.J., Jannin, J. and Salvatella, R. (2006) The future of Chagas disease control. *Trends in Parasitology* 21, 583–588.

Service, M.W. (ed.) (2001) *Encyclopedia of Arthropod-transmitted Infections.* CAB International, Wallingford, UK.

Stevens, L., Dorn, P.L., Schmidt, J.O., Klotz, J.H., Lucero, D. and Klotz, S.A. (2011) Kissing bugs. The vectors of Chagas. *Advances in Parasitology* 75, 169–192.

Torres-Montero, J., Lopez-Monteon, A., Dumonteil, E. and Ramos-Ligonio, A. (2012) House infestation dynamics and feeding sources of *Triatoma dimidiata* in Central Veracruz, Mexico. *American Journal of Tropical Medicine and Hygiene* 86, 677–682.

Usinger, R.L., Wygodzinsky, P. and Ryckman, R.E. (1966) The biosystematics of the Triatominae. *Annual Review of Entomology* 11, 309–330.

Yamagata, Y. and Nakagawa, J. (2006) Control of Chagas disease. *Advances in Parasitology* 61, 129–165.

Zapata, M.T.G., Schofield, C.J. and Marsden, P.D. (1985) A simple method to detect the presence of live triatomine bugs in houses sprayed with residual insecticides. *Transactions of the Royal Society of Tropical Medicine and Hygiene* 79, 558–559.

Zeledon, R. and Rabinovich, J.E. (1981) Chagas' disease: an ecological appraisal with special emphasis on its insect vectors. *Annual Review of Entomology* 26, 101–133.

Zeledon, R., Guardia, V.M., Zuniga, A. and Swartzwelder, J.C. (1970) Biology and ethology of *Triatoma dimidiata* (Latreille, 1811). I. Life cycle, amount of blood ingested, resistance to starvation and size of adults. *Journal of Medical Entomology* 7, 313–319.

Zeledon, R., Solano, G., Zuniga, A. and Swartzwelder, C. (1973) Biology and ethology of *Triatoma dimidiata* (Latreille, 1811). III. Habitat and blood sources. *Journal of Medical Entomology* 10, 363–370.

Zeledon, R., Alvarado, R. and Jiron, L.F. (1977) Observations on the feeding and defecation patterns of three triatomine species (Hemiptera: Reduviidae). *Acta Tropica* 34, 65–77.

Lice (Chewing) (Phthiraptera: Amblycera and Ischnocera)

1. INTRODUCTION

The lice (Phthiraptera) are wingless, dorso-ventrally flattened, permanent ectoparasites of birds and mammals, with a high degree of host specificity. Lice are clearly recognizable as insects since they have a segmented body divided into a head, thorax and abdomen. They have three pairs of jointed legs and a pair of short antennae. As they spend their life on the same animal, the sensory organs are poorly developed and the eyes are vestigial or absent.

Four suborders of Phthiraptera are readily recognizable: the chewing lice, Amblycera and Ischnocera (commonly grouped as the Mallophaga), the bloodsucking lice (Anoplura) and the Rhynchophthirina. The Rhynchophthirina includes just two species, *Haematomyzus elephantis*, found on the Indian and African elephant, and the closely related *Haematomyzus hopkinsi*, found on wart hogs. The Amblycera and Ischnocera will be discussed here; the Anoplura are dealt with elsewhere (see entry for 'Lice (Sucking)').

The Amblycera and Ischnocera have chewing mouthparts and feed on skin debris on birds and mammals. There are about 2600 species; most (about 85%) are ectoparasites of birds, but they also occur on 9 of the 18 orders of living mammals. They feed on fragments of feathers, hair and other epidermal products. The eyes are reduced or absent; there are no ocelli; the antennae are three- to five-segmented; the prothorax is free; the mesothorax and metathorax may be fused.

2. TAXONOMY

2.1 Amblycera

The 836 species of the suborder Amblycera are arranged in seven families, of which three (729 species) occur on birds and four (107 species) on marsupials and mammals in South America and Australia. The family Boopidae are parasites of marsupials and are distinguished from other domestic mammal-infesting lice by possessing two claws. One species of Boopidae, *Heterodoxus spiniger*, occurs on domestic dogs in many parts of Australia, Africa, Asia and the Americas. Guinea pigs are frequently infested with two species of the family Gyropidae, *Gyropus ovalis*, with an oval abdomen which is broad in the middle, and *Gliricola porcelli*, a slender louse with the sides of the abdomen somewhat parallel.

Several species of the family Menoponidae occur on domestic birds, of which the most important are *Menopon gallinae*, the shaft louse (Fig. 54), and *Menacanthus stramineus*, the chicken body louse. *M. gallinae* is about 2 mm long and lays its eggs singly at the base of a feather. It occurs on the thigh and breast feathers and is harmful to young fowl. *M. stramineus* is the commonest and most destructive louse found on chickens and has a worldwide distribution (Fig. 55). It is up to 3.5 mm long and deposits its eggs in masses at the base of feathers, especially around the vent. It occurs on the breast, thighs and around the vent, causing a marked reddening of the skin, and sometimes gnaws through the skin or punctures the soft quills near the base and consumes the blood that oozes out. Other species of Menoponidae occur on ducks, geese and pigeons, but heavy infestations are rarely seen on these birds and little harm results.

2.2 Ischnocera

Three families of the suborder Ischnocera are recognized, of which two are of veterinary importance: the Philopteridae (1460 species) on birds and the Trichodectidae (about 300 species) on mammals. The Philopteridae have five-segmented antennae and paired claws on the tarsi (Fig. 56) and the Trichodectidae have three-segmented antennae and single claws on the tarsi (Fig. 57). The antennae of the trichodectid *Bovicola ovis* show sexual dimorphism, but in both sexes the antennae contain olfactory and chemosensory pegs and a possible thermohygroreceptor.

3. MORPHOLOGY

Adult Amblycera and Ischnocera are usually about 2–3 mm in length. They have large, rounded heads on which the eyes are reduced or absent. In the Amblycera, the four-segmented antennae are protected in antennal grooves, so that only the last segment is visible (Fig. 54). In the Ischnocera, the antennae are three- to five-segmented and are not hidden in grooves (Fig. 56). At least the first two segments of the thorax are usually visible. The single pair of thoracic spiracles is on the ventral side of the mesothorax. Typically, there are six pairs of abdominal spiracles, but the number may be reduced. The three pairs of legs are weak and slender and end in either one or two claws, depending on the species. The lice of birds usually have two tarsal claws on each leg, enabling them to grip more efficiently to their highly mobile hosts (Fig. 55),

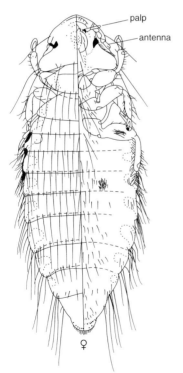

Fig. 54. Female *Menopon gallinae*, showing dorsal view on left and ventral view on right. *Source*: Ferris, G.F. (1924) *Parasitology* 16, 58.

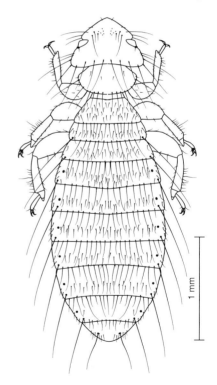

Fig. 55. Female *Menacanthus stramineus*. *Source*: Emerson *et al.* (1973).

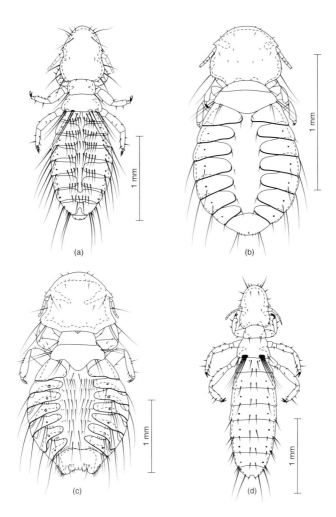

Fig. 56. Philopteridae which occur on poultry. (a) Female *Cuclotogaster heterographus*; (b) female *Goniocotes gallinae*; (c) *Goniodes dissimilis*; (d) *Lipeurus caponis*. *Source*: Emerson *et al.* (1973).

while only one tarsal claw is present on the lice of mammals (Fig. 57).

Both suborders have distinct, mandibulate mouthparts which are typical of chewing insects, composed of a labrum, a pair of mandibles and a pair of maxillae attached laterally to the labium, which is reduced to a simple broad plate. In the Amblycera, the mandibles lie parallel to the ventral surface of the head and cut in a horizontal plane. There is a pair of maxillary palps, which are two- to four-segmented. In the Ischnocera, the mandibles lie at right angles to the head and

cut vertically and there are no maxillary palps. Species from both these suborders usually feed on fragments of keratin in skin, hair or feathers and possess gut symbionts or specific enzymes to aid its digestion. However, they will take blood exuding from scratches in the skin and some are able to pierce the skin.

The thorax of both suborders consists of two, or occasionally three, visible segments. In the Amblycera, the mesothorax and metathorax are usually separate; in the Ischnocera, the mesothorax and metathorax are fused to form the pterothorax.

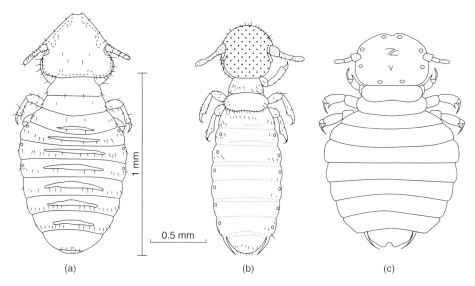

Fig. 57. (a) Female *Felicola subrostratus*; (b) *Bovicola ovis*; (c) *Trichodectes canis*. *Source*: Emerson *et al.* (1973); (b) from Calaby and Murray (1991).

4. LIFE CYCLE

Chewing lice firmly attach their eggs individually to the hairs or at the base of the feathers of their host. The nymph, which closely resembles a smaller but usually less sclerotized version of the adult, hatches from the egg within 1–2 weeks of oviposition. Over the course of between 1 and 3 weeks, the nymphs feed and moult through three to five stadia, eventually moulting to become a sexually mature adult. The entire egg-to-adult life cycle can be completed in as little as 4–6 weeks. Adults probably live for up to a month, although the normal lifespan of chewing lice on poultry may be several months.

Being permanent ectoparasites, lice are only removed accidentally from their host, and then their low powers of survival and high temperature thresholds for activity severely limit the probability of their finding another host. Transfer from one individual host to another occurs when animals are closely herded or penned and in the close contact of mother and young, with lice usually transferring from mother to young within a few hours of birth. Occasionally, lice may be transferred between animals by flies (phoresy).

5. BEHAVIOUR AND BIONOMICS

5.1 Philopteridae

Five species of Philopteridae occur on poultry and have virtually a worldwide distribution. The chicken head louse, *Cuclotogaster heterographus* (Fig. 56a), occurs on the skin and feathers of the head and neck, where the lice feed on tissue debris and occasionally ingest blood. Severe infestations in young chickens are sometimes fatal. The fluff louse, *Goniocotes gallinae* (Fig. 56b), is a small louse which occurs on the down feathers anywhere on the body and generally causes little irritation. *Goniodes dissimilis* (Fig. 56c) and *Goniodes gigas* are about 3 mm long and brown in colour. They are among the largest lice that are found on chickens and occur anywhere on the body. In small numbers, they have little effect on their host. *G. gigas* is more prevalent in the tropics and *G. dissimilis* in temperate regions. The fifth species, the wing louse, *Lipeurus caponis* (Fig. 56d), is not very active and occurs on the underside of the wing and tail feathers. Other philopterids occur on turkeys, ducks and pigeons and appear to do little harm.

5.2 Trichodectidae

Felicola subrostratus is the only louse that occurs on cats (Fig. 57a). The head is triangular, with the point directed forwards and notched at the apex. Ventrally, there is a median longitudinal groove on the head, which fits around the hair of the host. *F. subrostratus* is of minor importance, being found in large numbers only on elderly or sick cats, especially if they are long haired.

Trichodectes canis (Fig. 57c) is found on the domestic dog and wild canids throughout the world. The head is broader than long, being rectangular with rounded corners. It is found on the head, neck and tail, attached to the base of hairs. Infestations are commoner on very young, very old or sick dogs. This louse can act as an intermediate host of the tapeworm, *Dipylidium caninum*.

B. ovis is a small, pale species (Fig. 57b), which occurs on sheep worldwide. For oviposition, it requires both a suitable temperature and fibres of an appropriate diameter to which eggs can be attached. The temperature at the skin surface of sheep is usually around 37°C and this is the temperature at which maximum oviposition occurs in *B. ovis*. The distribution of the eggs of *B. ovis* on sheep is also governed by skin temperature. Low temperatures in certain areas of the body, for example legs and tail, inhibit egg laying. When the thickness of the fleece is 30–100 mm, most eggs are oviposited within 6 mm of the skin surface; even with long fleece (>100 mm), few eggs are laid more than 12 mm from the skin surface. Eggs develop and hatch over the range 33–39°C and are not affected strongly by humidity over the range 7–75% RH (relative humidity). However, very few eggs hatch at >90% RH. In fleeces where the temperature ranges from 38°C at the skin surface to 15°C near the tip of the fleece, 69% of the mobile population (nymphs and adults) may be found within 6 mm of the skin surface and only 15% more than 12 mm from the skin. When the tip of the fleece is shaded and warmed, adults and third-stage nymphs will come to the surface.

Populations of *B. ovis* are limited by a number of factors, including shearing when 30–50% of the population may be lost. Heavy rain can cause high mortality due to soaking the fleece, immersing all stages of the louse and maintaining a high humidity during the drying out period. In Australia in the summer, temperatures in a fleece exposed to the sun can reach 45°C at the skin surface in 5–10 min, with temperatures near the fleece tip being 65–70°C. Such temperatures will be lethal to all stages of the louse and help to explain why louse populations are low in summer.

Bovicola bovis is a small, reddish-brown louse on cattle, found particularly on dairy cattle. This louse is commonest on the head and midline of cattle, spreading more widely in heavy infestations. Its effect on the host is usually minimal. The intensity of infestation may be affected strongly by nutrition. *Bovicola equi*, which occurs on horses, cannot attach its eggs to the coarse hairs of the face, mane and tail and consequently it suffers a reduction in population size when the coat is shed. *Bovicola caprae* parasitizes goats.

6. MEDICAL AND VETERINARY IMPORTANCE

Chewing lice are of no significant medical concern (although occasionally humans are infested with the dog tapeworm, *D. caninum*, if its intermediate louse host *T. canis* is accidentally ingested) and are predominantly of importance in livestock and poultry farming.

For such animals, the effect of lice is usually a function of their density. A small number of lice may present no problem, and in fact may be considered to be a normal part of the skin fauna. However, louse populations can increase dramatically, reaching high densities. For example, the number of the louse *B. ovis* on a sheep has been recorded as increasing from about 4000 in autumn/fall to more than 400,000 in spring, with a rate of natural increase of about 6.7% per day. Such heavy louse infestations may cause pruritus, alopecia, excoriation and self-wounding. Birds attempt to remove lice when grooming, scratching the head with the feet and preening the body with the bill. Infestation can cause severe irritation, leading to feather damage, restlessness and cessation of feeding, and birds may pluck their feathers.

Heavy infestations of lice are associated with young animals or old animals in poor health and/or animals maintained in unhygienic conditions. Nevertheless, the irritation caused by even modest populations of lice leads to animals scratching and rubbing, causing damage to fleece and hides. There is conflicting evidence on the effect of lice on the production of milk or beef and this is likely to be dependent on other factors affecting the health of the host.

T. canis also act as intermediate hosts for a subcutaneous filarioid, Acanthocheilonema reconditum, which infests dogs worldwide.

7. PREVENTION AND CONTROL

There are a number of different ways to prevent and control lice infestations in livestock, involving both chemical intervention with pediculicides and hygiene. Pediculicides may be applied as dusts, powders, sprays, dips, ear and tail tags, resin strips, gut boluses, collars, pour-ons, lotions and injections. In addition, dust bags or back rubbers containing pediculicides can be placed at self-dosing stations for cattle treatment. It is important to note that systemic treatments with macrocyclic lactones, such as ivermectin, are not completely effective against chewing lice.

Spot treating infected individuals may be successful at controlling the overall spread of the infection, but treatment of all in-contact animals may be required for its elimination. Individuals should be treated twice a week, for between 2 and 4 weeks; however, the level of treatment required will vary depending on the severity of the infection. On sheep, treatments are usually applied immediately after shearing. The effectiveness of pediculicides is usually greatest when the application is carried out in late autumn/fall, as louse populations are typically at their highest during the winter months. However, in many countries, widespread resistance to these chemicals has developed. As a result, there is currently considerable interest in the use of alternatives to neurotoxic insecticides, such as plant-derived oils.

Simple preventative husbandry measures are also important in control. For example, avoiding physical contact between infected individuals and reducing overcrowded conditions will lower the risk of initial infection within a host population. Nutrition and the living conditions of livestock will affect their susceptibility to lice infestation. Providing cattle with a high-energy diet and keeping them in an uncrowded condition can reduce the risk of louse infestation and control its spread.

SELECTED BIBLIOGRAPHY

Clay, T. (1974) Geographical distribution of the avian lice (Phthiraptera): a review. Journal of the Bombay Natural History Society 71, 536–547.

Colebrook, E. and Wall, R. (2004) Ectoparasites of livestock in Europe and the Mediterranean region. Veterinary Parasitology 120, 251–274.

Crawford, S., James, P.J. and Maddocks, S. (2001) Survival away from sheep and alternative methods of transmission of sheep lice (Bovicola ovis). Veterinary Parasitology 94, 205–216.

Emerson, K.C. (1956) Mallophaga (chewing lice) occurring on the domestic chicken. Journal of the Kansas Entomological Society 29, 63–79.

Emerson, K.C. and Price, R.D. (1981) A host–parasite list of the Mallophaga of mammals. Miscellaneous Publications of the Entomological Society of America 12, 1–72.

Gawler, R., Coles, G.C. and Stafford, K.A. (2005) Prevalence and distribution of the horse louse, Werneckiella equi equi, on hides collected at a horse abattoir in south-west England. Veterinary Record 157, 419–420.

James, P.J. (1999) Do sheep regulate the size of their mallophagan louse populations? International Journal for Parasitology 29, 869–875.

Kettle, P.R. (1974) The influence of cattle lice (Damalinia bovis and Linognathus vituli) on weight-gain in beef animals. New Zealand Veterinary Journal 22, 10–11.

Khater, H.F., Ramadan, M.Y. and El-Madawy, R.S. (2009) Lousicidal, ovicidal and repellent efficacy of some essential oils against lice and flies infesting water buffaloes in Egypt. *Veterinary Parasitology* 164, 257–266.

Mencke, N., Larsen, K.S., Eydal, M. and Sigurðsson, H. (2004) Natural infestation of the chewing lice (*Werneckiella equi*) on horses and treatment with imidacloprid and phoxim. *Parasitology Research* 94, 367–370.

Mencke, N., Larsen, K.S., Eydal, M. and Sigurðsson, H. (2005) Dermatological and parasitological evaluation of infestations with chewing lice (*Werneckiella equi*) on horses and treatment using imidacloprid. *Parasitology Research* 97, 7–12.

Murray, M.D. (1957) The distribution of the eggs of mammalian lice on their hosts. II. Analysis of the oviposition behaviour of *Damalinia ovis* (L.) on the sheep. *Australian Journal of Zoology* 5, 19–29.

Murray, M.D. (1960) The ecology of lice on sheep. II. The influence of temperature and humidity on the development and hatching of the eggs of *Damalinia ovis* (L.). *Australian Journal of Zoology* 8, 357–362.

Murray, M.D. (1968) Ecology of lice on sheep. VI. The influence of shearing and solar radiation on populations and transmission of *Damalnia ovis*. *Australian Journal of Zoology* 16, 725–738.

Murray, M.D. and Gordon, G. (1969) Ecology of lice on sheep. VII. Population dynamics of *Damalinia ovis* (Schrank). *Australian Journal of Zoology* 17, 179–186.

Lice (Sucking) (Phthiraptera: Anoplura)

1. INTRODUCTION

The lice (Phthiraptera) are wingless, dorso-ventrally flattened, permanent ectoparasites of birds and mammals, with a high degree of host specificity. Lice are clearly recognizable as insects, since they have a segmented body divided into a head, thorax and abdomen. They have three pairs of jointed legs and a pair of short antennae. As they spend their life on the same animal, the sensory organs are poorly developed and the eyes are vestigial or absent.

Four suborders of Phthiraptera are readily recognizable: the sucking lice (Anoplura), the chewing lice (suborders Amblycera and Ischnocera, which are commonly grouped as the Mallophaga) and the Rhynchophthirina. All Anoplura are bloodsucking ectoparasites of mammals. Rhynchophthirina includes just two species, *Haematomyzus elephantis*, found on the Indian and African elephant, and the closely related *Haematomyzus hopkinsi*, found on wart hogs. The Anoplura will be discussed here, whereas the Amblycera and Ischnocera will be discussed elsewhere (see the entry for 'Lice (Chewing)').

2. TAXONOMY

There are probably more than 1000 species of Anoplura, of which only about half have been described so far. These have been arranged into 15 families and 47 genera. Two-thirds of the Anoplura are grouped into the families Hoplopleuridae and Polyplacidae and are parasites of rodents. Each species is highly host specific, with 63% being found on a single species of host and only 13% occurring on four or more host species.

The two medically important genera are *Pediculus* and *Pthirus*. For the first, there is some controversy as to whether the human lice known as the 'body louse' and the 'head louse' are subspecies of *Pediculus humanus* (viz. *Pediculus humanus humanus* (or *Pediculus humanus corporis*) and *Pediculus humanus capitis*, respectively) or whether they are separate species that should be called *Pediculus humanus* and *Pediculus capitis*, respectively. For simplicity, they will hereinafter be called *P.h. humanus* and *P.h. capitis*, and whatever their taxonomic relationship, they occur in one of two distinct habitats; body lice have evolved to attach their eggs to clothing fibres, whereas head lice attach their eggs to the base of head hairs. For the other important louse genus, *Pthirus*, there are two distinct species, which occur on separate hosts – humans and gorillas.

Three genera are of veterinary importance. *Haematopinus* (22 species) and *Linognathus* (51 species) parasitize Artiodactyla and Perissodactyla, with another four species of *Linognathus* parasitizing the Canidae. The majority of species of *Linognathus* are parasites of the Bovidae, again with a high level (72%) of host specificity. Species of the genus *Solenopotes* (10 species) parasitize Cervidae and Bovidae.

3. MORPHOLOGY

Anoplura are small insects, ranging from less than 0.5 mm to 8 mm in length in the adult; about 2 mm is an average length. The antennae are usually five-segmented; the eyes are reduced and usually absent, and there are no ocelli (Fig. 58). The mouth opening is terminal. The highly specialized mouthparts

© R.C. Russell 2013. *The Encyclopedia of Medical and Veterinary Entomology* (R.C. Russell *et al.*)

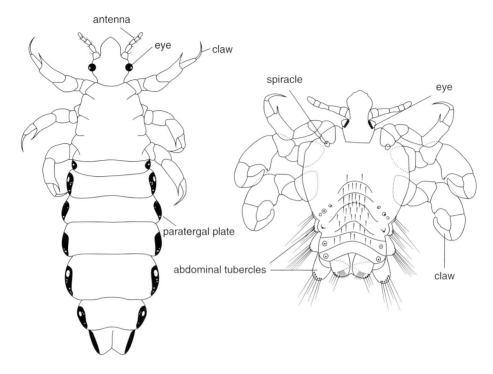

Fig. 58. Two common human lice. *Pediculus humanus capitis* (left) and *Pthirus pubis* (right) from dorsal view. Setae omitted from *P. humanus capitis* and only selected setae shown in *P. pubis*.

are not visible externally. The mouthparts are quite different to those of other insects and are highly adapted for piercing the skin of hosts. They are composed of three stylets in a ventral pouch, which form a set of fine-cutting structures. The true mouth, known as the prestomum, opens at the anterior extremity of the ventral pouch. The prestomum is usually lined with fine teeth. During feeding, the prestomum is inverted and the teeth help to secure the louse to the host's skin. The stylets are then used to puncture the skin and blood is sucked into the prestomum by a muscular cibarial pump. The mouthparts have no palps and are usually retracted into the head when not in use so that all that can be seen of them is their outline in the head or their tips protruding.

The three thoracic segments are fused. The legs have only a single tarsal segment and a single claw; when the claw is retracted, it makes contact with a thumb-like process on the tibia (the enclosed space having the diameter of the hairs of the host) and enables the louse to maintain itself firmly attached to an active host. There is one pair of spiracles (mesothoracic) on the thorax and six pairs (segments 3–8) on the abdomen, which has nine segments in all.

The sexes can be easily distinguished. In the male, the tip of the abdomen is rounded and, ventrally, the sclerotized genitalia are prominent in the midline posteriorly. In the female, the tip of the abdomen is bilobed and paired lateral gonopods and sternal plate of the eighth abdominal segment are sclerotized to varying degrees.

An important feature of the gut of anopluran lice is that on the ventral surface of the ventriculus there lies the mycetome, a cellular mass containing symbionts. In development, the mycetome arises as a pouch off the midgut and symbionts, which are in the gut of the embryo, enter the mycetome. In nymphs and males, they remain there throughout the life of the individual, but in females they migrate to the ovary and there is transovarian transmission of symbionts from one generation to another. In the absence of symbionts, nymphs live for only a few days and females are sterile.

4. LIFE CYCLE

Anopluran lice firmly attach their eggs individually to the hairs of their host or, in the case of *P.h. humanus*, to fibres of the clothing of its host (Fig. 59a). The nymph, which closely resembles a smaller, but usually less sclerotized, version of the adult, hatches from the egg within 1–2 weeks of oviposition. Over the course of between 1 and 3 weeks, the nymphs feed and moult through three to five stadia, eventually moulting to become a sexually mature adult. The entire egg-to-adult life cycle can be completed in as little as 4–6 weeks.

When on the skin permanently, the longevity of female *P.h. humanus* has been estimated as 34 ± 13 days and that of males as 31 ± 12 days. Under these conditions, a female lays 270–300 eggs, on average, at a rate of about 9–10 eggs/day. The individual nymphal stages last 3–5 days and the total nymphal life about 12–15 days. This is not an exceptionally high reproductive rate when

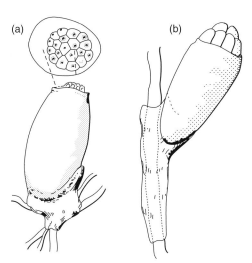

Fig. 59. (a) Egg of *Pediculus humanus humanus* with operculum viewed from above; (b) egg of *Pthirus pubis*. *Source*: Ferris (1951).

compared to other insects but, since eggs are cemented firmly to individual hairs or feathers when present, in a protected, favourable microclimate, the mortality of immature lice may be very low. This allows a rapid increase in louse populations. Some species of lice may be facultatively parthenogenetic, which increases greatly their potential rate of population growth. Using an observed hatching rate of 87.7% and survival of nymphs to adults of 86%, the rate of natural increase has been estimated at 11.7% per day.

Lice usually are unable to survive for more than 1–2 days off their host and tend to remain with a single host animal throughout their lives. Most species of louse are highly host specific and many species specialize in infesting only one part of their host's body. Transfer between hosts is most commonly brought by close physical contact between individuals. Occasionally, lice also may be transferred between animals by attachment to flies (phoresy).

5. BEHAVIOUR AND BIONOMICS

Lice respond to warmth, humidity and chemical odours. Many receptors are located on the antennae, but heat and humidity receptors are located over the entire body. Lice have a tightly defined band of humidity and temperature tolerance and respond to humidity or temperature gradients by showing increased rates of turning in favourable microclimates, which tends to keep them in suitable areas. The preferred temperature is 29–30°C and movement into areas of higher or lower temperature results in more frequent turning of the louse to bring it back to the preferred temperature. Being ectoparasitic on warm-blooded animals, lice live at a relatively high ambient temperature and *Pediculus* does not oviposit at temperatures below 25°C.

Lice are positively thigmotactic, moving less on rough surfaces. In addition, they usually move away from direct light and towards dark objects. The responses of lice to stimuli are kineses and not taxes; that is, lice are not attracted directly to the source of the stimulus but show increased turning when they move away from the source.

5.1 *Pediculus humanus/capitis*

The two 'forms' of *Pediculus* – body and head – are generally indistinguishable, except for their association with their particular habitat, and they will interbreed in the laboratory.

Adult *Pediculus* (Fig. 58) measure 2–3 mm in length in the male and 2.4–3.6 mm in length in the female. The simple, lateral eyes are well developed, for a louse. The legs are essentially the same size and shape. The margins of the abdomen are more or less strongly lobed, with lobes on segments 3–8 being covered by sclerotized paratergal plates. The thorax bears a sclerotized sternal plate.

The egg measures 0.8 × 0.3 mm and is glued to the hair in the case of *P.h. capitis* and to fibres of the inner clothing by *P.h. humanus* (Fig. 59a). The eggs of head lice, and sometimes the lice themselves, are often commonly referred to as 'nits', although that term strictly refers to the hatched eggshell remaining attached to the hair shaft. Eggs will hatch over the temperature range 24–37°C, with the highest hatching rate of 70–90% occurring at 29–32°C. Outside that range, the hatching rate declines, reaching 10% at a low of 24°C and a high of 37°C. The percentage of eggs hatching, but not the duration of the egg stage, is affected by humidity, with the highest rate occurring at 75% RH (relative humidity). The egg stage lasts 7–10 days at 29–32°C and the maximum time that eggs can survive unhatched is 3–4 weeks, which may be important when considering the survival of lice in infested clothing.

At hatching, the nymph swallows the amniotic fluid and air, forces the operculum off the egg and tears the vitelline membrane, using an elaborate hatching device on a ridged area of embryonic cuticle on the front of the head. When kept on the skin, the three nymphal stages are completed in 8–9 days, but when removed at night, the duration of the nymphal stages extends to 16–19 days. Adults mate frequently throughout life, beginning soon after the final moult. Human lice survive off the host for only a few days, the duration varying inversely with the temperature; at low temperatures, lice are inactive, which reduces their chance of finding another host.

5.2 *Pthirus pubis*

Pthirus pubis, the pubic or crab louse, has simple eyes and is shorter (1.5–2 mm) and broader than the more slender *P.h. humanus/capitis* (Fig. 58). Its body is less than twice as long as wide. The thorax is very wide and passes imperceptibly into the short abdomen. Compared to the first pair of legs, the second and third pairs are strongly developed. There is no thoracic sternal plate, the abdomen bears four pairs of lateral sclerotized tubercles and, as a result of compression, the first three pairs of abdominal spiracles are in an almost straight, transverse row.

P. pubis occurs on hair in the pubic and perianal regions of the body, and occasionally in the axillae, eyebrows, eyelashes and beard. The incidence of *P. pubis* has been reported to have increased during the latter decades of the 20th century in some regions (possibly associated with increasing sexual freedom in those societies); however, in more recent years, there have been reports suggesting that rates are now declining in some communities, as a result of trends towards body hair removal in both males and females, particularly those who are sexually active.

Transmission can also occur between individuals sleeping in the same bed, when one is infested. A common feature of infestation with *P. pubis* is the presence of blue or slate-grey macules on the skin, which are considered to be either altered patient blood pigments or substances excreted from the louse's salivary glands. Fortunately, *P. pubis* is not involved in the transmission of any pathogenic organisms. It has a similar life cycle to that of *P. humanus*, but it attaches its eggs to hairs in the area of infestation (Fig. 59b). *P.h. pubis* is considered to be less mobile than *P.h. humanus*, moving only 100 mm/day compared with 175 mm/h for the latter. It survives for even shorter periods off the host, dying in less than 48 h at 15°C.

5.3 Haematopinidae

Twenty-two species of *Haematopinus* have been described and they are all parasites of ungulates (Fig. 60). Those occurring on domestic animals include *Haematopinus suis* on pigs, *Haematopinus asini* on equines and three species on cattle: *Haematopinus eurysternus*, the short-nosed sucking louse, *Haematopinus quadripertusus*, the tail louse, and *Haematopinus tuberculatus*, the buffalo louse.

Species of *Haematopinus* are large lice, measuring about 4 mm in length, with prominent ocular points but without eyes. The thoracic sternal plate is well developed, the legs are all of similar size, the paratergal plates are strongly sclerotized on abdominal segments 2 or 3–8 and there is a sclerotized plate at the base of the tarsal segment, which is referred to as the pretarsal sclerite or discotibial process (Fig. 60c).

H. suis is the largest anopluran found on domestic animals and occurs in folds of the neck and jowl and around the ears of pigs. It causes severe irritation, resulting in a depressed growth rate. *H. asini* is about 3.5 mm in length and favours the roots of the mane, the forelock, round the butt of the tail and above the hooves. Louse populations in domestic animals are usually reduced at the beginning of summer by the shedding of the coat. This has least effect on populations of *H. asini* because the coarse hairs of the mane and tail, to which they attach their eggs, are not shed.

The female *H. eurysternus* measures about 3.5 mm in length and occurs on cattle worldwide. The main areas of infestation are the head and neck, spreading in heavy infestations to other parts of the body. The life cycle from egg to egg averages 4 weeks, with females living up to 16 days and laying 35–50 eggs. *H. quadripertusus* occurs on cattle in tropical areas. It is a rather larger louse, measuring 4.5 mm, and occurs mainly in the tail-switch, where the eggs are laid almost exclusively. The nymphs migrate to the soft skin around the anus, vulva and eyes. *H. tuberculatus* is the largest of the three, measuring 5.5 mm, and was originally described from the Indian buffalo. In Australia, it has been found infesting camels and cattle but is not considered of any great importance.

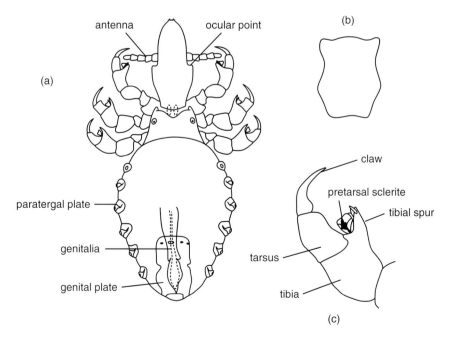

Fig. 60. Male *Haematopinus asini*. (a) In dorsal view; (b) sternal plate; (c) terminal segments of leg.

5.4 Linognathidae

There are two genera of veterinary importance in the family *Linognathidae*: *Linognathus* and *Solenopotes*. Members of this family are distinguished by the absence of eyes and ocular points and by the second and third pairs of legs, which end in large stout claws, being considerably larger than the first pair, and by the thoracic sternal plate being absent or weakly developed in *Linognathus* but distinct in *Solenopotes*. In addition, the latter does not have paratergal plates on the abdomen. Most species of *Linognathus* are found on Artiodactyla and a few on carnivores. More than 50 species of *Linognathus* have been described and six occur on domestic animals. *Linognathus setosus* parasitizes dogs, particularly long-haired breeds, on which it infests the neck and shoulders.

The eggs of both genera hatch over a relatively narrow range of temperatures around 35°C and few hatch at 38°C or higher temperatures. During the summer months in Australia, the temperature near the skin of the sheep may rise to over 45°C, and within the fleece the temperature may exceed 50°C. Such conditions have an adverse effect on louse populations.

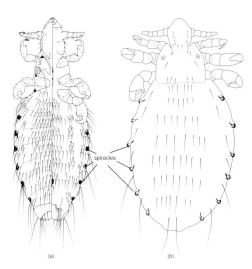

spiracles

(a)   (b)

Fig. 61. (a) Female *Linognathus ovillus* – dorsal view presented in left half and ventral view in the right half of the illustration; (b) dorsal view of female *Solenopotes capillatus*. Source: *L. ovillus* from Ferris (1951).

Two species, *Linognathus pedalis*, the foot louse, and *Linognathus ovillus*, the face louse (Fig. 61a), are most commonly found on sheep. Both frequent the hairy parts of the body, *L. ovillus* occurring on the face and lower jaw, from which it spreads to the body, and *L. pedalis* occurring on the lower hairy parts of the body, namely shanks, belly and scrotum. *L. pedalis* is able to survive for several days off the host, 50% being alive after 7 days at 12°C and 75% RH. In comparison, *L. ovillus* dies in 4 days under the same conditions.

The fact that *L. pedalis* can survive for several days off the host means that there is a possibility of infection being acquired from contaminated pasture and housing. On sheep, *L. pedalis* is more sedentary and congregates in clusters, a behaviour which does not occur in *L. ovillus*. *Linognathus stenopsis*, the goat louse, also occurs on sheep. It is larger than the other lice on sheep and can cause scabby, bleeding areas on the host. *Linognathus africanus* parasitizes domestic sheep and goats. *Linognathus vituli* is the long-nosed sucking louse of cattle and it is more common on calves and young stock, and more important among dairy cattle.

Solenopotes capillatus is the smallest of the sucking lice (Fig. 61b) on cattle and occurs in conspicuous clusters on the neck, head, shoulders, dewlap, back, anus and tail. In *Solenopotes*, the abdominal spiracles are borne on lightly sclerotized tubercles that project slightly from the body.

6. MEDICAL AND VETERINARY IMPORTANCE

Louse infestation is described as pediculosis; although, technically, this term originally was restricted to infestation by *Pediculus*, it has by now obtained wider currency.

6.1 Medical

Infestations of *Pediculus* on humans cause considerable irritation and scratching, leading to skin lesions and secondary infections if untreated. However, while head and pubic lice are of particular concern only in relation to disturbance, body lice are also well known as vectors of the important human diseases, epidemic typhus and relapsing fever, caused by *Rickettsia prowazekii* and *Borrelia recurrentis*, respectively, and also *Bartonella quintana*, which causes what is known as trench fever.

6.1.1 Epidemic typhus (*Rickettsia prowazekii*)

Epidemic typhus is a very severe febrile disease associated with muscle, joint, headache and other neurological symptoms, and generally a maculopapular rash; it may lead to a high mortality (30–60%), particularly in populations weakened by malnutrition. Treatment is with tetracyclines as the drugs of choice, and patients generally recover completely with prompt treatment, but supportive therapy may also be required. For this reason, epidemic typhus was widespread globally prior to the introduction of modern antibiotics in the middle of the 20th century. Epidemics of louse-borne typhus generally have been associated with overcrowded, unsanitary conditions (in England, typhus was known as gaol fever) and have been prevalent in times of war, occurring among both refugees and fighting men, when personal hygiene practices break down and body louse populations flourish.

The body louse orientates to human odours and the excreta of its own species. Both these responses will keep body lice in the vicinity of their host. They prefer rough to smooth surfaces, i.e. woollen cloth to cotton cloth and both to silk cloth, and have great powers of reproduction; under optimal conditions, a female louse will live for nearly 5 weeks, producing about 280 eggs, of which around 40% will become adults. Under normal conditions, an infested human will take corrective action as the louse population builds up. In practice, natural populations of head and body lice commonly consist of about 10 or 20 insects (nymphs and adults), though hundreds are not very rare and populations exceeding 1000 have been recorded. Lice exposed to a temperature gradient will move to a temperature range of 29–30°C. This has the effect that lice leave a person in a fever, e.g. suffering from typhus, and also leave a

corpse. When refugees are crowded together for warmth and shelter, lice spread rapidly throughout the human population.

Although *P.h. humanus* occurs widely in human populations, epidemic typhus is commoner in the temperate regions and in the cooler regions of the tropics above 1600 m, and is absent from the lowland tropics. It is present in mountainous regions where heavy clothing is worn continuously, favouring infestations with body lice. It occurs in Mexico, Guatemala, the Andean highlands, the Himalayan region including Pakistan and Afghanistan, the highlands of Ethiopia, Burundi, Rwanda and Lesotho in Africa, and northern China. The incidence of louse-borne typhus has been declining steadily in recent decades, with the majority of cases occurring in Africa, Peru and Ecuador.

Epidemics of typhus have changed the course of history and at the end of the First World War and the period immediately succeeding it (1917–1923), it has been estimated that 30 million cases of epidemic typhus occurred in Russia and Europe, with over 300,000 deaths. A little more than a century earlier, typhus had played a major role in the defeat of Napoleon's armies that invaded Russia. The potato famine in Ireland in the 1840s led to a major movement of the population to America, and of the 75,000 Irish who migrated in 1847, 30,000 (40%) contracted typhus and 20,000 (67%) died from the disease, reflecting the debilitated state of the health of the migrants.

An epidemic in Naples in 1943–1944 during the Second World War was the first time an epidemic of typhus had not exhausted itself but had been terminated by human action by dusting fully clothed individuals with effective anti-louse powders. Initially, a pyrethrum preparation and one containing derris and naphthalene were used, and later it was 10% DDT in talc. Nearly 3 million dust treatments were made to individuals and the epidemic brought to a halt.

The head louse, *P.h. capitis*, and the crab louse, *P.h. pubis*, can transmit *R. prowazekii* experimentally, but epidemics have always occurred in conditions where body lice, *P.h. humanus*, were particularly prevalent, and this species is the usual vector. *R. prowazekii* multiplies in the epithelial cells of the midgut and, when these burst, it is passed out with the faeces of the louse. *R. prowazekii* is pathogenic to the body louse and kills it in about 10 days. People become infected by scratching in response to the feeding of the louse, scarifying the skin and facilitating the entry of *R. prowazekii*. It is possible that *R. prowazekii* can gain entry into the human body by other routes, for example by inhalation of louse faeces or by penetrating mucosa or the conjunctiva of the eye. Fatalities among first research workers arose because they concentrated on the feeding of the louse and not on its faeces; *R. prowazekii* can survive 66 days in dry faeces at ambient temperatures. Hence, fresh cases of typhus can be contracted for 2 months after the conclusion of a successful body louse eradication programme.

Louse-borne typhus can be epidemic, endemic-epidemic or endemic and there is a question as to how *R. prowazekii* survives between outbreaks. There is no transovarial transmission, individual lice survive for only about 6 weeks and even less if infected, and *R. prowazekii* for only about 2 months in louse faeces. Humans are the primary host of *R. prowazekii* and individuals who recover from epidemic typhus often retain small numbers of organisms, presumably in their lymph nodes, giving rise to a milder form of typhus (i.e. Brill-Zinsser disease), which occurs in the absence of body lice.

R. prowazekii has been shown to be present as natural infections in flying squirrels (*Glaucomys volans*) in the USA, and has been recovered from the bloodsucking louse, *Neohaematopinus sciuropteri*, and the flea, *Orchopeas howardii*; this louse is host-specific but *O. howardii* has an extensive host range, including humans. While the significance of this finding to the epidemiology of epidemic typhus is not known, sporadic human cases have arisen in houses harbouring flying squirrels.

6.1.2 Epidemic relapsing fever (*Borrelia recurrentis*)

B. recurrentis can cause a severe febrile illness with non-specific generalized symptoms, including rash and neurological features in many patients. The first attack of relapsing fever lasts about 6 days and the first relapse about 2 days, with a 9-day period of remission in between. Case fatality rates can be above 50%, but are much reduced with antibiotic treatment with tetracycline or erythromycin, producing in most patients a short-lived, non-fatal severe reaction involving a sharp rise in body temperature and a fall in blood pressure while the borreliae are being cleared from the bloodstream.

Although in the first half of the 20th century there were major epidemics of louse-borne relapsing fever in Africa, eastern Europe, the Middle East and Russia, from the 1960s, louse-borne relapsing fever has been associated mainly with the highlands of Sudan, Somalia and Ethiopia.

B. recurrentis occurs in the circulating blood of an infected person, from where it can be acquired by the body louse when it feeds. Most of the spirochaetes die in the midgut of the louse, but a few survive to penetrate the midgut and reach the haemocoel, in which they multiply. From the 6th day after an infective feed, spirochaetes become increasingly abundant in the haemolymph of the louse. They invade the neural ganglia and the muscles of the head and thorax, but have never been observed in salivary glands or ovaries, indicating that *B. recurrentis* can neither be transmitted by the bite of the louse nor by transovarian transmission. Few *B. recurrentis* are passed in the faeces of the louse, but most of them are moribund so that transmission via the faeces is highly unlikely. Conversely, transmission of *B. recurrentis* does occur when a louse is crushed and the infected haemolymph released on to the skin, although there is also evidence that *B. recurrentis* can penetrate unbroken skin. Thus, the borrelia transmission involves the death of the louse, and an individual louse can only infect one person. Consequently, louse-borne relapsing fever epidemics are dependent on high louse populations occurring in human populations, as for epidemic typhus. Humans are the only known reservoir for *B. recurrentis*. This raises questions about the survival of *B. recurrentis* during non-epidemic periods, since in the absence of transovarian transmission, *B. recurrentis* cannot survive in the louse population.

6.1.3 Trench fever (*Bartonella quintana*)

Formerly known as *Rochalimaea quintana*, infection with *B. quintana* causes trench fever or 5-day fever in humans. The disease is a non-specific febrile illness that is fatal only rarely, but it has been reported that nearly half the convalescents recovering from trench fever were carriers of the pathogen for months or even years, forming a source of further cases.

Epidemics of trench fever occurred in the First World War (1914–1918), affecting at least 1 million military personnel. It was virtually unheard of during the interwar period, but reappeared in Germany in 1941–1942 and had become widespread by 1943. The disease seemed to disappear again after the Second World War, but in recent decades it has reappeared in immunocompromised and homeless people in Europe and North America, and has been also reported from North Africa, northern Asia, Mexico and Bolivia. Because of its relative rarity, there is no established treatment regimen, but the drugs of choice are likely to be tetracyclines and doxycycline.

B. quintana is spread among human populations by the body louse, which acquires the pathogen when feeding on the blood of an infected person. *B. quintana* multiplies in the cuticular margin of the midgut epithelial cells and appears after 6–10 days in the faeces of the louse, causing infection when faeces are scarified into the skin or possibly by inhalation. The longevity of the louse is not affected by the presence of *B. quintana* and it remains infective for the rest of its life (maximum 5 weeks in an adult louse). There is no transovarian transmission, so that newly emerged lice are free from infection. However, since transmission of the pathogen is by the faeces of the louse, it is possible for new cases to arise for some time after elimination of the louse population, since *B. quintana* remains viable in dry louse faeces for many months, and possibly in excess of a year.

6.2 Veterinary

Sucking lice associated with domestic animals have also been implicated in the transmission of disease. For example, the pig louse, *H. suis*, may spread pox virus and cattle lice may transmit rickettsial anaplasmosis. Some species of lice may act as intermediate hosts to the tapeworm, *D. caninum*. However, despite this, lice are predominantly of veterinary interest because of the direct damage they can cause to their hosts, rather than as vectors. The disturbance caused may result in lethargy and loss of weight gain or reduced egg production. Severe infestation with sucking lice may cause anaemia. Heavy infestations are usually associated with young animals or older animals in poor health, or those kept in unhygienic conditions.

In temperate habitats, louse populations exhibit pronounced seasonal fluctuations. In cattle and sheep, population numbers increase in late autumn/fall and throughout the winter months, and then decrease during the warm weather of spring and early summer. This decrease may be attributed to rising skin surface temperatures and, more significantly, to shedding of the winter coat and self-grooming. As the thicker winter coat is shed, adult lice, nymphs and eggs attached to the hairs are lost directly, while those remaining experience reduced regulation of the microclimate and increased exposure to atmospheric conditions. The seasonal increase in louse populations may be exacerbated by winter housing, if the animals are in poor condition and particularly if animals are deprived of the opportunity to groom themselves properly. Louse infestation may also be indicative of some other underlying problem, such as malnutrition, neglect or chronic disease.

Louse infestation is more common in cattle than other domestic animals. Cattle heavily infested with lice develop an unthrifty appearance and can show reduced vigour or weight loss. The hair coat of louse-infested animals becomes discoloured and will appear greasy. Dairy animals produce less milk when infested. Calves that become infested with lice in autumn/fall may not achieve normal weight gain rates during the winter and may remain stunted until spring. If populations of sucking lice are high, infested animals may become anaemic and may be predisposed to respiratory diseases, abortion and death. In sheep, transmission occurs through direct contact between ewe and lamb, ram and ewe during mating and during aggregation. Lice may be a problem in housed flocks and in heavily fleeced breeds, where there are increased transmission opportunities. After initial infestation of a sheep, it takes several months before the louse population has increased to the numbers that cause rubbing and fleece loss. If the burden is heavy, the fleece develops a characteristic tatty appearance, with snags of loose wool, and it develops a yellow stain. Shearing can remove up to 50% of the louse population of a sheep.

Louse infestation in pigs is very common and it occurs most often in the folds of the neck and jowl and around the ears. Light infestation causes only mild irritation. Pediculosis in pigs leads to scratching and skin damage. In horses, light infestations are most commonly found in the mane, base of the tail and submaxillary space. As the population of lice increases, the infestation may spread over the body. As with other animals, the lice spread by contact and their presence leads to irritation, restlessness and rubbing. Long-eared and long-haired breeds of dog and cat are especially prone to infestation, although heavy infestations are most usually seen in neglected, underfed animals.

7. PREVENTION AND CONTROL

There are a number of different ways to prevent and control lice infestations in humans and livestock, involving both chemical intervention and hygiene. Various chemicals, including organochlorines, carbamates, organophosphates, pyrethrins and pyrethroids, have been used specifically to kill lice over the past 50 years and they are known as pediculicides.

In humans, particularly children, head lice (*P.h. capitis*) infestation is common. Lice are usually transmitted by close physical contact, but they may also be transmitted by the transfer of personal objects from an infected individual to a non-infected one. Thus, the sharing of combs, hats, earphones and blankets should be avoided and infested children kept away from others until effectively treated. Washing hair may help to reduce the number of nymphs and adults in an infestation, but combing with a special finely toothed 'louse comb' will have a more marked effect in removing the lice and eggs and reducing the infestation (although the eggs are tightly cemented to the hair shaft and not easy to remove). Fine-combing hair containing a 'conditioner' applied after shampooing has been shown to be even more effective in removing lice (which become 'trapped' in the viscous liquid). Washing has the added advantage of removing the louse's faeces, which can be a source of irritation. Chemical pediculicides can be used effectively as lotions, shampoos and gels to kill adult and nymphal populations, but because it is relatively difficult to affect the egg, treatments must be repeated each week for 2–4 weeks, so that newly-hatched nymphs can be killed, and reinfestation can still occur quickly from contact with untreated people.

Unfortunately, there is widespread resistance in head lice to many of the available chemical insecticides, exacerbating problems of control, and while permethrin is probably the agent most commonly and widely used currently, advice on locally effective agents should guide treatment regimes. Body lice (*P.h. humanus*) can be combated by providing clean clothing and boiling infested clothing to destroy eggs and lice, or by heating clothing to a temperature in excess of 60°C for 15 min (e.g. in a tumble drier), but these methods can be inapplicable to refugee situations, where body lice can multiply quickly and spread unchecked. Pubic louse (*P. pubis*) infestations can be prevented by avoiding infested sexual partners and can be treated as for head lice, with the same insecticidal preparations, and with appropriate hot laundering of shared bed and bathroom linen.

In animals, pediculicides may be applied as dusts, powders, sprays, dips, ear and tail tags, resin strips, gut boluses, collars, pour-ons, lotions and injections. In addition to this, dust bags or back rubbers containing pediculicides can be placed at self-dosing stations for cattle treatment. Spot treating infected individuals is often successful at controlling the overall spread of the infection, but treatment of all in contact may be required for its elimination. Individuals should be treated twice a week, for between 2–4 weeks, but the level of treatment required will vary depending on the severity of the infection. In many countries, widespread resistance to these chemicals has developed and thus the development of new control agents for lice is essential for control in the future. Simple preventative husbandry measures are also important in control. For example, avoiding physical contact between infected individuals and reducing overcrowded conditions will lower the risk of initial infection within a host population. Nutrition and living conditions of livestock will affect their susceptibility to lice infection. Providing cattle with a high-energy diet and keeping them in

an uncrowded condition can reduce the risk of louse infestation and control its spread. Shearing sheep is a highly effective control strategy, removing up to 80% of the lice population.

The effectiveness of pediculicides is usually greatest when the application is carried out in late autumn/fall, as louse populations are typically at their highest during the winter months. Biological control agents have also been used to treat livestock. These include bacteria such as *Bacillus thuringiensis* and nematodes such as *Steinernema carpocapsae*.

SELECTED BIBLIOGRAPHY

Armstrong, N.R. and Wilson, J.D. (2006) Did the 'Brazilian' kill the pubic louse? *Sexually Transmitted Infections* 82, 265–266.

Azad, A.F. and Beard, C.B. (1998) Interactions of rickettsial pathogens with arthropod vectors. *Emerging Infectious Diseases* 4, 179–186.

Burgdorfer, W. (1976) The epidemiology of the relapsing fevers. In: Johnson R.C. (ed.) *The Biology of Parasitic Spirochetes*. Academic Press, New York, pp. 191–200.

Burgdorfer, W. and Hayes, S.F. (1990) Vector–spirochaete relationships in louse-borne and tick-borne borrelioses with emphasis on Lyme disease. *Advances in Disease Vector Research* 6, 127–150.

Burgess, I.F. (1995) Human lice and their management. *Advances in Parasitology* 36, 271–342.

Burgess, I.F. (2009) Current treatments for pediculosis capitis. *Current Opinion in Infectious Diseases* 22, 131–136.

Burgess, I.F. (2010) Do nit removal formulations and other treatments loosen head louse eggs and nits from hair? *Medical and Veterinary Entomology* 24, 55–61.

Buxton, P.A. (1950) *The Louse*. Edward Arnold, London.

Collins, R.C. and Dewhirst, L.W. (1965) Some effects of the sucking louse, *Haematopinus eurysternus*, on cattle on unsupplemented range. *Journal of the American Veterinary Medical Association* 146, 129–132.

Downs, A.M.R. (2004) Managing head lice in an era of increasing resistance to insecticides. *American Journal of Clinical Dermatology* 5, 169–177.

Gratz, N.G. (1997) *Human Lice: Their Prevalence, Control and Resistance to Insecticides. A Review 1985–1997*. World Health Organization/CTD/WHOPES/97.8, Geneva, Switzerland.

Heukelbach, J., Oliveira, F.A., Richter, J. and Häussinger, D. (2010) Dimeticone-based pediculicides: a physical approach to eradicate head lice. *The Open Dermatology Journal* 4, 77–81.

Ibarra, J. (1993) Lice (Anoplura). In: Lane, R.P. and Crosskey, R.W. (eds) *Medical Insects and Arachnids*. Chapman and Hall, London, pp. 517–528.

Kim, K.C. (1985) Evolution and host associations of Anoplura. In: Kim, K.C. (ed.) *Coevolution of Parasitic Arthropods and Mammals*. John Wiley, New York, pp. 197–231.

Leo, N.P. and Barker, S.C. (2005) Unravelling the evolution of head lice and body lice of humans. *Parasitology Research* 98, 44–47.

Leo, N.P., Campbell, N.J.H., Yang, X., Mumcuoglu, K. and Barker, S.C. (2002) Evidence from mitochondrial DNA that head and body lice of humans (Phthirapter: Pediculidae) are conspecific. *Journal of Medical Entomology* 39, 663–666.

Leo, N.P., Hughes, J.M., Yang, X., Poudel, S.K.S., Brogdon, W.G. and Barker, S.C. (2005) The head and body lice of humans are genetically distinct (Insecta: Phthiraptera, Pediculidae): evidence from double infestations. *Heredity* 95, 34–40.

Li, W., Ortiz, G., Fournier, P., Gimenez, G., Reed, D.L., Pittendrigh, B.R., *et al.* (2010) Genotyping of human lice suggests multiple emergences of body lice from local head louse populations. *PLoS Neglected Tropical Diseases* 4, e641.

Lyal, C.H.C. (1985) Phylogeny and classification of the Psocodea, with particular reference to the lice (Psocodea: Phthiraptera). *Systematic Entomology* 10, 145–165.

Meleney, W.P. and Kim, K.C. (1974) A comparative study of cattle-infesting *Haematopinus*, with redescription of *H. quadripertusus* Fahrenholz, 1916 (Anoplura: Haematopinidae). *Journal of Parasitology* 60, 507–522.

Mumcuoglu, K.Y., Magdassi, S., Miller, J., Ben-Ishai, F., Zentner, G., Helbin, V., *et al.* (2004) Repellency of citronella for head lice: double-blind randomized trial of efficacy and safety. *Israel Medical Association Journal* 6, 756–759.

Olds, B.P., Coates, B.S., Steele, L.D., Sun, W., Agunbiade, T.A., Yoon, K.S., *et al.* (2012) Comparison of the transcriptional profiles of head and body lice. *Insect Molecular Biology* 21, 257–268.

Sonenshine, D.E., Bozeman, F.M., Williams, M.S., Masiello, S.A., Chadwick, D.P., Stocks, N.I., *et al.* (1978) Epizootiology of epidemic typhus (*Rickettsia prowazekii*) in flying squirrels. *American Journal of Tropical Medicine and Hygiene* 27, 339–349.

Van der Stichele, R.H., Dezeure, E.M. and Bogaert, M.G. (1995) Systematic review of clinical efficacy of topical treatments for head lice. *British Medical Journal* 311, 604–608.

Waniek, P.J. (2009) The digestive system of human lice: current advances and potential applications. *Physiological Entomology* 34, 203–210.

Wisseman, C.L. (1991) Rickettsial infections. In: Strickland, G.T. (ed.) *Hunter's Tropical Medicine*. W.B. Saunders, Philadelphia, Pennsylvania, pp. 256–286.

World Health Organization (1997) A large outbreak of epidemic louse-borne typhus in Burundi. *Weekly Epidemiological Record* 21, 152–153.

Yang, Y.C., Lee, H.S., Lee, S.H., Marshall Clark, J. and Ahn, Y.J. (2005) Ovicidal and adulticidal activities of *Cinnamomum zeylanicum* bark essential oil compounds and related compounds against *Pediculus humanus capitis* (Anoplura: Pediculidae). *International Journal for Parasitology* 35, 1595–1600.

Zdrodovskii, P.K. and Golinevich, H.M. (1960) *The Rickettsial Diseases*. Pergamon Press, London.

Millipedes (Polydesmida and Spirobolida)

1. INTRODUCTION

Millipedes (class Diplopoda) are moisture-loving, metamerically segmented terrestrial arthropods, which feed on decaying vegetable matter (Fig. 62). They are distributed worldwide but are more abundant (and larger) in warmer countries. Many millipedes have repugnatorial glands that produce chemical secretions when disturbed and which can be exuded or squirted as a defensive mechanism; some of these body secretions can irritate and discolour human skin and affect the eye.

2. TAXONOMY

Millipedes are classified into three subclasses and 16 extant orders; more than 100 families and approximately 12,000 species are described (although it is estimated that as many as 80,000 species may exist). Several species in the orders Polydesmida and Spirobolida are of medical interest.

3. MORPHOLOGY

Millipedes have a head with one pair of short antennae and two pairs of appendages associated with the mouth (Fig. 62). Each apparent body segment behind the head bears two pairs of legs and two pairs of spiracles as respiratory openings (except the first which has none, the next few which have one and the last one or few which have none). Subcylindrical millipedes may be up to 30 cm in length and flat-backed millipedes up to 13 cm.

4. LIFE CYCLE

Millipedes have copulatory sexual contact, with the male transferring a sperm packet to the female. Eggs (up to some hundreds) are laid in moist soil and hatching immatures have no more than seven segments and three pairs of legs. They undergo up to seven moults (during which additional segments and legs are

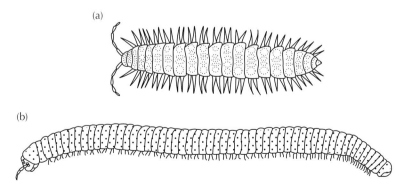

(a)

(b)

Fig. 62. (a) Dorsal view of a flat-back diplopod, *Polydesmus* sp.; (b) lateral view of a cylindrical diplopod, *Cylindroiulus* sp. Redrawn from Hopkin and Read (1992).

© R.C. Russell 2013. *The Encyclopedia of Medical and Veterinary Entomology* (R.C. Russell *et al.*)

added) before reaching sexual maturity after 2–5 years. Adults can live up to 10 years and, in cold climates, they overwinter in the soil.

5. BEHAVIOUR AND BIONOMICS

Millipedes are generally slow moving and essentially herbivorous; they feed on living and decomposing vegetation (although some are omnivorous or carnivorous and feed on soil invertebrates). They tend to be nocturnally active, seeking shelter during the day among leaf litter, under rocks or in the surface layers of friable soil. Some are attracted by lights and gather around dwellings, which they may invade when in large numbers.

The defensive secretions of most species are simply oozed out of lateral openings on most abdominal segments; however, some species (e.g. species of *Epibolus*, *Rhinocrichus*, *Spirobolus* and *Spirostreptus*) can squirt the fluids for some distance (up to 80 cm is recorded for *Rhinochrichus latespargor*).

6. MEDICAL AND VETERINARY IMPORTANCE

Millipedes do not possess biting jaws like those of the centipedes, and it is only the defensive secretions that are of medical interest. Species in the order Polydesmida can squirt the secretions, while others in the order Spirobolida simply exude them. The secretions from some species act as antifeedants, sedatives and toxins to predators; some species have been reported to discharge iodine and a range of organic compounds including benzoquinones, quinazolinones, *p*-cresol, phenols, alkaloids, terpenes and cyanogenic glycosides (which, for example, enables *Pachydesmus crassicutis* to repel fire ants and *Apheloria corrugata* to discourage predacious doryline army ants). These fluids may cause considerable discomfort (prickling and burning sensations and possibly blistering) to humans with sensitive skin or, if they get into the eyes, cause a burning sensation and possibly blindness. Skin 'burns' have been reported from Africa (*Epibolus pulchripes*), Indonesia (*Spirostrepus* spp.), Haiti (*R. latespargor*), Mexico (*Orthoporus* spp.) and Papua New Guinea (*Polyconoceras alokistus*). Episodes of temporary blinding in children and permanent blindness in dogs from Papua New Guinea have been reported.

7. PREVENTION AND CONTROL

Residual insecticides can be effective but removal of favourable harbourages and decaying vegetation will assist in preventing build-up of populations around homes.

Treatment of contacts with secretions involves thorough washing to remove the fluids from the skin with, for instance, alcohols. If the eye is involved, thorough irrigation should be carried out and expert medical treatment should be sought.

SELECTED BIBLIOGRAPHY

Eisner, T., Alsop, D., Hicks, K. and Meinwald, J. (1978) Defensive secretions of millipedes. In: Bettini, S. (ed.) *Arthropod Venoms*. Springer-Verlag, Berlin, pp. 41–72.

Eisner, T., Eisner, M. and Siegler, M. (2005) *Secret Weapons: Defenses of Insects, Spiders, Scorpions, and Other Many-legged Creatures*. Harvard University Press, Cambridge, Massachusetts.

Girardin, B.W. and Steveson, S. (2002) Millipedes – health consequences. *Journal of Emergency Nursing* 28, 107–110.

Hopkin, S.P. and Read, H.J. (1992) *The Biology of Millipedes*. Oxford University Press, Oxford, UK, 233 pp.

Hudson, B.J. and Parsons, G.A. (1997) Giant millipede 'burns' and the eye. *Transaction of the Royal Society of Tropical Medicine and Hygiene* 91, 183–185.

Mason, G.H., Thomson, H.D.P., Fergin, P. and Anderson, R. (1994) The burning millipede. *Medical Journal of Australia* 160, 718–726.

Radford, A.J. (1975) Millipede burns in man. *Tropical and Geographic Medicine* 27, 279–287.

Ratnapalan, S. and Das, L. (2011) Causes of eye burns in children. *Pediatric Emergency Care* 27, 151–156.

Sierwald, P. and Bond, J.E. (2007) Current status of the Myriapod Class Diplopoda (millipedes): taxonomic diversity and phylogeny. *Annual Review of Entomology* 52, 401–420.

Southcott, R.V. (1976) Arachnidism and allied syndromes in the Australian regions. *Records of the Adelaide Children's Hospital* 1, 97–187.

Mites (Arachnida: Acari)

1. INTRODUCTION

The Acari is a large group of arthropods, with over 30,000 described species and more than 2000 genera. However, this is a small proportion of the half a million species which, it is believed, exist today. Mites are widely distributed throughout the world, most being free-living and terrestrial predators, herbivores or detritivores, occupying a wide range of habitats from soil to oceans and from deserts to ice fields. However, the freshwater aquatic mites (Hydrachnellae) and the marine mites (Halacaroidea) form small distinct ecological groups.

A relatively small number of mite species are parasites, feeding on blood, lymph, skin debris or sebaceous secretions. They affect many classes of invertebrate and all classes of vertebrate, particularly birds and mammals. The majority of these mite species are ectoparasites and a few act as vectors of disease-causing pathogens, while a small number (about 500 species) are endoparasites, living in the lungs or nasal passages of various birds, mammals and reptiles. Some free-living mites associated with human dwellings and accoutrements can cause skin and respiratory allergies.

2. TAXONOMY

There is general agreement as to the main groups within the Acari, but not on their names or classification. The Acari are a subclass of the Arachnida, with two super-orders: Parasitiformes (Anactinotrichida) and Acariformes (Actinotrichida). The Acariformes are characterized by the absence of visible stigmata posterior to the coxae of the second pair of legs, and by coxae (epimeres) often fused to the ventral body wall. The Parasitiformes possess one to four pairs of lateral stigmata posterior to the coxae of the second pair of legs, and freely movable coxae. The Acariformes include three orders, the Acaridida (Astigmata), the Actinedida (Prostigmata) and the Oribatida (Cryptostigmata). The Parasitiformes comprise four orders, two of which include only a few species of no medical or veterinary importance and two, the Gamasida (Mesostigmata) and Ixodida (Metastigmata), which are of considerable medical and veterinary significance.

The Astigmata are a well-defined group of slow-moving, weakly sclerotized mites including the economically important families Sarcoptidae and Psoroptidae, which cause mange and scab in various vertebrates (including humans), a number of families containing parasites and commensals of mammals and birds and families associated with the infestation of stored foods.

The Prostigmata are the most heterogeneous of the acarine orders. Three of the 31 superfamilies included in this order are of medical or veterinary importance. The Trombidioidea includes the Trombiculidae, with some members parasitizing vertebrates (as the larval stage) and others acting as vectors of scrub typhus, a rickettsial disease of humans. The Cheyletoidea comprises eight families which are, with few exceptions, parasites of arthropods and vertebrates, including humans. The Pyemotidae includes a few species that are responsible for dermatitis in people and domestic animals. The Oribatida are soil- or humus-dwelling mites, which are of importance as intermediate hosts of certain tapeworms of domestic animals.

The Mesostigmata are a large and successful group of predatory mites, and some species are external or internal parasites of mammals, birds, reptiles and invertebrates. Most species of medical and veterinary importance belong to the Dermanyssoidea.

The Metastigmata, the ticks, are the most important group and include blood-feeding parasites that have the potential to act as vectors of bacterial, viral and parasitic diseases in animals and humans; they are dealt with elsewhere (see entries for 'Ticks (Hard)' and 'Ticks (Soft)').

3. MORPHOLOGY

Acarines are arachnids with a body typically composed of an anterior gnathosoma or capitulum and a posterior idiosoma, separated by a circumcapitular suture. The gnathosoma resembles the head of a generalized arthropod only in that the mouthparts are appended to it. The brain lies in the idiosoma. The idiosoma can be subdivided into the areas of legs, the podosoma, and that behind the fourth pair of legs, the opisthosoma (Fig. 63).

The gnathosoma bears, laterally, a pair of palps and, more medially, the chelicerae (Fig. 63). Typically, the palps are simple sensory appendages which aid the acarine in locating its food. However, in some predatory Meso-

stigmata, the palps are used to manipulate the prey. The palps are composed of one or two segments in most Astigmata and five segments in the Mesostigmata. The chelicerae are composed of two segments in the Ixodida and three in the remaining orders. The chelicerae are usually toothed, with the third segment being movable; however, some modifications occur according to the feeding habits of the different mite species.

The legs of a typical mite bear six segments, but seven if the pretarsus is included (Fig. 64). The other segments are the coxa, trochanter, femur, genu, tibia and tarsus. The tarsus carries the pretarsus, bearing the paired claws and/or a median empodium. A terminal membranous pulvillus may also be present. The empodium is variable in morphology and can be hair-like, pad-like, sucker-like or claw-like. The first pair of legs often differs from the other three pairs, since they may have a sensory function. Frequently, the first pair of legs is longer and more slender. In predatory species, they may also be modified for capturing prey, and in some Rhinonyssidae (Mesostigmata) they function as surrogate chelicerae.

In small mites, gas exchange may occur entirely through the cutaneous tissues; however, in larger species, gas exchange is

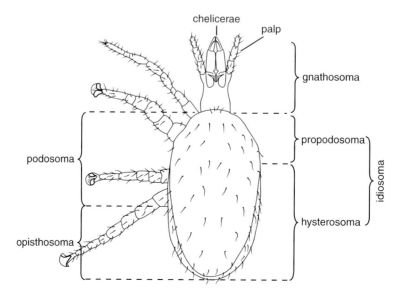

Fig. 63. Divisions of the body of a mite. *Source*: Savory, T.H. (1977) *Arachnida*. Academic Press, London.

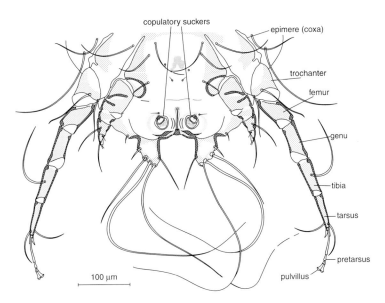

Fig. 64. Ventral view of hysterosoma of male *Psoroptes ovis.*

facilitated by a complex tracheal system opening to the exterior via the stigmata. The Mesostigmata are characterized by one pair of lateral stigmata; larvae of Trombiculidae lack stigmata but there is one pair located between the bases of the chelicerae in the nymphs and adults; stigmata are absent in the Astigmata.

Transfer of spermatozoa from the males to the females may occur directly (e.g. in the Astigmata) or through a complicated process known as podospermy (e.g. in the Mesostigmata); in the Trombiculidae, the male deposits a spermatophore on the ground and this is subsequently internalized by the female.

Astigmata, the deutonymph may be completely distinct from the preceding and succeeding stages, both in morphology and behaviour. Such heteromorphic nymphs (hypopus, plural hypopodes) are highly resistant to environmental stresses and some perform phoresy, where they are able to climb on to passing animals to aid their dispersal. True tritonymphs are known in the Oribatida and in certain Prostigmata and Astigmata. It is common practice to restrict the term deutonymph to the heteromorphic nymph of the Astigmata and refer to the two nymphal stages as protonymph and tritonymph, respectively.

4. LIFE CYCLE

Female mites produce relatively large eggs and only a few, sometimes one or two, are laid at each oviposition. When the egg hatches, a hexapod larva emerges. Later, the larva moults to an octopod nymph. One to three (usually two) nymphal stages may occur before the moult to the final adult mite form. The three nymphal stages are referred to as protonymph, deutonymph and tritonymph, respectively. One or more of these developmental stages may be inactive and non-feeding. In the

5. ASTIGMATA

The Astigmata are small mites with a relatively thin integument that lacks obvious shields. The coxae are sunk into the body and are referred to as epimeres. The empodium is claw-like and the membranous pulvillus is stalked or sessile. True paired claws are absent. Fertilized eggs are extruded through an anteroventral slit, the oviporus or genital opening. In both sexes, the genital opening may be reinforced anteriorly by a pregenital plate, or epigynium. Most of the astigmatic mites of medical and veterinary importance belong to the families Sarcoptidae and Psoroptidae within the division Psoroptoidea; the Knemidokoptidae, Cytoditidae and Laminosoptidae in the Analgoidea; and the Pyroglyphidae in the Pyroglyphoidea. The division Acaridia contains the stored food mites, which are important as sources of allergens.

5.1 Mites and human allergies

Mites within the Astigmata have been extremely successful in exploiting patchy or ephemeral food resources; their short life cycle allows a rapid build-up of the numbers of individuals within populations. Dispersal typically occurs through phoretic association with an arthropod or vertebrate, which carries the mite from habitat to habitat. Most households harbour two groups of mites: the 'house dust' and the 'storage' mites.

Of the house dust mite species, *Dermatophagoides pteronyssinus*, *Dermatophagoides*

farinae, *Blomia tropicalis* and *Euroglyphus maynei* are members of the Pyroglyphidae and they are sources of a range of cross-reactive allergens (Fig. 65). House dust mites thrive in humid dwellings, especially in human environments. Densities as high as 3500/g of house dust have been recorded. At relative humidities (RH) above 65–70%, the mites secrete a hygroscopic solution from the supracoxal glands and extract water from air that is subsequently ingested. Survival during dry periods is by means of a desiccation-resistant protonymphal stage. *D. pteronyssinus* can be reared under experimental conditions using shaved human beard growth as a culture substrate. Under natural conditions, *D. pteronyssinus* feeds on human skin scales and its life cycle (from egg to adult mite) can be completed within 21 days under optimum environmental settings. Laboratory populations of *D. pteronyssinus* grow rapidly at 25°C and 75–80% RH. A female is able to produce 100 eggs within its lifespan (10 weeks), of which about 90 are produced within the first 5 weeks.

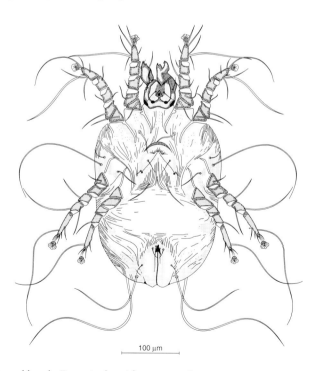

100 μm

Fig. 65. Ventral view of female *Dermatophagoides pteronyssinus.*

The allergens associated with house dust mites are present not only in the living mite itself but also in its secretions, body parts, dead skins and excreta. Although still controversial, there is increasing circumstantial evidence that house dust mite allergens play a role in atopic eczema; however, their association with atopic asthma is well accepted. To prevent allergic sensitization and asthma problems in the domestic environment, it is often recommended that mite populations should be reduced by maintaining a low relative humidity and dust-free surfaces, including limiting soft furnishings (particularly carpets) and regular vacuuming of those that exist. None the less, there is little evidence to demonstrate that this approach has clinical benefits.

The so-called 'storage' mites are commonly found infesting stored food or other materials of plant or animal origin, including nuts, grains and flour, cheeses and dried meats, and hay and straw; sensitive individuals handling heavily infested produce may develop dermatitis. Historically, the mites have had occupational associations; for example, *Tyrophagus putrescentiae* has been associated with 'copra itch', *Glycyphagus domesticus* with 'grocer's itch' and *Lepidoglyphus* (formerly *Glycyphagus*) *destructor* with 'hay itch', but they can be found in large numbers in warehouses and homes, causing skin reactions in workers and residents, and can be responsible for some of the allergic material associated with house dust.

Different allergens are produced by the stored food mites, *Acarus siro*, *T. putrescentiae*, *L. destructor* and *Blomia kulagini*. Sensitization to storage mites is not restricted to occupational exposure but occurs among people living in urban environments. Persons handling infested material contact the mites, which will crawl on to the skin and can penetrate the upper strata, resulting in a hypersensitivity dermatitis caused by the mites and their products (but they do not feed on blood or skin tissues). The allergens can also be inhaled or ingested and resultant anaphylactic reactions have been reported. Management is dependent on identifying the sources of the mites and instituting preventive and avoidance procedures.

5.2 Sarcoptidae

Sarcoptid mites are parasitic throughout their life, burrowing into the skin of mammals. They are globose mites with a flat ventral surface, the cuticle finely striated and the chelicerae adapted for cutting and paring. Within the Sarcoptidae, the species *Sarcoptes scabiei* is an important human and animal parasite, while *Notoedres cati* is of minor veterinary importance.

5.2.1 *Sarcoptes scabiei*

S. scabiei (Fig. 66), the itch mite, causes scabies in humans and sarcoptic mange in a wide range of domestic and wild mammals worldwide. Hosts include chimpanzees and gibbons within the Primates; horses and tapirs in the Perissodactyla; domestic and wild bovids, sheep, goats, kudu and hartebeest, Old World camels, South American llamas, pigs within the Artiodactyla; lions, foxes, wolves, dogs and ferrets within the Carnivora; rabbits and guinea pigs within the Rodentia; and koalas and wombats within the Marsupialia.

The only known species is *S. scabiei*. However, detailed morphological studies on *Sarcoptes* mites from a wide range of hosts have shown that some morphological characters, such as body size and the length of some setae, vary within the same population, between populations of different hosts or even between populations on the same host species but from different localities. The populations of *S. scabiei* infesting different mammalian species may differ physiologically as well as morphologically and populations show local adaptation to a particular host. Hence, populations from one host species do not usually establish themselves readily on another host species. In the latter case, they are often responsible for transitory infestation. Human infestations with *S. scabiei* acquired from infected dogs, horses or pigs are usually mild and self-limiting.

S. scabiei can be identified readily by its size, shape and morphology. The skin is striated and bears a central patch of raised scales dorsally (Fig. 66), which rarely posterolaterally. The dorsal surface bears three pairs of lateral spines about midway along the

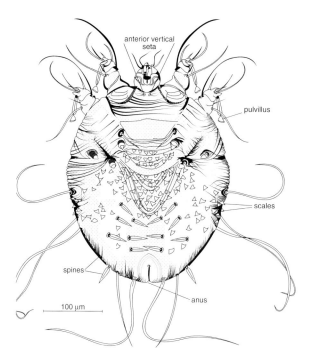

Fig. 66. Dorsal view of female *Sarcoptes scabiei.*

body and six or seven pairs of spines posteromedially. A pair of anterior vertical setae of taxonomic significance is located dorsally on the propodosoma, behind the gnathosoma. In both sexes, the pretarsi of legs I and II bear empodial claws and stalked pulvilli; the latter are sometimes referred to as suckers. Legs III and IV in the female end in long setae and lack stalked pulvilli; they are located on the ventral surface and are not visible in dorsal view. In the female, the genital opening is a transverse slit in the middle of the ventral surface of the body and the copulatory bursa, in which the spermatozoa are deposited, is on the dorsal side just anterior to the anus, which is terminal and slightly dorsal.

The male is morphologically similar to, but smaller than, the female and is characterized by the presence of stalked pulvilli on the fourth pair of legs (Fig. 67); between the latter is the obvious sclerotized genital apparatus. The nymphs are similar but smaller than the female and lack a genital opening, while the larvae resemble nymphs but have only one pair of legs posteriorly.

LIFE CYCLE

Female mites are found at the end of burrows in the horny layer of the skin. The burrows contain faeces and the relatively large eggs, which are laid singly. The egg of *S. scabiei* hatches within 50–53 h, giving rise to a hexapod larva, which moves rapidly to the surface of the skin. Larvae find shelter, and presumably also food, by entering hair follicles. Within 2–3 days, the larva moults into an octopod protonymph, which is also found in hair follicles. This moults to become a tritonymph and, after a final moult, to become an adult male or immature female. At this stage, both genders are about 250 µm in length, they make burrows (<1 mm in length) into the skin and mate. Subsequently, the females generate longer permanent burrows. As the ovaries develop, the females increase in size so that the mature female is 300–500 µm in length. The female never leaves the burrow but, if removed undamaged, will generate another. Each female buries herself in the skin within 1 h, using the chelicerae and empodial claws on the pretarsi of the first two pairs of

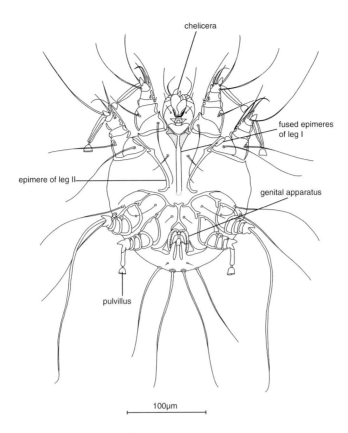

chelicera

fused epimeres
of leg I

epimere of leg II

genital apparatus

pulvillus

100µm

Fig. 67. Ventral view of male *Sarcoptes scabiei.*

legs. Each burrow is extended by the female by 0.5–5 mm each day. The females become mature within 3–4 days and each lays from 1–3 eggs/day during a reproductive cycle which last about 2 months. The total length of the life cycle from egg to egg is about 10–14 days, during which mortality may be as high as 90%. It has been estimated that within about 2 months the ovigerous female population will increase by 17-fold.

SCABIES IN HUMANS
Under most circumstances, transmission of *S. scabiei* mites from one individual to another is highly unlikely. Cases of transmission usually occur during close bodily contact under warm conditions, such as in bed, when the mobile *S. scabiei* exhibit a strong thermotaxis, showing a tendency to move to the warmest part of a temperature gradient and responding positively to both the temperature and the

odour of a host. Most new infestations are initiated by a newly inseminated female mite, which starts to establish a permanent burrow.

Survival of *S. scabiei* off the host is dependent on a high relative humidity restricting dehydration. Survival is greatest at low temperatures and high relative humidities, but *S. scabiei* is virtually immobile at temperatures below about 20°C. Survival beyond 48 h, at normal ambient temperatures, is rare. However, live *S. scabiei* have been recovered from the home environment of scabietic patients, indicating that transmission via fomites is possible.

The incidence of scabies in the human population generally peaks through the autumn/fall into winter and declines in spring. In humans, symptoms of scabies appear one or more months after infestation, during which sensitization occurs but without reactions. Following sensitization, the host reacts strongly by scratching due to intense itchiness; this facilitates the establishment of secondary infections, which complicate the pathogenesis of mite infestation. People who have received treatment with an acaricide retain sensitivity and may continue to experience itchiness for another month or so. However, when reinfected, reaction is immediate. Following a persistent infection, ultimately a stage of tolerance may be reached, which may result in crusted scabies (formerly called Norwegian scabies) in which large numbers of mites occur within hyperkeratotic skin of a host who does not itch or scratch. This condition frequently involves mites infesting the nail beds or nail plates on hands or feet. The existence of asymptomatic carriers with large numbers of mites impacts severely on control strategies of outbreaks of scabies and emphasizes the importance of adequate treatment for both symptomatic and asymptomatic individuals.

A firm diagnosis of scabies or mange must rely on visualization and morphological identification of mites from the affected host. In heavy infestations, skin scrapings cleared in alkali (10% potassium hydroxide) and viewed under the light microscope can reveal mites, their eggs aligned in the tunnels and faecal pellets throughout. When diagnosing scabies clinically and selecting sites for skin biopsies, it is important to consider that the distribution of the rash on the body may not relate to the distribution of the mites. Nearly two-thirds of the mites infecting a host are found on the hands and wrists, whereas the remaining third is more or less distributed equally between the elbows, feet and genital area. The rash often develops bilaterally, being concentrated on the axillae, waist and inner and posterior parts of the upper thighs and buttocks. The use of contact microscopy allows the rapid examination of a large number of lesions without discomfort to the patient.

To treat a human infestation with *S. scabiei*, topical acaricides should be applied thoroughly to all areas of the adult body from the chin downwards and including the face, head and neck in babies. All members living within the same household should be treated together, regardless of whether or not they show symptoms of infection. Applications should be repeated once. Older skin treatments, such as

benzyl benzoate, have been replaced largely by permethrin-based products (although resistance by the mite to permethrin has been reported). Crotamiton can also be effective (and is suitable for younger children), but may require repeated applications, and oral ivermectin (often along with topical permethrin) has been used successfully in difficult cases, such as with communal/institutional infestations and immunocompromised individuals with crusted scabies. Sharing of bed linen and clothes between infected and non-infected individuals should be avoided. Clothing and bed linen which has been in contact with the infected individual should be washed at a high temperature (>60°C) and left untouched for a minimum of 3 days.

SARCOPTIC MANGE IN ANIMALS

On animals, *S. scabiei* is found more frequently on the sparsely-haired regions of the body, such as the face and ears of goats, sheep and rabbits; the hock, muzzle and root of the tail of dogs and foxes; the inner surface of the thighs, underside of neck and brisket and around the base of the tail of cattle, from which it may spread to the whole body surface within 6 weeks; the head and neck of equines and the trunk, head and ears of pigs. The burrowing and feeding of mites in the skin cause irritation and consequential scratching, which leads to inflammation and exudations that form crusts on the skin. If left untreated, the skin wrinkles and thickens with proliferation of the connective tissue. This is followed by depilation (loss of hair). Small foci of infection do not usually impact on the health of an animal; however, under certain conditions the infestation may spread all over the body and, if untreated, cause death. Mange is a disease associated with animals in poor condition and is therefore more common at the end of winter or early spring.

The spread of *S. scabiei* among animals is by close contact; therefore, it is frequent in herds of domestic animals and in families or groups of wild animals. Large numbers of mites can be found in the ears of infected sows and transmitted to piglets after farrowing. Environmental conditions affect the survival of *S. scabiei* off its host. Housing previously occupied by infected animals should either be

Plate 1. Dorsal view of a bull ant, *Myrmecia* species (Hymenoptera, Formicidae). Courtesy of S.L. Doggett.

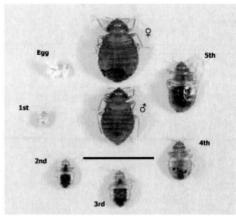

Plate 2. Life cycle of the common bed bug, *Cimex lectularius* (Hemiptera, Cimicidae), showing the five different instars, both adults and the egg stage. Bar = 5 mm. Courtesy of S.L. Doggett.

Plate 3. Side view of a near replete adult female common bed bug, *Cimex lectularius* (Hemiptera, Cimicidae). Courtesy of S.L. Doggett.

Plate 4. Side view of a wasp (Hymenoptera, Vespidae). Courtesy of S.L. Doggett.

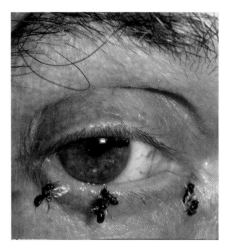

Plate 5. Bees at the human eye. Seven workers of *Lisotrigona cacciae* (Hymenoptera, Apidae, Meliponini) (one airborne, one on top of another) imbibing tears. Courtesy of H. Bänziger.

Plate 6. Rove beetles of the genus *Paederus* (Coleoptera, Staphylinidae). Courtesy of S.L. Doggett.

Plate 7. Larvae of a carpet beetle, *Anthrenocerus* species (Coleoptera, Dermestidae). Courtesy of S.L. Doggett.

Plate 8. Side view of a biting midge, *Culicoides* species (Diptera, Ceratopogonidae). Courtesy of S.L. Doggett.

Plate 9. A black fly, *Austrosimulium* species (Diptera, Simuliidae). Courtesy of S.L. Doggett.

Plate 10. A blow fly, *Calliphora* species (Diptera, Calliphoridae). Courtesy of S.L. Doggett.

Plate 11. *Lucilia* flowflies (Diptera, Calliphoridae) ovipositing on carrion. Courtesy of Richard Wall.

Plate 12. Larvae of *Lucilia sericata* (Diptera, Calliphoridae) on the skin of a black-fleeced sheep. Courtesy of the University of Bristol.

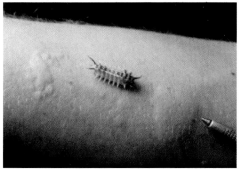

Plate 13. An urticating caterpillar of the white cedar moth, *Leptocneria reducta* (Lepidoptera, Lymantridae). Courtesy of S.L. Doggett.

Plate 14. Skin lesions due to a caterpillar, *Doratifera* species (Lepidoptera, Limacodidae). Courtesy of R.C. Russell.

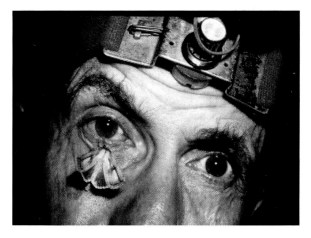

Plate 15. Male *Chaeopsestis ludovicae* (Lepidoptera, Thyatiridae) drinking tears on human eyes in North Thailand. Courtesy of H. Bänziger.

Plate 16. The house centipede, *Allothereua maculata* (Chilopoda, Scutigeridae). Courtesy of S.L. Doggett.

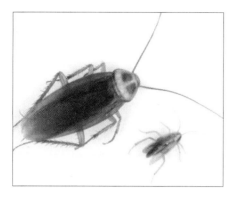

Plate 17. The cockroaches *Periplaneta americana* (left) and *Blatella germanica* (right) (Blattodea, Blatellidae). Courtesy of R.C. Russell.

Plate 18. Eye fly, *Phortica variegata* (Diptera, Drosophilidae), the vector of *Thelazia callipaeda*, feeding on human eye in Southern Italy. Courtesy of D. Otranto.

Plate 19. SEM of the cat flea, *Ctenocephalides felis* (Siphonaptera, Pulicidae). Courtesy of A.R. Walker.

Plate 20. A plague flea, *Xenopsylla* species (Siphonaptera, Pulicidae). Courtesy of A.R. Walker.

Plate 21. Dog massively infested by the cat flea, *Ctenocephalides felis* (Siphonaptera, Pulicidae). Courtesy of R.P. Lia.

Plate 22. An adult flesh fly (Diptera, Sarcophagidae). Courtesy of S.L. Doggett.

Plate 23. Severe traumatic myiasis in the vulva region of a sheep caused by flesh fly, *Wohlfahrtia magnifica* (Diptera, Sarcophagidae). Courtesy of A. Giangaspero.

Plate 24. An adult horn fly, *Haematobia irritans* (Diptera, Muscidae). Courtesy of A. Giangaspero.

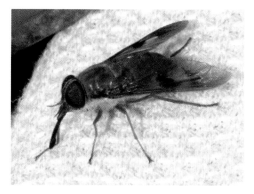

Plate 25. An adult horse fly (Diptera, Tabanidae).
Courtesy of S.L. Doggett.

Plate 26. An adult house fly, *Musca domestica.*
(Diptera, Muscidae). Courtesy of S.L. Doggett.

Plate 27. Different larval stages of the human bot fly, *Dermatobia hominis* (Diptera, Oestridae). Scale in mm. Courtesy of A.R. Walker.

Plate 28. An adult ked or louse fly (Diptera, Hippoboscidae) from
the subfamily Ornithomyiinae (probably *Ortholfersia* spp.).
Courtesy of S.L. Doggett.

Plate 29. Kissing bug, *Triatoma dimidiata* (Hemiptera, Reduviidae), vector of Chagas' disease. Courtesy of Guanyang Zhang.

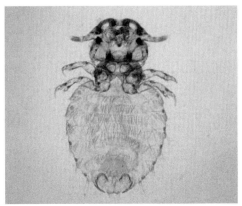

Plate 30. Female of the chewing louse of dogs, *Trichodectes canis* (Phthiraptera, Mallophaga). Courtesy of D. Otranto.

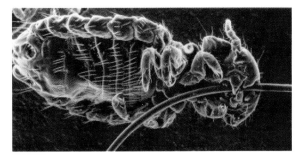

Plate 31. SEM of the chewing louse of sheep, *Bovicola bovis* (Phthiraptera, Mallophaga). Courtesy of A.R. Walker.

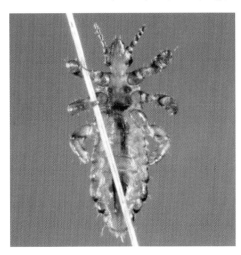

Plate 32. An adult human head louse, *Pediculus humanus capitis* (Phthiraptera, Anoplura). Courtesy of S.L. Doggett.

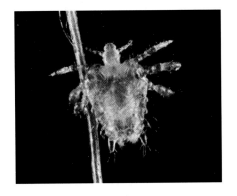

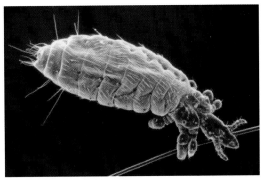

Plate 33. An adult human pubic louse, *Phthirus pubis* (Phthiraptera, Anoplura). Courtesy of S.L. Doggett.

Plate 34. SEM of the long-nosed sucking louse of cattle, *Linognathus vituli* (Phthiraptera, Anoplura). Courtesy of A.R. Walker.

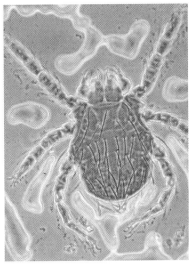

Plate 35. An adult millipede (Diplopoda). Courtesy of S.L. Doggett.

Plate 36. A chigger or harvest mite (Acari, Trombiculidae). Courtesy of R.C. Russell.

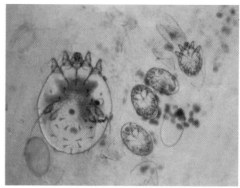

Plate 37. Furfuraceous dermatitis by the fur mite, *Cheyletiella yasguri* (Acari, Cheyletiellidae). Courtesy of R.P. Lia.

Plate 38. The scabies mite, *Sarcoptes scabiei* (Acari, Sarcoptidae). Females, embryos and empty eggs within a cleared skin scraping. Courtesy of M.J. Geary.

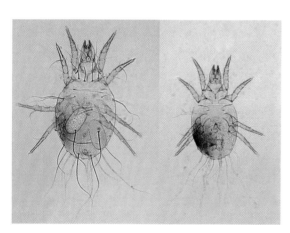

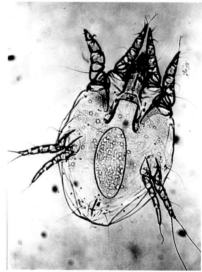

Plate 39. The stored product mite, *Tyrophagus putresecentiae*, female with eggs on the left and male on the right. Courtesy of S.L. Doggett.

Plate 40. Adult female of the sheep scab mite, *Psoroptes ovis* (Acari, Psoroptidae), with a single egg visible in the opisthosoma. Courtesy of the University of Bristol.

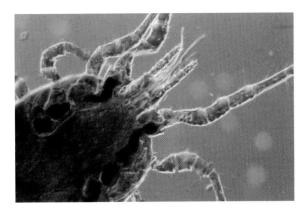

Plate 41. The fowl mite, *Dermanyssus gallinae* (Acari, Dermanyssidae). Courtesy of R.P. Lia.

Plate 42. Larvae of the harvest mites, *Neotrombicula autumnalis* (Acari, Trombiculidae), on the face of a sheep. Courtesy of R.P. Lia.

Plate 43. The yellow fever (and dengue) mosquito, *Aedes aegypti* (Diptera, Culicidae). Courtesy of S.L. Doggett.

Plate 44. The Asian tiger mosquito, *Aedes albopictus* (Diptera, Culicidae). Courtesy of S.L. Doggett.

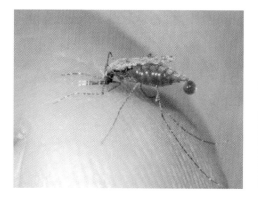

Plate 45. A malaria mosquito, *Anopheles farauti* (Diptera, Culicidae). Courtesy of S.L. Doggett.

Plate 46. Larvae of the filariasis vector, *Culex quinquefasciatus* (Diptera, Culicidae). Courtesy of S.L. Doggett.

Plate 47. Late instar larvae of the sheep bot fly, *Oestrus ovis* (Diptera, Oestridae). Courtesy of S.L. Doggett.

Plate 48. Larvae of the nostril fly, *Oestrus ovis* (Diptera, Oestridae), at different developmental stages in the nasal cavity of a sheep. Courtesy of R.P. Lia.

Plate 49. Female of a sand fly, *Phlebotomus* spp. (Diptera, Psychodidae). Courtesy of D. Otranto.

Plate 50. An adult marbled scorpion, *Lychas* spp. (Scorpiones, Buthidae). Courtesy of S.L. Doggett.

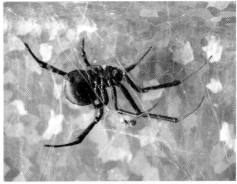

Plate 51. An adult male Sydney funnel web, *Atrax robustus* (Arachnida, Hexathelidae). Courtesy of S.L. Doggett.

Plate 52. An adult female red back spider, *Latrodectus hasseltii* (Arachnida, Theridiidae). Courtesy of S.L. Doggett.

Plate 53. Side view of a stable fly, *Stomoxys calcitrans* (Diptera, Muscidae). Courtesy of S.L. Doggett.

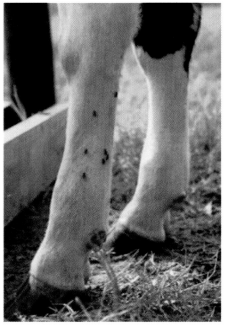

Plate 54. Stable flies, *Stomoxys calcitrans* (Diptera, Muscidae), feeding on cow's leg. Courtesy of A. Giangaspero.

Plate 55. Massive gastric infestation by larvae of the stomach bot fly, *Gasterophilus intestinalis* (Diptera, Oestridae). Courtesy of D. Otranto.

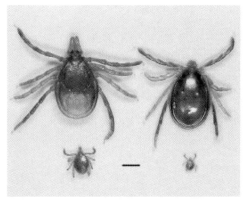

Plate 56. Life stages of the paralysis tick, *Ixodes holocyclus* (Acari, Ixodidae). From top left clockwise, adult female, adult male, larvae, nymph. Bar = 1 mm. Courtesy of S.L. Doggett.

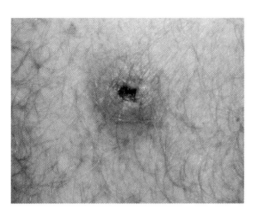

Plate 57. Rickettsial eschar following infected ixodid tick bite. Courtesy of R.C. Russell.

Plate 58. Female of *Amblyomma variegatum* (Acari, Ixodidae). Courtesy of A.R. Walker.

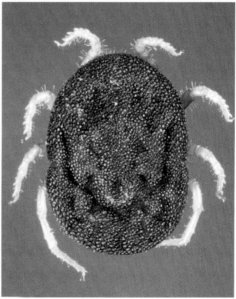

Plate 59. Dog heavily infested by the brown dog tick, *Rhipicephalus sanguineus* (Acari, Ixodidae). Courtesy of D. Otranto.

Plate 60. The soft tick, *Ornithodoros savignyi* (Acari, Argasidae). Courtesy of A.R. Walker.

Plate 61. An adult tumbu fly, *Cordylobia anthrophaga* (Diptera, Oestridae). Courtesy of S.L. Doggett.

Plate 62. Adult female of the tsetse fly, *Glossina morsitans* (Diptera, Glossinidae). Courtesy of the Tsetse Research Laboratory, University of Bristol.

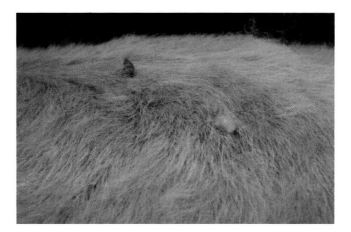

Plate 63. Cattle warble fly infestation by *Hypoderma lineatum* (Diptera, Oestridae). Courtesy of D. Otranto.

Plate 64. Larvae of warble fly, *Hypoderma bovis* (Diptera, Oestridae). Scale in mm. Courtesy of A.R. Walker.

left in a dry state for 3 weeks or be treated with an acaricide.

When a case of mange is diagnosed, treatment should be administered to all animals that have been in contact with the infected subjects, because the early stages of infection may be asymptomatic. The external application of organophosphorus insecticides has proven effective in controlling *S. scabiei*, as has the administration of macrocyclic lactones, such as ivermectin, given subcutaneously or orally. The latter is now the treatment of choice for sarcoptic mange, a single dose being completely effective in pigs. A second treatment may be necessary in cases of heavy infestations in goats.

For diagnosis, a skin scraping of the infected area is necessary; the material may be examined directly or preferably after disrupting the keratin by boiling for a few minutes in 10% caustic soda or potassium hydroxide. The fluid is then centrifuged and the sediment examined. In skin scrapings, males are rarer than females, probably as a consequence of their shorter lifespan.

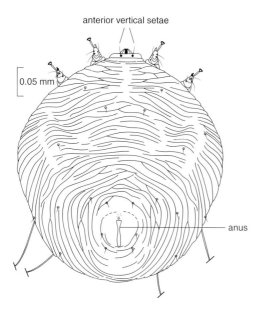

Fig. 68. Dorsal view of a female *Notoedres muris*. *Source*: Lavoipierre, M.M.J. (1964) *Journal of Medical Entomology* 1, 11. Reproduced by permission of the editor.

5.2.2 *Notoedres*

More than 20 species of *Notoedres* have been described, most of them being parasites of tropical bats. Three species are of interest to the veterinary entomologist and one, *N. cati*, is important. *Notoedres muris* (Fig. 68) occurs on rats throughout the world, including laboratory colonies, and *Notoedres musculi* infests the house mouse in Europe.

N. cati is a mange mite of cats that, on occasions, may also infest dogs, and it can cause a transient dermatitis in humans. A morphologically identical mite also occurs on rabbits, and the unlikelihood of a cat parasite being transmitted to rabbits led to the hypothesis that two sibling species may exist; however, there is as yet no firm evidence to support this suggestion. In both hosts, *N. cati* localize to the head and ears and, more rarely, in advanced cases, to legs, genitalia and perineum. The original lesion is similar to a pinhead in size, but as it spreads, a crust develops and hair falls out. Notoedric mange is highly contagious and intensely pruritic among cats.

Diagnosis is effected by recovering mites from skin scrapings. *N. cati* is morphologically similar to *S. scabiei*, with stalked pulvilli on legs I and II in all stages and on leg IV in the male. However, it is considerably smaller than *S. scabiei*, the female being 225 μm and the male 150 μm in length. In *N. cati*, the anus is located on the dorsal surface, as it is in *N. muris* (Fig. 68); there are no projecting scales but, mid-dorsally, the striae are broken into a scale-like pattern and stout setae replace the lanceolate spines of *S. scabiei*.

The female burrows into the stratum corneum, in which eggs are laid. Larvae and nymphs may stay in the female's burrow or move on to the surface of the skin, where they form small pits in which they moult. All moults may occur in the pit made by the larva, or each developmental stage may form a separate pit. The immature females remain in the moulting pit until inseminated, after which they form a permanent burrow. The life cycle from egg to adult can be completed within 17 days, and maturation of the adult female and deposition of the first egg 4–5 days after the final moult. Hence, a new generation appears within 3 weeks. The ovigerous female lays about 60 eggs in a lifetime of 2–3 weeks at a rate of 3–4 eggs/day.

5.2.3 *Trixicarus caviae*

Trixicarus caviae is a burrowing mite of guinea pigs which superficially resembles *S. scabiei* and has a very similar life cycle. Its burrowing activity results in irritation, inflammation and pruritus, causing biting, scratching and rubbing of the infested areas and leading to alopecia. The dorsal striations of the idiosoma of *T. caviae* are similar to those of *S. scabiei*. However, the dorsal scales, which interrupt the striations, are more sharply pointed and the dorsal setae are simple and not spine-like. Like *N. cati*, the anus is located on the dorsal surface. *T. caviae* is also smaller than *S. scabiei* and similar in size to *N. cati*; females are about 240 μm in length and 230 μm in width. It originated in South America but it is now spread worldwide. The infestation spreads quickly from the initial lesions to cause generalized mange. Death may occur within

3–4 months after infestation. Transmission is by close physical contact and from mother to offspring. Treatment involves the administration of macrocyclic lactones such as ivermectin, twice at 7- to 10-day intervals. All infested bedding must be replaced and the guinea pig's local environment thoroughly cleaned.

5.3 Psoroptidae

Members of the Psoroptidae are oval, non-burrowing mites which are parasites of the skin of mammals. The third and fourth pairs of legs are usually visible from above, the epimeres of the first pair of legs are not fused and no vertical setae are present on the propodosoma. On the ventral surface of the ovigerous female, just posterior to the second pair of legs, an obvious inverted U-shaped genital opening is present, through which the eggs are passed (Fig. 69). Two nymphal stages

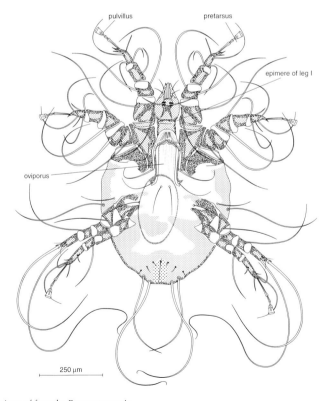

Fig. 69. Ventral view of female *Psoroptes ovis*.

occur during the life cycle. The male has prominent adanal copulatory suckers (see Fig. 64) which engage with copulatory tubercles on the female tritonymph (Fig. 70). Three genera, *Psoroptes*, *Chorioptes* and *Otodectes*, are of veterinary importance, and mites within the latter genus are also of occasional medical significance.

5.3.1 *Psoroptes*

Psoroptes is a cosmopolitan genus of obligate ectoparasites which cause a debilitating dermatitis involving hair or wool loss and pruritic scab formation. All stages of *Psoroptes* are distinguished by sucker-like pulvilli borne at the end of long, jointed pretarsi (peduncles) (Figs. 64 and 69). Five species have been acknowledged: *Psoroptes ovis*, causing body mange in sheep, cattle and horses; *Psoroptes equi*, a body mite of horses; *Psoroptes cuniculi*, an ear mite of rabbits, goats, horses, sheep, bighorn sheep and deer; *Psoroptes natalensis*, a body mite of domestic cattle occurring mainly in the southern hemisphere (South Africa, South America and New Zealand); and *Psoroptes cervinus*, an ear mite on the American bighorn sheep. However, it has proved difficult to separate the species based on morphological characters, host specificity or genetic markers; hence, the evidence supports the hypothesis that, like *S. scabiei*, one single species, *P. ovis*, exists, which exhibits host-adapted populations.

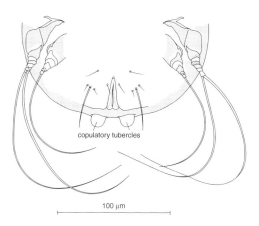

copulatory tubercles

100 µm

Fig. 70. Ventral view of hysterosoma of female tritonymph (female) of *Psoroptes ovis*.

The ovigerous *P. ovis* is about 750 µm long and lays oval, white eggs which are about 250 µm long. The larva which emerges has pulvilli on legs I and II and leg III terminates in two long setae. The larva is about 330 µm in length; it moults to become a protonymph, with pulvilli on legs I, II and IV and long setae on leg III, while legs III and IV are shorter and less stout than legs I and II. The female protonymph and tritonymph are similar but bear copulatory tubercles posterodorsally (Fig. 70).

The adult male is readily identifiable by its paired adanal copulatory suckers and paired posterior lobes, which carry two long and three short setae (see Fig. 64). Leg III is the longest, leg IV the shortest and legs I and II are markedly stouter than legs III and IV. Pulvilli are borne on legs I, II and III. Female tritonymphs attach to males until they moult to ovigerous females, which is when insemination occurs. The legs of the ovigerous females are more or less similar, with pulvilli on all except leg III, which bears two long setae. A female lives 11–42 days and lays between 30 and 40 eggs at a rate of 1–5 eggs/day.

In each developmental stage of the life cycle, a period of active feeding is followed by a quiescent immobile phase prior to moulting. Under optimal conditions, the quiescent phase lasts about 1 day and the active feeding phase about 2 days. The minimum duration of the life cycle from egg to egg is about 11 days.

VETERINARY IMPORTANCE

Psoroptes mites are non-burrowing and feed superficially on a lipid emulsion of skin cells, bacteria and lymph on the host skin, produced as a result of a hypersensitivity reaction to the presence of antigenic mite faecal material. This hypersensitivity causes inflammation, surface exudation, scale and crust formation, with excoriation (scratching) due to self-trauma. Infestation is described as psoroptic mange or 'sheep scab'. The serous exudate produced in response to the mites dries on the skin to form a dry, yellow crust, surrounded by a border of inflamed skin covered in moist crust. Mites are found on the moist skin at the edge of the lesion, which extends rapidly and may take as little as 6–8 weeks to cover three-quarters of the host's skin. Eventually, the crust lifts off as the new fleece grows.

In sheep, lesions may occur on any part of the body but are particularly obvious on the neck, shoulders, back and flanks. In cattle, lesions usually appear on the withers, neck and around the root of the tail, from which, in severe cases, they may spread to the rest of the body. Populations of *Psoroptes* may also be found localized in the ears of sheep, causing chronic irritation, often associated with haematomas, head shaking and scratching.

Sheep scab can affect sheep of all ages but may be particularly severe in young lambs and sheep in poor condition. The incidence of the disease varies according to season. In warm weather, mite populations may decline, with residual populations left in sites such as the axilla, groin, infra-orbital fossa and inner surface of the pinna and auditory canal during spring, summer and early autumn.

Transmission is primarily through physical contact and the majority of sheep become infected while the mites are active and multiplying. However, transmission may also be acquired from the environment. The survival time off host is strongly dependent on environmental temperature and humidity and, at low temperatures (<15°C) and high humidity (>75%), survival may extend beyond 18 days, allowing transmission from housing, bedding or contaminated machinery such as shearing equipment. Periods of decline in the mite population, either as a response to environmental conditions or to the establishment of the host's immune response, affect the epidemiology of the disease significantly. Sheep that appear not to be infested but carry small populations of mites may be responsible for the introduction of the disease to healthy flocks during summer and autumn, and subsequently for outbreaks.

Sheep scab was eradicated from Australia and New Zealand, but it remains common in sheep-rearing areas of northern Europe and South America. The main source of infection of a flock is through the introduction of new animals. Every new individual must be inspected thoroughly and undergo a quarantine period, if possible. Two treatments with injectable macrocyclic lactone, such as ivermectin at 200 μg/kg at an interval of 7 days, have proven successful in clearing *P. ovis* infestation. Alternatively, single injections of doramectin (300 μg/kg) or moxidectin (200 μg/kg) are effective in controlling infestations. All are now licensed in several countries for this purpose. The short population turnover period facilitates rapid spread and, because of this, legislative control of the parasite occurs in many countries, since the economic burden of uncontrolled sheep scab can be severe.

5.3.2 *Chorioptes*

Chorioptic mange is the commonest form of mange in horses and cattle; it may also be common in llamas and alpacas. Detailed studies of *Chorioptes* have suggested that two distinct species exist: *Chorioptes bovis* and *Chorioptes texanus*, separated by differences in the lengths of the posterior setae of adult males. Both are found infesting the body of their host, but no clear host preference appears to exist in the two putative species. No behavioural differences in their parasitic behaviour have been recorded. A third, as yet unnamed, species found within the ear canal of reindeer and moose may also exist.

Adult female *C. bovis* are about 300 μm in length, considerably smaller than *P. ovis*. *Chorioptes* (Fig. 71) do not have jointed pretarsi; their pretarsi are shorter than in *Psoroptes* and the sucker-like pulvillus is more cup-shaped, as opposed to trumpet-shaped in *Psoroptes*. In the adult female, tarsi I, II and IV have short-stalked pretarsi and tarsi III have a pair of long, terminal, whip-like setae. The first and second pairs of legs are stronger than the others and the fourth pair has long, slender tarsi. In the male, all legs possess short-stalked pretarsi and pulvilli. However, the fourth pair is extremely short, not extending beyond the body margin. Male *C. bovis* have two broad, flat setae and three normal setae on well-developed posterior lobes (Fig. 71). The mouthparts are distinctly rounder and the abdominal tubercles of the male are noticeably more truncate than those of *Psoroptes*. The life cycle of *C. bovis* is similar to that of *P. ovis*. *Chorioptes* mites are thought to feed on skin debris; however, this still remains to be demonstrated.

In horses, the mites occur on the lower parts of the legs and are rarely found on other parts of the body. They are a source of irritation and reduce the performance of

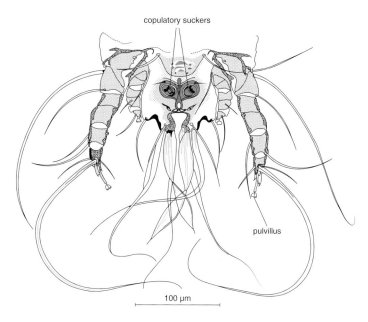

copulatory suckers

pulvillus

100 µm

Fig. 71. Ventral view of hysterosoma of male *Chorioptes bovis*.

infested horses. In cattle, *Chorioptes* mites most commonly cause lesions at the base of the tail, on the perineum and the udder. Chorioptic mange occurs most commonly during winter, whereas during summer the mites are found on the area above the hooves on the hindlegs. In cattle, most cases are characterized by little damage; however, occasional severely pathogenic cases may occur. In sheep, *Chorioptes* mites may affect the wool-less areas, particularly the lower parts of the hindlegs and scrotum, and the latter may affect the fertility of rams. Infection is passed from animal to animal by contact and possibly by grooming tools.

Infestation by *Chorioptes* can be controlled with the dips used for psoroptic mange, repeated at 2-week intervals. Ivermectin, doramectin, eprinomectin and moxidectin applied topically as a pour-on are also effective against chorioptic mange. Regular checks of livestock and quarantining of infected animals will aid the control of the frequency and extent of infestations.

5.3.3 *Otodectes* cynotis

Otodectes mites generally live deep in the ear canal, near the eardrum, of dogs, foxes, cats and ferrets; however, lesions caused by these mites may also be seen on other regions of the body. These mites are mainly parasites of carnivores. They resemble *Chorioptes*, as they are of similar size and also have unjointed pretarsi (Fig. 72). They can be reared *in vitro* on epidermal debris and hair from the ear canal of carnivores. In addition to the hosts listed above, the life cycle has been completed *in vitro* on ear debris from other carnivores including coyote, timber wolf and black bear.

The life cycle of *Otodectes cynotis* is similar to that of other psoroptids. In the tritonymph, copulatory tubercles occur in both genders. The adult male has pulvilli on all four pairs of legs, copulatory suckers, weakly developed posterior processes and a slightly emarginate hind margin to the body (Fig. 72). Males attach to female tritonymphs and copulation occurs as the adult female emerges. Females that are not attached are not inseminated at ecdysis and are infertile. In the ovigerous female, leg IV is reduced and lacks a pulvillus.

In heavily infested cats and dogs, convulsions may occur. This condition requires treatment with a suitable acaricide. In kennels and catteries, since the mites can survive for

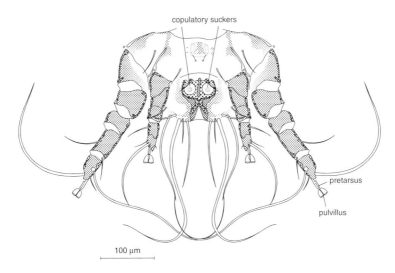

Fig. 72. Ventral view of hysterosoma of male *Otodectes cynotis.*

some time off their hosts (i.e. up to 17 days at 95% RH and 10°C), treatment of the premises (weekly for 4 weeks) with a residual acaricide such as diazinon or malathion is necessary. Individual animals may be treated with one or two subcutaneous injections of a macrocyclic lactone such as ivermectin or by applying Amitraz in mineral oil as ear drops twice a week.

5.4 Knemidokoptidae

Twelve species of Knemidokoptidae have been described, of which three are of veterinary importance. *Knemidokoptes mutans* and *Knemidokoptes gallinae* infest poultry and *Knemidokoptes pilae* (Fig. 73) is common on caged parakeets. Female knemidokoptid mites are about 400 μm long, but have neither spines, sharp-pointed scales nor anterior vertical setae. Anteriorly on the mid-dorsal surface, there are two sclerotized, more or less parallel, longitudinal bands, which are connected posteriorly by a less well-developed transverse band (Fig. 74). In the female, the epimeres of the first pair of legs are concave laterally and do not meet in the midline. In the male, the epimeres of the first pair of legs fuse in the midline and have a posteromedian extension (Fig. 75). Stalked pulvilli are present

on all legs of the male and larva but are absent in the nymphal stages and female. The female is viviparous and there are one larval and two nymphal stages.

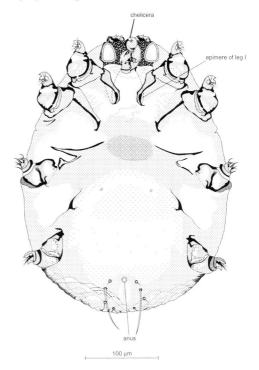

Fig. 73. Ventral view of female *Knemidokoptes pilae.*

5.4.1 *Knemidokoptes mutans*

K. mutans causes a condition known as 'scaly leg' in domestic poultry. At first, the infestation localizes on the legs to the lower ends of the tarsus and digits, where the epidermal scales swell up and exude a whitish, floury powder. This may develop into a thick, nodular, spongy crust and, in advanced cases, the comb and neck may also be affected. The disease develops slowly over many months, while the affected bird loses its appetite and considerable weight loss occurs. Diagnosis relies on the observation of adult mites on the underside of the crust where ovigerous females are found, usually surrounded by a proliferation of epidermal cells.

Female *K. mutans* can be distinguished from *K. gallinae* by the presence, mid-dorsally, of rounded or oval plaques resembling smooth scales. In *K. gallinae*, the dorsal surface presents regular striations, some of which may be very finely toothed (Fig. 76).

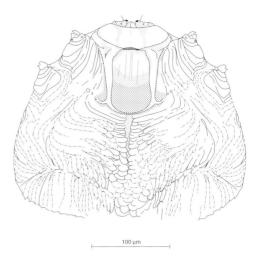

100 µm

Fig. 74. Dorsal view of the anterior half of a female *Knemidokoptes pilae*.

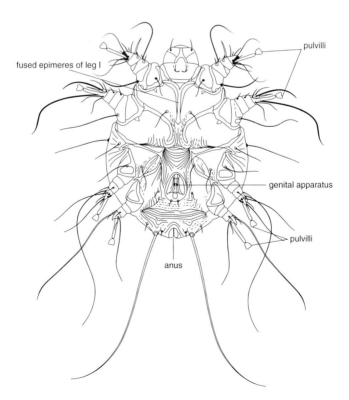

Fig. 75. Ventral view of a male *Knemidokoptes mutans*. *Source*: Hirst (1922).

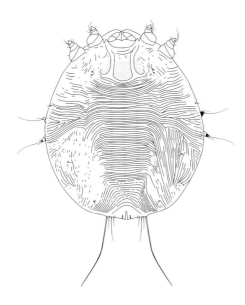

Fig. 76. Dorsal view of a female *Knemidokoptes gallinae*. *Source*: Hirst (1922).

5.4.2 *Knemidokoptes gallinae*

K. gallinae infests poultry, pheasants and geese, in which it is responsible for a condition known as 'depluming itch', characterized by the loss of feathers over extended areas of the body. The mite localizes at the base of the feathers on the back, head, neck, around the vent and on the breast and thighs. The condition can be diagnosed by plucking a few feathers from these areas and the mites will be found embedded in the tissue or scales at the base of the quill. This disease is more noticeable during spring and summer.

5.5 Cytoditidae

5.5.1 *Cytodites nudus*

This oval mite is usually longer than 500 μm. It infests the linings of the air sacs and air passages of chickens and, less frequently, of other poultry. Small numbers of mites have no noticeable effects on the host, but in large numbers they may cause death. These mites have sometimes been found within the peritoneal and thoracic cavities. Diagnosis is

only possible based on clinical symptoms or by observing the mites at the post-mortem examination. The mites have a smooth cuticle, largely devoid of striations (Fig. 77). They are characterized by a few short setae but no anterior vertical setae. The chelicerae are absent and the pedipalps are fused to form a soft sucking organ through which fluids exuded by the host are imbibed. The epimeres of the first pair of legs are fused into Y-shaped structures. In the female, all legs have long pretarsi bearing subglobose pulvilli, while the male is characterized by the presence of pulvilli and short pretarsi.

5.6 Laminosoptidae

5.6.1 *Laminosoptes cysticola*

Laminosoptes cysticola is present in many regions of the world, including North and South America, Europe and Australia. Large numbers of mites (millions) may occur in the cellular tissue of turkeys and chickens, where they destroy the fibres. The mites are responsible for the formation of nodules, which become calcified when the mites die, thus reducing the market value of the carcass and meat intended for human consumption.

Active mites occur deeply in the tissues; they are small, elongate mites, about 250 μm long, with smooth cuticle and few long setae (Fig. 78). The gnathosoma is recessed on the ventral surface. The epimeres of the first pair of legs are fused into a Y-shaped structure and those of the second pair of legs meet in the midventral line and then diverge posteriorly. Legs I and II bear claw-like tarsi and legs III and IV have been observed to end in long spatulate pretarsi, though some descriptions state that all legs end in a 'pedunculate sucker'.

The mode of transmission of this mite is unknown, but it has been estimated that around 1% of free-living urban pigeons harbour the mite. For treatment, macrocyclic lactones such as ivermectin may be effective, but euthanasia may be preferred for the rapid elimination of infected birds. Destroying or quarantining infected birds may aid reducing infestations within flocks.

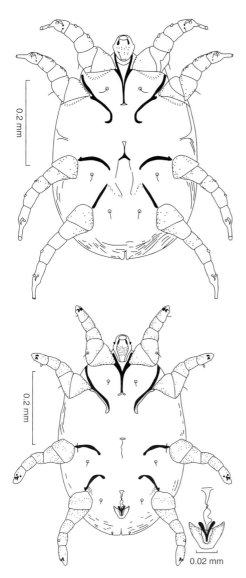

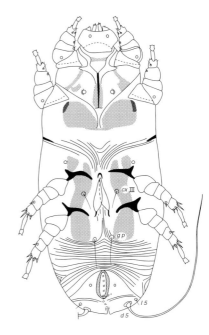

Fig. 78. Ventral view of female *Laminosoptes cysticola*. *Source*: Fain (1981).

Fig. 77. Ventral view of *Cytodites nudus* female (top) and male (bottom). *Source*: Fain (1960).

5.7 Prevention and control of mange mites in animals

A wide range of neurotoxic insecticides may be used to eradicate mange mites. Newer and more convenient products with a better residual effect are the systemic organophosphates, pyrethroids and the macrocyclic lactones. Avoiding close contact with infected individuals is the main way in which infection can be controlled. Preventing infection of livestock requires a number of different husbandry practices. For example, quarantining new untreated individuals before they enter the herd should be carried out for a period of 3 weeks to ensure no infection is brought into the herd. Communal grazing between different herds should be kept to a minimum and a closed herd policy maintained. Any animals suffering severe infestations should be culled and the most resistant individuals selected for. Diagnosis of mange is made by finding the mites in skin scrapings made from the moist areas at the periphery of the scabs.

6. PROSTIGMATA

The Prostigmata are a large, diverse and complex group of mites which vary in size from 100 μm to about 1 mm (although the trombiculid African red velvet mites, *Dinothrombium* spp., may grow up to 10 mm in length). Most species are free-living and survive by sucking the juices of animals and plants. Predatory prostigmatid mites occur in terrestrial, freshwater and marine habitats and include the Hydrachnidia or freshwater mites. Plant-feeding Prostigmata include the Tetranychidae or spider mites and the Eriophyoidea or bud mites, many of which are economically important pests in orchards and horticulture. Other Prostigmata are parasites of invertebrates, including insects, and vertebrates, including humans. The parasitic developmental stages are of medical and veterinary importance.

The typical prostigmatid mite is weakly sclerotized and, where there is an internal respiratory system, the stigmata open on the gnathosoma or anterior part of the propodosoma. It is this feature that gives rise to the name Prostigmata. The chelicerae are either blade-like, as in *Trombicula*, or styletiform, as in *Pyemotes*; they are rarely chelate. Specialized sensillae, the trichobothria, are often present on the propodosoma, such as in the larval stage of *Trombicula*. The coxae of the legs may be incorporated into the ventral surface of the body, as in *Trombicula*, and may be fused, as in *Demodex*.

6.1 Trombiculidae

Trombidioids are generally parasitic in the larval stage and free-living predators in the adult and nymphal stages, feeding on the eggs and juvenile stages of other arthropods. The Trombidioidea includes eight families, the most important of which is the Trombiculidae. More than 1200 species of trombiculid mites have been described and about 50 of these have been known to attack humans or livestock. They are widely distributed throughout the world. In Europe, they are known as harvest mites (e.g. *Neotrombicula autumnalis*), in the Americas as chiggers (e.g. *Eutrombicula alfreddugesi*) and in Asia and Australia as scrub itch mites (e.g. *Eutrombicula sarcina*). Infestation by the larvae causes a pruritic dermatitis in humans. In eastern and southeastern Asia, New Guinea and northern Australia, however, some species (e.g. *Leptotrombidium akamushi* and *Lepto-*

trombidium deliense) act as vectors of the rickettsial disease, scrub typhus. In the southern states of North America, *Neoschongastia americana* is an important pest of turkeys and a minor pest of chickens, and attacks free-ranging birds; however, infestations are sporadic and localized.

The prominent gnathosoma of the larva bears strong, blade-like chelicerae anteromedially (Fig. 79). The chelicerae are flanked by stout, segmented palps, the penultimate of which (palpal tibia) bears a claw. The terminal segment (palpal tarsus) is located opposite to the claw and bears ciliated setae. There are no true stigmata or tracheae and respiration is cutaneous. Behind the gnathosoma on the dorsal surface there is a scutum, which represents an important feature for the taxonomic classification of these mites. The scutum bears a pair of sensilla, or trichobothria, which are setaceous in many genera and inflated in some, e.g. *Schoengastia*. Typically, the scutum also bears five ciliated setae, one at each corner of the more or less rectangular scutum and a single anteromedian seta (paired in *Apolonta*). Both the ventral and dorsal body surfaces carry moderately long ciliated setae, the number and distribution of which are specific characters. Nymphal and adult trombiculids are known as velvet mites, a term which refers to the dense covering of pilose setae covering their legs, body and, to some

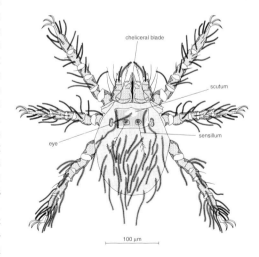

Fig. 79. Dorsal view of larva of *Leptotrombidium deliense.*

degree, palps. In these free-living stages, the stigmata open at the base of the chelicerae.

The larva attaches itself to the host by its chelicerae and feeds on the host's tissues by partially digesting them with saliva, which it pumps to and fro, leading to the formation of a feeding tube or stylostome in the host at the point of larval attachment. The larva feeds for several days before falling from the host and entering a quiescent phase prior to moulting to the protonymph. There are three nymphal stages but only one, the second, is active. The adult mite is about 1 mm long and its body is 'waisted', producing a figure-of-eight shape.

In the warmer regions of the world, trombiculid mites probably breed throughout the year; however, in the cooler regions, the number of generations per year is limited and infestations with trombiculid mites are seasonal; for instance, larvae of N. autumnalis are abundant in late summer and early autumn and hence are referred to as 'harvest bug'.

The female trombiculid deposits its spherical eggs in damp but well-drained soil. Larvae are about 250 μm long. They climb grass stems to a height of 60–80 mm to await the passage of a suitable host, to which they cling. This habit leads to the fact that the faces and legs of grazing animals are often the first sites of infestation by trombiculid larvae and the faces of horses and cattle are often affected. Infestations with E. sarcina can cause severe pruritus in horses and sheep.

6.1.1 Scrub typhus (Orientia tsutsugamushi)

Orientia tsutsugamushi (formerly known as Rickettsia tsutsugamushi and Rickettsia orientalis) is the aetiological agent of scrub typhus, an acute febrile disease, transmitted by trombiculid mites. O. tsutsugamushi occurs in an area bounded by Pakistan and Tadzhikistan in the west to Japan and south-east Siberia in the east, and to Indonesia, New Guinea and tropical northern Australia in the south. It is also present on certain islands in the Pacific and Indian Oceans, including Diego Garcia, midway between Madagascar and the Oriental region, where it is believed it was introduced by infected vectors on birds or flying foxes.

When an infected mite feeds on a susceptible human, an ulcer-like eschar usually forms at the site of the mite's attachment.

Typically, symptoms of disease include fever, severe headache, rash and lymphadenopathy lasting for 2–3 weeks, with severity according to strain of the pathogen and a mortality which may exceed 30% in untreated cases. Prompt treatment with tetracycline antibiotics (especially doxycycline) and also chloramphenicol can prevent mortality cases. The multiple serotypes of O. tsutsugamushi are responsible for the establishment of transient cross-immunity in humans, which is not protective against subsequent infections.

The association between scrub typhus and mites was reported in a 16th century Chinese manuscript on natural history and has been known for at least 200 years by the Japanese, who named the disease tsutsugamushi, which means dangerous mite. More than 1000 species of trombiculid mites have been described, but only about 10 species in the genus Leptotrombidium are considered to be major vectors of O. tsutsugamushi to humans. They include L. akamushi, L. deliense, Leptotrombidium pallidum and Leptotrombidium scutellare. Species of Ascoschoengastia are involved in transmitting O. tsutsugamushi among rodents, and other genera of chiggers such as Blankaartia, Gahrliepia, Eutrombicula, Microtrombicula and Odontocarus have also been found to carry the pathogen and may have a role in its transmission.

Scrub typhus is a zoonosis and humans may become infected when they enter an enzootic focus, where the pathogen, Leptotrombidium spp., and wild rodents, especially of the subgenus Rattus (Rattus), are present. Such foci tend to be characterized by the presence of transitional vegetation, an ideal habitat for Rattus species. L. akamushi and L. deliense have a wide geographical distribution within which they are patchily distributed, being abundant in ecologically favourable 'mite islands' and absent from other habitats. Only trombiculid mites which parasitize rodents can become infected with O. tsutsugamushi, while species of Eutrombicula and Schoengastia, which parasitize birds and reptiles, do not usually carry this pathogen. About 15 species of these last two genera cause scrub itch but not scrub typhus in humans. This explains why the distributions of scrub typhus and scrub itch do not necessarily overlap.

The environment in which scrub typhus occurs is essentially man-made. In Malaya, Sumatra, New Guinea and tropical Queensland, scrub typhus is associated with a coarse, rasping, fire-resistant grass, *Imperata cylindrica*, known in some areas as kunai grass. It provides a suitable habitat for field rodents and it is prevalent in areas where bush fires are common, as the latter prevent the establishment of shrubs and trees, whose shade would control the grass. In areas where the forest is re-established, scrub typhus is either absent or its occurrence is limited. There is also an ecotone effect, with scrub typhus being more likely to occur at the fringes of the grassland habitat.

As the trombiculid larvae feed on one individual host only, they have the potential to acquire the infection from the host or transmit it to the host. The latter case occurs when *O. tsutsugamushi* is transmitted transovarially from the adult female mites to the offspring (e.g. as in various species of *Leptotrombidium*). *O. tsutsugamushi* has been detected in almost all organs and tissues of larval and adult *L. pallidum*, especially the salivary glands, epidermal cells and reproductive organs. While rodents may act as reservoirs, *O. tsutsugamushi* survives mainly in infected mites, as they maintain infections for longer (by transovarian transmission) than the rodent host; therefore, mites serve as both the reservoir and the vector of this pathogen.

Preventive measures involve treatment of clothing and exposed skin with topical repellents (e.g. diethyltoluamide – DEET) or pyrethroid toxicants (e.g. permethrin), particularly the lower legs, ankles and feet (parts of the body more likely to encounter larval mites). Even if practical, control of the rodent population is usually ineffective as a short-term measure to reduce scrub typhus; indeed, removing the rodent hosts from the ecosystem may result in trombiculid larvae seeking alternative hosts such as humans.

6.2 Psorergatidae (family)

6.2.1 *Psorergates (Psorobia)*
Two species of *Psorergates* (= *Psorobia*) have been recovered from domestic livestock.

Psorergates bos has been found on cattle in the USA, where it has little pathological effect, whereas *Psorergates ovis* occurs on sheep, mainly in Australia, New Zealand, South Africa, South America and the USA, and may be a persistent pest. Adult *Psorergates* can be identified by their legs, which are radially arranged around a more or less circular body (Fig. 80).

Female *P. ovis* lay only a few eggs in their lifetime; from these, larvae with reduced legs hatch. There are three nymphal stages; in these the legs become progressively larger until, in the adult stage, the legs are well developed and the mites are mobile. Adults of both sexes are very small, measuring only 200 µm in length. The coxae of all the legs are sunk into the body as epimeres. Those of the first pair of legs are relatively broad and reflected laterally to be hook-shaped (Fig. 80b and c). The female has two pairs of long setae posteriorly and the genital opening opens posteriorly on the ventral surface. The male is more oval than the rounded female, it has only a single pair of long setae on a small posteromedian process and the genital opening is located anterodorsally (Fig. 80a and b), behind which an elongated penis may be seen. The complete life cycle of *P. ovis* takes 4–5 weeks.

P. ovis affects mainly merino sheep, which react to the irritation by chewing the fleece, resulting in a condition known as fleece derangement and leading to the wool clip being downgraded. Affected sheep may become tolerant after 1 or 2 years, albeit still being infested. The mite spreads very slowly through a flock and may affect up to 15% of sheep in a neglected flock. Only the adults are mobile and they are very sensitive to desiccation; they survive for only a short time in the fleece and die within 24–48 h when removed from their host. Consequently, transmission from one host to another occurs during the brief period which follows shearing. Most mites are located under the stratum corneum in the superficial layers of the skin and infestation with *P. ovis* is diagnosed by detecting mites in skin scrapings. Peak numbers of *P. ovis* can be observed during spring and treatment is best applied after shearing, by administering two dippings of an

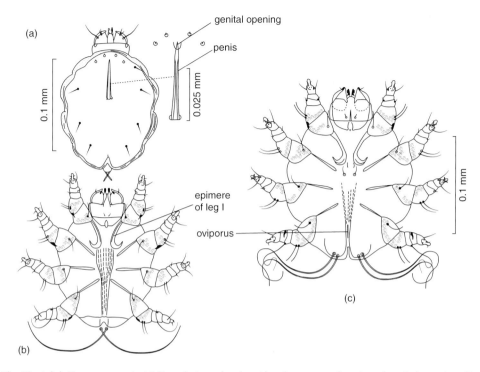

Fig. 80. Adult *Psorergates ovis*. (a) Dorsal view of male with enlargement of penis and genital opening; (b) ventral view of male; (c) ventral view of female. *Source*: Fain (1961).

insecticide such as an organophosphate or pyrethroid, 1 month apart.

6.3 Demodicidae

6.3.1 *Demodex*

Members of the genus *Demodex* are minute, annulate, worm-like parasitic mites (Fig. 81), which live head-down in hair follicles and in the sebaceous and Meibomian glands of the skin. Demodicid mites occur in humans and a wide range of wild and domestic animals, including bats, insectivores, carnivores, rodents, horses and ruminants.

The taxonomy of *Demodex* mites is not well defined; however, it is accepted they form a group of sibling species, with different species occurring on different hosts, although more than one species may occur on the same host, e.g. *Demodex folliculorum* and *Demodex brevis* on humans. The relationship between demodicid mites on a host and clinical disease is unclear, as they may occur in low numbers on most healthy, asymptomatic hosts. However, in humans, *Demodex* mites, most often detected on the cheeks, forehead and nose, have been implicated as an aetiological agent of the facial erythema known as 'rosacea' and have also been associated with the facial rough eruption known as 'pityriasis folliculorum'.

The female lays eggs from which larvae with short legs ending in a single trifid claw hatch (Fig. 82b). An unusual feature of the life cycle is the emergence of a second hexapod form, designated as a protonymph, in which each leg terminates in a pair of trifid claws. The deutonymph stage which follows is octopod. Both protonymph and deutonymph have a pair of crescent-shaped sternal scutes on the ventral surface between each pair of legs (Fig. 82c). The deutonymph moults into an adult in which the coxal epimeres are fused to form a median longitudinal bar (Fig. 81). The female genital opening is on the ventral

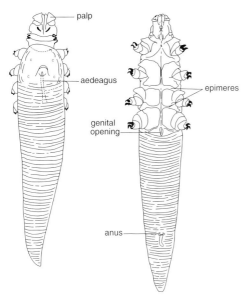

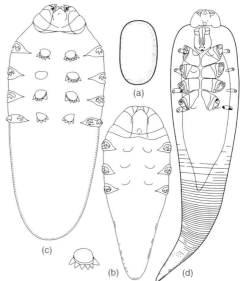

Fig. 81. Left: dorsal view of male *Demodex*. Right: ventral view of female *Demodex*. *Source*: Krantz (1978).

Fig. 82. Stages in life cycle of *Demodex*. (a) Egg; (b) larva; (c) deutonymph of *Demodex muscardinus* with, below, an enlarged sternal scute; (d) deutonymph of *Demodex bovis* containing fully formed adult within. *Source*: Hirst, S. (1922) Economic Series No 13. *Mites Injurious to Domestic Animals*. British Museum (Natural History), London.

surface just posterior to the fourth pair of legs, while the male genital opening is located dorsally at the level of the second and third pair of legs. In *Demodex*, the legs are closely associated anteriorly so that the striated opisthosoma forms at least half the total body length. Adults measure 250–300 μm in length and 40 μm in width.

Among domestic animals, demodectic mange is of greatest importance in dogs, goats and pigs, of lesser importance in cattle and horses, rare in sheep and very rare in cats. The appearance of small nodules and pustules is a sign of demodicosis in domestic animals. In pigs, *Demodex phylloides* is found on the face, spreading down the ventral surface to the neck and chest to the belly. In cattle (*Demodex bovis*) and goats (*Demodex caprae*), the lesions occur most commonly on the brisket, lower neck, forearm and shoulder and dorsally behind the withers. In pigs and goats, pustules may develop into large abscesses and, in some instances, cause death. Species of *Demodex* are host specific and interspecific transmission of any *Demodex* species has never been reported.

Transmission of *Demodex* within a host species occurs very early in life, probably while

the youngsters are suckling. Infestation by *Demodex canis* is acquired during the first 2–3 days of neonatal life. Two distinct clinical conditions can occur: a localized and a generalized demodicosis. In young dogs (3–12 months), localized demodicosis is characterized by a small number of squamous patches developing on the face, especially around the eyes and mouth. In most cases, the condition is self-limiting and relapses are rare. However, when occurring, relapses can develop into the generalized form, which is commoner in pure-bred and short-haired dogs. A hereditary specific T-cell defect for *D. canis* is involved in the occurrence of the generalized form. In dogs of more than 5 years old, generalized demodicosis is a rare but severe condition underlying immunosuppression. Generalized demodicosis usually begins on the face and spreads to the head, legs and trunk. Secondary infections with bacteria follow and crusted, pyogenic, haemorrhagic lesions develop on most of the body surface. This condition is difficult to treat and may cause fatalities. Diagnosis of demodectic mange is based on the detection of mites from skin scrapings and morphological identification of *Demodex* mites is easy; however, identification of *Demodex* specimens to species requires specialist expertise.

6.4 Cheyletiellidae

The Cheyletiellidae includes nine genera of mites which are parasites of birds and small mammals. They are characterized by stiletiform chelicerae, which are used for piercing the host, and strong, curved, palpal claws for holding the fur or feathers of the host (Fig. 83). Species of *Cheyletiella* are large mites, about 350 μm in length, which cause a mild, non-suppurative dermatitis in dogs, cats and rabbits and a transitory dermatitis in humans; species have also been described from wild animals. *Cheyletiella* is an obligate parasite which lives on the keratin layer of the epidermis and is not associated with hair follicles,

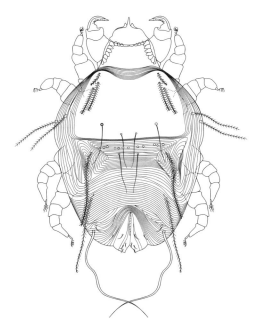

Fig. 83. Dorsal view. *Cheyletiella parasitivorax* female. Leg setae omitted.

although periodically it may pierce the skin and engorge on a clear, colourless fluid.

The life cycle is completed on the host. The large eggs (230 × 100 μm) are attached to the hairs of the host 2–3 mm above the skin; the prelarva and larva develop within the eggshell. The larva is hexapod, but the prelarva possesses only rudimentary gnathosomal appendages. There are two nymphal stages before the adult stage is reached. Females can survive for approximately 10 days in cool temperatures, whereas males and immature stages die within 48 h after they are removed from the host. The mobility of these mites and their presence throughout the hair of the host facilitates relatively easy transmission between hosts by physical contact. However, females have been found attached to fleas (Siphonaptera) and louse flies (Hippoboscidae) and it has been suggested that phoresy plays a role in transmission of the mite from host to host.

The mites move rapidly and this behaviour gives rise to the term 'walking dandruff'. *Cheyletiella yasguri* causes a highly contagious infection in puppies; it usually begins on the rump, from where it may spread over the back to the head, and older dogs may be asymptomatic carriers of small populations of mites. *Cheyletiella blakei* causes a mild dermatitis in cats, and *Cheyletiella parasitivorax* occurs in the scapular region of rabbits. Human infestations with *Cheyletiella* are transitory and reactions are highly variable.

Cheyletiella infestation is diagnosed by detecting mites in the hair of the host. Treatment involves alleviation of the symptoms and eradication of the mites on the infested pet. Cats, dogs and rabbits can be treated safely with pyrethrins, carbaryl powders or lime sulfur dips and dogs with malathion or carbaryl. Three treatments are needed at weekly intervals. Two subcutaneous injections of a macrocyclic lactone, 2–3 weeks apart, may also be effective.

6.5 Myobiidae

The Myobiidae is a cosmopolitan family of ectoparasitic mites which occur on marsupials, rodents, bats and insectivores. In small numbers, they appear to have little effect on the health of the host. Myobiids can be recognized readily by the first pair of legs, which are highly modified for clinging to a single hair (Fig. 84). *Myobia musculi* causes a mild dermatitis in mice; *Radfordia ensifera* and *Radfordia affinis* infest rats and mice, respectively. They feed at the base of hairs and ingest extracellular tissue fluid and sometimes blood.

6.6 Pyemotoidea

Species of *Pyemotes* mites are predators of insects infesting stored grains, straw and similar products and can attack humans and domestic animals that come in contact with the infested materials. The inseminated female *Pyemotes tritici* attaches itself to an insect by its chelicerae. The eggs are fertilized and hatch within the female and all the immature stages develop inside the female's swollen opisthosoma (Figs 85 and 86). Adult males emerge 2 days before the females and remain in the vicinity of the female's opisthosoma, assisting in delivery of virgin females by pulling them through the birth pore. Copulation then takes place and the inseminated female moves away to find a host. *P. tritici* males are haploid, while females are diploid; at birth, the progeny is 92% female. Males are fully potent for their first 15 matings. On average, a female produces 250 offspring; a population of mites increases by 0.63-fold each day and doubles within 1.1 days.

P. tritici (the grain or hay itch mite) is the cause of pruritic dermatitis in humans in many parts of the world, but species of *Pyemotes* vary in their pathogenicity to humans; *P. tritici* is highly pathogenic, as is *Pyemotes beckeri*, while *Pyemotes scolyti* is not. In susceptible humans, a vesicle develops in the centre of an erythematous weal, and rubbing and scratching can cause the vesicle to burst and the occurrence of secondary infections. The mites do not establish breeding populations on

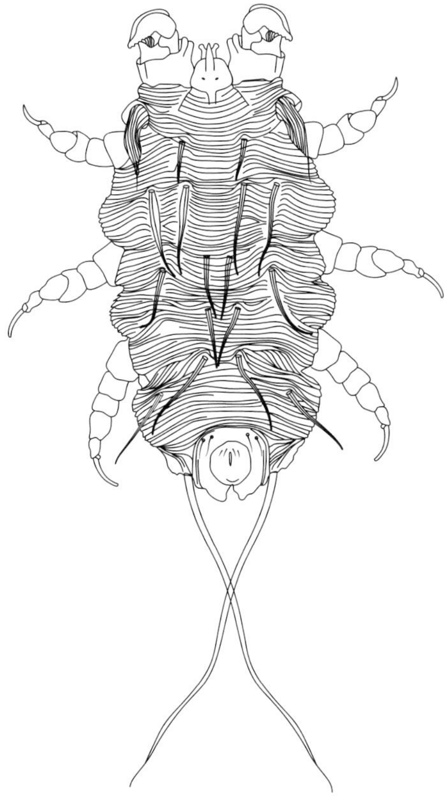

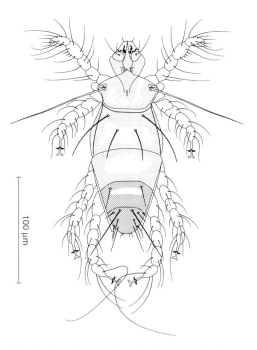

100 µm

Fig. 84. *Myobia musculi.* Minor setae omitted.

Fig. 85. Dorsal view of female *Pyemotes tritici.*

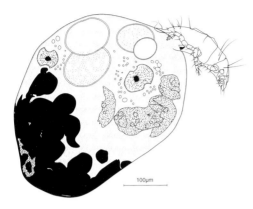

Fig. 86. Lateral view of gravid *Pyemotes tritici*. All of the swollen area is the opisthosoma.

mammals and, if reinfestation is prevented, the condition subsides within a few days. Infestations of *Pyemotes* are associated mainly with grains, straw, hay and grasses, but also occur with pulses and other crops. The insect hosts of *P. tritici* are various, but include larvae of the Angoumois grain moth (*Sitotroga cerealella*), the saw-toothed grain beetle (*Oryzaephilus surinamensis*), the cowpea weevil (*Callosobruchus maculatus*) and the rice weevil (*Sitophilus oryzae*). The venom of *P. tritici* causes immediate paralysis and eventual death of its insect host.

7. ORIBATIDA (CRYPTOSTIGMATA)

7.1 Oribatidae

The Oribatida comprises the four families Ceratozetidae, Galumnidae, Oribatulidae and Schleroribatidae, and includes over 7000 species. They are a very large group of free-living, soil-dwelling mites that are important detritivores and help to enhance soil fertility. Also known as 'moss' or 'beetle' mites, they are often the most numerically dominant and diverse arthropods inhabiting the organic layer of soil of temperate regions. None are parasitic. However, some species can act as intermediate hosts of tapeworms in humans and domestic animals. More than 125 species of oribatid mites act as intermediate hosts of anoplocephalid tapeworms.

Oribatid mites are free-living, dark-coloured mites, often covered with large sclerotized plates, from which the common name of 'beetle mite' derives. They range in size from 0.2 mm to over 2 mm in length. Besides some rare exceptions, they possess prominent, club-like sensillae (pseudostigmatic or bothridial organs), which arise from large pits on the posterolateral margins of the propodosoma. In heavily sclerotized species, respiration occurs through tracheal tubes, with stigmata opening at the bases of the legs. The alternative name Cryptostigmata refers to the fact that the stigmata are poorly defined and difficult to observe.

The life cycle of oribatid mites consists of six postembryonic instars (prelarva, larva, protonymph, deutonymph, tritonymph and adult). The prelarval instar takes place within the egg. This is an inactive form, with no visible appendages or setae. The hexapod larva hatches from the egg and precedes the octopod nymphal and adult instars, which are all active and feed normally. In temperate environments, the life cycle is usually completed within several months, whereas in cold climates it may take several years to complete. Development may be interrupted by periods of quiescence. Usually, one single generation occurs per year. Because of their low metabolic rate, oribatid mites are well adapted to periods of starvation and this, along with low rates of reproduction, results in only small fluctuations of their population densities from one year to the next.

Oribatid mites are iteroparous, producing offspring in successive (annual or seasonal) cycles. In comparison with other mite groups, the fecundity and reproduction rate of oribatid mites is low, with females usually producing 50–100 eggs/lifetime or less. Fertilization in most cases is indirect, with the males releasing free-standing spermatophores and having no direct contact with the females. One exception exists within the genus *Collohmannia*, where fertilization is direct and a well-defined courtship behaviour exists, involving the provision of nuptial food for the female by the male. Many species are obligatorily parthenogenetic. Oribatid mites are generally slow moving and do not travel more than a few centimetres to seek food or favourable sites for oviposition. Hence, populations tend to be highly aggregated.

The medical and veterinary importance of oribatid mites derives from their ability to act as intermediate hosts of various cestode parasites. Infected hosts pass tapeworm segments via their faeces. These segments release tapeworm eggs in the dung and surrounding grass, which the mites ingest as they graze on the decaying organic material. Once inside the mite gut, the tapeworm eggs hatch and the oncospheres penetrate into the haemocoel and develop to the cysticercoid stage. This stage may persist for up to 4 months. As the mites move through grass, they are ingested by herbivores and the tapeworm cysticercoids are released. In the vertebrate host, the cysticercoids develop into mature tapeworms and the cycle repeats. It takes as little as 15–20 weeks following ingestion of an infected oribatid mite for the tapeworm to begin shedding segments containing eggs into the vertebrate host's faeces.

Oribatid mites are particularly important in the transmission of the tapeworms *Moniezia benedeni* and *Moniezia expansa*, which inhabit the small intestines of cattle, sheep and goats, and the tapeworms *Anoplocephala perfoliata* and *Anoplocephala magna*, which infest horses. Oribatid mites may also transmit *Bertiella* spp. tapeworms. A wide range of Old World monkeys such as rhesus monkeys, Japanese macaques, baboons, chimpanzees and orang-utans are infested by *Bertiella studeri*, and although the potential of European oribatids to act as intermediate hosts of these pathogen has been demonstrated under experimental conditions, the species involved in the transmission of this cestode under natural conditions is still unknown. However, *Bertiella mucronata*, which parasitizes monkeys in South America, can develop in *Dometorina suramerica*, *Scheloribates atahualpensis* and other oribatid species. In Asia and Africa, *B. studeri* may also inhabit the small intestine of humans, similarly to tapeworms of the genus *Mesocestoides*, which are believed to be transmitted by oribatid mites infesting carnivorous mammals and charadriiform birds in Europe, Africa, Asia and North America.

Because of their ubiquity, control of the mites themselves is seldom possible. Maintenance of pasture hygiene, such as removing

faeces, may contribute towards reducing the potential for tapeworm transmission. Grazing in areas of high mite contamination should be avoided.

8. MESOSTIGMATA

Of the 12 suborders of Mesostigmata, only the Dermanyssina includes species of medical and veterinary importance and all of them are ranked in the superfamily Dermanyssoidea.

In the Dermanyssoidea, the transition between free-living organisms and obligatory parasites is well represented by existing species. A single genus, for example *Haemogamasus*, includes non-parasitic polyphagous nest-dwellers, facultative parasites and obligatory haematophagous parasites. The early stages of this progression towards parasitism are evident in the family Laelapidae, in which one subfamily, the Hypoaspidinae, does not include any parasites of vertebrates, albeit a number of hypoaspidines are typically found in the nests of small mammals and birds. Two other subfamilies, the Laelapinae and Haemo-gamasinae, are largely parasitic but include some polyphagous nest-dwelling species. The main families of medical importance are the bloodsucking Dermanyssidae and Macro-nyssidae, the Rhinonyssidae and the subfamily Halarachninae of the Halarachnidae that infest the respiratory passages of birds and mammals, respectively, and the Raillietiinae (another halarachnid subfamily), which parasitize the ears of bovids.

Most of the damage caused by haema-tophagous dermanyssoids results from the direct effect of large numbers feeding on the host rather than from their role as vectors of pathogens. However, various pathogens have been detected in dermanyssoids collected from the environment and the transmission of some has been proven under experimental conditions. However, it is generally accepted that, with the exception of the agent of rickettsial pox in humans (*Rickettsia akari*), bloodsucking dermanyssoids play little role in the epidemiology of vertebrate disease-causing pathogens. Bloodsucking dermanyssoids may be host specific or parasitize a range of related hosts; however, in certain circumstances, such as when birds desert an infested nest, these mites can attack unusual hosts, such as humans.

8.1 Dermanyssidae

Dermanyssid mites are haematophagous ectoparasites of birds and mammals. The adults are 750–1000 μm in length and in the unfed state are greyish-white in colour, becoming bright red after feeding and darker as the meal is progressively digested. When not feeding, the mites spend most of their time in the nest. Their eggs are deposited in the nest or in associated crevices. These mites have a high engorgement capacity, which enables them to withstand starvation for months. Females are often found on the host, which favours dispersal of the species.

The chelicerae are chelate with minute, weakly dentate digits (Fig. 87). In the nymphs and female, the chelicerae are elongate and stylet-like, with the second segment considerably lengthened. In the male, the second segment is of normal length and a long, grooved spermadactyl is fused with the movable digit, which is considerably longer than the fixed digit. The corniculi are membranous and there are nine or more deutosternal denticles in a single file. In the life cycle, the larva is non-feeding, while both the protonymph and deutonymph are actively feeding stages. Two species are of medical and veterinary importance, *Dermanyssus gallinae* and *Liponyssoides sanguineus*.

8.1.1 *Dermanyssus gallinae*

D. gallinae, the chicken mite, is a cosmopolitan ectoparasite of poultry and a range of wild and commensal birds, including pigeons, sparrows and starlings. It attacks mammals when other hosts are not available. Unfed *D. gallinae* are approximately 700 μm long and 400 μm wide, but increase to more than 1 mm long when fully engorged. The female has a single large dorsal shield, which is truncate posteriorly. The setae on the dorsal shield are shorter than those on the adjacent body surface. There are only two pairs of setae on the sternal plate, the posterior pair being far from the plate. There are also two pairs of pores on the sternal plate.

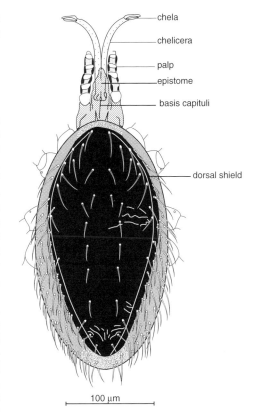

Fig. 87. Dorsal view of female *Ornithonyssus bursa*.

The genito-ventral plate is rounded posteriorly and bears one pair of setae. The anal plate is large and bears three setae. In the male, there is a single holoventral shield.

Very large populations of *D. gallinae* can build up rapidly in poultry houses and bird nests. In North America, *D. gallinae* is rare in modern commercial caged layer operations; however, they are seen frequently in broiler breeder farms, with populations peaking during summer. The mites are nocturnal, when they feed on roosting birds, and during daytime, they hide away in cracks and crevices. Under optimal conditions in the presence of suitable hosts, the life cycle is completed rapidly. The eggs hatch within 2–3 days and both nymphal stages moult 1–2 days following a blood meal; adult females are ready to oviposit 12–24 h following blood feeding. Fed adults can survive 34 weeks without feeding;

therefore, infestations are difficult to eliminate, even when the hosts are removed. The control measures described below for *Ornithonyssus sylviarum* are also applicable to the control of *D. gallinae*, provided that greater attention is paid to the surroundings of the birds.

Large populations of *D. gallinae* have severe effects on domestic poultry. Egg production is reduced, hens may leave incubating eggs and death may occur from exsanguination. A number of pathogens have been detected in *D. gallinae* (e.g. *Salmonella* spp., *Escherichia coli*, *Shigella* sp. and *Staphylococcus* spp.) and transmisson of some has been shown experimentally; however, *D. gallinae* is not considered to play a significant role in their transmission.

8.1.2 *Liponyssoides sanguineus*

L. sanguineus (formerly known as *Allodermanyssus sanguineus*) is a widely distributed ectoparasite of small rodents. It is very similar to *D. gallinae* but with two dorsal shields, a larger tapering anterior shield and a very small posterior one, bearing one pair of setae. The sternal plate has three pairs of setae and two pairs of pores. The genital plate is slender and tapering and there is a small anal plate. The life cycle of *L. sanguineus* (18–23 days) is slower than that of *D. gallinae* (7 days). *L. sanguineus* is normally a nest dweller and only feeds on the host. It is the vector of *R. akari*, which causes rickettsial pox in humans.

RICKETTSIAL POX *(RICKETTSIA AKARI)*

R. akari is included with the agents of tick typhus in the spotted fever group within the genus *Rickettsia*. Rickettsial pox was first recognized as a new human disease in the mid-1940s, when it was described from Boston and New York (USA), and a few years later from the former Soviet Union. In 1949–1950, the disease occurred in epidemic form in the Ukraine, causing approximately 1000 cases in the Donets Basin. Rickettsial pox has since been identified in South Korea, South Africa and French Equatorial Africa and it is also suspected to occur in Yugoslavia and Italy (Sicily). The disease is comparatively benign and symptoms include non-specific fevers and sweats, with head and muscle aches and a papular rash; often, an eschar at the bite site indicates a rickettsial infection.

In the vector, *L. sanguineus*, *R. akari* is transmitted transovarially to the offspring. Under experimental conditions, the tropical rat mite, *Ornithonyssus bacoti*, can harbour *R. akari* and transovarial transmission can occur. There is little information on the circulation of *R. akari* in wild rodents; however, the pathogen has been detected in various rats, voles and gerbils. Treatment of patients usually involves the administration of antibiotics such as tetracycline, while chloramphenicol can also be effective.

8.2 Macronyssidae

Members of the Macronyssidae are bloodsucking ectoparasites of mammals, birds and reptiles. It is believed that these mites evolved primarily on bats and subsequently they have adapted to other hosts. In the feeding protonymph and female, the chelicerae are chelate and edentate, and in the inactive, non-feeding larva and deutonymph, the cheliceral digits are rudimentary. Two genera are of medical interest, *Ornithonyssus*, which occurs on birds and mammals, and *Ophionyssus*, an ectoparasite of reptiles. Female *Ornithonyssus* have the genital setae inserted on the genital shield and only a single dorsal shield, while *Ophionyssus* presents two dorsal shields and genital setae inserted on the integument adjoining the genital shield. In male *Ornithonyssus*, the anal shield is fused with the other ventral shields, while in male *Ophionyssus* the anal shield is discrete. In female *Ornithonyssus*, the dorsal shield tapers posteriorly, whereas in the male it is more extensive. Three separate shields are present on the ventral surface. The sternal shield bears three pairs of setae (two pairs in *O. sylviarum*); the genital shield, which tapers posteriorly, bears one pair of setae and the anal shield is oval, tapering posteriorly and bearing three setae.

8.2.1 *Ornithonyssus bacoti*

Although *O. bacoti* is referred to as the tropical rat mite, it is cosmopolitan, occurring in both tropical and temperate areas of the world, especially in seaports. *O. bacoti* is a pest of mice, rats, hamsters and small marsupials. It is characterized by the setae on the dorsal shield being of similar length to those on the adjoining integument, and by three pairs of setae on the sternal shield.

The life cycle is completed rapidly under optimal conditions, within 11–16 days from egg to adult. As with other haematophagous mesostigmatic mites, large populations can cause the death of their host by exsanguination. *O. bacoti* is the vector of the filarial worm, *Litomosoides carinii*, which infests rodents. This mite has also been shown to transmit a number of other pathogens under experimental conditions; however, its vectorial role under natural condition is considered marginal.

8.2.2 *Ornithonyssus sylviarum*

O. sylviarum, the northern fowl mite, is a serious pest of poultry and wild birds throughout the northern temperate regions of Europe and North America, and it is common in southern Australia. It has also been described from wild birds in South Africa. The setae on the dorsal shield are shorter than those on adjoining integument and only two pairs of setae are present on the sternal plate, the third pair being on the integument. A peculiar feature of the life cycle of *O. sylviarum* is that the adults remain on the host, giving birds with heavy infestation a greyish to blackish appearance.

Oviposition occurs on the host, primarily in the area of the vent, and egg development is completed within 1–2 days. The protonymph feeds at least twice before moulting to the non-feeding deutonymph. The entire life cycle can be completed within a week, enabling populations of *O. sylviarum* to grow rapidly. In North America, populations peak during winter. Poultry infestations cause losses to commercial producers, as a consequence of a decline in egg production and higher feed costs.

O. sylviarum can survive for 3–4 weeks off the host and is introduced into new settings through contaminated egg crates and packaging, and sometimes personnel. The introduction of *O. sylviarum* into a poultry facility can be minimized by careful examination of all introduced materials to ensure that they

are mite free. *O. sylviarum* has developed resistance to a range of insecticides.

8.2.3 *Ornithonyssus bursa*

Ornithonyssus bursa (Figs 87 and 88) is the tropical poultry mite, which occurs also on pigeons, sparrows and mynah birds, as well as, on occasions, humans. It causes temporary irritation on humans because of its inability to survive for long periods away from its bird hosts. In *O. bursa*, as in *O. sylviarum*, the setae on the dorsal shield are shorter than those on the integument (Fig. 87); however, *O. bursa* differs from the latter in that it possesses three pairs of setae on the sternal plate. *O. bursa* occurs either on the bird or in its nest. Attacks on humans result from the dispersal of mites from infested nests deserted by the breeding birds and their nestlings.

8.2.4 *Ophionyssus natricis*

Ophionyssus natricis is an ectoparasite of reptiles; it is rare in the wild but can be troublesome in zoos. This species has two dorsal plates, an anterior lemon-shaped plate on the podosoma and a posterior plate, which is immediately dorsal to the anal plate but considerably smaller. The posterior plate bears no setae. There are two pairs of setae and two pairs of pores on the sternal plate, with the third pair of each located behind the plate on

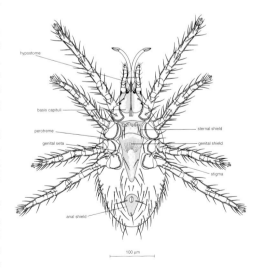

Fig. 88. Ventral view of female *Ornithonyssus bursa*.

the integument. The anal plate bears the usual three setae.

The adult mite feeds under the scales on the snake and leaves its host to oviposit in dark, moist crevices. Its life cycle can be completed in 2–3 weeks when suitable hosts are abundant; in their absence, some females can survive unfed for 5–6 weeks. Heavy infestations affect the health of snakes severely, causing anaemia, which may lead to the snake's death. *O. natricis* is believed to be a mechanical vector of a haemorrhagic septicaemia of snakes caused by *Aeromonas hydrophila hydrophila* (= *Proteus hydrophilus*), a motile, facultatively anaerobic bacillus. Infestations of snakes in captivity can be reduced by allowing them to immerse in water.

8.3 Halarachnidae

Two subfamilies are included within the Halarachnidae. The Halarachninae are obligatory parasites of the respiratory tract of mammals, whereas the Raillietiinae are obligatory parasites occurring in the external ear of mammals. As adaptations to an 'endoparasitic' mode of life, the dorsal and ventral shields in the Halarachninae are delicate and reduced and the genital plate is rudimentary. In addition, the tritosternum is reduced or absent and the peritremes, associated with the stigmata, are reduced or vestigial. These mites inhabit the respiratory system and, consequently, the ambulacral apparatus at the extremity of the legs is well developed. In the Raillietiinae, the genital shield is well developed, a bifid tritosternum is present and the peritremes are elongate and well developed. In both subfamilies, the life cycle is compressed. The hexapod larva is active, whereas the protonymph and deutonymph are non-feeding, non-motile, ephemeral stages with rudimentary claws.

8.3.1 Halarachninae

The Halarachninae includes about 35 species which parasitize primates, terrestrial carnivores, phocid and otariid seals, rodents, hyraxes and artiodactyls. Some species are larviparous and others oviparous. The larva

has well-developed ambulacra and is the only developmental stage capable of moving between hosts. Transmission seems to occur easily from one host species to another. Two species are of veterinary importance: the dog parasite, *Pneumonyssoides caninum*, and the monkey parasite, *Pneumonyssus simicola*.

P. caninum occurs in the sinuses and nasal passages of dogs in Australia, South Africa and the USA. Adult *P. caninum* are pale yellow, oval mites with few body setae (Fig. 89). The chelicerae are well developed with opposable digits. The dorsal plate is small, irregularly shaped and covered with microscopic spines. The first pair of legs is equipped with a pair of heavily sclerotized brown claws, while the other three pairs are tipped with a long, stalked pulvillus and two slender claws. Infections by *P. caninum* are mild and are relatively free from marked symptoms. However, in some cases, the mites penetrate tissues and migrate throughout the body. They have

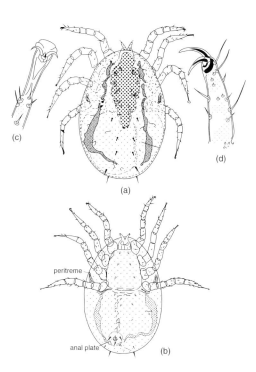

Fig. 89. Female *Pneumonyssoides caninum*. (a) Dorsal view; (b) ventral view; (c) detail of tarsus, typical of legs II, III and IV; (d) tarsus of leg I. *Source*: Chandler and Ruhe (1940).

been retrieved from the bronchi, the renal fat and the liver.

P. simicola is a common parasite of the lungs of the rhesus monkey, *Macaca mulatto*, in which prevalence may reach up to 100%. This species also occurs in a range of Old World primates, albeit less commonly. *P. simicola* is similar to *P. caninum* but smaller. The mites live and feed in the lung, where they may be grouped in nodules which superficially resemble tubercles. The nodules contain a characteristic golden brown to black pigment, which may be faecal material resulting from the mite feeding on blood. Clinical signs are usually absent. The mite spreads readily through susceptible animals through coughing and sneezing. Rhesus monkeys taken from their mothers at birth and reared in isolation are free from infection.

8.3.2 Raillietiinae

Species of *Raillietia* occur in the ears of bovids. *Raillietia auris* is an oval mite about 1 mm long, with a small oval dorsal plate (Fig. 90). The second pair of legs in the male is modified for grasping the female. The movable digit of the chelicera is entirely fused to the hypertrophied spermadactyl, which collects sperm from the male genital opening and deposits it into the sperm induction pores of the female. *R. auris* occurs in the ears of cattle in North America, Europe and Australia, and

of sheep in Iran. The mite is considered to feed on epidermal cells and wax but not on blood. Infestations are usually benign, lacking obvious symptoms, but in northern Queensland, *R. auris* has been associated with otitis media. *Raillietia caprae* occurs in goats and *Raillietia amis* in domestic bovids.

8.4 Rhinonyssidae

Most members of the Rhinonyssidae are parasites of the nasopharynx of birds. Rhinonyssids are weakly sclerotized, elongate mites with well-developed legs, reduced or absent peritreme and usually without a tritosternum. They are an extraordinarily successful group, closely related to the Macronyssidae. In both families, the protonymph is a feeding stage and the deutonymph a non-feeding stage. One species of minor veterinary importance is the canary lung mite, *Sternostoma tracheacolum*.

8.4.1 *Sternostoma tracheacolum*

Most species of *Sternostoma* are nasal mites, but *S. tracheacolum* has been recorded from the respiratory tract of a range of domestic and wild birds, including canaries and budgerigars. The species is widely distributed throughout the world, occurring in Africa, North and South America, Europe, Australia and New Zealand.

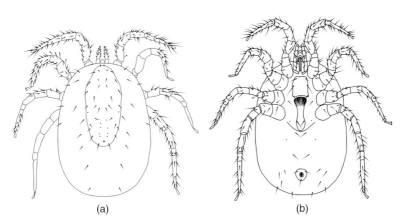

(a)   (b)

Fig. 90. Female *Raillietia auris*. (a) Dorsal view; (b) ventral view. Redrawn from Hirst, (1922) *Mites Injurious to Domestic Animals*. British Museum (Natural History), London.

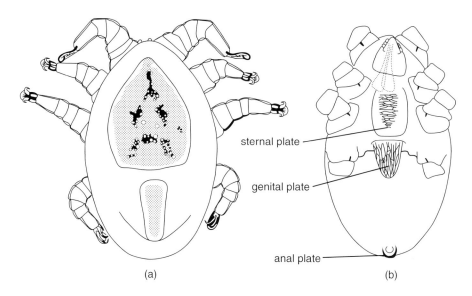

Fig. 91. Female *Sternostoma tracheacolum*. (a) Dorsal view; (b) ventral view. *Source*: Lawrence (1948).

S. tracheacolum is a yellowish-brown mite about 0.5 mm in length, with two dorsal plates, a pentagonal anterior plate and a narrower posterior one (Fig. 91); there are distinct sternal and genital plates and a reduced anal plate, strong, thickset palps and large, mobile and chelate chelicerae. On legs II–IV, the ambulacral apparatus consists of paired claws and a pulvillus, modified on leg I.

S. tracheacolum has been isolated in the tracheae, air sacs, bronchi and parenchyma of the lung and also on the surface of the liver, but rarely from the nasal cavities. In canaries, the mites are attached firmly to the inner walls of the tracheae, from which they ingest blood. Mites have been detected in the lungs as well as the tracheae and air sacs of Gouldian finches in aviaries; in infected birds, a wasting disease may develop and result in their death. In northern Australia, *S. tracheacolum* is considered to be the main factor endangering wild populations of Gouldian finches and prevalence may reach 62% in wild populations.

SELECTED BIBLIOGRAPHY

Akrami, M.A., Saboori, A. and Eslami, A. (2007) Observations on oribatid mites (Acari: Oribatida) serving as intermediate hosts of *Moniezia expansa* (Cestoda: Anoplocephalidae) in Iran. *International Journal of Acarology* 33, 365–369.

Arlian, L.G. (1989) Biology, host relations, and epidemiology of *Sarcoptes scabiei*. *Annual Review of Entomology* 34, 139–161.

Arlian, L.G. and Platts-Mills, T.A. (2001) The biology of dust mites and the remediation of mite allergens in allergic disease. *Journal of Allergy and Clinical Immunology* 107, S406–S413.

Baker, E.W. and Wharton, G.W. (1952) *An Introduction to Acarology*. Macmillan, New York.

Baker, E.W., Camin, J.H., Cunliffe, F., *et al.* (1958) *Guide to the Families of Mites*, Contribution No. 3. Institute of Acarology, University of Maryland.

Baker, E.W., Evans, T.M., Gould, D.J., Hull, W.B. and Keegan, H.L. (1956) *A Manual of Parasitic Mites of Medical or Economic Importance*. National Pest Control Association, New York.

Barnes, R.D. (1974) *Invertebrate Zoology*. W.B. Saunders, London.

Bates, P. (1999) Inter- and intra-specific variation within the genus *Psoroptes* (Acari: Psoroptidae). *Veterinary Parasitology* 83, 201–217.

Bisdorff, B., Milnes, A. and Wall, R. (2006) Prevalence and regional distribution of scab, lice and blowfly strike in sheep in Great Britain. *Veterinary Record* 158, 749–752.

Bronswijk, J.E.M.H. and de Kreek, E.J. (1976) *Cheyletiella* (Acari: Cheyletiellidae) of dog, cat and domesticated rabbit, a review. *Journal of Medical Entomology* 13, 315–327.

Burgess, I. (1994) *Sarcoptes scabiei* and scabies. *Advances in Parasitology* 33, 235–293.

Bush, R.K. (2011) Does allergen avoidance work? *Immunology and Allergy Clinics of North America* 31, 493–507.

Cameron, M.M. and Hill, N. (2002) Permethrin-impregnated mattress liners: a novel and effective intervention against house dust mites (Acari: Pyroglyphididae). *Journal of Medical Entomology* 39, 755–762.

Colloff, M.J. and Spieksma, F.T. (1992) Pictorial keys for the identification of domestic mites. *Clinical and Experimental Allergy* 22, 823–830.

Debboun, M. and Strickman, D. (2012) Insect repellents and associated personal protection for a reduction in human disease. *Medical and Veterinary Entomology*, doi: 10.1111/j.1365-2915.2012.01020.x

Denegri, G.M. (1993) Review of oribatid mites as intermediate hosts of tapeworms of the Anoplocephalidae. *Experimental and Applied Acarology* 17, 567–580.

Desch, C.E. (1984) Biology of biting mites (Mesostigmata). In: Nutting, W.B. (ed.) *Mammalian Diseases and Arachnids*, Vol I. CRC Press, Florida, pp. 83–109.

Duncan, S. (1957) *Dermanyssus gallinae* (De Geer 1779) attacking man. *Journal of Parasitology* 43, 637–643.

Fain, A. (1994) Adaptation, specificity and host–parasite coevolution in mites (Acari). *International Journal for Parasitology* 24, 1273–1283.

Frances, S.P. (2005) Potential for horizontal transmission of *Orientia tsutsugamushi* by chigger mites (Acari: Trombiculidae). *International Journal of Acarology* 31, 75–82.

Frances, S.P., Yeo, A.E.T., Brooke, E.W. and Sweeney, A.W. (1992) Clothing impregnations of dibutylphthalate and permethrin as protectants against a chigger mite, *Eutrombicula hirsti* (Acari: Trombiculidae). *Journal of Medical Entomology* 29, 907–910.

Golant, A.K. and Levitt, J.O. (2012) Scabies: a review of diagnosis and management based on mite biology. *Pediatrics in Review/American Academy of Pediatrics* 33, e1–12.

Kelly, D.L., Fuerst, P.A., Ching, W.M. and Richards, A.L. (2009) Scrub typhus: the geographic distribution of phenotypic and genotypic variants of *Orientia tsutsugamushi*. *Clinical Infectious Diseases* 48, S203–230.

Krantz, G.W. (1978) *A Manual of Acarology*. Oregon State University Book Stores, Corvallis, Oregon.

Krinsky, W.L. (1983) Dermatoses associated with the bites of mites and ticks (Arthropoda: Acari). *International Journal of Dermatology* 22, 75–91.

Kuo, C.C., Huang, J.L., Ko, C.Y., Lee, P.F. and Wang, H.C. (2011) Spatial analysis of scrub typhus infection and its association with environmental and socioeconomic factors in Taiwan. *Acta Tropica* 120, 52–58.

Lassa, S., Campbell, M.J. and Bennett, C.E. (2011) Epidemiology of scabies prevalence in the UK from general practice records. *British Journal of Dermatology* 164, 1329–1334.

Leung, V. and Miller, M. (2011) Detection of scabies: a systematic review of diagnostic methods. *Canadian Journal of Infectious Diseases and Medical Microbiology* 22, 143–146.

Lusat, J., Bornstein, S. and Wall, R. (2011) *Chorioptes* mites: a re-evaluation of species integrity. *Medical and Veterinary Entomology* 25, 370–376.

Meana, A., Pato, N.F., Martin, R., Mateos, A., Perez-Garcia, J. and Luzon, M. (2005) Epidemiological studies on equine cestodes in central Spain: infection pattern and population dynamics. *Veterinary Parasitology* 130, 233–240.

Nurmatov, U., van Schayck, C.P., Hurwitz, B. and Sheikh, A. (2012) House dust mite avoidance measures for perennial allergic rhinitis: an updated Cochrane systematic review. *Allergy* 67, 158–165.

Nutting, W.B. (1976) Pathogenesis associated with hair follicle mites (Acari: Demodicidae). *Acarologia* 17, 493–506.

Otranto, D., Milillo, P., Mesto, P., De Caprariis, D., Perrucci, S. and Capelli, G. (2004) *Otodectes cynotis* (Acari: Psoroptidae): examination of survival off-the-host under natural and laboratory conditions. *Experimental Applied Acarology* 32, 171–179.

Pegler, K.R., Evans, L., Stevens, J.R. and Wall, R. (2005) Morphological and molecular comparison of host-derived populations of parasitic *Psoroptes* mites. *Medical and Veterinary Entomology* 19, 392–403.

Rehacek, J. (1979) Spotted fever group rickettsiae in Europe. In: Rodriguez, J.G. (ed.) *Recent Advances in Acarology II*. Academic Press, New York, pp. 245–255.

Russell, R.C. (2001) The medical significance of Acari in Australia. In: Halliday, R.B., Walter, D.E., Proctor, H.C., Norton, R.A. and Colloff, M.J. (eds) *Proceedings of the 10th International Congress of Acarology*. CSIRO Publishing, Melbourne, Australia, pp. 535–546.

Schuster, R. and Murphy, O.W. (1991) *The Acari*. Chapman and Hall, London.

Schuster, R., Coetzee, L. and Putterill, J.F. (2000) Oribatid mites (Acari, Oribatida) as intermediate hosts of tapeworms of the Family Anoplocephalidae (Cestoda) and the transmission of *Moniezia expansa* cysticercoids in South Africa. *Onderstepoort Journal of Veterinary Research* 67, 49–55.

Scott, D.W., Schultz, R.D. and Baker, E.B. (1976) Further studies on the therapeutic and immunological aspects of generalized demodectic mange in the dog. *Journal of the American Animal Hospital Association* 12, 202–213.

Scott, G.R. and Chosidow, O. (2011) European guideline for the management of scabies, 2010. *International Journal of STD and AIDS* 22, 301–303.

Sengbusch, H.G. and Hauswirth, J.W. (1986) Prevalence of hair follicle mites, *Demodex folliculorum* and *D. brevis* (Acari: Demodicidae), in a selected human population in western New York, USA. *Journal of Medical Entomology* 23, 384–388.

Sinclair, A.N. and Gibson, A.J.F. (1975) Population changes of the itch mite *Psorergates ovis*, after shearing. *New Zealand Veterinary Journal* 23, 14.

Sweatman, G.K. (1958) On the life-history and validity of the species in *Psoroptes*, a genus of mange mite. *Canadian Journal of Zoology* 36, 905–929.

Tovey, E.R. and Marks, G.B. (2011) It's time to rethink mite allergen avoidance. *Journal of Allergy and Clinical Immunology* 128, 723–727.

Traub, R. and Wisseman, C.L. (1974) The ecology of chigger-borne rickettsiosis (scrub typhus). *Journal of Medical Entomology* 11, 237–303.

Varma, M.G.R. (1993) Ticks and mites (Acari). In: Lane, R.P. and Crosskey, R.W. (eds) *Medical Insects and Arachnids*. Chapman and Hall, London, pp. 597–658.

Walker, A. (1994) *Arthropods of Humans and Domestic Animals*. Chapman and Hall, London.

Wall, R. (2007) Psoroptic mange: rising prevalence in UK sheep flocks and prospects for its control. In: Takken, W. and Knols, B.G.J. (eds) *Emerging Pests and Vector-Borne Diseases in Europe. Ecology and Control of Vector-Borne Diseases* 1, pp. 227–239.

Wisseman, C.L. (1991) Rickettsial infections. In: Strickland, G.T. (ed.) *Hunter's Tropical Medicine*. W.B. Saunders, Philadelphia, Pennsylvania, pp. 256–286.

Woolley, T.A. (1961) A review of the phylogeny of mites. *Annual Review of Entomology* 6, 263–284.

Wraith, D.G., Cunnington, A.M. and Seymour, W.M. (1979) The role and allergenic importance of storage mites in house dust and other environments. *Clinical Allergy* 9, 545–561.

Mosquitoes (Diptera: Culicidae)

1. INTRODUCTION

Mosquitoes (Culicidae) are globally distributed, nuisance biting pests and vectors of a number of pathogenic protozoans, nematodes and viruses, which cause serious diseases such as malaria, lymphatic filariasis and yellow fever in humans and related diseases in domestic animals.

2. TAXONOMY

At the end of 2010, at least 3520 species were recognized; however, there is considerable debate about the ranking of some genera. Three subfamilies are generally recognized among the Culicidae: Anophelinae, Culicinae and the primitive Toxorhynchitinae. However, some authors suggest that the Toxorhynchitinae should be placed as a tribe within the Culicinae. The currently recognized number of genera is uncertain because of contention surrounding recent and ongoing reviews.

The Toxorhynchitinae contains only a single genus and 76 species of largely forest-dwelling insects that do not blood feed, are therefore not involved in disease transmission and will not be discussed further here.

Three genera of mosquitoes are included in the Anophelinae but only one, *Anopheles*, is widely distributed in tropical and temperate regions. *Bironella* is confined to New Guinea and tropical Australia and *Chagasia* to the Neotropical region, and neither is of medical importance. Some *Anopheles* are of great medical importance, being the sole vectors of human malaria where it occurs, and they also play a substantial role in transmitting lymphatic filariasis in some tropical regions.

The Culicinae are generally widespread, although the genus *Eretmapodites* occurs only in the Afrotropical region, *Haemagogus* only in the Neotropical region, *Psorophora* only in the Nearctic and Neotropical regions, *Heizmannia* only in the Oriental region (with the exception of one species in the Australasian region (Moluccas)) and *Opifex* only in the Australasian region (in New Zealand). The Culicinae can be classified ecologically into five groups, which are more or less self-contained. The aedine genera, which include *Aedes* and *Psorophora*, are globally distributed and have desiccation-resistant eggs that enable them to breed in temporary ground pools and small containers. The sabethines are predominantly Neotropical in distribution, with seven of the twelve genera being restricted to that region; the *Wyeomyia* genus includes four species in the adjoining Nearctic region, *Topomyia* is largely confined to the Oriental region (with the exception of two species in the Australasian region (New Guinea) and one in the Palaearctic region (Ryukyu Islands)) and *Tripteroides* to the Oriental and Australian regions, while the small genus *Malaya* has a wide distribution in the Old World tropics. Collectively, the sabethines breed in small collections of water associated with plants (e.g. leaf axils of epiphytic bromeliads, pitcher plants, tree holes and bamboo). The quasi-sabethines exhibit both aedine and sabethine features, breed in containers such as bamboo, tree holes and leaf axils and include such genera as *Eretmapodites*, *Haemagogus* and *Armigeres*. A fourth ecological group is associated with dense aquatic vegetation and includes *Ficalbia*, *Coquillettidia* and *Mansonia*. The fifth group is an assemblage of miscellaneous genera, of which the largest is *Culex*, which is globally distributed and tends to be associated with permanent and semi-permanent aquatic habitats, and another, *Deinocerites*, is associated with crab holes in the Nearctic and Neotropical regions.

Some Culicinae species are important vectors of human disease-causing pathogens such as arboviruses (e.g. *Aedes aegypti* for yellow fever and dengue, *Culex tritaeniorhynchus* for Japanese encephalitis) and also of filarioids (e.g. *Culex quinquefasciatus* and *Aedes polynesiensis* for *Wuchereria bancrofti*, and *Mansonia uniformis* for *Brugia malayi*).

3. MORPHOLOGY

Adults of the Culicidae have scales on their wings and body. The wings are long and narrow and have a characteristic wing venation in which the longitudinal veins run more or less parallel to the long axis of the wing and end at the wing tip, with veins R_{2+3} and M_{1+2} forked. A fringe of narrow scales extends along the posterior border of the wing. There is a long, forwardly directed proboscis, which is longer than the head and thorax combined, and the palps are porrect (Fig. 92).

Adults can be sexed by examination of the antennae and palps. Most male mosquitoes of all three subfamilies have plumose antennae and females have pilose antennae with fewer, shorter hairs. This difference is a functional one, the antennae being sound receptors enabling the male to locate the female by the sound of her wingbeat. In the female anopheline, the palps are as long and straight as the proboscis, while the palps of the female culicine are considerably shorter, usually about one-quarter of the length of the proboscis. In the anopheline male, the palps are as long as the proboscis and often laterally directed and clubbed at the distal end, with the last two segments swollen. In the typical culicine male, the palps are as long as the proboscis, often upturned and tapering distally, although the tapering is sometimes obscured by the development of tufts of hair on the distal segments (Figs 93 and 94). However, in a number of culicine genera, including *Sabethes*, *Uranotaenia* and *Wyeomyia*, the palps of the

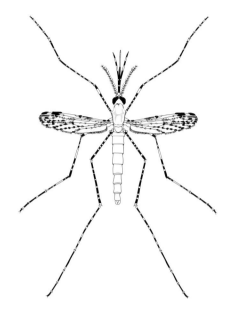

Fig. 92. Female *Anopheles annulipes.*

male are short and similar to those of the female.

There are also differences between the subfamilies in the distribution of scales on the body, the shape of the scutellum and the number of spermathecae. In the Anophelinae, the abdominal sterna, and usually also the terga, are completely or largely devoid of scales, but in the Culicinae the abdomen is covered with a uniform layer of scales. The scutellum is evenly curved in the Anophelinae and there is a regular row of setae on the posterior border, while in the Culicinae the scutellum is trilobed and the setae are grouped on the lateral and median lobes (Fig. 95). There is only one spermatheca in the Anophelinae but there are three in most Culicinae (although only two in *Mansonia* and one in *Uranotaenia* and *Aedeomyia*).

4. LIFE CYCLE

4.1 Egg

Typically, the female *Anopheles* mosquito lays a batch of 100–150 eggs singly, usually at night on the surface of the water. They are

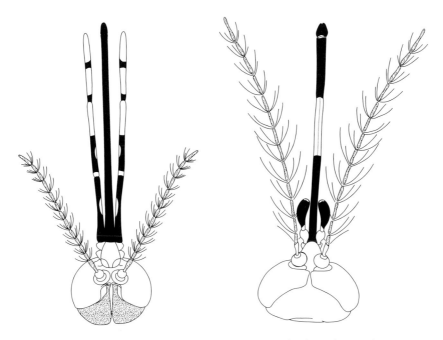

Fig. 93. Heads of female mosquitoes. Left: *Anopheles annulipes*; and right: *Culex annulirostris*.

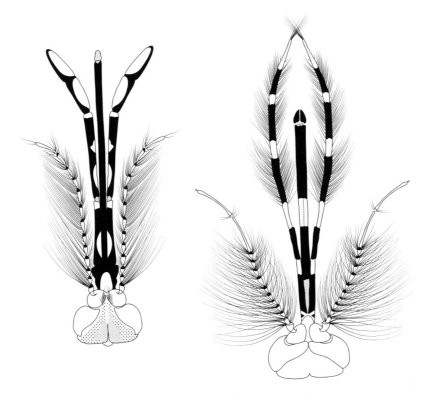

Fig. 94. Heads of male mosquitoes. Left: *Anopheles annulipes*; and right: *Culex annulirostris*.

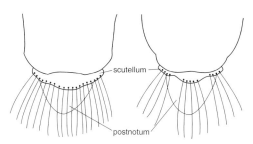

Fig. 95. Scutella of an *Anopheles* (left) and of a culicine (right).

'boat-shaped' and about 1 mm long, with a flattened upper surface (the deck) and a keel-shaped lower surface (which is submerged). A usual feature of anopheline eggs is the presence of paired, lateral, air-filled floats, but these are missing in some species. Scanning electron microscopy of the fine structure of the egg surface reveals differences between closely related species. The surface structure of the egg is such that eggs will attach end-to-end or side-to-side, but never end-to-side. They become attached by surface forces to objects such as vegetation and muddy pool edges projecting from the water; this property prevents eggs from drifting into open stretches of water, where they would be subject to predation or downstream flushing. Some species have characteristic behaviours; for example, *Anopheles multicolor* eggs are laid side-by-side in a row resembling cartridges in a bandolier. Anopheline eggs develop directly into larvae and only rarely undergo diapause (although *Anopheles lesteri* overwinters in the egg stage). Under optimal conditions, anopheline eggs hatch in 1–2 days.

The eggs of *Aedes* are also laid singly in a batch and not attached to each other. They differ from those of *Anopheles* by the absence of floats. They are laid on the moist surface at the water's edge or in the soil (e.g. flood-water mosquitoes) and not on the water itself. When the eggs are first laid, they are permeable to water and increase their weight rapidly. At this stage, they are susceptible to desiccation and collapse, and die if dried. Later, when the serosal cuticle is formed, the eggs can withstand desiccation and remain viable in the dried state for many months, depending on the species. The production of eggs resistant

to desiccation makes *Aedes* species ideal colonizers of temporary collections of water (e.g. salt marshes subject to tidal inundation, rock pools and tree holes, containers, etc.). When the eggs are flooded, most of them hatch immediately, but some will remain dormant and hatch at the second or third flooding. The early emergence of larvae is essential if the life cycle is to be completed before the habitat dries up. Similar dormant eggs are produced by other aedine genera, including *Haemagogus* and *Psorophora*.

The egg raft, which is regarded erroneously as typical of the Culicinae, is produced in only six genera: *Armigeres*, *Coquillettidia*, *Culex*, *Culiseta*, *Trichoprosopon* and *Uranotaenia*. Rafts commonly measure 3–4 mm long and 2–3 mm wide, but the shape is a specific attribute. The lower side of the raft on the water is convex and the upper concave. The eggs are orientated at right angles to the water surface but arranged so that the larva is head down in the egg and emerges from the underside of the raft within a few days from oviposition. The eggs cannot withstand desiccation and, if dried, they collapse and the embryos die.

The eggs of *Mansonia* are laid in clusters on the undersurface of floating leaves of aquatic plants such as *Pistia*, *Salvinia* and *Eichhornia*. The eggs are tapered at the free end, giving the cluster of eggs the appearance of a miniature pincushion.

4.2 Larva

Mosquito larvae emerge from the egg by using a small egg tooth (the egg buster), placed posterodorsally on the head. Three regions can be differentiated in the body of the larva, which is apodous (legless): a well-developed sclerotized head, a broad thorax in which the three segments are fused and a segmented abdomen (Fig. 96). Culicid larvae breathe atmospheric air through a pair of spiracles on the eighth abdominal segment at the posterior end of the abdomen. The opening of the spiracles is in the floor of the spiracular apparatus, which emerges from the water surface. To prevent water from entering the spiracles, a film of oil is secreted by the perispiracular glands.

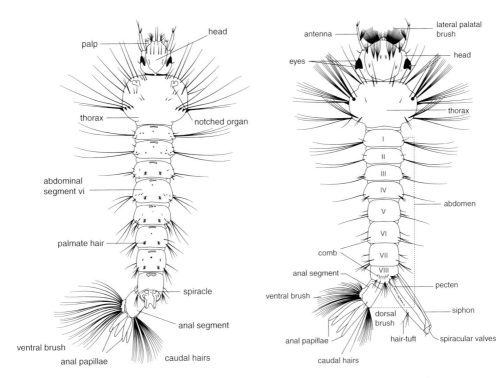

Fig. 96. Larva of *Anopheles maculipennis* viewed from above, with anal segment twisted round to display the ventral brush and caudal hairs. *Source*: Marshall (1938) *The British Mosquitoes*. British Museum (Natural History), London.

Fig. 97. Main features of a culicine larva. *Source*: Marshall (1938) *The British Mosquitoes*. British Museum (Natural History), London.

When anopheline larvae lie parallel to the water surface, supported by the paired prothoracic notched organ, the posterior spiracular apparatus and paired palmate hairs (unique to anopheline larvae and present on most abdominal segments and occasionally on the thorax) spread out in contact with the water surface and help to support the larva (Fig. 96). At the water surface, its dorsal side is uppermost and the mouthparts are directed downwards. Most *Anopheles* larvae feed by collecting/filtering at the air–water interface. The larva rotates its head through 180° and uses its lateral palatal brushes, often referred to as mouth brushes, to create currents in the water surface, which are directed medially to bring food particles between the brushes into the feeding groove.

Culicine larvae differ from those of *Anopheles* by the possession of a siphon on the penultimate segment of the abdomen (Fig. 97). The tracheae are continued into the siphon and the spiracles open at its tip. Within the subfamily, there is great variety in the shape and size of siphons. In *Aedes*, the siphon is typically short and stout. In *Culex*, the siphon is typically long and slender, and when these larvae are submerged, the valves of the siphon close the opening and prevent the entry of water. In *Mansonia* and *Coquillettidia*, the siphon is short and highly specialized (knife-like and serrated) for penetrating aquatic plant tissue to access air.

Possession of the typical culicine siphon enables larvae to hang head down from the surface film and simultaneously respire and feed below the surface. Such collecting/filtering feeding is used by most species of *Culex*, *Culiseta* and some *Aedes* to feed within the water column or, in the case of

Coquillettidia and *Mansonia* species, within the plant root zone. Species of *Psorophora*, *Haemagogus*, *Wyeomyia* and most *Aedes* feed more widely in the water column, on the surface of sediments and at the air–water interface while suspended from the surface. They are classified as collector-gatherers. Many *Aedes* predominantly feed by scraping microorganisms from mineral and organic surfaces, and others are classified as shredders, feeding on detritus and dead invertebrates. The larvae of *Aedes* (*Mucidus*), *Lutzia* (*Metalutzia*), *Psorophora* and *Toxorhynchites* are obligatory predators on other larvae (including their own species). *Eretmapodites* larvae, inhabiting shallow waters in leaf axils, have short siphons but elongated abdomens, enabling them to breathe at the surface and feed on the bottom at the same time. In addition to filter feeding, culicine larvae also drink; for larvae living in a saline environment, this intake is required to counteract the loss of water through the integument, but freshwater species also drink and the rate is increased by phagostimulants and the presence of particulate food.

At intervals, the larva moults and the head capsule is inflated rapidly to the size characteristic of the next instar. During the first three instars, the head capsule increases in length by the addition of a collar, but in the fourth instar the collar remains a narrow band.

Larval identification is usually predicated on the descriptions of fourth instars, but the only efficient criteria for identifying fourth instar larvae from previous ones are head size and the extent of the collar. Other body dimensions, such as length, increase steadily throughout larval life and there is an overlap between instars. Under optimal conditions, the fully-grown fourth instar larva will pupate in 7–10 days, but the duration of the larval stage is temperature dependent and, in the cool season in subtropical areas, the larval stage may last several weeks to months. The effect of seasonal changes in temperature in summer is to produce smaller larvae with smaller head capsules and smaller adults with shorter wings.

4.3 Pupa

The head and thorax of the pupa are combined into a single division, the cephalothorax, which is joined posteriorly to a segmented abdomen (Fig. 98). At rest, the pupa floats at the water surface, with its abdomen curled under the cephalothorax. The pupa does not feed, but it breathes through a pair of trumpets dorsally placed on the cephalothorax. The ninth segment of the abdomen carries a pair of broad, flat plates, the paddles. The pupa remains quiescent unless disturbed, when the abdomen is straightened out, the paddles

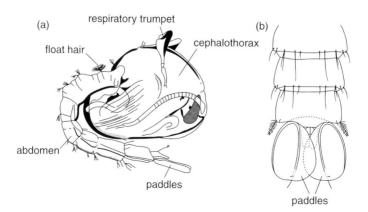

Fig. 98. (a) Pupa of *Anopheles maculipennis*; (b) dorsal view of terminal segments of abdomen of pupa of *Anopheles hilli. Source*: (a) Nuttall, G.H.F. and Shipley, A.E. (1901) *Journal of Hygiene* 1.

spread widely and the abdomen is flexed rapidly. Depending on the orientation of the cephalothorax, this movement of the abdomen serves to drive the pupa forward or to dive below the surface, where it may merely float to the surface again or stay down without any apparent further effort. Pupae can complete their development out of water and, on a dry surface, pupae can jump about, possibly as a defence mechanism against predators.

The culicine pupa is very similar to that of the anopheline, apart from the differences in the shape of the respiratory horn, which is tubular in the Culicinae compared with the distally expanded horn in the Anophelinae. In *Mansonia* and *Coquillettidia*, the respiratory horns of the pupa are modified (knife-like) for penetrating plant tissue and, on pupation, the larval exuviae is not shed until the pupa has attached itself firmly to the plant. In these genera, at adult emergence, the pupa frees itself from the plant, leaving behind the tips of the respiratory horns, and rises to the surface, where it respires in the same manner as other culicine pupae.

4.4 Development time and adult emergence

In general, the immature development times for culicines are as short as or shorter than those of *Anopheles*, but the life cycles of *Mansonia* and *Coquillettidia* are particularly protracted; *Coquillettidia aurites* takes 40–50 days to develop from egg to adult at 23°C, whereas *Ae. aegypti* can complete its development in 10 days at the same temperature.

In preparation for emergence, gas appears between the pupal cuticle and the enclosed pharate adult. Then gas appears in the midgut, thus increasing the buoyancy of the pupa. The abdomen is raised into a horizontal position and a split appears mid-dorsally in the pupal cuticle. Pumping in of air raises the adult out of the pupal exuviae. The newly emerged adult expands its wings and legs by increasing haemolymph pressure. Eclosion may take about 15 min, at the end of which the newly emerged adult is distended by air and fluid. Diuresis contributes to eliminate excess fluid and air is dispersed from the midgut, enabling the adult to be able to fly normally after about an hour. There follows a period of rest in which intense metabolic activity occurs involving the breakdown of larval muscles in the abdomen.

5. BEHAVIOUR AND BIONOMICS

When adults emerge from the pupae of a single egg batch, the males emerge first. During this initial 24 h period of adult life, the male terminalia rotate through 180° before the male is ready for mating. Hence, by the time the females emerge, the males are competent for mating. The males obtain additional energy for flight by taking nectar from plants, as do the females. Both sexes of the salt-marsh mosquito, *Aedes taeniorhynchus*, show a marked circadian pattern of nectar feeding, with peaks in the early morning and from late afternoon until nearly midnight. Both sexes of *Ae. aegypti* and *Anopheles arabiensis* respond to odours of composite flowers, with the *Ae. aegypti* response being biphasic with dawn and dusk peaks and the nocturnal *An. arabiensis* exhibiting a peak at 20:00 h.

Female mosquitoes are referred to as endophilic or exophilic, depending on whether they rest indoors or outdoors, and as endophagic or exophagic, depending on whether they feed indoors or outdoors. *Anopheles balabacensis* is both exophilic and exophagic; *Anopheles maculatus* is exophilic and endophagic, while *Anopheles minimus* is both endophilic and endophagic.

5.1 Mating

Mating is often preceded or accompanied by swarming, in which the males associate over a marker and fly in a particular manner. Swarming occurs commonly at sunset, when the wind tends to die away and the rapidly fading light has a stimulating effect on mosquito activity. Markers are visually prominent objects such as the tops of bushes or trees, or ground objects that contrast with the background. Male swarms form over the markers and virgin females orientating to markers will encounter swarming males. Males recognize a female from a short distance and fly directly towards her and attempt to mate and, if successful, the pair leaves the swarm *in copula*, with insemination being completed elsewhere, although in some species, insemination occurs in flight, with the couples separating when they reach the ground. Swarming occurs for a short period at dusk at 24 h intervals, and such temporal discontinuity of swarming and mating increases and maintains the genetic diversity on which natural selection and survival depend. In *Anopheles stephensi*, males become active at dusk and form mating swarms, where they are joined by virgin females; following mating, females cease activity and engage in nocturnal host seeking and oviposition.

Other mosquitoes, such as *Ae. aegypti*, mate without swarming. Males aggregate around items attractive to females, such as water-filled containers, dark objects and blood hosts, where they intercept females. The male locates females by responding to the sound of the female's wingbeat, which in 3- to 4-day-old females ranges from 450 to 600 cycles/s, with an average of 500 cycles/s. Male *Ae. aegypti* can detect females up to a distance of 25 cm and respond to a sound of approximately 20 decibels. Male *Ae. aegypti* have higher wingbeat frequencies than females, but as both sexes have lower frequencies early in adult life, the wingbeat of young males and old females is similar.

In some species, the male antennae are pilose, like the females', and sexual recognition is based on other than auditory stimuli. Males of *Deinocerites cancer* search the surface of breeding sites for female pupae and mate with the female when it emerges. Female pupae are recognized through a contact pheromone, which the male detects with its antennae. Males of *Opifex fuscus* are capable of mating within an hour of emergence and will copulate with females that are still within the pupal exuviae. In *Sabethes chloropterus*, the male hovers over the resting female and tests her receptivity by tapping the hindlegs with its midtarsi.

During mating, spermatozoa are deposited in the bursa copulatrix of the female, from which they move to the spermathecae, and

secretions from the accessory glands of the male produce a mating plug in the female. In *Ae. aegypti*, it has been shown that the plug contains matrone, a substance which stimulates oviposition and inhibits mating.

5.2 Host finding

In some species, for example *Ae. taeniorhynchus*, *Anopheles hilli* and *Culex pipiens pipiens* biotype *molestus*, the first egg batch may be matured autogenously, i.e. without a blood meal, but most mosquitoes require a blood meal for ovarian development. The source of the blood meal is a major factor in determining the potential of a species to be a nuisance pest and/or vector of disease. A number of basic feeding patterns have been recognized among mosquitoes, including species that feed almost entirely on mammals (e.g. *Anopheles gambiae*), almost entirely on birds (e.g. *Culiseta melanura*), readily on both birds and mammals (e.g. *Cx. p. quinquefasciatus*), almost exclusively on amphibians (e.g. *Culex territans*) or predominantly on reptiles (e.g. *Deinocerites dyari*). Within the mammal-feeding mosquitoes, the feeding habits can be further subdivided into those which are anthropophilic (feeding primarily on humans), for example *Ae. aegypti* and *An. gambiae*, and those which are zoophilic (feeding primarily or exclusively on other mammals), such as *Anopheles quadriannulatus* in Africa and *An. maculatus* in Nepal.

There are several stages in the feeding behaviour of mosquitoes, which are designated activation, orientation, landing and probing. Most mosquitoes have a crepuscular or nocturnal circadian cycle of activity, which is stimulated by rapidly fading illumination. In its upwind movement towards a host, orientation can be both visual and chemical, usually occurring at less than 20 m. Host-seeking mosquitoes are more attracted to low-intensity colours, such as blue, black and red, than they are to high-intensity colours, such as white and yellow. Convection currents given off by a warm body enable mosquitoes to orientate to a host very effectively over short distances. Because many mosquitoes are active on the darkest nights when visibility is minimal, it has been proposed that mosquitoes mostly orientate themselves upwind by adopting an undulating flight pattern seeking chemical cues.

Carbon dioxide (CO_2) acts as an attractant to mosquitoes, which fly upwind in response, and vertebrate body odours are also known to be attractive. Responses to octenol (1-octen-3-ol) by host-seeking mosquitoes have been recognized and, in combination with CO_2, it increases greatly the catch of some species of *Aedes*, *Anopheles*, *Coquillettidia*, *Mansonia* and *Psorophora*, with the effect being a synergistic one; however, species of *Culex* show little response to octenol in combination with CO_2 (probably because species of *Culex* are generally ornithophilic and octenol is a mammalian emanation). Other chemicals, such as lactic acid (the product of muscular activity) and fatty acids of various chain length (e.g. caproic acid), acetone, ammonia and dimethyl disulfide, have also been shown to be attractive to some mosquitoes when used in various combinations with/without CO_2.

5.3 Biting cycle

Overall, mosquitoes are not limited in their activity to ground level. In forests, potential blood hosts are to be found from ground up to the canopy, the domain of birds and primates. Generally, blood feeding follows a circadian rhythm, with most species of mosquitoes being nocturnal or crepuscular, and a smaller number diurnal. Most species of *Anopheles* are nocturnal and many, including major malaria vectors such as *An. gambiae* and *An. minimus*, have their peak biting activity in the late night hours around midnight. *Culex* mosquitoes are also mostly crepuscular or nocturnal, with many of the arbovirus vectors such as *Culex annuliorstris*, *Culex tarsalis* and *Cx. tritaenirohynchus* biting early in the evening hours, although the filariasis vector *Cx. quinquefasciatus* bites late at night. Species of *Aedes* (*Stegomyia*), including the virus and filariasis vectors *Ae. aegypti*, *Aedes albopictus*, *Ae. polynesiensis* and *Aedes scutellaris*, are diurnal, with *Ae. aegypti* having two or three peaks of activity, one just

after dawn and one just before sunset, 17:00–18:00 h, with another sometimes in the late morning.

These biting patterns have practical implications in that mosquito nets can be effective against nocturnal species, especially those which feed after midnight, whereas they give little protection against crepuscular or diurnal species, which require house screening or the use of repellents and protective clothing when outdoors. However, although circadian rhythms are characteristic of species, they can be modified by environmental conditions; in particular, heavy cloud shade and higher humidity levels often promote feeding activity outside normal hours, while low temperatures can inhibit such activity.

5.4 Blood feeding and ovarian development

Blood feeding takes only a few minutes. When pool feeding, the female waits until sufficient blood has collected, before it is ingested rapidly and the gut becomes visibly distended. The amount of blood ingested varies with species and size of the female. Further, some species, for example *Ae. aegypti*, pass drops of clear fluid from the anus while they are feeding, and others, for example *An. stephensi*, pass apparently unchanged blood from the anus, thereby increasing or concentrating the overall volume ingested. *Ae. aegypti* may ingest 4–5 µl of blood, while the larger *Cx. quinquefasciatus* may ingest at least 10 µl of blood.

In the tropics, the gonotrophic cycle of blood ingestion, digestion and ovarian development leading to oviposition may take no more than 48 h (or sometimes less) with *Anopheles* species, but with *Aedes* and *Culex* spp., it may take at least 3–4 days. Following nocturnal activity, four categories of resting females can be recognized in the early morning: those which have fed the previous night and are full of blood, those which have fed the night before and contain some dark blood and developing eggs, fully gravid females that will oviposit during the following night and empty females (either nulliparous or parous) that have oviposited but not yet fed again. The feeding and oviposition cycles have a minimum duration of 2–3 days, but are dependent on mosquito species and environmental conditions.

Most mosquitoes show gonotrophic concordancy in which there is strict alternation of blood feeding and oviposition, with each blood meal being followed by the maturation and oviposition of a batch of eggs. Some species, such as *An. gambiae* and *Anopheles funestus*, require two blood meals to mature the first batch of eggs, but very few feed twice in subsequent ovarian cycles. *Ae. aegypti*, however, will feed twice during the maturation of its first batch of eggs and will take multiple blood meals during later ovarian cycles, particularly if interrupted during feeding attempts.

In each ovarian cycle, only one egg is developed in each ovariole and therefore the number of eggs matured in an egg batch depends on the number of ovarioles, and also with the species and the source of the blood meal. For instance, *Anopheles melanoon* can lay up to 500 eggs in its first gonotrophic cycle, while *Cx. p. pipiens* can develop about 250 eggs following a meal on bird blood but only 120 eggs from the same amount of human blood.

5.5 Oviposition

Oviposition is thought to follow a circadian rhythm in many mosquitoes. In the laboratory, peak oviposition of *Ae. aegypti* occurs just before sunset and remains so even when the period of daylight is varied substantially; the factor which controls oviposition is the onset of darkness and not the length of daylight. In Trinidad, *Ae. aegypti* has a biphasic cycle of oviposition, with peaks occurring after dawn between 06:00 and 08:00 h and before dusk at 16:00–18:00 h. Under simulated tropical conditions, newly blood-fed, inseminated female *An. stephensi* are inactive for 2 nights and on the 3rd night, when their eggs are mature, they become active at dusk and seek out oviposition sites during the night; following oviposition, females resume nocturnal host seeking. Gravid female *Cx. tarsalis* are attracted to oviposit in water containing egg rafts of the same species or to water in which

egg rafts have been laid previously; the attractant is an ether-soluble pheromone that is seemingly species specific. Female *Cx. quinquefasciatus* respond in a similar manner to egg rafts of the same species and deposit their eggs nearby on the water surface; they respond to the presence of a volatile pheromone secreted in the droplet found at the apex of each egg. Egg rafts that have hatched 24 and 48 h previously continue to attract ovipositing females and, in Kenya, gravid female *Cx. quinquefasciatus* have been shown to distinguish between sites with and without pheromone from a distance of up to 10 m. Species that breed in small containers can have modified morphology and/or behaviour. For instance, *Armigeres dolichocephalus* has a long, narrow thorax that enables the adult female to enter small holes in bamboo for oviposition, while ovipositing *Sa. chloropterus* hover in front of vertical holes in bamboo and flick eggs through the hole, one at a time.

5.6 Selection of breeding site

Gravid females locate a suitable site for oviposition that may be more or less characteristic for a genus/species. Although it is possible to describe the typical breeding site characteristic of a species, there is a good deal of variation and species might be found in unusual habitats. Breeding sites can be classified in a number of ways and a major division into aboveground and subterranean waters has been proposed. These breeding sites include flowing streams, ponded streams, lake edges, swamps and marshes, shallow permanent ponds, shallow temporary pools, intermittent ephemeral puddles, natural and artificial containers. Subterranean waters are classified as natural or artificial.

In considering different habitats, species will be cited as examples of mosquitoes which breed in such habitats, but each species has a limited geographical distribution. Still edges of running water habitats are important sources of *Anopheles* mosquitoes. The grassy edges of sunlit, flowing water represent breeding sites for species such as *An. maculatus* and

Anopheles fluviatilis, while *An. minimus* and *Anopheles umbrosus* are found in the grassy edges of shaded streams and larvae of *Anopheles superpictus* hide among pebbles at the edges of shallow, bare, hill streams.

Permanent ground habitats, such as large swamps and vegetated lakes, are the breeding sites of species of *Culex*, *Mansonia* and *Coquillettidia*, as well as some anophelines such as *An. funestus* and *Anopheles hyrcanus*. Rice fields are semi-permanent man-made swamps suitable for a number of species, including *Anopheles culicifacies* and *Cx. tritaeniorhynchus*, but some species occur in rice fields only when the plants are small and disappear as the rice develops into tall plants – not because the water is no longer suitable but because, as the plants grow taller, the female's oviposition behaviour is at first inhibited and then prevented. Permanent saltwater fishponds that carry a luxuriant growth of floating green algae, which form the food of the herbivorous fish, provide a breeding site for *Anopheles sundaicus*.

Water quality can be important: *An. funestus* occurs only where the water is clear and unpolluted, but *Mansonia* species and *Cx. tritaeniorhynchus* can thrive in moderately polluted water, *Cx. quinquefasciatus* colonizes highly polluted sewage treatment ponds, *Cs. melanura* breeds in oligotrophic waters of swamps and *Aedes communis* in pools made by melting snow on the tundra.

Temporary ground pools of salt water are formed in coastal salt marshes and in the tropics in mangrove communities, and are a prolific source of anthropophilic *Aedes* mosquitoes in almost all parts of the world. Salt marshes represent the breeding sites for *Ae. taeniorhynchus* and *Aedes sollicitans* in the Caribbean and in the eastern coast of the USA, for *Aedes caspius* and *Aedes detritus* in the Palaearctic region, and for *Aedes vigilax* and *Aedes camptorhynchus* in the Australasian region, respectively. Temporary freshwater ground pools are the main breeding sites of various *Aedes* species, *An. gambiae* and *Anopheles punctulatus*. Early exploitation of temporary pools enables the immature stages to minimize predation.

Container habitats are prolific sources of mosquitoes. All the sabethines breed in them

and many culicines, but few anophelines. Container habitats include tree holes, a source of *Anopheles plumbeus*, *Aedes africanus*, *Aedes notoscriptus*, *Aedes triseriatus* and various *Toxorhynchites* species; collections of water in bamboo internodes breed *Sa. chloropterus*; leaf axils are a source of *Aedes bromeliae*; pitchers of plants breed species of *Tripteroides* and *Wyeomyia*; epiphytic bromeliads are a source of *Anopheles bellator*, fruits and husks of *Ae. albopictus* and species of *Armigeres* and *Eretmapodites*; artificial containers, such as vases, water jars, tyres, etc., are breeding sites of *Ae. aegypti*, and snail shells of *Eretmapodites*. Movement of such artificial containers (particularly water barrels and automobile tyres) containing desiccation-resistant eggs has been responsible for the wide geographic spread of *Ae. aegypti* and *Ae. albopictus*.

Underground water habitats are mostly artefacts, such as storage tanks and wells. In the Mediterranean area, *Anopheles claviger* breeds in wells and cisterns, and its counterpart in the Oriental region is *An. stephensi*, but neither species is restricted to underground water and *An. claviger* breeds in cool waters in other habitats and *An. stephensi* occurs also in irrigation ditches. *Cx. p. pipiens* breeds in storage tanks of clean water and *Cx. quinquefasciatus* in highly polluted water, such as that found in soakage pits, flooded latrines and septic tanks. Crab holes are natural underground water habitats and in the Australasian region breed *Aedes daliensis*, in the Pacific region *Ae. polynesiensis* and in the Neotropical region *De. cancer*.

5.7 Dispersal

In the field, species vary greatly in the extent to which they disperse from their breeding sites. The flight range of a species can be considered in terms of probability; it is not a limit beyond which a species will not fly but an indication of the distance beyond which the species will be present only in insignificant numbers. This approach emphasizes the fact that the greater the population emerging from the site, the larger the number of mosquitoes that will reach a fixed distance and the greater the

maximum distance reached by some members
of the population. Dispersal is influenced by
the prevailing wind, by longevity of the species
and by the presence of suitable hosts. However,
as a general feature of mosquitoes, males do
not disperse as far as females, and if males are
found, it is an indication that the breeding site
is close by, often within 200 m.

Forest mosquitoes appear to have a more
restricted flight range than those that breed in
open situations. Where suitable domestic
breeding sites are available, Ae. aegypti is
unlikely to disperse more than 0.5 km in
significant numbers. In the tropics, larval
control against Anopheles mosquitoes is
usually effective if extended for 3 km. Many
species occur abundantly up to 1 km from
their breeding site, but rarely reach 5 km.
Studies have shown that only 20% of a
population of An. funestus disperses further
than 0.8 km from the breeding site, with a
practical limit of dispersal of 7 km, while few
female An. gambiae s.s. disperse further than
3 km (with the average being 1.0–1.6 km).
However, Anopheles pharoensis is a large
mosquito that breeds abundantly in the Nile
Delta and regularly disperses 6 km from its
breeding site and, on occasion, for distances of
100 km with the wind.

Salt-marsh mosquitoes seem to be
particularly likely to undertake long dispersal
flights. The New World species Ae.
taeniorhynchus and Ae. sollicitans emerge in
vast numbers after their breeding sites have
been inundated; they can disperse tens of
kilometres, with females recovered up to
40 km from their site of origin, and they
disperse upwind at low wind velocities
(0.25 m/s) but downwind at high velocities
(1.5–2.0 m/s). In Australia, Ae. vigilax
behaves in a comparable manner and has been
recorded at least 70 km downwind from
source.

5.8 Hibernation, diapause, aestivation and seasonal cycles

In tropical regions, breeding of most species
continues throughout the year and seasonal
fluctuations in numbers are related to the rainy

and dry seasons. In the severe dry season,
breeding either continues in relation to reduced
bodies of permanent water or the population
survives this hostile period as dormant eggs or
aestivating adults. In the valley of the White
Nile in Sudan, An. arabiensis maintains itself
through the dry season by low-level breeding,
but in the more arid areas, 20 km from the
Nile valley, there is evidence that the females
aestivate. In the USA, Culiseta inornata
enters into winter diapause in the north, and in
the south, it aestivates during the hot, dry
season.

There is much more information available
on the effect of low temperature on mosquito
biology. In the subtropics and adjoining areas
of the warmer temperate regions, the main
response to lower temperature in the cool
season is to slow down the rate of biological
processes. Ovarian development may take
10–14 days, compared with 2–3 days under
optimum conditions, and development from
egg to adult takes 2–3 months. In the
temperate high-veld region of South Africa,
Cx. p. pipiens and Culex theileri overwinter
by quiescence, as does Cx. annulirostris in
southern Australia and Cx. tarsalis in southern
California (but in the cooler central California,
Cx. tarsalis enters hibernation/diapause).

Hibernation is induced in a developing
generation by exposure to decreasing hours of
daylight, reinforced by lower temperatures. In
species with an extensive north–south
distribution, for example Anopheles freeborni,
Ae. triseriatus and Wyeomyia smithii, there
is clinal variation in the length of the critical
photoperiod to induce diapause. Diapause in
the egg stage is found in all Aedes, Psorophora
and Haemogogus, in which it may be a
response to dryness as well as a photoperiod-
induced diapause, and it also occurs in
Anopheles walkeri and Culiseta morsitans.
Larval diapause in winter is a feature of
Culiseta species, for example Cs. melanura,
and is also found in Anopheles barberi, Ae.
triseriatus and Wy. smithii, with the latter
able to survive freezing, but not for prolonged
periods. Other species pass the winter as
larvae in waters that are protected from
freezing, for example An. plumbeus and Ae.
triseriatus in tree holes, and Cs. melanura in

bogs. Diapausing larvae are as active as non-diapausing larvae, but feed less.

The commonest way of surviving winter in species that do not produce dormant eggs is by overwintering as inseminated, nulliparous females. Two mechanisms have been recognized: (i) gonotrophic dissociation, where the female feeds on blood at intervals during the winter but does not develop eggs, and (ii) gonotrophic concordancy, where the female feeds on plant juices, develops body fat and its ovaries remain undeveloped. Dissociation is well recognized in the Californian population of *An. freeborni* but not elsewhere in the western USA. Similarly, throughout much of its range in south-eastern USA, overwintering *An. quadrimaculatus* take infrequent blood meals. The development of body fat and the survival rate over the winter is the same for female *An. freeborni*, independent of whether they followed gonotrophic concordancy or dissociation. Diapausing *An. freeborni* and *Cx. tarsalis* develop body fat during the autumn and utilize the stored lipids for energy during winter. High humidity is necessary for most overwintering adult mosquitoes. Female *Culiseta alaskensis* have been found at the bases of dense stands of grass under snow, and in France several species have been recovered from dense vegetation in reed swamps. Some species, for example *Anopheles sacharovi*, *An. freeborni* and *Cx. tarsalis*, undertake a prehibernation flight from their breeding sites to their winter resting places, or hibernacula. It is considered that diapause decreases in intensity with time and may end in midwinter, with the insect remaining quiescent until favourable conditions return in spring. Diapause is under hormonal control and that of *Ae. triseriatus* larvae is broken by exposure to ecdysone and for female *An. freeborni* by juvenile hormone.

In the humid tropics, breeding may be continuous throughout the year, with little variation in the size of the adult populations. In the subtropics, with hot dry summers, mosquito populations are commonly bimodal, with peaks in spring and autumn, and lower populations during summer. In more temperate regions, one or two generations may occur during the warmer months of each year.

5.9 Longevity

In the epidemiology of vector-borne disease, two features of a species' behaviour and ecology are particularly relevant: the host on which the females commonly feed and their survival between feeds. For some time, it was a common assumption that mortality of female mosquitoes was independent of age, but in many important vectors of malaria, mortality rate has been shown to increase with age. As there is usually one blood meal per ovarian cycle, survival between blood meals is the same as that between ovarian cycles, and mosquito dissection for investigating ovarian morphology (such as expansion in ovarian tracheation and presence/number of ovariolar dilatations), as well as determining the presence/absence of gut meconium and the number of growth lines on thoracic phragma, have been used to indicate longevity. More recently, biochemical and molecular methods (including analysing changes to cuticular hydrocarbons and assaying gene transcription) have become available. Depending on the species in question, some of these are more useful than others (e.g. ovariolar dilatations have been investigated more with *Anopheles* species) and accurate methods that are applicable across genera are still awaited. In general, little is known about the longevity of males, although with some species at least, it seems to be similar to that of the conspecific females.

6. MEDICAL AND VETERINARY IMPORTANCE

Mosquitoes commonly cause inflammatory skin reactions and mild allergic responses to their salivary secretions when they bite, and while more severe allergic reactions have been reported, anaphylaxis is rare. They can also cause serious adverse impacts on animal herds attacked by extraordinary numbers of blood-seeking mosquitoes. However, the main medical and veterinary importance relies on their role as vectors of disease-causing pathogens.

Some pathogens can be transmitted mechanically by mosquitoes, the principal disease example being the myxoma virus that is spread among rabbits primarily by mosquitoes in Australia (although in Europe, the principal vector of myxomatosis is the flea, *Spilopsyllus cuniculi*). With respect to mechanical transmission of other organisms, mosquitoes have been implicated in the transmission of some bacteria, for example *Francisella tularensis* (causing tularaemia) and *Mycobacterium ulcerans* (causing Buruli, and other-named, ulcers), although direct evidence for these is weak.

Mosquitoes are more important as biological vectors of pathogenic protozoans, nematodes and viruses, which cause a range of diseases, including malaria, filariasis and encephalitis in humans and similar diseases in domestic animals.

6.1 Arboviruses

The term 'arbovirus' means 'arthropod-borne-virus'. These multiply in vertebrate hosts and are moved between these hosts by arthropod vectors. A range of viruses within the families Flaviridae, Togaviridae, Bunyaviridae and Rhabdoviridae are transmitted by mosquitoes and these will be addressed in turn.

6.1.1 Flaviviridae
The type species of Flavivirus is the yellow fever virus, which explains the generic name Flavivirus. Other significant members of the genus include the dengue viruses, West Nile virus and several encephalitis viruses including Japanese, Murray Valley and St Louis. These viruses are all mosquito-borne, but other flaviviruses are tick-borne (e.g. Omsk, Powassan and Kyasanur Forest disease viruses – see the entry for 'Ticks (Hard)').

YELLOW FEVER VIRUS (YFV)
This disease has historically been of great international concern as a major quarantinable infection. Epidemics occurred regularly in urban areas of tropical and subtropical America during the 17th, 18th and 19th centuries, and outbreaks of various sizes are still frequent in Africa, particularly in western Africa. Fortunately, a fully effective vaccine (17D) is available, which probably protects for life and is currently valid for 10 years.

The virus is endemic in Africa from 15°N to 10°S, especially in West Africa, and in parts of tropical Central and South America, from which it may spread to other areas. Three epidemiological patterns are recognized: sylvatic, intermediate (rural) and interhuman. In Africa, sylvatic activity occurs in rainforest, where it is enzootic among monkeys, and a major vector is *Ae. africanus*. Intermediate (endemic) activity occurs in the humid, rural savannah adjoining rainforest, where the vectors are *Ae. bromeliae* (formerly *Aedes simpsoni*), *Aedes furcifer* and *Aedes taylori* (the last two species being indistinguishable in the adult female). Interhuman (epidemic) outbreaks, with mortalities of 20–50%, occur in village and urban communities, where the vector is the domestic *Ae. aegypti*.

Ae. aegypti becomes infected with the YFV when it feeds on an infected person in the early stages of the disease, from about 6 h before the onset of clinical signs to about 4 days later. The virus undergoes a temperature-dependent cycle in the mosquito which takes 2 days at 30°C and 12 days at 18°C, during which time the virus multiplies in the cells of the midgut and then the salivary glands. When the cycle is complete, the mosquito remains infective for the rest of its life, secreting the virus in its saliva. Multiple feeding during a single gonotrophic cycle by *Ae. aegypti* increases greatly the possibility of it acquiring and transmitting the virus.

In a susceptible individual, the virus will be incubated for 3–6 days before the onset of clinical symptoms of disease. Infection can follow a rapid course, often terminating fatally within a week. Although some infections are subclinical, the acute phase is represented by fever, muscle pain, shivers, headache, appetite loss, nausea and/or vomiting. Most cases improve, but 15% enter a 'toxic' phase wherein fever reappears and body systems are affected, with jaundice, abdominal pain and vomiting, internal and orifice bleeding and kidney failure. Approximately 50% of toxic patients die in 10–14 days, but the remainder recover without significant organ damage.

In South America, up to a few hundred cases of yellow fever are reported annually, which may underestimate the actual number of cases. Monkeys are the principal wild vertebrate hosts and deaths of howler monkeys (*Alouatta* spp.) are an early sign of a yellow fever epizootic in progress. Infection in African monkeys rarely results in illness or death, indicating a balanced long-term host–parasite relationship and suggesting that YFV is a relatively recent introduction into the Neotropical region, possibly with the slave trade when ships crossed the Atlantic regularly from West Africa to the New World. Water storage containers on board would have provided ideal breeding sites for *Ae. aegypti*, which may have been introduced into the Americas at the same time.

Humans become involved in sylvatic yellow fever when they enter an enzootic area and are bitten by infected mosquitoes. Individuals incubating the virus can then introduce it into a human ecosystem infested with *Ae. aegypti* and the scene is set for an outbreak of yellow fever. Interchange between the sylvatic and humid savannah ecosystems can be achieved by monkeys moving from one to the other. In western Uganda, yellow fever circulates among the forest monkeys through *Ae. africanus*. Monkeys are attracted into groves of cultivated bananas, where they are exposed to attack by day-biting *Ae. bromeliae*, which breeds in water accumulating in the axils of plants (including bananas). *Ae. bromeliae* is susceptible to infection with YFV and readily bites both monkeys and humans, making it potentially a very important vector.

At least 14 species of mosquitoes are able to transmit YFV in Africa, and some have been captured in large numbers during epidemics. *Ae. africanus* is a vector of sylvatic yellow fever, *Ae. bromeliae*, *Ae. furcifer* and *Ae. taylori* are involved in the transmission of YFV in rural situations, *Aedes vittatus* has been considered to be a vector in Sudan and *Aedes luteocephalus* in Nigeria. With two exceptions, the species of *Aedes* mentioned as vectors of YFV belong to the subgenus *Stegomyia*, a group of black mosquitoes with silvery markings. The exceptions, *Ae. furcifer* and *Ae. taylori*, belong to the subgenus *Diceromyia*, in which the tarsi are all dark and the broad wings have pale and light scales. In South America, the vectors are various species of *Haemagogus*, including *Haemagogus*

spegazzinii, *Haemagogus leucocelaenus* and *Haemagogus janthinomys*; *Sa. chloropterus* is a relatively inefficient vector of YFV, but may play a role in virus survival because of it being relatively drought resistant.

Monkeys and humans have been considered to be reservoirs of YFV but, in fact, they probably play more of a temporary role as amplifiers of the virus. In humans, the viraemia declines from the fourth day after the onset of clinical disease and, on recovery, people are immune, as are monkeys after infection. The virus arguably survives longer in infected female mosquitoes, which pass it on through transovarian transmission to a very small proportion of their offspring. Such transmission has been shown to occur in species of *Aedes* and *Haemagogus*. In both of these genera, mature eggs can survive desiccation for several months in a state of diapause and it is possible that this is an important way in which the virus survives, especially if transovarian transmission occurs in successive generations of mosquitoes.

There is no specific treatment for YFV, with supportive therapy given because of the risk of renal failure and shock, but there is a very effective vaccine available that provides protection for at least 10 years. Overall, prevention of urban disease outbreaks is dependent on controlling the vector *Ae. aegypti*, a mosquito that breeds in containerized water (including domestic receptacles). Control therefore depends generally on eliminating or treating the breeding sites of *Ae. aegypti* and, since these are often small collections of water in domestic situations, requires the cooperation of the general public. During urban epidemics, adulticiding with knock-down and residual pyrethroids in and around homes will help to interrupt transmission and reduce the likelihood of a continuing epidemic.

In recent years, a concerted mass vaccination programme, known as the 'Yellow Fever Initiative', has been instigated in Africa, with a focus on countries where the disease is most prominent. All age groups are targeted and, between 2007 and 2010, ten countries (i.e. Benin, Burkina Faso, Cameroon, Central African Republic, Guinea, Liberia, Mali, Senegal, Sierra Leone and Togo) completed their preventive campaigns.

DENGUE VIRUSES

Dengue disease is caused by one or more of four viruses, known as dengue 1, 2, 3 and 4 (DENV 1–4), and it occurs in nearly all tropical countries where the major vector, *Ae. aegypti*, is present and abundant in urban situations. Other vectors of lesser importance in different areas are *Ae. albopictus, Ae. scutellaris* and *Ae. polynesiensis*. An ancestral sylvatic cycle exists among non-human primates and *Aedes* mosquitoes in forests in South-east Asia and West Africa, and occasional sylvatic infections occur in humans; however, humans are generally thought to be the critical reservoir and amplification hosts for the human cycle, and urban outbreaks generally are the result of human distribution of the viruses to places with abundant *Ae. aegypti*. The incubation period in humans is 2–7 days, with the individual becoming infective to mosquitoes 6–18 h before the onset of fever and during the fever period, which lasts about 6 days. In the mosquito, the virus replicates in the cells of the midgut epithelium and then the salivary gland, where the virus is passed with saliva during feeding. This extrinsic cycle takes a minimum of 8 days, but more often 11–14 days.

In classical dengue (dengue fever), symptoms include fever, headache, rash and joint pains, but there is no mortality; recovery is complete, but weakness and depression may be severe and last several weeks. Epidemics of dengue fever are noted for affecting a large proportion of the human community. The four different serotypes of dengue virus offer only cross-immunity of short (a few months) duration, although long-term immunity is afforded to the infecting serotype.

The dengue haemorrhagic fever (DHF) syndrome, first recognized in the Philippines in 1953, which shows as internal bleeding and circulatory failure, and the associated dengue shock syndrome (DSS), with its acute plasma leakage from vascular permeability and a mortality rate of 1–5% (particularly in children), are now widespread in South-east Asia and have occurred in the Pacific and American regions. The causation of DHF is not completely understood and both virus virulence and host immunity factors may be involved, but it often follows sequential exposure to two different serotypes of the virus, indicating an antibody

enhancement effect. It affects infants born to dengue-immune mothers during their first infection and children more than 1 year old who acquire a second infection, but it is rare in adults and children above 14 years of age.

In the 1950s and 1960s, a major effort was made to eradicate *Ae. aegypti* from the Americas. It was successful in Mexico and most of Central and South America, but not in the USA, Venezuela and the Caribbean. In the 1970s, surveillance was reduced, most of the region was reinfested and the progress made lost. Subsequently, since 1977, there has been an increased incidence of dengue in the Americas, with DHF occurring in a number of regions, including Mexico, Cuba, Puerto Rico, Central America, Venezuela and Brazil.

There is no specific treatment with drugs, but supportive therapy can be important and, with haemorrhagic patients, hypovolaemic shock should be treated with fluid replacement. There is no vaccine available for dengue (although a number are under development and field trial) and control of the disease depends on controlling the vector. *Ae. aegypti* is a domestic mosquito and therefore not readily exposed to aircraft- or ground vehicle-based insecticide applications, so adulticiding within homes may be required to counter urban epidemics. Prevention of outbreaks generally depends on eliminating or treating (larviciding) the breeding sites of *Ae. aegypti* and, since these are often small collections of water in domestic situations, requires the cooperation of the general public.

JAPANESE ENCEPHALITIS VIRUS (JEV)

This is the most common cause of epidemic encephalitis in the world. In humans, symptoms range from fever and headache to meningitis and encephalitis, coma and death; mortality varies with the age group, but it is always considerable (20–30%) and is 50% for people over 50 years of age. It has been known from epidemics in Japan since 1870 and it has 'spread' southwards through China to Thailand and Malaysia, westwards through Burma to India and Pakistan and eastwards to Indonesia, New Guinea and Australia. The natural cycle involves wading aquatic birds, including herons and egrets, as the maintenance hosts. In the Japanese spring, there is intense virus transmission among young herons by *Cx. tritaeniorhynchus*, a rice field breeding mosquito. The virus then spreads to local pigs, in which it causes abortion. These are the main amplifier hosts from which the virus is transmitted to humans and horses (both dead-end hosts). Other vectors are *Culex gelidus*, species of the *Culex vishnui* group, and *Cx. annulirostris*. As with other arboviruses, there is no specific treatment for JEV infection. Control of virus activity and prevention of major outbreaks by insecticides has been inhibited by the logistics of treating extensive areas of rice fields and by the widespread resistance to insecticides in the major vector, *Cx. tritaeniorhynchus*. Vaccines are available and, in a number of countries, protection against JEV infection is provided by mass immunization programmes and, along with improved management of rice cultivation (via strategic manipulation of water levels, e.g. intermittent irrigation) to reduce vector populations and pig husbandry (via immunization of pigs and relocation away from domestic premises), to reduce virus carryover to humans, many communities are free of the disease, although it remains a constant threat in many rural areas of southern and south-eastern Asia.

ST LOUIS ENCEPHALITIS VIRUS (SLEV)

This is the most important native arbovirus in North America and it extends to parts of Central America. In humans, the majority of infections is inapparent or produces only a mild influenza-like illness, but there can be central nervous system involvement, from headache to encephalitis, coma and death; there is a high (30%) mortality in the elderly and long-term sequelae in many that recover.

This is a bird virus, with natural cycles involving urban birds (e.g. sparrows and pigeons), and members of the *Cx. pipiens* group (including *Cx. quinquefasciatus*) are involved in maintenance, enzootic and epidemic transmission cycles. In the mid-west and eastern urban areas of the USA, SLEV is associated with periods of drought and poor drainage, which provide urban breeding sites for the vectors. However, in western rural areas, the main vector is *Cx. tarsalis* and epidemics are associated principally with periods of high rainfall (although the other

Culex species may be involved, particularly in/near urban areas). In rural Florida, the main vector is *Culex nigripalpus*. Overall, major outbreaks are uncommon.

There is no vaccine to prevent infection with SLEV and prevention and control efforts are dependent on active surveillance of virus activity. Mosquito control directed at urban *Culex* larval populations needs particularly to target drainage systems; immature stages of rural vector populations can be logistically difficult to manage and adulticiding operations might be required in the event of outbreaks.

MURRAY VALLEY ENCEPHALITIS VIRUS (MVEV)

This has caused occasional epidemics in south-eastern Australia, especially in the Murray–Darling Basin, but in 1974 an epidemic involved all mainland states and both tropical and temperate regions. Since then, virtually all cases have occurred in northern Western Australia and the Northern Territory, where the virus is considered to be endemic. The virus is considered to have wading water birds as its maintenance hosts and the principal epidemic vector is *Cx. annulirostris*, which breeds in semi-permanent and temporary vegetated breeding sites created by rainfall or flooding. Endemic activity is related to seasonal monsoon conditions in northern Australia. There appear to be associations of MVEV outbreaks in south-eastern Australia with favourable climatic conditions, particularly extraordinary rainfall in the catchment areas, which generates enhanced bird breeding and mosquito production in the riverine wetlands. As with other arboviral encephalitides, MVEV can cause a range of symptoms, from headache to encephalitis and coma; mortality is approximately 20% and almost 50% of survivors experience residual neurological sequelae. Horses are also severely affected. No control is undertaken in endemic areas, but active surveillance indicates activity and generates public health warnings advising personal protection. As widespread epidemic activity has been relatively rare, there is no experience with operational interventions.

WEST NILE VIRUS (WNV)

This disease, associated with wild birds and bird-feeding mosquitoes, has been known from

Africa, southern Europe, the Middle East and India, and from 1999 became established in North America (with intrusions into Central and South America). WNV disease in humans is often asymptomatic but, otherwise, fever, headache, rash and a range of symptoms have been reported; meningoencephalitis can occur and neurological problems occur mostly in the young and elderly; cases occasionally end fatally (range 2.7–4%), but recovery is usually complete without permanent sequelae. Horses can develop encephalitis. Epidemics usually occur during the summer months, when populations of culicine mosquitoes are large. In Egypt, Israel and South Africa, the vector has been *Culex univittatus*, in Europe *Cx. p. molestus*, in Israel and France *Culex modestus* and in India *Cx. vishnui*. In North America, various species have been involved, although *Cx. pipiens* and *Cx. tarsalis* have been most important. The virus circulates widely in birds (pigeons and crows), especially in the nesting season, and infects humans via 'bridge vectors' that feed on both birds and humans. Horses can be protected by vaccination but there is no vaccine for humans, and personal protection measures along with urban mosquito control (targeting *Culex* spp. in particular) provide the best protection against infection during periods of WNV activity.

KUNJIN VIRUS (KUNV)

Recently recognized to be a subtype of West Nile Virus, KUNV is known principally from Australia, where it is endemic in northern tropical regions but occurs occasionally in southern areas. It is a bird virus, transmitted predominantly by *Culex* species, particularly *Cx. annulirostris*. It can cause serious disease in horses, and in humans causes symptoms ranging from mild (fever, rash and aching joints) to serious (encephalitis), but it has not been shown to cause human fatalities. There are no specific management strategies, but active surveillance allows the distribution of public health warnings when virus activity is detected.

ZIKA VIRUS (ZIKV)

This virus has been reported from various countries in Africa (e.g. Central African Republic, Egypt, Gabon, Nigeria, Sierra Leone, Tanzania and Uganda) and Asia (e.g. India, Indonesia, Malaysia, the Philippines, Thailand and Vietnam), but also in the western Pacific (Yap Island). It has caused relatively mild illness, characterized by fever, rash and arthralgia (and conjunctivitis in the Yap outbreak).

It is thought that monkeys are the likely vertebrate reservoir host for ZIKV and the virus has been isolated from a number of different mosquitoes across its range, mostly forest *Aedes*, including *Ae. africanus*, *Ae. luteocephalus* and *Ae. furcifer* in Africa, but also *Ae. aegypti* in Asian urban situations. The outbreak on Yap in 2007 was considered to be an example of viruses being spread by travel or commerce.

6.1.2 Togaviridae

This family contains the genus Alphavirus. The type species is Sindbis virus. Other significant mosquito-borne viruses in the genus include Chikungunya, Barmah Forest, O'Nyong-nyong and Ross River, which cause various symptoms including arthritis, and three equine encephalomyelitis (or encephalitis) viruses (eastern, western and Venezuelan) that occur in the New World and cause disease in humans as well as horses.

SINDBIS (SINV)

This causes a mild febrile illness (with vesicular rash, mild fever and joint pain) in humans that in many cases goes undetected. Human cases have been recorded in northern Europe (where it is known as Ockelbo disease or Karelian fever), parts of Africa (including Egypt, Uganda and South Africa), Asia and Australia (where the illness appears to be particularly mild). Mosquitoes acquire infection from feeding on viraemic wild birds, and the vectors vary with region. In Sweden, *Cx. p. pipiens*, *Culex torrentium* and *Cs. morsitans* are involved enzootically with passerine birds, and the bridge vector to humans appears to be *Aedes cinereus*, while in the former Soviet Union, *Aedes* species (e.g. *Ae. communis*) have been incriminated. In Africa, isolations have been made from *Cx. univittatus* and *Cx. theileri*, while probable vectors in other countries include *Cx. annulirostris* (Australia), *Cx.*

tritaeniorhynchus (Malaysia), *Culex bitaeniorhynchus* (Philippines) and *Culex antennatus* (Egypt). There is no vaccine, and prevention of infection, if required, would need mosquito control tailored to suit the particular local vector/s.

ROSS RIVER VIRUS (RRV)

This is endemic throughout Australia and causes a syndrome with painful joints, rash and low-grade fever in humans. It is often referred to as epidemic polyarthritis. The arthralgia may be prolonged for some months, but no permanent damage is suffered. The virus is thought to be maintained in native macropods (kangaroos and wallabies), but fruit bats and possums may be involved in urban areas, and it infects various animals (including horses and cattle) as well as humans. The vectors of RRV in Australia vary with region and seasonal conditions, but *Ae. vigilax* and *Ae. camptorhynchus* are important in northern and southern coastal regions, respectively, *Cx. annulirostris* is the major vector in most inland regions, various species of *Aedes* (*Ochlerotatus*), including *Aedes normanensis*, can be significant in semi-arid and arid regions, and in urban and peri-urban areas *Ae. notoscriptus* and *Cx. annulirostris* appear to be involved. RRV is also known from Papua New Guinea and the Solomon Islands and, in 1979–1980, it spread eastwards into the Pacific, occurring in epidemic form in Fiji (up to 90% of the population infected), American Samoa (44%), the Cook Islands (69%), New Caledonia (33%) and probably also in Tonga, Kiribati and Western Samoa (where *Ae. polynesiensis* was arguably the principal vector). There has been no such regional outbreak outside Australia since that time, although some further instances of infection have been reported for Fiji during the past decade. Mosquito control (particularly larviciding on salt marshes) is undertaken to reduce seasonal vector populations and thus reduce infection potential in some coastal areas, but such measures in inland areas are often not practical.

BARMAH FOREST VIRUS (BFV)

This virus is known only from Australia and causes a disease similar to that caused by RRV, although while the rash may be more florid, the arthralgia is usually less pronounced. Overall, national case reports are fewer than for RRV, but they have been increasing significantly, particularly in coastal eastern Australia. The vectors are generally similar to those regionally associated with RRV, but otherwise there is little definitive information concerning the natural vertebrate hosts, and the natural history of the two viruses appears to be different, as activity periods are not necessarily coincident. Vector control is as above for RRV.

O'NYONG-NYONG VIRUS (ONNV)

This was an alphavirus that caused a widespread epidemic in 1959–1962 in East Africa, affecting an estimated several million people. Antibodies have been found more widely in the Afrotropical region. It is the only well-documented example of an anopheline-borne epidemic viral disease of humans. The vertebrate hosts are unknown. The vectors are said to be *An. gambiae* s.l. and *An. funestus*, the main vectors of malaria in tropical Africa, and consequently outbreaks of ONNV coincide with outbreaks of malaria. Symptoms include fever, headache, pruritic rash, eye pain, myalgia and arthralgia. Prevention and control can be effected by routine indoor residual spraying and other methodologies appropriate to the *Anopheles* vectors.

CHIKUNGUNYA VIRUS (CHIKV)

This is an alphavirus found in much of sub-Saharan Africa (Senegal to South Africa) and South-east Asian (India to Indonesia) regions. In Africa, there is a sylvatic cycle involving primates and forest-dwelling *Aedes* species, and urban epidemics are sustained by a human–mosquito–human cycle, typically involving *Ae. aegypti* as the vector. In a rural epidemic in the wooded savannah of the eastern Transvaal, South Africa, the vectors were *Ae. furcifer* and *Ae. taylori*, which were transmitting the virus among baboons and humans, and baboons were regarded as the primary vertebrate host from which the virus extended into the human population. Elsewhere in Africa, there are similar enzootic cycles involving transmission between non-human primates and mosquitoes. The virus has been isolated from *Ae. africanus* in Uganda and the Central African Republic, and from *Ae. luteocephalus* and *Ae. vittatus* in Senegal; these have been presumed to be responsible for monkey–monkey, monkey–human and human–human transmission in rural areas, but *Ae. aegypti* has been accepted as the usual urban vector in African situations. The existence of a similar forest cycle in Asia has yet to be demonstrated, with CHIKV appearing to be more urban-based with human–human cycles and *Ae. aegypti* accepted as the major urban vector, although *Ae. albopictus* is suspected of also being involved. Recently, an outbreak in islands off the coast of East Africa (Comoros, Mayotte, Mauritius, Reunion) that started in 2005 was characterized by *Ae. albopictus* being the vector; a mutation in the CHIKV strain made it more infectious for *Ae. albopictus*, which was particularly abundant at the time of the epidemic, whereas *Ae. aegypti* was limitedly distributed. Subsequently, the outbreak spread to India in 2006, where activity was widely associated with *Ae. aegypti*, but with *Ae. albopictus* in some regions, and also to Indonesia and Italy in 2007. In endemic areas, adults are immune and outbreaks occur at 5- to 10-year intervals, depending on the build-up of a population of susceptible children. In humans, the illness can be symptomless, but fever, rash and arthralgia can occur and the disease can be confused with dengue; in children, haemorrhagic manifestations are recorded and there have been rare fatalities, with the virus being associated with deaths in elderly persons. There is no specific treatment, and prevention and control are vector based, as for dengue.

MAYARO VIRUS (MAYV)

This is an alphavirus that is endemic in northern countries of South America (including Bolivia, Brazil, French Guiana, Panama, Peru Suriname, Trinidad and Venezuela). The vertebrate reservoirs are thought to be forest primates, with *Haemagogus* mosquitoes as vectors (similar to YFV), and human cases are generally associated with sylvan exposure. Illness in humans usually involves fever and

headache, rash, and myalgia and arthralgia that may persist for months, but is non-fatal. Although infections typically are sylvan related, that *Ae. aegypti* and *Ae. albopictus* are competent laboratory vectors of MAYV indicates the potential for urban outbreaks.

EASTERN EQUINE ENCEPHALOMYELITIS VIRUS (EEEV)

This virus (also known as eastern equine encephalitis virus) has been recorded from the eastern USA, the Caribbean and Central and South America to Argentina. Different subtypes are recognized in North America and Central and South America. The virus causes serious mortality in horses and various birds; during outbreaks of EEEV, epizootics have occurred among exotic birds, for example pheasants and emus, causing many deaths. In North America, human cases are rare but the majority of symptomatic cases proceed to encephalitis. The mortality rate in humans and horses is very high (50–90%) and those that recover often suffer permanent brain damage. The virus circulates among passerine birds, mainly through *Cs. melanura*, an ornithophilic, nocturnal species, which breeds in forested freshwater swamps. The mosquito species that transfer the virus from the enzootic cycle to humans and horses have not been identified definitively, but *Coquillettidia perturbans*, *Ae. sollicitans*, *Aedes vexans* and various other *Aedes* species have been implicated as bridge vectors, although their importance varies from year to year with environmental conditions. In Central and South America, there is little evidence of human disease and it is likely that the virus variant is less virulent, but cases in horses are common in some regions. The natural cycles are not well understood, but *Culex taeniopus* is recognized as an enzootic vector in various locations, with other mosquitoes also being involved. Prevention and control of EEEV infections is problematic, because the virus is maintained in zoonotic cycles in swamp and forest habitats. Surveillance detecting activity of the virus in horses or sentinel birds can provide an impetus for mosquito control in the locality and public health warnings (e.g. advising personal protection measures for people entering high-risk environments). A vaccine is available for horses and domestic animals at risk but is not used for humans, except for laboratory workers.

WESTERN EQUINE ENCEPHALOMYELITIS VIRUS (WEEV)

This virus (also known as western equine encephalitis virus) is another avian arbovirus, which on occasions spills over into the human and equine populations. Human cases occur in western and central USA (west of the Mississippi River), western Canada and in Central and South America (where human cases are unknown but equine epizootics occur). The virus causes both apparent and inapparent infections (1:100–1:500), with the former ranging from an influenza-like illness to encephalitis and death; the most severe cases occur in the young and many surviving infants develop severe debilitating neurological sequelae. Horses exhibit classical neurological symptoms and recorded case fatality rates have been up to 50%. In western and central USA, the virus is enzootically active in wild birds and the primary vector is *Cx. tarsalis*, populations of which show considerable variation in their ability to transmit the virus, although there appears to be a secondary cycle involving rabbits and *Aedes melanimon*. In the eastern USA, the primary vector of WEEV among its avian hosts is the ornithophilic *Cs. melanura*, but the virus is not a public health or veterinary problem in that region. In South America, the host and vector cycles have not been well investigated and the complexity of virus subtypes with different virulence for horses has been confounding. Surveillance programmes in North America lead to vector control by local authorities when the virus is detected, although increased water (particularly snow melt and irrigation) management and vaccination of horses have resulted in less activity threatening humans in recent years.

VENEZUELAN EQUINE ENCEPHALITIS VIRUS (VEEV)

This virus (also known as Venezuelan equine encephalomyelitis virus) is more tropical in distribution in the New World, being recorded from Colorado and Florida in the USA, Mexico, tropical Central and South America to northern Argentina and Trinidad. Several subtypes of VEEV are enzootic in small rodents and subtype 1 infects horses and occasionally humans. Although some strains are not virulent, VEEV in horses can cause symptoms ranging from inapparent to fever to acute encephalitis and death (with mortality rates up to 80%). Likewise in humans, with fever, headache and myalgia, through to nervous system involvement (most often in children and the elderly); there have been fatalities, but neurological disease occurs in less than 20% of symptomatic cases and the mortality rate is low (<3%) in adults, although it is 10–20% in children less than 1 year old.

The principal enzootic vertebrate hosts are forest rodents, although opossums have also been implicated. Enzootic vectors are thought to be various *Culex* species, such as *Culex portesi*, *Culex cedecei*, *Culex ocossa*, *Culex panocossa* and *Cx. taeniopus* in different regions. Epizootics and epidemics involve horses (and donkeys and mules) acting as amplification hosts, and a wide variety of species may be involved as epidemic vectors: *Psorophora confinnis* and *Aedes taeniorhynchus* are confirmed vectors, but other species are almost certainly involved, depending on locality.

For prevention and control, a vaccine is available for horses, which not only protects the vaccinated animal but also eliminates a potential source of virus to the vector. Otherwise, mosquito control is critical (if it is feasible in the local circumstances) and personal protection with clothing and topical repellents is appropriate for people living or working near horses during outbreaks.

6.1.3 Bunyaviridae

The Bunyaviridae includes the Orthobunya-viruses (including California and La Crosse encephalitis viruses and Oropouche virus) and others not yet placed in a genus (including GanGan, Mapputta and Trabanaman viruses) that are primarily mosquito-borne, and the Phleboviruses (including the mosquito-borne Rift Valley Fever virus) that are otherwise transmitted mainly by phlebotomine sand flies (see the entry for 'Sand Flies').

CALIFORNIA GROUP VIRUSES (LA CROSSE VIRUS LACV)

The 17 Bunyaviruses in the California serogroup are found in the USA, Canada, Central and South America, Finland and Central Europe. California encephalitis virus was isolated in 1943 from *Ae. melanimon* and three children. The more important, and more common, La Crosse virus was not isolated until 1963. It circulates in chipmunks, squirrels and rabbits, among which it is transmitted by *Ae. triseriatus*, a tree-hole breeder, which overwinters in the larval stage. It was the first arbovirus to be shown to be transmitted transovarially in mosquitoes, producing infected larvae which give rise to adults capable of transmitting the virus at their first feed, and cases are thus likely to occur in spring. Amplification of the virus also occurs by venereal transmission between infected males and uninfected females. LACV is associated with most USA states east of or contiguous with the Mississippi River. In humans, LACV may produce an inapparent infection, a mild fever and headache in adults, and in children an involvement of the central nervous system leading to meningitis and encephalitis. Most cases are in children less than 10 years old. Mortality is less than 0.5% and recovery is usually complete. There is no vaccine for LACV and prevention is dependent on vector management (e.g. source reduction through removing tyres and other artificial containers and filling tree-holes with cement) and personal protection from mosquito bites (e.g. with clothing and repellents)

OROPOUCHE VIRUS (OROV)

This was first isolated in Trinidad and is now also known from Brazil, Panama and Peru, having caused many epidemics (mostly in Brazil). Infection with OROV causes an acute febrile illness with general aches and pains, and meningitis has been reported. No deaths have been recorded, although a proportion of patients become severely ill. It is likely that there is a sylvatic cycle with non-human primates (antibodies against OROV have been found in several genera of monkeys, and isolations of virus have been made from the three-toed sloth), but humans seem to be the main host during urban epidemics. The virus has been isolated occasionally from mosquitoes, including *Cx. quinquefasciatus*, and frequently from the biting midge, *Culicoides paraensis*, with the latter proving to be the more efficient vector in the laboratory and raising questions about the role of mosquitoes as major vectors. Vector control targeted against *Cx. quinquefasciatus* may be achievable in some circumstances, but targeting the *Culicoides* vector or its larval habitats (decomposing vegetation) would be problematic in many situations.

RIFT VALLEY FEVER VIRUS (RVFV)

This occurs mainly in Egypt and sub-Saharan Africa, but it has been reported also from Saudi Arabia and Yemen. It causes an acute, febrile disease of cattle, sheep and humans, characterized by high mortality in calves and lambs and abortion and some deaths in adult sheep and cattle; in humans, ocular disease with retinitis is severe and fatalities have been associated with the virus causing encephalitis and haemorrhagic fever. Transmission of RVFV can be by contamination from infected animal tissues but is normally by mosquito bites, most commonly from members of the *Cx. pipiens* complex in some areas such as Egypt, but in South Africa the major vectors have been reported as *Aedes caballus* and *Cx. theileri*, in Uganda *Eretmapodites chrysogaster*, and in Kenya *Aedes lineatopennis* and *Cx. antennatus*. The virus survives during inter-epidemic periods by transovarial transmission, and this has been demonstrated in *Ae. lineatopennis* in Kenya and *Ae. vexans* in Senegal, and there is a suggestion that rodents (e.g. the Namaqua Rock mouse) might be involved as a reservoir host in South Africa. In enzootic areas, annual vaccinations can protect livestock. It is likely that advanced warning of possible epizootics of RVFV may be provided by satellite remote sensing to forecast the flooding of grassland depressions (dambos), which are the breeding habitat of *Ae. lineatopennis* and other significant vector species.

6.1.4 Rhabdoviridae

The mosquito-borne arboviruses in this family are bovine ephemeral fever virus and vesicular stomatitis virus (although the role of mosquitoes in the latter may be incidental).

BOVINE EPHEMERAL FEVER VIRUS (BEFV)

This is a disease of cattle and water buffalo (*Bubalus bubalis*), which is enzootic in Africa, the Middle East, Asia and Australia and causes epizootics in countries that are partially free of the virus. With infection, there is an initial onset of fever, generalized inflammation and toxaemia, followed by a short-term paralysis, which may resolve itself suddenly or result in death. Mortality rates have been reported to be in the order of 1–3%. Recovered animals are considered to have a lifelong sterile immunity. Economic losses come from a sharp drop in milk production, deaths of dairy and beef animals and an abnormally delayed pregnancy. The evidence for the involvement of insect vectors in transmission is largely circumstantial and has not yet been demonstrated. Virus has been recovered from *Anopheles bancrofti* and a mixed pool of culicine species in Australia, but also from pools of *Culicoides* spp. in Africa and *Culicoides brevitarsis* in Australia. The virus needs to be injected into the circulatory system for disease transmission; intradermal, subcutaneous and intramuscular inoculations of BEFV do not infect cattle, so transmission by capillary-feeding mosquitoes rather than by pool-feeding biting midges is perhaps more likely. Vaccines are available and used to protect herds, but since the vectors have not been identified, there is little reason to propose any vector control programmes.

6.2 Anaplasmataceae

Eperythrozoon species are small organisms that appear as rings or cocci on the erythrocytes, or free in the plasma, of vertebrates. They are distributed worldwide and have been found in domestic and wild mammals. *Eperythrozoon suis* causes icteroanaemia of swine, a disease of some economic importance in the USA. In cattle, clinical disease due to *Eperythrozoon wenyonii* is uncommon. In sheep, *Eperythrozoon ovis* causes anaemia, which is more severe in young sheep and may be the principal cause of ill thrift in lambs, and it has been shown experimentally to be transmitted by *Ae. camptorhynchus* and *Cx. annulirostris* in Australia.

6.3 Malaria (*Plasmodium*) and other Haemosporidia (Sporozoa)

The order Haemosporidia contains three families: the Plasmodiidae, the Haemoproteidae and the Leucocytozoidae. The most important of these is the Plasmodiidae, which contains the genus *Plasmodium* and includes the species responsible for human (and other vertebrate) malarias. In *Plasmodium* spp., schizogony occurs in the blood; gametocytes develop in mature erythrocytes; the end product of the digestion of haemoglobin is a dark pigment, haemozoin, and the vectors are mosquitoes. In the other two families, schizogony does not occur in the blood and the vectors are Diptera other than mosquitoes.

6.3.1 Human malaria

Malaria is the most widespread and persistent disease that affects human populations throughout the world and is the most prevalent and devastating disease in the tropics, with 40% of the world's population being at risk of infection. It is estimated that out of a world population of 5 billion people, 110 million (2%) develop clinical disease each year and 280 million (5.6%) carry the parasite. The World Health Organization (WHO) estimated >650,000 deaths from malaria in 2010, with >80% being children under 5 years of age. The disease is becoming more difficult to control due to the spread of resistance of vectors to insecticides, and of the parasite to antimalarial drugs, and an increase in the number of foci of intense malarial transmission due to changing environmental conditions.

THE PARASITE

More than 100 species of *Plasmodium* have been described from vertebrates (about 20 species are known from non-human primates, a similar number in other mammals and about

40 each in birds and reptiles). Classically, four species (*Plasmodium falciparum*, *Plasmodium vivax*, *Plasmodium malariae* and *Plasmodium ovale*) have been recognized to infect humans; however, in recent years, a number of cases of human infection with non-human primate malaria parasites, such as *Plasmodium knowlesi* (from macaques in South-east Asia) and *Plasmodium simium* (from howler and spider monkeys in Brazil), has been reported. Because of the increasing reports, *P. knowlesi* is now recognized as a fifth species causing human malaria, albeit as a zoonosis because, to date, there is no evidence of human–human transmission via a mosquito vector in areas without simian reservoirs.

PLASMODIUM LIFE CYCLE
Sporozoites injected into a human by an *Anopheles* female develop in the liver either into latent hypnozoites, of which some may delay development (with *P. vivax* and *P. ovale*) to cause relapses later, or undergo immediate schizogony and release merozoites. The latter forms enter red blood cells and become feeding trophozoites and, in the early stages of an infection, the fully-grown trophozoite becomes a schizont, producing a small number of new merozoites. Release of the merozoites from the erythrocytes brings on an attack of malaria (characterized by chills and fever). Some merozoites develop into male or female gametocyctes, which only develop further when ingested by an *Anopheles* vector. In the vector's gut, male gametocytes undergo exflagellation to produce male gametes, one of which will fuse with a female gamete to form a zygote. The zygotes become motile ookinetes, which pass between the cells of the midgut to form oocysts. The oocyst enlarges, ending in the formation of motile sporozoites, which invade the haemocoel and penetrate the mosquito's salivary glands, from which they are passed into the host with the saliva when the mosquito next feeds and, if the host is susceptible, the cycle is repeated. The time taken to complete sporozoite development in the vector is temperature dependent; for *P. falciparum* at 30°C it is 9 days, doubling to 20 days at 20°C, and taking longer than 30 days at 19°C, while for *P. vivax* at 30°C it is 7–8 days, doubling to 15–16 days at 20–21°C,

and at 19°C it is still less than 20 days (although it cannot complete development below 16–17°C). These limitations on extended low temperature development, and the unlikely survival of the vectors over those long periods, explain the association between the distribution of malaria and the summer isotherm of 16°C for *P. vivax* and 20°C for *P. falciparum*. Once infective, anopheline mosquitoes remain so for up to 12 weeks, and for all practical purposes that means that mosquitoes are infected for life.

THE DISEASE
Clinical malaria in humans is associated with the bursting of infected erythrocytes, and a typical episode has an abrupt onset with chill, which turns within an hour into profuse sweating with headache and high temperature lasting for 2–6 h, after which the temperature falls rapidly to normal and the patient may feel 'well'. In endemic areas, infants have passive immunity from antibodies acquired from their mothers across the placenta and in breast milk, which protect them for about the first 3 months of life, after which they suffer repeated bouts of malaria, which may prove fatal if untreated, especially when associated with measles, severe gastroenteritis or malnutrition. In later life, the mortality rate from malaria declines from around 1% for children 1–4 years old, to 0.1% in adolescents (10–14 years old) and to 0.03% in adults.

Although they produce a similar illness with attacks recurring at 48 h or 72 h intervals and are grouped under the one heading 'malaria', the four species are different pathologically and clinically. Infection with *P. falciparum* brings the most severe form of the disease and in the absence of treatment may kill up to 25% of non-immune adults within 2 weeks. After the initial series of attacks have passed, malaria may recur from the activation of latent erythrocytic forms (recrudescences). In contrast, relapses, which occur in *P. vivax* and *P. ovale* infections, arise from liver hypnozoites (dormant tissue forms) from the initial exoerythrocytic cycle. *P. falciparum* has a higher temperature threshold for development than *P. vivax* and is commoner in the warmer areas of the world, being limited by a summer isotherm of 20°C. *P. vivax* causes a milder disease, but is more persistent than *P. falciparum*. It is widely distributed throughout the world, being limited by the 16°C summer isotherm and is therefore often the only species present in the cooler temperate regions. *P. ovale* is the rarest of the human malaria parasites and was not described until 1922. It produces hypnozoites, which may cause relapses at 3-monthly intervals for a period of up to 4 years. It is the least pathogenic and produces a tertian fever after a longer incubation period than either *P. vivax* or *P. falciparum*. In West Africa, it replaces *P. vivax* among the indigenous people. It is also present in Papua New Guinea, Thailand, Kampuchea and Vietnam. *P. malariae* is a slow-growing parasite, which has a worldwide but patchy distribution. Although widespread, it is usually less common than either *P. falciparum* or *P. vivax*. However, it is next to *P. falciparum* in pathogenicity, with death resulting from kidney failure. *P. malariae* does not produce hypnozoites and its merozoites invade ageing erythrocytes. It has remarkable powers of persistence, with recrudescences occurring for up to 50 years. *P. malariae* is limited by the 16°C summer isotherm. The primate parasite *P. knowlesi* is very similar to *P. malariae* and morphological differentiation is problematic. Human infections are now known to be widely distributed in Malaysia and the species has caused fatalities.

THE VECTORS
The vectors of the mammalian species of *Plasmodium* are invariably species of *Anopheles* mosquitoes. More than 400 species of *Anopheles* have been described, of which 68 have been associated with malaria, 40 as main vectors and 28 as subsidiary vectors in one or more epidemiological zones. Some species are important vectors wherever they occur, for example *Anopheles albimanus*, *Anopheles aquasalis* and *Anopheles darlingi* in the Central and South American zones; *An. fluviatilis* in the Indo-Iranian and Indo-Chinese hills zones; and *Anopheles dirus* in the Indo-Chinese hills and Malaysian zones. Some species, for example *Anopheles melas* in the Afrotropical region, are important local vectors but are not classified as main vectors because of their limited geographical distribution.

Four of the eight species associated with stable malaria occur in the Afrotropical region, where three (*An. arabiensis*, *An. funestus* and *An. gambiae*) are widespread; two (*An. minimus* and *An. fluviatilis*) occur in the Oriental region and two (*Anopheles labranchiae* and *An. sacharovi*) in the Palaearctic region. Another eight species are associated with intermediate malaria, including *An. quadrimaculatus* and *An. darlingi* in the Nearctic and Neotropical regions, respectively; *Anopheles farauti* in the Australasian region; three species (*An. balabacensis*, *Anopheles sinensis* and *An. sundaicus*) in the Oriental region and two (*Anopheles atroparvus* and *Anopheles sergenti*) in the Palaearctic region. Thirteen species are associated with unstable (epidemic) malaria, and of these, seven are in the Oriental region; two each in the Neotropical and Palaearctic regions and one in each of the Afrotropical and Australian regions.

However, many of the traditionally attributed vector 'species' have been revealed to be not a single entity but a 'species complex', comprising morphologically indistinguishable sibling species that are genetically distinct and have different distributions, behaviours and vectorial capacities (transmission potentials), thus presenting different risks for malaria transmission in different areas. The principal examples include the *Anopheles maculipennis* complex in Europe, the *An. gambiae* complex in Africa, the *An. punctulatus* complex in Melanesia, the *An. hyrcanus* complex in China and the *An. culicifacies* complex in India, as well as the *An. dirus*, *Anopheles barbirostris*, *An. maculatus* and *An. minimus* complexes in South-east Asia. Further complexities may be realized in the future within the currently identified siblings; for instance, within *An. gambiae* s.s., two races known as *M* and *S*, which are identical in appearance but differ chromosomally, have been recognized.

TREATMENT AND PREVENTION
Any attack on malaria must have the full support of the local community. Measures may be directed against either the parasite, with drugs and vaccines when available, or the vector, with measures of personal protection

and insecticide applications, and in the latter case, against the immature stages in the breeding sites or against emerged adults in houses and animal shelters.

For some decades, chloroquine was the drug of choice for chemoprophylaxis of infection with P. vivax, P. malariae, P. ovale and strains of P. falciparum sensitive to the drug. Unfortunately, resistance to chloroquine has developed in P. falciparum in most parts of the world and, in these areas, while other drugs such as doxycycline and mefloquine might be appropriate, the use of insecticide-treated bed nets has become more important.

For treatment of malaria infections, quinine is effective against chloroquine-resistant P. falciparum and quinidine against both chloroquine- and quinine-resistant P. falciparum. Artemisinin (qinghaosu), a drug extracted from the herb, Artemisia annua (sweet wormwood), and its analogues, is effective against chloroquine-resistant P. falciparum – particularly when used with other drugs as artemisin-based combination therapies (aka ACTs), although some evidence of artemisin resistance has been reported recently from South-east Asia and East Africa. Other drugs effective against multidrug-resistant P. falciparum are halofantrine and mefloquine. Primaquine is an excellent gametocytocidal and sporontocidal drug which is used to prevent relapses of P. vivax and P. ovale by its action against exoerythrocytic hypnozoites. Pyrimethamine and proguanil act slowly against blood schizonts but are effective against the exoerythrocytic schizonts of P. falciparum. Resistance to both compounds is extensive, but pyrimethamine together with sulfadoxine is widely used for treatment, although they are less active against the latent exoerythrocytic stages of P. vivax and P. ovale. Proguanil in conjunction with chloroquine has been used extensively for prophylaxis against chloroquine-resistant P. falciparum in Africa but is currently being replaced in many areas by mefloquine.

There has been much research investigating the development of a vaccine to protect against malaria; however, Plasmodium species exist as antigenically distinct strains so that infected humans develop an effective immunity against the specific strain but only a weaker immunity to heterologous strains. Therefore,

an effective vaccine will necessarily incorporate critical antigens which cannot readily be dispensed with or altered by the parasite. Various candidate vaccines have been developed and some have undergone field trials but, to date, none is ready for distribution or general use.

In areas where ineffective or no control measures are in operation, the individual must rely on personal protection, using screening of living accommodation (especially bedrooms), sleeping under nets, wearing protective clothing and using repellents. The use of bed nets impregnated with a residual insecticide, such as a pyrethroid (e.g. permethrin, deltamethrin), has been a relative success, significantly reducing mortality of infants in many endemic areas, but chemicals that are less irritant (to vectors) than those presently used are being sought. For travellers or workers from non-endemic regions visiting highly endemic areas, chemoprophylaxis is recommended as a short-term measure, but the drugs must be chosen according to the known resistance profiles of the local parasites.

Adult mosquito control involving the indoor spraying of residual insecticides is preferred where the vector is endophilic and endophagic, but is ineffective against an exophilic vector such as An. balabacensis balabacensis. The aim of these anti-adult measures is to increase the daily mortality of female Anopheles that are likely to contact humans inside the house, so that few live long enough to become infective. Larval control is only practical where the breeding sites are well defined and limited in space and/or time. Breeding sites can be eliminated by filling holes (a permanent solution) or draining (may require regular maintenance to remain free of breeding). The use of chemicals to control breeding of Anopheles is rarely feasible in rural areas, but may be practical in urban and peri-urban situations.

Before the advent of synthetic insecticides, oil and Paris green (a copper–arsenic compound) were the main larvicides, while pyrethrum sprays were virtually the only adulticide available. Rigorously applied, such simple materials could be highly effective, but the discovery in the 1940s of the insecticidal properties of DDT and other chlorinated hydrocarbons dramatically changed the approach to malaria control by focusing on adult *Anopheles*. The outstanding property of DDT and related compounds was persistence, and deposits applied to resting places of mosquitoes in houses and animal shelters remained insecticidally active for up to 6 months. However, resistance to DDT developed in many regions and, over the past decades, for that and other reasons, DDT was replaced by newer compounds that were at first successful but in time suffered the same fate of vector resistance. One further problematic aspect has been that the compounds replacing DDT have been inherently more expensive and also have much shorter periods of activity in the field (e.g. 1 month for malathion), thus increasing programme costs substantially on three fronts, including labour costs for more frequent applications, and these financial pressures can lead to breakdowns in full coverage or cessation of programmes. In some regions (e.g. a number of countries in Africa), DDT spraying has been resumed because of resistance problems with other compounds.

OTHER PLASMODIIDAE

More than 30 valid species of malaria parasites have been described from about 500 species of birds. Five species, *Plasmodium durae*, *Plasmodium elongatum*, *Plasmodium gallinaceum*, *Plasmodium juxtanucleare* and *Plasmodium relictum*, are of veterinary importance.

These are mostly infections of wild birds that can be highly pathogenic in various domestic poultry (chickens, ducks and turkeys) and cage birds (particularly canaries). *P. durae* is found in sub-Saharan Africa, associated with wild francolins, and has caused epizootics with high mortality among domestic turkeys and pheasants in both East and West Africa. *P. elongatum* is found in Europe and the Americas and has a wide host range (>50 bird species), infecting ducks, pigeons and various 'cage birds'. *P. gallinaceum* is native of Southeast Asia (India and Sri Lanka, through Malaysia to Indonesia), with the wild red jungle fowl as its natural host, and it is highly pathogenic in domestic chickens (particularly the young birds). *P. juxtanucleare* is more widely distributed in eastern Asia from Japan south to Malaysia, in southern Africa and in Latin America, where it is associated with wild birds and causes severe epizootics in domestic chickens. *P. relictum* has a worldwide distribution and wide host range (>350 species of wild birds); it is highly pathogenic in canaries and other 'cage birds' but it usually does not kill wild birds – however, in areas where it has been newly introduced (e.g. Hawaii), it can decimate populations of local birds that have lost their resistance.

The natural histories of these bird *Plasmodium* species and of their vectors are less well known than with human malaria; although culicine mosquitoes, particularly *Culex* species, are proven natural vectors in some situations, many species across the genera *Anopheles*, *Aedes*, *Armigeres*, *Culex*, *Culiseta* and *Mansonia* have been shown capable of transmission of the various bird *Plasmodium* species in the laboratory.

6.4 Filarioidea

In the vertebrate host, the Filarioidea are parasites of the blood or lymphatic system, muscles or connective tissue, or of the serous cavities of their host. The Filarioidea contains two families, the Onchocercidae and the Filariidae. All the medically important mosquito-borne species are in the Onchocercidae, including the mosquito-borne *W. bancrofti*, *B. malayi* and *Brugia timori*, the causative organisms of bancroftian and brugian filariasis, respectively.

Filarial parasites of veterinary importance in the Onchocercidae transmitted by mosquitoes include species of *Onchocerca* and *Dirofilaria* and species of *Setaria* and *Elaeophora*. In the Filariidae, species of *Parafilaria* and *Stephanofilaria* are of minor veterinary importance.

6.4.1 Lymphatic filariasis (*Wuchereria bancrofti*, *Brugia malayi*, *Brugia timori*)

The World Health Organization has estimated that there are 120 million cases of lymphatic filariasis infection (40 million of which have clinical disease) in 83 countries globally, of which 90% are caused by *W. bancrofti*. The diseases caused by *W. bancrofti* and the periodic form of *B. malayi* are anthroponoses, and the nocturnal subperiodic form of *B. malayi* is an anthropozoonosis. Subperiodic *B. malayi* is common in wild monkeys and wild and domestic carnivores (dogs and cats) and is limited to foci in swamp forests in Southeast Asia, where people and domestic animals are surrounded by virgin forest with wild animals and mosquitoes. The closely related *Brugia pahangi* is sympatric with this form of *B. malayi* and occurs in a similar range of hosts. *B. pahangi* is primarily a parasite of carnivores, with primates and other vertebrates as incidental hosts, whereas subperiodic *B. malayi* is primarily a parasite of humans and leaf monkeys (*Presbytis* spp.), with wild and domestic carnivores as incidental hosts. There has been no proven natural human infection with *B. pahangi*.

THE PARASITES

W. bancrofti is widely distributed throughout tropical Africa and the Indo-Pacific region. In the latter, it occurs in India, Bangladesh, Myanmar, Vietnam, New Guinea and Polynesia. Its incidence has been much reduced in China, Indonesia, Malaysia, Sri Lanka and Thailand, and it has been eradicated from the Solomon Islands. In Africa, *W. bancrofti* is endemic in the Nile Delta and in much of the Afrotropical region. In the latter, its northern limit is a line from Senegal to Somalia (Mogadishu) and its southern limit a line from Angola (Benguela) to Mozambique (Beira). There is high prevalence of *W. bancrofti* in the coastal regions of East Africa, Madagascar and the islands off the East African coast and in the Gulf of Guinea. The distribution of *W. bancrofti* in tropical America is much reduced, being present in the Guianas and areas around Belém and Recife in coastal north-east Brazil. *B. malayi* occurs in Malaysia, Indonesia and Mindanao in the Philippines. *B. timori* occurs on Timor, Flores and other islands of the Savu Sea.

FILARIA LIFE CYCLE

In the human body, the female worm is viviparous, liberating microfilariae into the lymphatic system in large numbers (50,000/female/day), and these appear in the per-

ipheral blood. The microfilariae are taken up with the blood ingested by the vector. In the midgut, they shed their sheaths and penetrate the epithelium to reach the haemocoel, through which they migrate to the thoracic flight muscles. Here they develop into thicker, shorter 'sausage' forms, which undergo two moults before developing into elongate, snakelike mature infective larvae (L_3) measuring about 1.5×0.02 mm. Mature L_3 larvae leave the thoracic musculature and enter the haemocoel, in which they move around actively and accumulate in the head. When the mosquito is feeding, they enter the labium and escape by rupturing the labella. They are deposited in a drop of haemolymph and enter the host through the puncture made by the feeding mosquito. In the human host, they become adult and lodge in the lymphatic vessels and start producing microfilariae that reach the peripheral blood and become available to feeding mosquitoes. In the vector, the development is filarial species dependent, mosquito species dependent and temperature dependent, and can take 7–21 days. The adult worms are long-lived (10–14 years) and can be reproductively active, producing millions of microfilariae over a period of 5–8 years.

Populations of microfilariae show variation in their abundance in the circulating blood at different times of the day. With *W. bancrofti*, these are generally referred to as the nocturnal periodic (and subperiodic) and the diurnal subperiodic (or aperiodic) forms, which are geographically separated in the Pacific region at the 170° east longitude; this distinction reflects the adaptation of the parasite to the feeding behaviour of its local mosquito vectors (see discussion below). With *B. malayi*, there are two principal forms (nocturnal periodic and subperiodic) that are associated with different vectors, and there is a localized diurnally subperiodic form of *B. malayi* in west Malaysia (see discussion of vector below). With *B. timori*, there is only a nocturnal periodic form, which equates with it having only one known vector, the nocturnal *An. barbirostris*.

Both the nocturnal periodic *W. bancrofti* and *B. malayi* have similar periodicities, with peak numbers of microfilariae being in the circulating blood from 23:00 to 03:00 h (but in east Thailand, there is a nocturnal subperiodic form of *W. bancrofti* which peaks at 21:00 h). The diurnal subperiodic form of *W. bancrofti* in the Pacific peaks at 14:00–17:00 h. The nocturnal subperiodic form of *B. malayi* is associated with dense swamp forest in South-east Asia and its microfilariae have a peak in the early evening. The diurnally subperiodic form of *B. malayi* in west Malaysia has a periodicity similar to that of the diurnal subperiodic *W. bancrofti* in the Pacific.

THE DISEASE

Adult worms inhabit the lymphatic vessels and nodes of the human host, causing local inflammation of the lymphatic vessels (lymphangitis), swelling of the lymphatic nodes (lymphadenitis) and destruction of the lymphatics. The disease develops slowly, with recurrent episodes of fever and adeno-lymphangitis in the first decade, which, left untreated, can lead to genital lesions and a reversible lymphoedema of the extremities in the second decade that can lead to irreversible elephantiasis. These symptoms become more frequent and severe in the third and fourth decades, after which symptoms may remain steady or decline. Elephantiasis is a long-term result of chronic infections and more commonly found in the legs and scrotum than in the arms, breasts and labia. Other complications include chyluria due to the rupture of the lymphatics into the urinary tract and, in males, hydrocoele and lymph scrotum, chronic epididymitis and inflammatory swelling of the spermatic cord, some of which can be relieved by surgical treatment. Genital lesions are rare in brugian filariasis; however, gross manifestations of the disease can be unusually severe in infections with *B. timori* – in some communities, elephantiasis, which is usually below 5%, can be present in 35% of the adult population.

THE VECTORS

The most widespread cause of lymphatic filariasis is the nocturnal periodic form of *W. bancrofti*, of which the main vector in urban areas is *Cx. quinquefasciatus*, a highly anthropophilic species which feeds readily both indoors and outdoors, and has its peak biting period between midnight and 03:00 h, coinciding with the peak microfilarial abun-

dance in the peripheral blood. In rural areas, the vectors include many species of *Anopheles*, and many of these are also local vectors of malaria. At least 16 species of *Aedes* have been listed as vectors of diurnal subperiodic *W. bancrofti*, of which 10 are in the subgenus *Stegomyia* (9 in the *Ae. scutellaris* group) and 6 in the subgenus *Finlaya*. The main vector is the day-biting, exophilic *Ae. polynesiensis*, which breeds in a wide range of small water containers, including coconut shells, tins, tyres, drums, tree holes, crab holes, canoes and the axils of *Pandanus*, and it has a minor peak of feeding at 08:00 h and a major one just before sunset at 17:00–18:00 h, which more or less corresponds with the time of microfilarial maximum abundance at 16:00 h.

Near the coast of west Malaysia, the main vectors of nocturnal subperiodic *B. malayi* are four species of *Mansonia* (*Mansonioides*): *Mansonia annulata*, *Mansonia bonneae*, *Mansonia dives* and *Ma. uniformis*. The first two species breed in dense swamp forest, where the larvae attach to the pneumatophores of trees. The adults are exophilic and exophagic, and largely zoophilic. At ground level in swamp forest, biting occurs all day and night, with a peak after sunset. In more open areas around houses, there is a sharp peak in biting after sunset, and within houses, the peak of biting occurs after midnight. These species act as vectors because of their large numbers and the range of hosts on which they will feed. Although largely exophilic and zoophagic, they will enter houses and feed on humans. On the coastal rice plains in west Malaysia, nocturnal periodic *B. malayi* is largely transmitted by *Anopheles campestris*, an anthropophilic, endophilic, endophagic species which breeds in ditches, wells and 'borrow pits' under semi-shade. *Anopheles* species are poor hosts for the nocturnal subperiodic form of *B. malayi*. The reverse response is shown by nocturnal periodic *B. malayi*, of which few or no larvae develop in the *Mansonia* vectors of the subperiodic form. In China and Korea, the vector of nocturnal periodic *B. malayi* is *Aedes togoi*, which breeds in brackish water in rock holes and also in rain-filled artificial containers, and is endophilic, with the peak biting rate occurring after sunset.

The actual infection of vector species is influenced by internal physical and other factors once microfilariae are ingested. Some mosquitoes (e.g. *An. gambiae* and *An. farauti*) have cibarial and pharyngeal armatures that influence the level of vector infections because they damage microfilariae during feeding, while in others (e.g. *Cx. quinquefasciatus*) the armatures are only weakly developed and in others still (e.g. *Ae. polynesiensis* and *Ae. togoi*) they are absent. Further, with some species, the numbers of microfilariae escaping the gut and developing to L_3 infective larvae are proportional with the numbers of microfilariae ingested; this is called 'facilitation' and applies particularly with *Anopheles* vectors of *W. bancrofti*. In others, as the number of microfilariae ingested increases, the percentage escaping the gut and developing to L_3 decreases; this response is called 'limitation' and occurs with subperiodic *W. bancrofti* in *Ae. polynesiensis* in Samoa and Tahiti, with periodic *W. bancrofti* in *Cx. p. quin-quefasciatus* in India, Sri Lanka, Tanzania and the Americas and with periodic *W. bancrofti* in *Cx. p. pipiens* biotype *molestus* in Egypt (and apparently also with *B. malayi* and *Ma. dives*), making these vectors more efficient at lower human infection rates and thus decreasing the effectiveness of drugs that reduce microfilaraemias.

PREVENTION AND CONTROL

Because the worms are long-lived and produce microfilariae for many years, control measures against filariasis must be maintained at an appropriate level over a long time. Measures can be directed against the parasite or against the vector, or to minimizing the human–vector contact. For many years, the drug of choice for treatment has been diethylcarbamazine (DEC), which causes rapid disappearance of microfilariae from the circulation and also kills adult worms, but some microfilariae and some adults survive repeated treatments. Mass treatment of populations with DEC regimens has been effective in some areas but less so in others, and it has failed where there have been deficiencies in coverage and/or compliance of the populations. More recently, the drugs albendazole and ivermectin have been added to mass drug administration programme

strategies. Ivermectin can clear microfilariae from the blood more rapidly than DEC, but it is not very effective against adult worms. Albendazole, which can kill adult worms, is now recommended for use in combination with DEC (except where onchocerciasis or loiasis is endemic, and there the recommended combination is albendazole and ivermectin). A recent novel approach has been the investigation of the antibiotic doxycycline as a macrofilaricide, in as much as it is an anti-*Wolbachia* drug and can exploit the symbiosis between the filarial nematode and its *Wolbachia* bacteria.

Vector control strategies are dependent on the species involved. House spraying and insecticide-treated nets for malaria control have been effective against *Anopheles* transmitted *W. bancrofti* in the Solomon Islands and West African countries, respectively, and bed nets should be protective in areas where indoor transmission by *Cx. quinquefasciatus* is important. However, control measures have usually been directed against the immature stages of the vectors of lymphatic filariasis, particularly where *Cx. quinquefasciatus* is involved, and there is access to its larval habitats, although the species has become widely resistant to many insecticides. Rural and forest vectors, particularly *Mansonia* species associated with large swamps, are problematic, but the bacterial product *Bacillus sphaericus* has been used to control *Mansonia* breeding in smaller ponds and is an appropriate agent for *Cx. quinquefasciatus* in stagnant polluted waters, where it can persist and recycle. Physical control by using polystyrene beads to cover stagnant, confined waters, such as is found in cesspits, has been widely used and has been highly effective, persisting for at least 5 years in the absence of flooding. Raising living standards, in particular by the introduction of piped water supply with accompanying drainage and sewerage, removes breeding sites of *Cx. quinquefasciatus* and has given control of filariasis in the southern USA, Puerto Rico and the Mediterranean.

When the vector is exophilic and diurnal, control is more difficult. Mosquito proofing of houses offers little protection; neither do mosquito nets. This is the situation with *Ae.*

polynesiensis and others in the *Ae. scutellaris* group, which are vectors of the diurnal subperiodic *W. bancrofti* in the Pacific. They breed in a wide range of small containers both natural and artificial, which are too numerous to locate and deal with individually. Here and elsewhere, where neither adult nor larval control is practical, mass chemotherapy offers the best prospect of control, but issues related to treatment coverage, sustainability and compliance have to be addressed for it to be successful in the long term.

6.4.2 Other filarioid parasites of domestic animals

DIROFILARIA IMMITIS

The canine heartworm occurs mainly in the tropics and subtropics, where it infests dogs, other canids and rarely cats or humans. Adult worms, measuring 12–20 cm in the male and 25–31 cm in the female, are found in the right ventricle of the heart and in the pulmonary artery. They restrict the circulation, leading to a loss of exercise tolerance, chronic cardiac insufficiency and heart failure. Dogs living in infected areas can be protected by daily doses of DEC or monthly/annual doses of ivermectin.

The microfilariae of *D. immitis* show a nocturnal periodicity in the circulating blood. When they are ingested by mosquitoes, the microfilariae escape from the midgut into the haemocoel and develop in the Malpighian tubes, in which development is completed in 15–16 days in temperate regions and in 8–10 days in tropical regions. Infective larvae move into the head and enter the labium, from which they escape when the mosquito is feeding. Mature worms reach the heart in 3–4 months and microfilariae are produced in 6–8 months. A range of mosquito species of various genera, but mostly *Aedes* and *Culex*, have been reported as actual or potential vectors, being generally nocturnal feeders that coincide with the nocturnal periodicity of the microfilariae.

Human dirofilariasis is a rare condition, but infections with *D. immitis* have been reported from various countries, in the Americas, Africa, Asia and Australia. While *D. immitis* is the causative agent of human dirofilariasis in the New Word, in Europe human cases of dirofilariasis are caused by *Dirofilaria repens*, a less pathogenic species for dogs in which it causes subcutaneous infestation. In addition, human subcutaneous infestations by *Dirofilaria tenuis* (from raccoons), *Dirofilaria ursi* (from bears), *Dirofilaria subdermata* (from porcupines) and *Dirofilaria striata* (from wild cats) have been recorded less frequently in North America. Human intraocular dirofilariasis by a nematode morphologically and phylogenetically close to *D. immitis* has been reported from northern Brazil, suggesting that other, yet unknown, *Dirofilaria* species of wild animals may have potential to cause zoonotic infections. Zoonotic dirofilariases are generally asymptomatic, with the filariae never becoming mature, and no microfilariae appear in the circulation. The parasites may appear as subcutaneous nodules and/or encystations in various organs, particularly the lungs and eyes.

7. PREVENTION AND CONTROL

Strategies and methodologies for mosquito control associated with particular disease situations have been covered above, and covered here are more general aspects of mosquito control that are mainly dependent on the larval habitat and adult behaviour of the target species, with different approaches for coastal salt marsh, freshwater swamp, irrigated agriculture, urban drainage and domestic container species. Typically, larval control is the preferred option (unless the larval habitats are cryptic, inaccessible or unmanageable), with malaria control by indoor application of residual insecticide being the principal exception. However, when a mosquito-borne disease is epidemic, larval control (while serving to reduce the adult populations in the longer term) will not bring rapid interruption of transmission and adult control (designed to kill the infective active females) should be the frontline approach.

7.1 Control of the immature stages

As a first principle, source reduction, where the larval habitat is eliminated by filling, draining or physical removal, should be the initial consideration; otherwise, approaches which physically modify the habitat to make it unsuitable for the target species have been effective in various situations. However, various concerns for environmental impacts can make this approach untenable. When such approaches have failed, or are considered to be not appropriate or relevant, the use of anti-larval chemical or biological agents is the normal recourse. In extensive breeding sites, control of the immature stages of mosquitoes traditionally has consisted of the application of insecticides in solution in oils, as emulsions, wettable powders, granules or dusts. Light oils (such as kerosene and diesel) and oil-based insecticides will kill larvae and pupae of all surface-breathing culicids. Surface-applied dusts would selectively target *Anopheles* larvae. With many organic insecticides having been prohibited in recent years for use in water bodies in and near human communities, one remaining successful compound for many situations is temephos, an organophosphate with low mammalian and other non-target toxicities that has proved to be a particularly efficient larvicide for both surface water and container habitats (although its effectiveness is much reduced in colloidal and polluted waters, and chemical resistance and potential impacts on non-target species have restricted its use in some situations). It is often applied in a granular formulation that disintegrates in water, releasing insecticide slowly and thus prolonging its effect, but preparations that require ingestion of the toxic agent do not affect the non-feeding pupal stage.

In some situations, non-toxic barriers can be effective control agents. Culicid larvae and pupae can be killed by the application of a monolayer of a water-insoluble surfactant, for example plant oil-, lecithin- or silicon-based monomolecular films. The larvae and pupae are unable to pierce the monolayer to make contact with the air and die from lack of oxygen. Other aquatic creatures that are dependent, for respiration, on oxygen dissolved in the water can be unaffected, but those that are dependent on surface tension for aspects of their life cycle could be impacted adversely. In a somewhat similar fashion, highly effective control of *Cx. quinquefasciatus* breeding in wet pit latrines has been achieved

by the application of expanded polystyrene beads to produce a layer about 7 mm deep over the water surface and prevent the emergence of adults and oviposition.

The introduction of vertebrate (fish) and invertebrate (insects and crustaceans) predators into water storage reservoirs, ponds or vessel water containers that provide breeding sites for important disease vectors is an option when the water is used for household purposes and/or there is consumer resistance to it being treated with insecticides. The only notable successes here have been with the use of various fish in a range of situations and copepods in some Asian communities (e.g. in Vietnam, where predatory *Mesocyclops* copepods have been highly successful in controlling *Ae. aegypti* larvae in water tanks in some communities). The contributions which various nematode, fungal, viral and protozoan parasites can make to the control of mosquitoes have been investigated, but there have been no commercially practical outcomes to date.

Two bacteria, *B. sphaericus* (*Bs*) and *Bacillus thuringiensis israelensis* (*Bti*), provide fermentation culture products that are highly toxic to some mosquito larvae but not to non-target organisms. *Bs* is toxic to larvae of *Culex* and *Anopheles* species but not to *Aedes*, while *Bti* has a broader spectrum and is effective against *Aedes* larvae as well. The bacterial toxins are ingested by feeding larvae and *Bti* is most effective in clear, shallow, warm water, while *Bs* has been developed principally for *Culex* species breeding in polluted water habitats.

Insect growth regulators (IGRs) (juvenile hormone analogues and chitin synthesis inhibitors) – methoprene, pyriproxyfen and diflubenzuron – are relatively widely used in marshland situations, large and small ground pools and container habitats, where they can prevent larval development to the adult stage effectively without endangering local non-target non-mosquito species.

A relatively new larval control agent, spinosad, with the active ingredients spinosyn A and D, is derived as a fermentation product from a naturally occurring soil actinomycete, *Saccharopolyspora spinosa*; it acts as a neurotoxin with a novel mode of action (and no indications of cross-resistance with other larvicides), has been shown to be effective against *Aedes*, *Anopheles* and *Culex* larvae and is applicable to both clean and polluted waters.

7.2 Control of the adult stages

Prevention of adult mosquito attack, by screening windows and doors in houses, using bed nets (with or without insecticide impregnation) when sleeping, wearing protective clothing (long sleeves and long trousers) and applying topical repellents (containing diethyl toluamide (DEET) or picaridin), is often advised for personal protection. Mosquito coils, or their modern equivalent of an electrically heated liquid or impregnated pad containing a pyrethroid insecticide, can be effective in still air conditions.

For community protection, adult mosquito control is generally achieved with the application of residual insecticides to surfaces on which the target mosquitoes will rest or the production of insecticidal mists and fogs to contact adult females during their periods of flight activity. The principal example of this residual approach is the use of indoor residual spraying against malaria vectors (from the introduction of DDT through various organophosphates and carbamates to modern pyrethroids), although the use of residual pyrethroids on leaf litter, vegetation or perimeter fences is increasing, to provide what is called 'barrier protection' around homes or communities. Examples of adult control through fogging for public health tend to be associated primarily with outbreaks of mosquito-borne diseases where interruption of transmission is required as a matter of urgency, and pyrethrum/pyrethrins or quick knock-down pyrethroids are usually the chemicals of choice, although the technique is widely used commercially to protect residents and/or visitors in some residential and tourist locations against pest mosquitoes.

SELECTED BIBLIOGRAPHY

Addison, D.S. and Ritchie, S.A. (1993) Cattle fatalities from prolonged exposure to *Aedes taeniorhynchus* in southwest Florida. *Florida Scientist* 56, 65–69.

Al-Afaleq, A.I. and Hussein, M.F. (2011) The status of Rift Valley fever in animals in Saudi Arabia: a mini review. *Vector-Borne and Zoonotic Diseases* 11, 1513–1520.

Amarasinghe, A., Kuritsky, J.N., Letson, G.W. and Margolis, H.S. (2011) Dengue virus infection in Africa. *Emerging Infectious Diseases* 17, 1349–1354.

Arlian, L.G. (2002) Arthropod allergens and human health. *Annual Review of Entomology* 47, 395–433.

Barrett, A.D.T. and Higgs, S. (2007) Yellow fever: a disease that has yet to be conquered. *Annual Review of Entomology* 52, 209–229.

Becker, N., Petric, D., Zgomba, M., Boase, C., Madon, M., Dahl, C., *et al.* (2010) *Mosquitoes and Their Control*, 2nd edn. Springer, Heidelberg.

Beier, J.C. (1998) Malaria parasite development in mosquitoes. *Annual Review of Entomology* 43, 519–543.

Bockarie, M.J., Tavul, L., Kastens, W., Michael, E. and Kazura, J.W. (2002) Impact of untreated bednets on prevalence of *Wuchereria bancrofti* transmitted by *Anopheles farauti* in Papua New Guinea. *Medical and Veterinary Entomology* 16, 116–119.

Bockarie, M.J., Pederson, E.M., White, G.B. and Michael, E. (2009) Role of vector control in the global program to eliminate lymphatic filariasis. *Annual Review of Entomology* 54, 469–487.

Boreham, P.F.L. and Atwell, R.B. (eds) (1988) *Dirofilariasis*. CRC Press, Boca Raton, Florida.

Bowen, M.F. (1991) The sensory physiology of host-seeking behavior in mosquitoes. *Annual Review of Entomology* 36, 139–158.

Brenner, R.J., Hayes, J. and Daniels, E. (2000) Transmission thresholds for dengue in terms of *Aedes aegypti* pupae per person with discussion of their utility in source reduction efforts. *American Journal of Tropical Medicine and Hygiene* 62, 11–18.

Bryan, J.H. and Southgate, B.A. (1976) Some observations on filariasis in Western Samoa after mass administration of diethylcarbamazine. *Transactions of the Royal Society of Tropical Medicine and Hygiene* 70, 39–48.

Buck, A.A. (1991) Filarial infections. General principles. Filariasis. In: Strickland, G.T. (ed.) *Hunter's Tropical Medicine*. Saunders, Philadelphia, Pennsylvania, pp. 711–727.

Burt, F.J., Rolph, M.S., Rulli, N.E., Mahalingam, S. and Heise, M.T. (2012) Chikungunya: a re-emerging virus. *Lancet* 379, 662–671.

Burkot, T.R. and Ichimori, K. (2002) The Pacific program for the elimination of lymphatic filariasis: will mass drug administration be enough? *Trends in Parasitology* 18, 109–115.

Campbell, G.L., Marfin, A.A., Lanciotti, R.S. and Gubler, D.J. (2002) West Nile virus. *Lancet Infectious Diseases* 2, 519–529.

Campbell, G.L., Hills, S.L., Fischer, M., Jacobson, J.A., Hoke, C.H., Hombach, J.M., *et al.* (2011) Estimated global incidence of Japanese encephalitis: a systematic review. *Bulletin of the World Health Organization* 89, 766–774.

Charlwood, J.D. (2011) Studies on the bionomics of male *Anopheles gambiae* Giles and male *Anopheles funestus* Giles from southern Mozambique. *Journal of Vector Ecology* 36, 382–394.

Christophers, S.R. (1960) *Aedes aegypti* (L.). Cambridge University Press, London, 739 pp.

Clements, A.N. (1992a) *The Biology of Mosquitoes, Vol 1. Development, Nutrition and Reproduction*. Chapman and Hall, London, 509 pp.

Clements, A.N. (1992b) *The Biology of Mosquitoes, Vol 2. Sensory Reception and Behaviour*. CAB International, Wallingford, UK, 740 pp.

Coetzee, M., Craig, M. and le Sueur, D. (2000) Distribution of African malaria mosquitoes belonging to the *Anopheles gambiae* complex. *Parasitology Today* 16, 74–77.

Collins, F.H. and Paskewitz, S.M. (1995) Malaria: current and future prospects for control. *Annual Review of Entomology* 40, 195–219.

Coluzzi, M., Sabatini, A., della Torre, A., Di Deco, M.A. and Petrarca, V. (2002) A polytene chromosome analysis of the *Anopheles gambiae* species complex. *Science* 298, 1415–1418.

Conn, J.E., Wilkerson, R.C., Segura, M.N.O., de Souza, R.T.L., Schlichting, C.D., Wirtz, R.A., *et al.* (2002) Emergence of a new Neotropiocal malaria vector facilitated by human migration and changes in land use. *American Journal of Tropical Medicine and Hygiene* 66, 18–22.

Curtis, C.F. (ed.) (1990) *Appropriate Technology in Vector Control*. CRC Press, Boca Raton, Florida.

Curtis, C.F. (2002) Restoration of malaria control in the Madagascar highlands by DDT spraying. *American Journal of Tropical Medicine and Hygiene* 66, 1.

Curtis, C.F. and Mnzava, A.E.P. (2000) Comparison of house spraying and insecticide-treated nets for malaria control. *Bulletin of the World Health Organization* 78, 1389–1400.

Curtis, C.F. and Townson, H. (1998) Malaria: existing methods of vector control and molecular entomology. *British Medical Journal* 54, 311–325.

Dale, P.E.R. and Hulsman, K. (1990) A critical review of salt marsh management methods for mosquito control. *Aquatic Sciences* 3, 281–311.

Dale, P.E.R., Ritchie, S.A., Territo, B.M., Morris, C.D., Muhar, A. and Kay, B.H. (1998) An overview of remote sensing and GIS for surveillance of mosquito vector habitats and risk assessment. *Journal of the Society for Vector Ecology* 23, 54–61.

Debboun, M. and Strickman, D. (2013) Insect repellents and associated personal protection for a reduction in human disease. *Medical and Veterinary Entomology*. 27, 1–9.

Dietz, K. (1988) Mathematical, models for transmission and control of malaria. In: Wernsdorfer, W.H. and McGregor, I. (eds) *Malaria: Principles and Practice of Malariology*, Vol 2. Churchill Livingstone, Edinburgh, UK, pp. 1091–1133.

Doggett, S.L., Russell, R.C., Clancy, J., Haniotis, J. and Cloonan, M.J. (1999) Barmah Forest virus epidemic on the south coast of New South Wales, Australia, 1994–95: viruses, vectors, human cases and environmental factors. *Journal of Medical Entomology* 36, 861–868.

Eisen, L., Beaty, B.J., Morrison, A.C. and Scott, T.W. (2009) Proactive vector control strategies and improved monitoring and evaluation practices for dengue prevention. *Journal of Medical Entomology* 46, 1245–1255.

Enayati, A. and Hemingway, J. (2010) Malaria management: past, present and future. *Annual Review of Entomology* 55, 569–591.

Erlanger, T.E., Weiss, S., Keiser, J., Utzinger, J. and Wiedenmayr, K. (2009) Past, present, and future of Japanese encephalitis. *Emerging Infectious Diseases* 15, 1–7.

Farajollahi, A., Fonseca, D.M., Kramer, L.D. and Kilpatrick, A.M. (2011) 'Bird biting' mosquitoes and human disease: a review of the role of *Culex pipiens* complex mosquitoes in epidemiology. *Infection, Genetics and Evolution* 11, 1577–1585.

Fonseca, D.M., Keyghobadi, N., Malcolm, C.A., Mehmet, C., Schaffner, F., Mogi, M., *et al.* (2004) Emerging vectors in the *Culex pipiens* complex. *Science* 303, 1535–1538.

Fradin, M.S. and Day, J.F. (2002) Comparative efficacy of insect repellents against mosquito bites. *New England Journal of Medicine* 347, 13–18.

Garnham, P.C.C. (1988) Malaria parasites of man: life-cycles and morphology (excluding ultrastructure). In: Wernsdorfer, W.H. and McGregor, I. (eds) *Malaria: Principles and Practice of Malariology*, Vol 1. Churchill Livingstone, Edinburgh, UK, pp. 61–96.

Genchi, C., Kramer, L.H. and Rivasi, F. (2011) Dirofilarial infections in Europe. *Vector-Borne and Zoonotic Diseases* 11, 1307–1317.

Gilles, H.M. and Warrell, D.A. (1994) *Bruce-Chwatts's Essential Malariology*, 3rd edn. Little, Brown, and Co, Boston, Massachusetts.

Gillett, J.D. (1971) *Mosquitoes*. Weidenfeld and Nicolson, London.

Gillies, M.T. (1980) The role of carbon dioxide in host-finding by mosquitoes (Diptera: Culicidae): a review. *Bulletin of Entomological Research* 70, 525–532.

Gillies, M.T. (1988) Anopheline mosquitoes: vector behaviour and bionomics. In: Wernsdorfer, W.H. and McGregor, I. (eds) *Malaria: Principles and Practice of Malariology*, Vol 1. Churchill Livingstone, Edinburgh, UK, pp. 453–485.

Graham, K., Mohammad, N., Rehman, H., Nazari, A., Ahmad, M., Kamal, M., *et al.* (2002) Insecticide-treated plastic tarpaulins for control of malaria vectors in refugee camps. *Medical and Veterinary Entomology* 16, 404–408.

Gratz, N.G. and Pal, R. (1988) Malaria vector control: larviciding. In: Wernsdorfer, W.H. and McGregor, I. (eds) *Malaria: Principles and Practice of Malariology*, Vol 2. Churchill Livingstone, Edinburgh, UK, pp.1213–1226.

Graves, P.M., Burkot, T.R., Saul, A.J., Hayes, R.J. and Carter, R. (1990) Estimation of anopheline survival rate, vectorial capacity and mosquito infection probability from malaria vector infection rates in villages near Madang, Papua New Guinea. *Journal of Applied Ecology* 27, 134–147.

Gubler, D.J. (1989) *Aedes aegypti* and *Aedes aegypti*-borne disease control in the 1990s – top down or bottom up. *American Journal of Tropical Medicine and Hygiene* 40, 571–578.

Gubler, D.J. (1998a) Resurgent vector-borne diseases as a global health problem. *Emerging Infectious Diseases* 4, 442–450.

Gubler, D.J. (1998b) Dengue and dengue hemorrhagic fever. *Clinical Microbiological Reviews* 11, 480–496.

Gubler, D.J. (2004) The changing epidemiology of yellow fever and dengue, 1900 to 2003: full circle? *Comparative Immunology, Microbiology and Infectious Diseases* 27, 319–330.

Gubler, D.J. (2012) The economic burden of dengue. *American Journal of Tropical Medicine and Hygiene* 86, 743–744.

Gubler, D.J. and Kuno, G. (1997) *Dengue and Dengue Hemorrhagic Fever*. CAB International, Wallingford, UK, 478 pp.

Guzman, M.G. and Kouri, G. (2002) Dengue: an update. *Lancet Infectious Diseases* 2, 33–42.

Halstead, S.B. (2008) Dengue virus – mosquito interactions. *Annual Review of Entomology* 53, 273–291.

Harbach, R.E. (2004) The classification of genus *Anopheles* (Diptera: Culicidae): a working hypothesis of phylogenetic relationships. *Bulletin of Entomological Research* 94, 537–553.

Harbach, R.E. (2011) Classification within the cosmopolitan genus *Culex* (Diptera: Culicidae): the foundation for molecular systematics and phylogenetic research. *Acta Tropica* 120, 1–14.

Harbach, R.E. and Kitching, I.J. (1998) Phylogeny and classification of the Culicidae (Diptera). *Systematic Entomology* 23, 327–370.

Harinasuta, T. and Bunnag, D. (1988) The clinical features of malaria. In: Wernsdorfer, W.H. and McGregor, I. (eds) *Malaria: Principles and Practice of Malariology*, Vol 1. Churchill Livingstone, Edinburgh, UK, pp. 709–734.

Harley, D., Sleigh, A. and Ritchie, S. (2001) Ross River virus transmission, infection, and disease: a cross-disciplinary review. *Clinical Microbiology Reviews* 14, 909–932.

Harrington, L.C., Edman, J.D. and Scott, T.W. (2001) Why do female *Aedes aegypti* (Diptera: Culicidae) feed preferentially and frequently on human blood? *Journal of Medical Entomology* 38, 411–422.

Hawley, W.A. (1988) The biology of *Aedes albopictus*. *Journal of the American Mosquito Control Association* 4, Supplement 1, 1–39.

Hay, S.I., Guerra, C.A., Tatem, A.J., Atkinson, P.M. and Snow, R.W. (2005) Urbanization, malaria transmission and disease burden in Africa. *Nature Reviews Microbiology* 3, 81–90.

Hayes, E.B., Komar, N., Nasci, R.S., Montgomery, S.P., O'Leary, D.R. and Campbell, G.L. (2005) Epidemiology and transmission dynamics of West Nile Virus disease. *Emerging Infectious Diseases* 11, 1167–1173.

Hemingway, J., Hawkes, N.J., McCarroll, L. and Ranson, H. (2004) The molecular basis of insecticide resistance in mosquitoes. *Insect Biochemistry and Molecular Biology* 34, 653–665.

Hiwat, H. and Bretas, G. (2011) Ecology of *Anopheles darlingi* Root with respect to vector importance: a review. *Parasites and Vectors* 4, 177.

Hubalek, Z. and Halouzka, J. (1999) West Nile fever – a reemerging mosquito-borne viral disease in Europe. *Emerging Infectious Diseases* 5, 643–650.

Hulsman, K., Dale, P.E.R. and Kay, B.H. (1989) The runnelling method of habitat modification: an environment-focussed tool for salt marsh mosquito management. *Journal of the American Mosquito Control Association* 5, 226–234.

Ijumba, J.N. and Lindsay, S.W. (2001) Impact of irrigation on malaria in Africa: paddies paradox. *Medical and Veterinary Entomology* 15, 1–11.

Johnson, P.D.R., Azuolas, J., Lavender, C.J., Wishart, E., Stinear, T.P., Hayman, J.A., *et al.* (2007) *Mycobacterium ulcerans* in mosquitoes captured during outbreak of buruli ulcer, Southeastern Australia. *Emerging Infectious Diseases* 13, 1653–1660.

Jupp, P.G., Kemp, A., Grobbelaar, A., Leman, P., Burt, F.J., Alahmed, A.M., *et al.* (2002) The 2000 epidemic of Rift Valley fever in Saudi Arabia: mosquito vector studies. *Medical and Veterinary Entomology* 16, 245–252.

Kay, B.H., Nam, V.S., Tien, T.V., Yen, N.T., Phong, T.V., Diep, V.T.B., *et al.* (2002) Control of *Aedes* vectors of dengue in three provinces of Vietnam by use of *Mesocyclops* (Copepoda) and community-based methods validated by entomologic, clinical, and serological surveillance. *American Journal of Tropical Medicine and Hygiene* 66, 40–48.

Kettle, D.S. (1995) *Medical and Veterinary Entomology*, 2nd edn. CAB International, Wallingford, UK.

Killeen, G.F., Smith, T.A., Ferguson, H.M., Mshinda, H., Abdulla, S., Lengeler, C., *et al.* (2007) Preventing childhood malaria in Africa by protecting adults from mosquitoes with insecticide-treated nets. *PLOS Medicine* 4, 1246–1258.

King, C.L. and Freedman, D.O. (2000) Filariasis. In: Strickland, G.T. (ed.) *Hunter's Tropical Medicine*, 8th edn. W.B. Saunders, Philadelphia, Pennsylvania, pp. 740–753.

Knight, K.L. (1978) *Supplement to 'A Catalog of the Mosquitoes of the World' (Diptera: Culicidae)*. The Thomas Say Foundation Supplement to Volume VI, Entomological Society of America, Lanham, Maryland, pp. 1–107.

Knight, K.L. and Stone, A. (1977) *A Catalog of the Mosquitoes of the World*. The Thomas Say Foundation 6, Entomological Society of America, Lanham, Maryland, pp. 1–611.

Knox, T.B., Yen, N.T., Nam, V.S., Gatton, M.L., Kay, B.H. and Ryan, P.A. (2007) Critical evaluation of quantitative sampling methods for *Aedes aegypti* (Diptera: Culicidae) immatures in water storage containers in Vietnam. *Journal of Medical Entomology* 44, 192–204.

Kramer, L.D., Styer, L.M. and Ebel, G.D. (2008) A global perspective on the epidemiology of West Nile virus. *Annual Review of Entomology* 53, 61–81.

Kuno, G. (1995) Review of the factors modulating dengue transmission. *Epidemiologic Reviews* 17, 321–335.

Kyle, J.L. and Harris, E. (2008) Global spread and persistence of dengue, *Annual Review of Microbiology* 62, 71–92.

Laird, M. (1988) *The Natural History of Larval Mosquito Habitats*. Academic Press, London.

Lambrechts, L., Paaijmans, K.P., Fansiri, T., Carrington, L.B., Kramer, L.D., Thomas, M.B., *et al.* (2011) Impact of daily temperature fluctuations on dengue virus transmission by *Aedes aegypti*. *Proceedings of the National Academy of Sciences of the United States of America* 108, 7460–7465.

Lundström, J.O. (1999) Mosquito-borne viruses in western Europe: a review. *Journal of Vector Ecology* 24, 1–39.

Lundström, J.O., Andersson, A.C., Backman, S., Schafer, M.L., Forsman, M. and Thelaus, J. (2011) Transstadial transmission of *Francisella tularensis holarctica* in mosquitoes, Sweden. *Emerging Infectious Diseases* 17, 794–799.

Macdonald, G. (1957) *The Epidemiology and Control of Malaria*. Oxford University Press, London.

Maciel-de-Freitas, R. and Lourenco-de-Oliveira, R. (2011) Does targeting key-containers effectively reduce *Aedes aegypti* population density? *Tropical Medicine and International Health* 16, 965–973.

Mackenzie, J.S., Lindsay, M.D., Coelen, R.J., Broom, A.K., Hall, R.A. and Smith, D.W. (1994) Arboviruses causing human disease in the Australasian Region. *Archives of Virology* 136, 47–67.

Mackenzie, J.S., Gubler, D.J. and Petersen, L.R. (2004) Emerging flaviviruses: the spread and resurgence of Japanese encephalitis, West Nile and dengue viruses. *Nature Medicine Supplement* 10, S98–109.

Marshall, I.D. (1988) Murray Valley and Kunjin encephalitis. In: Monath, T. (ed.) *The Arboviruses: Epidemiology and Ecology*, Vol III. CRC Press, Boca Raton, Florida, pp. 151–190.

Mattingly, P.F. (1969) *The Biology of Mosquito-Borne Disease*. Allen and Unwin, London.

Maxwell, C.A., Curtis, C.F., Haji, H., Kisumku, S., Thalib, A.T. and Yahya, S.A. (1990) Control of Bancroftian filariasis by integrating therapy with vector control using polystyrene beads in wet pit latrines. *Transactions of the Royal Society of Tropical Medicine and Hygiene* 84, 709–714.

McGreevy, P.B., Bryan, J.H., Oothuman, P. and Kolstrup, N. (1978) The lethal effects of cibarial and pharyngeal armatures of mosquitoes on microfilariae. *Transactions of the Royal Society of Tropical Medicine and Hygiene* 72, 361–368.

McHaffie, J. (2012) *Dirofilaria immitis* and *Wolbachia pipientis*: a thorough investigation of the symbiosis responsible for canine heartworm disease. *Parasitology Research* 110, 499–502.

Mcintosh, B.M., Jupp, P.G., Dos Santos, I. and Meenehan, G.M. (1976) Epidemics of West Nile and Sindbis viruses in South Africa with *Culex (Culex) univittatus* Theobald as vector. *South African Journal of Science* 72, 295–300.

Mcintosh, B.M., Jupp, P.G. and Dos Santos, I. (1977) Rural epidemic of Chikungunya in South Africa with involvement of *Aedes (Diceromyia) furcifer* (Edwards) and baboons. *South African Journal of Science* 73, 267–269.

Medlock, J.M., Hansford, K.M., Schaffner, F., Versteirt, V., Hendrickx, G., Zeller, H., *et al.* (2012) *Vector-Borne and Zoonotic Diseases* 12, 435–447.

Merritt, R.W., Dadd, R.H. and Walker, E.D. (1992) Feeding behavior, natural food, and nutritional relationships of larval mosquitoes. *Annual Review of Entomology* 37, 349–376.

Monath, T.P. (ed.) (1988) *The Arboviruses: Ecology and Epidemiology*, Vols I–IV. CRC Press, Boca Raton, Florida.

Monath, T.P. (ed.) (1989) *The Arboviruses: Ecology and Epidemiology*, Vol V. CRC Press, Boca Raton, Florida.

Mutebi, J.P. and Barrett, A.D.T. (2002) The epidemiology of yellow fever in Africa. *Microbes and Infection* 4, 1459–1468.

Nam, V.S., Yen, N.T., Kaym, B.H., Marten, G.G. and Reid, J.W. (1998) Eradication of *Aedes aegypti* from a village in Vietnam using copepods and community participation. *American Journal of Tropical Medicine and Hygiene* 59, 657–660.

Neafie, R.C. and Meyers, W.M. (1991) Dirofilariasis. In: Strickland, G.T. (ed.) *Hunter's Tropical Medicine*. Saunders, Philadelphia, Pennsylvania, pp. 748–749.

Omeara, G.F., Evans, L.F., Gettman, A.D. and Cuda, J.P. (1995) Spread of *Aedes albopictus* and decline of *Aedes aegypti* (Diptera, Culicidae) in Florida. *Journal of Medical Entomology* 32, 554–562.

Ooi, E.E., Goh, K.T. and Gubler, D.J. (2006) Dengue prevention and 35 years of vector control in Singapore. *Emerging Infectious Diseases* 12, 887–893.

Otranto, D. and Eberhard, M.L. (2011) Zoonotic helminths affecting the human eye. *Parasites and Vectors* 4, 41.

Otranto, D., Diniz, D.G., Dantas-Torres, F., Casiraghi, M., de Almeida, I.N., de Almeida, L.N., *et al.* (2011) Human intraocular filariasis caused by *Dirofilaria* sp. nematode, Brazil. *Emerging Infectious Diseases* 17, 863–866.

Ottesen, E.A. and Ramachandran, C.P. (1995) Lymphatic filariasis infection and disease: control strategies. *Parasitology Today* 11, 129–131.

Ottesen, E.A., Duke, B.O.L., Karam, M. and Behbehani, K. (1997) Strategies and tools for the control/elimination of lymphatic filariasis. *Bulletin of the World Health Organization* 75, 491–503.

Ottesen, E.A., Hooper, P.J., Bradley, M. and Biswas, G. (2008) The global programme to eliminate lymphatic filariasis: health impact after 8 years. *PLoS Neglected Tropical Diseases* 2, e317.

Pant, C.P. (1988) Malaria vector control: imagociding. In: Wernsdorfer, W.H. and McGregor, I. (eds) *Malaria: Principles and Practice of Malariology*, Vol 2. Churchill Livingstone, Edinburgh, UK, pp. 453–485.

Pates, H. and Curtis, C. (2005) Mosquito behaviour and vector control. *Annual Review of Entomology* 50, 53–70.

Rafatjah, H.A. (1988) Malaria vector control: environmental management. In: Wernsdorfer, W.H. and McGregor, I. (eds) *Malaria: Principles and Practice of Malariology*, Vol 2. Churchill Livingstone, Edinburgh, UK, pp. 1135–1172.

Phillips, R.S. (2001) Current status of malaria and potential for control. *Clinical Microbiology Reviews* 14, 208–226.

Pialoux, G., Gauzere, B.A., Jaureguiberry, S. and Strobel, M. (2007) Chikungunya, an epidemic arbovirosis. *Lancet Infectious Diseases* 7, 319–327.

Pichon, G. (2002) Limitation and facilitation in the vectors and other aspects of the dynamics of filarial transmission: the need for vector control against *Anopheles*-transmitted filariasis. *Annals of Tropical Medicine and Parasitology* 96 (Supplement 2), S143–152.

Rapley, L.P., Johnson, P.H., Williams, C.R., Silcock, R.M., Larkman, M., Long, S.A., *et al.* (2009) A lethal ovitrap-based mass trapping scheme for dengue control in Australia: II. Impact on populations of the mosquito *Aedes aegypti*. *Medical and Veterinary Entomology* 23, 303–316.

Reinert, J.F., Harbach, R.E. and Kitching, I.J. (2004) Phylogeny and classification of Aedini (Diptera: Culicidae), based on morphological characters of all life stages. *Zoological Journal of the Linnean Society* 142, 289–368.

Reisen, W.K., Meyer, R.P., Presser, S.B. and Hardy, J.L. (1993) Effect of temperature on the transmission of western equine encephalomyelitis and St Louis encephalomyelitis viruses by *Culex tarsalis* (Diptera, Culicidae). *Journal of Medical Entomology* 30, 151–160.

Reiter, P. (2001) Climate change and mosquito-borne disease. *Environmental Health Perspectives* 109, 141–161.

Reiter, P. (2010) West Nile virus in Europe: understanding the present to gauge the future (http://www.eurosurveillance.org/ViewArticle.aspx?ArticleId=19508, accessed June 2012). *Euro surveillance* 15, 19508.

Rishikesh, N., Dubitskij, A.M. and Moreau, C.M. (1988) Malaria vector control: biological control. In: Wernsdorfer, W.H. and McGregor, I. (eds) *Malaria: Principles and Practice of Malariology*, Vol 1. Churchill Livingstone, Edinburgh, UK, pp. 453–485.

Ritchie, S.A. and Rochester, W. (2001) Wind-blown mosquitoes and introduction of Japanese encephalitis into Australia. *Emerging Infectious Diseases* 7, 900–904.

Ritchie, S.A., Long, S., Hart, A., Webb, C. and Russell, R.C. (2003) An adulticidal sticky ovitrap for sampling container-breeding mosquitoes. *Journal of the American Mosquito Control Association* 19, 235–242.

Riviere, F., Kay, B.H., Klein, J.M. and Sechan, Y. (1987) *Mesocyclops aspericornis* (Copepoda) and *Bacillus thuringiensis* var. *israelensis* for the biological control of *Aedes* and *Culex* vectors (Diptera: Culicidae) breeding in crab holes, tree holes, and artificial containers. *Journal of Medical Entomology* 24, 425–430.

Roberts, D.R., Laughlin, L.L. and Legters, L.J. (1997) DDT, global strategies and a malaria control crisis in South America. *Emerging Infectious Diseases* 3, 295–302.

Rogers, D.J. and Randolph, S.E. (2006) Climate and vector-borne disease. *Advances in Parasitology* 62, 345–381.

Rogers, D.J., Wilson, A.J., Hay, S.I. and Graham, A.J. (2006) The global distribution of yellow fever and dengue. *Advances in Parasitology* 62, 181–220.

Romi, R., Razaiarimanga, M.C., Raharimanga, R., Rakotondraibe, E.M., Ranaivo, L.H., Pietra, V., *et al.* (2002) Impact of the malaria control campaign (1993–1998) in the highlands of Madagascar: parasitological and entomological data. *American Journal of Tropical Medicine and Hygiene* 66, 2–6.

Russell, R.C. (2002) Ross River virus: ecology and distribution. *Annual Review of Entomology* 47, 1–31.

Russell, R.C. and Dwyer, D.E. (2000) Arboviruses associated with human disease in Australia. *Microbes and Infection* 2, 1693–1704.

Russell, R.C. and Geary, M.J. (1996) The influence of microfilarial density of dog heartworm *Dirofilaria immitis* on infection rate and survival of *Aedes notoscriptus* and *Culex annulirostris* from Australia. *Medical and Veterinary Entomology* 10, 29–34.

Russell, R.C., Webb, C.E. and Davies, N. (2005a) *Aedes aegypti* (L.) and *Aedes polynesiensis* Marks (Diptera: Culicidae) in Moorea, French Polynesia: a study of adult population structures and pathogen (*Wuchereria bancrofti* and *Dirofilaria immitis*) infection rates to indicate regional and seasonal epidemiological risk for dengue and filariasis. *Journal of Medical Entomology* 42, 1045–1056.

Russell, R.C., Webb, C.E., Williams, C.R. and Ritchie, S.A. (2005b) Mark–release–recapture study to measure dispersal of the mosquito *Aedes aegypti* in Cairns, Queensland, Australia. *Medical and Veterinary Entomology* 19, 451–457.

Sasa, M. (1976) *Human Filariasis*. University of Tokyo Press, Tokyo.

Service, M.W. (1993a) Mosquitoes (Culicidae). In: Lane, R.P. and Crosskey, R.W. (eds) *Medical Insects and Arachnids*. Chapman and Hall, London, pp. 120–240.

Service, M.W. (1993b) *Mosquito Ecology. Field Sampling Methods*, 2nd edn. Elsevier, Barking, UK, 988 pp.

Service, M.W. (1997) Mosquito (Diptera: Culicidae) dispersal – the long and short of it. *Journal of Medical Entomology* 34, 579–588.

Service, M.W. (ed.) (2001) *Encyclopedia of Arthropod-transmitted Infections*. CAB International, Wallingford, UK.

Sheppard, P., Macdonald, W.W., Tonn, R.J. and Grab, B. (1969) The dynamics of an adult population of *Aedes aegypti* in relation to dengue haemorrhagic fever in Bangkok. *Journal of Animal Ecology* 38, 661–702.

Sinka, M.E., Bangs, M.J., Manguin, S., Coetzee, M., Mbogo, C.M., Hemingway, J., *et al.* (2010a) The dominant *Anopheles* vectors of human malaria in Africa, Europe and the Middle East: occurrence data, distribution maps and bionomic précis. *Parasites and Vectors* 3, 117.

Sinka, M.E., Rubio-Palis, Y., Manguin, S., Patil, A.P., Temperley, W.H., Gething, P.W., *et al.* (2010b). The dominant *Anopheles* vectors of human malaria in the Americas: occurrence data, distribution maps and bionomic précis. *Parasites and Vectors* 3, 72.

Sinka, M.E., Bangs, M.J., Manguin, S., Chareonviriyaphap, T., Patil, A.P., Temperley, W.H., *et al.* (2011) The dominant *Anopheles* vectors of human malaria in the Asia-Pacific region: occurrence data, distribution maps and bionomic précis. *Parasites and Vectors* 4, 89.

Sinka, M.E., Bangs, M.J., Manguin, S., Rubio-Palis, Y., Chareonviriyaphap, T., Coetzee, M., *et al.* (2012) A global map of dominant malaria vectors. *Parasites and Vectors* 5, 69.

Snow, L.C., Bockarie, M.J. and Michael, E. (2006) Transmission dynamics of lymphatic filariasis: vector-specific density dependence in the development of *Wuchereria bancrofti* infective larvae in mosquitoes. *Medical and Veterinary Entomology* 20, 261–272.

Tadei, W.P., Thatcher, B.D., Santos, J.M.H., Scarpassa, V.M., Rodrigues, I.B. and Rafael, M.S. (1998) Ecologic observations on anopheline vectors of malaria in the Brazilian Amazon. *American Journal of Tropical Medicine and Hygiene* 59, 325–335.

Takken, W. and Knols, B.G.J. (1999) Odor-mediated behavior of Afrotropical malaria mosquitoes. *Annual Review of Entomology* 44, 131–157.

Tan, C.H., Wong, P.S.J., Li, M.Z.I., Tan, S.Y.S., Lee, T.K.C., Pang, S.C., *et al.* (2011) Entomological investigation and control of a Chikungunya cluster in Singapore. *Vector-Borne and Zoonotic Diseases* 11, 383–390.

Taylor, M.J., Bandi, C. and Hoerauf, A. (2005) *Wolbachia* bacterial endosymbionts of filarial nematodes. *Advances in Parasitology* 60, 245–284.

Thomas, S.J. and Endy, T.P. (2011) Critical issues in dengue vaccine development. *Current Opinion in Infectious Diseases* 24, 442–450.

Tun-Lin, W., Kay, B.H. and Barnes, A. (1995) Understanding productivity, a key to *Aedes aegypti* surveillance. *American Journal of Tropical Medicine and Hygiene* 53, 595–601.

Van den Hurk, A.F., Ritchie, S.A. and Mackenzie, J.S. (2009) Ecology and geographical expansion of Japanese encephalitis virus. *Annual Review of Entomology* 54, 17–35.

Vasilakis, N., Cardosa, J., Hanley, K.A., Holmes, E.C. and Weaver, S.C. (2011) Fever from the forest: prospects for the continued emergence of sylvatic dengue virus and its impact on public health. *Nature Reviews Microbiology* 9, 532–541.

Walker, K. (2000) Cost-comparison of DDT and alternative insecticides for malaria control. *Medical and Veterinary Entomology* 14, 345–354.

Walker, K. and Lynch, M. (2007) Contributions of *Anopheles* larval control to malaria suppression in tropical Africa: review of achievements and potential. *Medical and Veterinary Entomology* 21, 2–21.

Ward, R.A. (1984) Second supplement to 'A catalog of the mosquitoes of the world' (Diptera: Culicidae). *Mosquito Systematics* 16, 227–270.

Ward, R.A. (1992) Third supplement to 'A catalog of the mosquitoes of the world' (Diptera: Culicidae). *Mosquito Systematics* 24, 177–230.

Watts, D.M., Pantuwatana, S., Yuill, T.M., DeFoliart, G.R., Thompson, W.H. and Hanson, R. (1975) Transovarial transmission of La Crosse virus in *Aedes triseriatus*. *Annals of the New York Academy of Sciences* 266, 135–143.

Weaver, S.C. and Barrett, A.D.T. (2004) Transmission cycles, host range, evolution and emergence of arboviral disease. *Nature Reviews Microbiology* 2, 789–801.

Weaver, S.C. and Vasilakis, N. (2009) Molecular evolution of dengue viruses: contributions of phylogenetics to understanding the history and epidemiology of the preeminent arboviral disease. *Infection, Genetics and Evolution* 9, 523–540.

Weaver, S.C., Ferro, C., Barrera, R., Boshell, J. and Navarro, J.C. (2004) Venezuelan equine encephalitis. *Annual Review of Entomology* 49, 141–174.

Webb, C.E. and Russell, R.C. (2009) Insect repellents and sunscreen: implications for personal protection strategies against mosquito-borne disease. *Australian and New Zealand Journal of Public Health* 33, 485–490.

Webber, R.H. (1979) Eradication of *Wuchereria bancrofti* infection through vector control. *Transactions of the Royal Society of Tropical Medicine and Hygiene* 73, 722–724.

Wharton, R.H. (1963) Adaptations of *Wuchereria* and *Brugia* to mosquitoes and vertebrate hosts in relation to the distribution of filarial parasites. *Zoonoses Research* 2, 1–12.

White, G.B. (1974) *Anopheles gambiae* complex and disease transmission in Africa. *Transactions of the Royal Society of Tropical Medicine and Hygiene* 68, 278–298.

White, G.B. (1978) Systematic reappraisal of the *Anopheles maculipennis* complex. *Mosquito Systematics* 10, 13–44.

Wilder-Smith, A. and Gubler, D.J. (2008) Geographic expansion of dengue: the impact of international travel. *Medical Clinics of North America* 92, 1377–1390.

Williams, C.R., Ritchie, S.A., Long, S., Dennison, N. and Russell, R.C. (2007) Impact of a bifenthrin-treated lethal ovitrap on *Aedes aegypti* oviposition and mortality in north Queensland, Australia. *Journal of Medical Entomology* 44, 256–262.

Wirtz, R.A. and Burkot, T.R. (1991) Detection of malarial parasites in mosquitoes. *Advances in Disease Vector Research* 8, 77–106.

World Health Organization (2011) World Malaria Report (http://www.who.int/malaria/world_malaria_report_2010/en/index.html, accessed December 2011).

Yohannes, M. and Boelee, E. (2012) Early biting rhythm in the afro-tropical vector of malaria, *Anopheles arabiensis*, and challenges for its control in Ethiopia. *Medical and Veterinary Entomology* 26, 103–105.

Nasal Bot Flies (Diptera: Oestridae, Oestrinae)

1. INTRODUCTION

The nasal bot flies are obligate larval parasites that develop in the nasopharyngeal cavities of ungulate mammals. Most species show a high degree of host specificity. The species of primary importance is *Oestrus ovis*, a parasite of sheep and goats. Three other species are also of economic significance: *Rhinoestrus purpureus*, *Cephalopina titillator* and *Cephenemyia trompe*, which respectively parasitize equines, camels and reindeer and caribou. Other species of minor importance include *Tracheomyia macropi*, which parasitizes the red kangaroo in Australia, and *Pharyngobolus africanus*, which is a parasite of the African elephant. The larvae of all these species have well-developed mouth-hooks to facilitate attachment and feeding, whereas the adults have primitive, usually non-functional mouthparts and are short-lived. Human infestations do occur, most commonly presenting as external ophthalmomyiasis, and allergic reactions against larval antigens have also been reported.

2. TAXONOMY

All the flies that act as important agents of medical and veterinary myiasis are calypterate Diptera in the superfamily Oestroidea. Within this superfamily, there are three major families of myiasis-producing flies: Oestridae (see also the entry for 'Stomach Bot Flies'), Calliphoridae (see the entry for 'Blow Flies') and Sarcophagidae (see the entry for 'Flesh Flies'). All the nostril flies belong to the Oestridae and the subfamily Oestrinae, which includes 34 species in nine genera.

3. MORPHOLOGY

Adult oestrines have a distinct postscutellum, large squamae and vein M bent towards and joining vein R_{4+5} before the margin (Fig. 99). The frons is enlarged, and the frons, scutellum and dorsal thorax bear small wart-like protuberances. The eyes are small and the abdomen is brownish-black.

Adult *O. ovis* have black pits dorsally between the eyes on the frons and black tubercles among the yellow hairs on the yellow-brown scutum and scutellum (Fig. 100). The abdomen is black, with an irregular pattern of lighter marking, which varies with the angle of illumination.

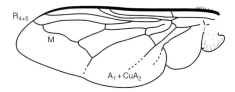

Fig. 99. The wing of *Oestrus ovis*.

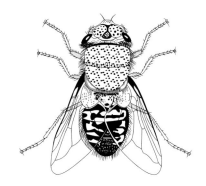

Fig. 100. Female *Oestrus ovis*. Source: Cameron, A.E. (1942) *Transactions of the Highland and Agricultural Society*.

The first-stage larvae of *O. ovis* are about 1 mm long and have gently curved mouth-hooks and 22–25 terminal spines arranged in two groups. These characters enable this larva to be separated from that of the first-stage *R. purpureus*, which can cause nasal myiasis in horses and donkeys and, occasionally, ocular myiasis in humans. The mature third-stage larva of *O. ovis* is about 25 mm long, white or yellowish in colour, with darker transverse dorsal bands and transverse rows of spines on the ventral surface of each segment, with well-developed mouth-hooks and the posterior spiracles are exposed, flat, D-shaped plates with the button enclosed by numerous small openings (Fig. 101).

4. LIFE CYCLE

Unlike other Oestroidea, the eggs of nostril flies develop and hatch *in utero*. The first-stage larvae are then squirted into the nose or eye of the host as the female hovers in the air. When

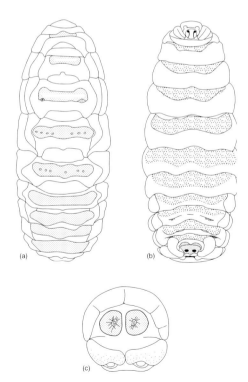

Fig. 101. Third-stage larva of *Oestrus ovis*. (a) Dorsal view; (b) ventral view; (c) posterior view.

in contact with a suitable vertebrate host, *O. ovis* deposits batches of about 20 larvae into the nasal cavities, where they undergo development for a month before moving into the frontal, and sometimes the maxillary, sinuses, which represents the final site for larval development. The larvae of *R. purpureus* may enter the tracheal branches of the lungs from the pharynx and then return to the host nasal sinuses or pharyngeal region. When fully developed, the larvae migrate back towards the nasal cavities, from which they are expelled by sneezing or coughing; subsequently, larvae burrow into the ground, pupate and eventually the next generation of adults emerge.

5. BEHAVIOUR AND BIONOMICS

5.1 *Oestrus*

O. ovis is considered to be a Palaearctic species whose worldwide distribution has been facilitated by human movement and trade.

Adults flies are particularly active during hot, dry weather, and this activity often leads to considerable disturbance and panic in a flock, with sheep typically keeping their nostrils pressed against the ground in an attempt to protect themselves from the adult flies. In areas with warm winters, year-round breeding of *O. ovis* is possible, but in many geographical areas where sheep are present, two generations per annum are considered to occur. The usual pattern in the northern hemisphere sees adults emerge in late spring in June, mate and larviposit, with the larvae developing rapidly and mature third-stages leaving the host in July and August. These pupate to produce an autumn generation, larvipositing in September and October. Overwintering of the first-stage larvae occurs within the host. These overwintering larvae may undergo arrested development for a long period before developing to become mature larvae in March, after which they leave the host, pupate and remain dormant until adult flies emerge in June. Mortality among the immature stages has been calculated at 90–94% in the first generation and 99% in the second generation.

High burdens of infestations with *O. ovis* in sheep are commonly recorded; for example,

73% in Pretoria, South Africa, and more than 90% in Kentucky, USA, with an average of 15 and 22 larvae per sheep, respectively. Annoyance by adult *O. ovis* causes sheep to lose valuable grazing time, whereas the presence of larvae in the nostrils irritates the mucosa, resulting in a mucopurulent discharge and impaired respiration. Control can be achieved in late summer and/or winter with various drugs, including macrocyclic lactones such as ivermectin. In one study, Merino rams treated against infestation showed reduced nasal discharge and increased weight gain.

Although sheep and goats are the main hosts for *O. ovis*, infestation occasionally occurs in dogs and humans; however, in these hosts, larvae do not complete their development. In humans, *O. ovis* infestations have been responsible for ocular myiasis.

5.2 *Cephenemyia*

The genus *Cephenemyia* is restricted to the Holarctic, with larvae developing exclusively in deer. Parasitism may affect an excess of 70% of a herd. Of particular interest, *C. trompe* and *Cephenemyia auribarbis* are found in reindeer and caribou, respectively. *Cephenemyia phobifer* and *Cephenemyia stimulator* are found in red and roe deer, respectively; *Cephenemyia pratti* may occur in the mule deer, *Cephenemyia jellisoni* in the whitetail and Pacific blacktail deer and *Cephenemyia apicata* in Californian deer. Females deposit live first-stage larvae in the nostrils of the hosts and the larvae subsequently move to the pharyngeal and nasal cavities. Mature larvae migrate to the anterior pharynx before leaving the host. The activity of the larvae causes nasal discharge, sneezing, coughing and restlessness.

5.3 *Rhinoestrus*

The genus *Rhinoestrus* is highly host specific and infests equines and large African mammals. Eleven species have been described, four of which are restricted to equines and seven to a wide range of wild animals, including giraffe, warthog, bushpig and antelope. One species, *Rhinoestrus usbekistanicus*, parasitizes horses and donkeys in the Palaearctic region and occurs in zebras in Africa. *Rhinoestrus latifrons* parasitizes domestic horses in Central Asia and adjoining regions. The most important species is *R. purpureus*, which parasitizes horses, donkeys and their cross-breeds. Larvae localize in the throat region, at the base of the tongue, in the pharyngeal cavity and in the turbinates.

Adult *Rhinoestrus* resemble *Oestrus* but have more conspicuous tubercles on the head and thorax (Fig. 102). In a study in the Caspian region of the former USSR, 97–100% of horses were found to be infested with *Rhinoestrus*, specifically *R. latifrons* and *R. purpureus*, with a mean of 154 larvae found in the head cavities of each infested horse and 899 larvae recovered from a single horse. Both *R. purpureus* and *R. usbekistanicus* are endemic in southern Europe, with prevalence up to 21% in native horses and an intensity rate as high as 65 larvae in examined animals. There is one generation a year; the female produces 700–800 eggs, which are deposited as larvae in batches of 8–40 at a time into the nostrils or eyes of a horse. Like *O. ovis*, *Rhinoestrus* spp. can cause ocular myiasis in humans; the first stage can be distinguished from that of *O. ovis* in that the former possess more strongly curved mouth-hooks and only 8–12 terminal hooklets in a single row. In the third-stage larva, the spiracles are crescent-shaped and do not surround the button completely. As with *O. ovis*, first-stage larvae

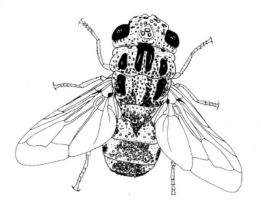

Fig. 102. Female *Rhinoestrus purpureus*. Redrawn from Zumpt (1965).

may remain in the nasal cavity for several months before moving to the sinuses to complete their development.

5.4 *Cephalopina*

Cephalopina is a monotypic genus with one species, *C. titillator*, the camel nasal bot fly. It is a parasite of camels in dry parts of the Palaearctic and Oriental regions and Australia. Infestations of up to 90% of camel herds have been recorded. The presence of larvae may cause considerable irritation and breathing difficulty. In Sudan, all camels are usually infested, harbouring up to 250 larvae/camel. Female *C. titillator* hover above the host before larvipositing rapidly in the nostrils. Mature larvae leave the nostril towards sunset and pupate in the soil. Heavy infestations cause nasopharyngeal lesions, providing a suitable environment for pathogens, leading to pyogenic infections and pneumonia.

6. MEDICAL AND VETERINARY IMPORTANCE

As they deposit larvae, the activity of adult *O. ovis* may disturb sheep, leading to a loss of grazing time, reduced weight gain in lambs and loss of condition. However, in general, infestations are relatively light, with an average of 2–20 larvae only infesting the frontal sinus of animals at any one time. The parasitic rhinitis caused by the larvae of *O. ovis* is characterized by a sticky, mucoid nasal discharge, which at times may be haemorrhagic. Histopathological changes in the nasal tissues of infected sheep include catarrh, infiltration of inflammatory cells and squamous metaplasia, characterized by the conversion of secretory epithelium to a stratified squamous type. Host immune responses against infestation by *O. ovis* have been recorded as a consequence of the mechanical action of spines and hooks on mucosal membranes during larval migration and as allergenic reactions provoked by molecules excreted/secreted by the larvae.

Clinical symptoms of infestation include mild discomfort, nasal discharge, sneezing, nose rubbing or head shaking. Dead larvae in the sinuses can cause allergic and inflammatory responses, followed by bacterial infection and occasionally death. Larvae may rarely penetrate the olfactory mucosa of turbinates and enter the brain throughout the lamina cribrosa of the ethmoid bone, leading to ataxia, circling and head pressing.

Infestation prevalence tends to be highly localized. In individual sheep flocks, infestation rates of up to 44–88% have been recorded in France and as low as 0.75% in Britain. Infestation rates of 6–52% have been recorded in Zimbabwe, 69% in India and 100% in Morocco, South Africa and Brazil. Infestation by *O. ovis* has been associated with reduced weight gain (between 1 and 4.5 kg), losses in wool production of up to 200–500 g and a reduction in milk production of up to 10%.

7. PREVENTION AND CONTROL

Where the numbers of larvae are small, it may not be economically viable to treat infected animals. However, in heavy infestations, closantel, nitroxynil and the macrocyclic lactone endectocides, ivermectin, doramectin and moxidectin, are highly effective, as are the organophosphates, trichlorphon and dichlorvos. Should a control scheme be necessary, flock treatment should be given twice a year, i.e. at the beginning of summer to kill newly acquired larvae and in midwinter to kill any overwintering larvae. Fly repellents may be used, but these have limited effect.

Human infestations do occur, most commonly presenting as external ophthalmomyiasis, but maggots have been reported from the interior of the eyeball. External infestation can be managed by physical removal of the larvae (with anaesthesia of the eyeball by lidocaine, which also paralyses the larvae and facilitates their removal), but management of internal infestations is dependent on the clinical situation and expert assistance should be sought.

SELECTED BIBLIOGRAPHY

Anderson, J.B. (1989) Use of deer models to study larviposition by wild nasopharyngeal bot flies (Diptera: Oestridae). *Journal of Medical Entomology* 26, 234–236.

Angulo-Valadez, C.E., Ascencio, F., Jacquiet, P., Dorchies, P. and Cepeda-Palacios, R. (2011) Sheep and goat immune responses to nose bot infestation: a review. *Medical and Veterinary Entomology* 25, 117–125.

Baron, R.W. and Colwell, D.D. (1991) Mammalian immune responses to myiasis. *Parasitology Today* 7, 353–355.

Breev, K.A., Zagretdinov, R.G. and Minar, J. (1980) Influence of constant and variable temperatures on pupal development of the sheep bot fly (*Oestrus ovis* L.). *Folia Parazitologica* 27, 359–365.

Bukshtynov, V.l. (1978) Determining the time of development of *Oestrus ovis*. *Veterinariya Moscow* 9, 60–62.

Colwell, D.D., Hall, M.J.R. and Scholl, P.J. (2006) *Oestrid Flies: Biology, Host–Parasite Relationships, Impact and Management*. CAB International, Wallingford, UK.

Dar, M.S., Amer, M.B., Dar, F.K. and Papazotos, V. (1980) Ophthalmomyiasis caused by the sheep nasal bot, *Oestrus ovis* (Oestridae) larvae, in the Benghazi area of eastern Libya. *Transactions of the Royal Society of Tropical Medicine and Hygiene* 74, 303–306.

Gracia, M.J., Lucientes, J., Peribanez, M.A., Castillo, J.A., Calvete, C. and Ferrer, L.M. (2010) Epidemiology of *Oestrus ovis* infection of sheep in northeast Spain (mid-Ebro Valley). *Tropical Animal Health and Production* 42, 811–813.

Hall, M.J.R. and Smith, K.G.V. (1993) Diptera causing myiasis in man. In: Lane, R.P. and Crosskey, R.W. (eds) *Medical Insects and Arachnids*. Chapman and Hall, London, pp. 429–469.

Hall, M.J.R. and Wall, R. (1994) Myiasis of humans and domestic animals. *Advances in Parasitology* 35, 258–334.

Horak, I.G. (1977) Parasites of domestic and wild animals in South Africa. 1. *Oestrus ovis* in sheep. *Onderstepoort Journal of Veterinary Research* 44, 55–63.

Horak, I.G. and Snijders, A.J. (1974) The effect of *Oestrus ovis* infestation on merino lambs. *Veterinary Record* 94, 12–16.

Masoodi, M. and Hosseini, K. (2003) The respiratory and allergic manifestations of human myiasis caused by larvae of the sheep botfly (*Oestrus ovis*): a report of 33 pharyngeal cases from southern Iran. *Annals of Tropical Medicine and Parasitology* 97, 75–81.

Musa, M.F., Harrison, M., Ibrahim, A.M. and Taha, T.O. (1989) Observations on Sudanese camel nasal myiasis caused by the larvae of *Cephalopina titillator*. *Revue d'Elevage et de Medecine Veterinaire des Pays Tropicaux* 42, 27–31.

Otranto, D., Colwell, D.D., Milillo, P., Di Marco, V., Paradies, P., Napoli, C., *et al.* (2004) Report in Europe of nasal myiasis by *Rhinoestrus* spp. in horses and donkeys: seasonal patterns and taxonomical considerations. *Veterinary Parasitology* 122, 79–88.

Otranto, D., Milillo, P., Traversa, D. and Colwell, D.D. (2005) Morphological variability and genetic identity in *Rhinoestrus* spp. causing horse nasal myiasis. *Medical and Veterinary Entomology* 19, 96–100.

Papadopoulos, E., Prevot, F., Diakou, A. and Dorchies, P. (2006) Comparison of infection rates of *Oestrus ovis* between sheep and goats kept in mixed flocks. *Veterinary Parasitology* 138, 382–385.

Reingold, W.J., Robin, J.B., Leipa, D., Kondra, L., Schanzlin, D.J. and Smith, R.E. (1984) *Oestrus ovis* ophthalmomyiasis externa. *American Journal of Ophthalmology* 97, 7–10.

Rogers, C.E. and Knapp, F.W. (1973) Bionomics of the sheep bot fly *Oestrus ovis*. *Environmental Entomology* 2, 11–23.

Zumpt, F. (1965) *Myiasis in Man and Animals in the Old World*. Butterworths, London.

Non-biting Midges (Diptera: Chironomidae)

1. INTRODUCTION

Chironomids are a large cosmopolitan family. They are mostly small, delicate flies that cannot bite (Fig. 103). None the less, adult midges can occur in plague-like numbers, are attracted to lights around dwellings and other buildings and can cause severe allergic responses in some people.

2. TAXONOMY AND MORPHOLOGY

The biodiversity of chironomids is probably underestimated because they are notoriously difficult to identify, usually being recorded by species groups. More than 5000 species have been described from 11 subfamilies and more than 400 genera, and there are probably more than 10,000 species worldwide. Overall, less than 100 species have been reported as nuisances.

Larger species may be confused with mosquitoes (Culicidae – see entry for 'Mosquitoes'), but lack their forwardly directed long proboscis, and smaller ones may be confused with biting midges (Ceratopogonidae – see entry for 'Biting Midges'), from which they can be separated by the unbranched wing vein M (iv) and non-piercing mouthparts.

3. LIFE CYCLE

The immature stages are aquatic, with the vermiform larvae being found in almost any aquatic or semi-aquatic habitat, including soil, rotting vegetation, tree holes, rice fields, drainage ditches, flood channels and sewage ponds, and also some artificial containers. There are four larval instars and development from egg to adult may take from 1 month to over a year, depending on species and environmental conditions.

4. BEHAVIOUR AND BIONOMICS

In aquatic sites, they are usually benthic organisms feeding on detritus (but some are filter feeders in the water column, while others are predators), with many living in tunnels they construct in or on the substrate. Some species have adapted to virtually anoxic conditions and are dominant in polluted waters. Larvae of some species (e.g. in genus *Chironomus*) are bright red in colour, due to a haemoglobin

Fig. 103. Chironomidae. (a) Adult female; (b) larva of *Chironomus tentans*. Larva redrawn from Johansen, O.A. (1934) *Aquatic Diptera*. Cornell University Agriculture Experiment Station, Memoir 164.

analogue, and these are known as 'blood-worms'.

Adults may emerge in massive numbers and rest by day and fly during the evening, night and early morning hours, with the males forming vast swarms. The adults may feed on plant sugars or honeydew, but some may not feed at all. Often, they may not live beyond a few days.

5. MEDICAL AND VETERINARY IMPORTANCE

When they emerge in large numbers (particularly from water reservoirs, ornamental ponds, rice fields, newly flooded basins and stored polluted waters associated with sewage treatment plants), some chironomids (e.g. species of *Chironomus* and *Tanypus*) can act as nuisance pests for both humans and livestock, because of the time spent trying to keep them out of the eyes and ears and to avoid inhaling them. Being strongly attracted to lights and pale-coloured buildings, they can create problems around houses and businesses, marking painted surfaces and interfering with the operation of equipment. When large numbers of adults die around buildings, their dead bodies can produce a pungent odour.

Further, though, chironomids are unusual insects in that many species contain low molecular weight haemoglobin molecules known as erythrocurorins and these have been shown to be potent allergens in humans that can sensitize and induce allergic disease. Testing of 14 chironomid species, belonging to five different genera, from Africa, Australia and Europe revealed that, in spite of considerable variation in the electrophoretic patterns of their larval proteins, there was considerable immunological cross-reactivity, mainly, or exclusively, due to common antigenic determinants of their haemoglobins. This cross-reactivity between the antigens of different chironomid species from different parts of the world translates into the fact that sensitized people may be at risk anywhere in the world where chironomids are a nuisance. Some of the species involved are *Chironomus*

(= *Cladotanytarsus*) *lewisi* along the Nile in the Sudan, *Chironomus salinarius* and *Chironomus thummi thummi* in Europe, *Glyptotendipes paripes*, *Chironomus decorous* and *Chironomus plumosus* in the USA, *Chironomus yoshimatsui* and *Tokunagayusurica akamushi* in Japan and *Kieferulus* (formerly *Cateronica*) *longilobus*, a saltwater breeding chironomid in the Indo-Pacific region.

In recent years, there have been reports of human pathogens (*Vibrio cholerae* and *Aeromonas* species) associated with chironomid egg masses. It has been hypothesized that adult midges can transfer such organisms mechanically in some circumstances, which may result in human infections; however, no direct evidence for this has been provided.

6. PREVENTION AND CONTROL

Inhalant allergy sensitive persons should take personal protection measures, such as wearing dust-type face masks when outdoors and encountering midge swarms. They may also need to seek medical treatment if highly sensitive.

The fact that large numbers can be a nuisance problem when attracted to domestic lights and that they cause allergic responses in some people has resulted in control measures being taken against some species. Attacking the adults with knock-down insecticides provides a temporary relief and residual insecticides applied to building walls where they rest can be more long-lasting. However, control of the larval populations is a preferable approach. Reducing the nutrient inflow to habitats will reduce the population output. Draining and drying of habitats such as ponds and detention basins has been used effectively in some situations. Fish have been useful in some situations, but only local fish species known to be effective should be used. Insecticides and insect growth regulators also have been used successfully against larvae, but recommendations should be based on local investigations that give due consideration to adverse impacts on non-target species in the habitats.

SELECTED BIBLIOGRAPHY

Ali, A. (1981) Nuisance chironomids and their control. *Bulletin of the Entomological Society of America* 26, 3–16.

Ali, A. (1991) Perspectives on management of pestiferous Chironomidae (Diptera), an emerging global problem. *Journal of the American Mosquito Control Association* 7, 260–281.

Ali, A. (1994) Nuisance, economic impact and possibilities for control. In: Armitage, P., Cranston, P.S. and Pinder, L.C.V. (eds) *The Chironomidae: Biology and Ecology of Non-biting Midges*. Chapman and Hall, London, 572 pp.

Ali, A. (1996) A concise review of chironomid midges (Diptera: Chironomidae) as pests and their management. *Journal of Vector Ecology* 21, 105–121.

Arlian, L.G. (2002) Arthropod allergens and human health. *Annual Review of Entomology* 47, 395–433.

Armitage, P., Cranston, P.S. and Pinder, L.C.V. (eds) (1995) *The Chironomidae: Biology and Ecology of Non-biting Midges*. Chapman and Hall, London, 572 pp.

Ballesteros, S.C., de Barrio, M., Baeza, M.L. and Sotes, M.R. (2006) Allergy to chironomid larvae (red midge larvae) in non professional handlers of fish food. *Journal of Investigational Allergology and Clinical Immunology* 16, 63–68.

Brasch, J., Bruning, H. and Paulke, E. (1992) Allergic contact-dermatitis from chironomids. *Contact Dermatitis* 26, 317–320.

Brown, H.M., Merrett, J. and Merrett, T.G. (2000) Fish food allergy. *Allergy* 55, 901–902.

Broza, M., Halpern, M., Gahanma, L. and Inbar, M. (2003) Nuisance chironomids in waste water stabilization ponds: monitoring and action threshold assessment based on public complaints. *Journal of Vector Ecology* 28, 31–36.

Cranston, P.S. (1988) Allergens of non-biting midges (Diptera: Chironomidae): a systematic survey of chironomid haemoglobins. *Medical and Veterinary Entomology* 2, 117–127.

Cranston, P.S. (1995) Medical significance. In: Armitage, P., Cranston, P.S. and Pinder, L.C.V. (eds) *The Chironomidae: Biology and Ecology of Non-biting Midges*. Chapman and Hall, London, pp. 365–384.

Cranston, P.S., Gad-El-Rab, M.O. and Kay, A.B. (1981) Chironomid midges as a cause of allergy in the Sudan. *Transactions of the Royal Society of Tropical Medicine and Hygiene* 75, 1–4.

Ferrington, L.C. Jr (2008) Global diversity of non-biting midges (Chironomidae; Insecta-Diptera) in freshwater. *Hydrobiologia* 595, 447–455.

Freeman, P. (1973) Chironomidae ('non-biting midges'). In: Smith, K.G.V. (ed.) *Insects and Other Arthropods of Medical Importance*. British Museum (Natural History), London, pp. 189–191.

Galindo, P.A., Melero, R., Garcia, R., Feo, E., Gomez, E. and Fernandez, F. (1996) Contact urticaria from chironomids. *Contact Dermatitis* 34, 297.

Hirabayashi, K., Kubo, K., Yamaguchi, S., Fjimoto, K., Murakami, G. and Nasu, Y. (1997) Studies of bronchial asthma induced by chironomid midges (Diptera) around a hypereutrophic lake in Japan. *Allergy* 52, 188–195.

Jeong, K.I., Yum, H.Y., Lee, I.Y., Ree, H.I., Hong, C.S., Kim, D.S., *et al.* (2004) Molecular cloning and characterization of tropomyosin, a major allergen of *Chironomus kiiensis*, a dominant species of nonbiting midges in Korea. *Clinical and Diagnostic Laboratory Immunology* 11, 320–324.

Jose Figueras, M., Beaz-Hidalgo, R., Senderovich, Y., Laviad, S. and Halpern, M. (2011) Re-identification of *Aeromonas* isolates from chironomid egg masses as the potential pathogenic bacteria *Aeromonas aquariorum*. *Environmental Microbiology Reports* 3, 239–244.

Kawai, K. and Konishi, K. (1986a) Fundamental studies on chironomid allergy. I. Culture methods of some Japanese chironomids (Chironomidae, Diptera). *Medical Entomology and Zoology* 37, 47–58.

Kawai, K. and Konishi, K. (1986b) Fundamental studies on chironomid allergy. II. Analysis of larval antigens of some Japanese chironomids (Chironomidae, Diptera). *Japanese Journal of Allergology* 35, 1088–1098.

Kawai, K., Murakami, G., Kasaya, S., Teranishi, H. and Muraguchi, A. (1997) Allergenic importance of 22 species of Japanese chironomid midges. *Allergology International* 46, 43–49.

McHugh, S.M., Credland, P.F., Tee, R.D. and Cranston, P.S. (1988) Evidence of allergic hypersensitivity to chironomid midges in an English-village community. *Clinical Allergy* 18, 275–285.

Morsy, T.A., Saleh, W.A., Magied, A., Farrag, A.K. and Rifaat, M.A. (2000) Chironomid potent allergens causing respiratory allergy in children. *Journal of the Egyptian Society of Parasitology* 30, 83–92.

Tee, R.D., Cranston, P.S., Dewair, M., Prelicz, H., Baur, X. and Kay, A.B. (1985) Evidence for hemoglobins as common allergenic determinants in IgE-mediated hypersensitivity to chironomids (non-biting midges). *Clinical Allergy* 15, 335–343.

Vaughan, I.P., Newberry, C., Hall, D.J., Liggett, J.S. and Ormerod, S.J. (2008) Evaluating large-scale effects of *Bacillus thuringiensis* var. *israelensis* on non-biting midges (Chironomidae) in a eutrophic urban lake. *Freshwater Biology* 53, 2117–2128.

Vezzulli, L., Pruzzo, C., Huq, A. and Colwell, R. (2010) Environmental reservoirs of *Vibrio cholerae* and their role in cholera. *Environmental Microbiology Reports* 2, 27–33.

Wirtz, R.A. (1984) Allergic and toxic reactions to non-stinging arthropods. *Annual Review of Entomology* 24, 47–69.

Sand Flies (Diptera: Psychodidae, Phlebotominae)

1. INTRODUCTION

The true sand flies (family Psychodidae, subfamily Phlebotominae) are distinct from the biting midges (family Ceratopogonidae) and blackflies (family Simuliidae), which, in some countries, are referred to as 'sand flies'. Phlebotomines are vectors of several diseases, of which the most important is leishmaniasis, caused by infection with various species of the protozoan *Leishmania*. The other subfamily of the Psychodidae, the Psychodinae (known as moth flies), are of little or no medical or veterinary importance (although there are some reports of their body and wing hairs being responsible for respiratory allergies in situations when they are particularly abundant).

Phlebotomines are primarily inhabitants of the warmer areas of the world, although they extend as far as 50°N in central Asia. Phlebotomines are a geologically old group, being identified from the Lower Cretaceous, about 120–135 million years ago. They therefore originated before the mammals and must have fed originally on reptiles.

2. TAXONOMY

The phlebotomine sand flies are a well-defined group of species, which are sometimes accorded family rank as the Phlebotomidae but more frequently are regarded as a subfamily (Phlebotominae) of the Psychodidae. More than 800 species of phlebotomines have been described and included in six genera. About half of the species are contained in the genus *Lutzomyia*, one-third in *Sergentomyia* and the majority of the remainder in the genus *Phlebotomus*, with a small number of species in *Brumptomyia*, *Warileya* and *Chinius* (one cave-dwelling species). *Phlebotomus*, *Ser-gentomyia* and *Chinius* are confined to the Old World and *Lutzomyia*, *Brumptomyia* and *Warileya* to the New World (mainly the Neotropical region).

3. MORPHOLOGY

Phlebotominae sand flies are small, brownish, long-legged flies with narrow, hairy bodies and long antennae (Fig. 104). The wings are narrow, hairy and lanceolate, characterized by more or less parallel longitudinal veins and held erect above the body. The radial sector is four-branched and the fork of R_{2+3} and R_4 occurs about the middle of the wing (Fig. 105). The mouthparts are moderately long, with functional mandibles in females but none in the males, which are not bloodsucking. The pendulous palps are five-segmented and, as in the Ceratopogonidae, the third segment bears sensillae. The maxillae are hooked at the tip in mammal-feeding *Phlebotomus* and *Lutzomyia* and ridge-tipped in reptile-feeding *Sergentomyia*. In the female, the cibarium often bears teeth, which are valuable in identification.

3.1 Major genera

The Old World genera *Phlebotomus* and *Sergentomyia* are separated from the New World genera by having the fifth palpal segment the longest, no postspiracular setae and no posterior bulge to the cibarium. In the New World genera, palpal segment three is usually the longest, postspiracular setae are present and there is a posterior bulge to the cibarium.

Phlebotomus includes mammal-biting sand fly species that reach their maximum

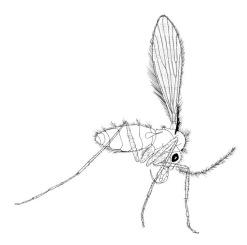

Fig. 104. Lateral view of a female *Phlebotomus* spp.

development in the warmer temperate and subtropical regions with hot summers and cold winters. It is characterized by the absence of cibarial teeth in the female and by the possession of erect hairs on the hind borders of abdominal tergites two to six (Fig. 104).

Sergentomyia is the dominant genus in the Old World tropics of Africa, India and Australia,

where its species feed on reptiles and amphibians. The genus is characterized by the possession of a posterior transverse row of cibarial teeth in the female and recumbent setae on abdominal tergites two to six.

Lutzomyia is mainly Neotropical in distribution but a few species occur in the south of the Nearctic region. They feed on both mammals and reptiles. Species of *Lutzomyia* are characterized by having a transverse row of hind teeth and one or more rows of fore teeth on the cibarium of the female.

4. LIFE CYCLE

The phlebotomine sand flies differ from the biting midges and black flies in that their immature stages are considered to be generally terrestrial rather than aquatic; however, they require a moist microhabitat with high humidity and, in some species (e.g. *Lutzomyia vexator occidentis*), both eggs and larvae need contact with water and are unable to survive even in a saturated atmosphere.

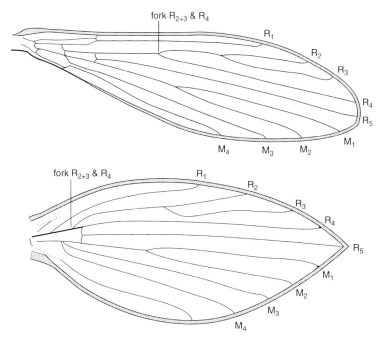

Fig. 105. Wing venation of a phlebotomine sand fly (above) and of a psychodine (below).

4.1 Egg

Phlebotomine eggs measure 300–400 µm long × 90–150 µm wide and they have one side flat, the other convex and the poles rounded. They are white when laid, but in a few hours darken to various shades of brown to black, according to the species. In the laboratory, *Phlebotomus longipes* lays an average of 52 eggs in a batch, with a range of 11–95; *L. vexator* matures 70 eggs and *Lutzomyia longipalpis* an average of 80 eggs (maximum 146). The eggs of 13 species, examined by SEM, have sculptured chorions of taxonomic importance, which probably act as plastrons when the eggs are covered by water (Fig. 106).

4.2 Larva

The larva which emerges from the egg passes through four stadia before pupating and, when mature, it is greyish-white with a dark head and no secondary annulations on the body (Fig. 107a). The antennae are small and leaf-like. The thorax is not differentiated from the abdomen, although the abdominal segments bear ventral pseudopods (i.e. unjointed evaginations from the body used for progression). The body segments bear characteristic pinnate hairs, with an unknown function (Fig. 107a). A diagnostic feature of phlebotomine larvae is the possession of two or four long caudal setae (Fig. 107c). First-stage larvae have only one pair of caudal setae, but second- to fourth-stage larvae have two pairs. The larva is amphipneustic, with spiracles opening on the prothorax and the eighth abdominal segment (Fig. 107c). The head (Fig. 107b) bears chewing mouthparts, which the larva uses to feed on decaying organic matter, leaf mould, insect bodies and, when living in animal burrows, faeces of the host animal.

4.3 Pupa

The pupa stands upright, being secured to the substratum by the larval exuviae, which is retained at the end of the abdomen (Fig. 108). Therefore, all the larval setal characters are available for identification of the pupa. The pupa is exarate, with legs and wings free from the body, and it has short prothoracic respiratory horns. Although phlebotomines have terrestrial larvae and pupae, they are very sensitive to desiccation. Pupae of *L. vexator* are independent of free water but require a relative humidity of 75–100% for survival.

4.4 Adult

In the laboratory, *P. longipes* reaches the adult stage from the egg in 6–7 weeks at 25°C, and mass-reared *Phlebotomus papatasi* and *L.*

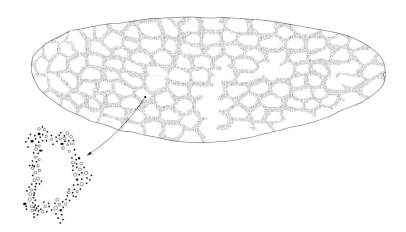

Fig. 106. Phlebotomine egg showing ornamentation on the chorion. *Source*: Abonnenc (1972).

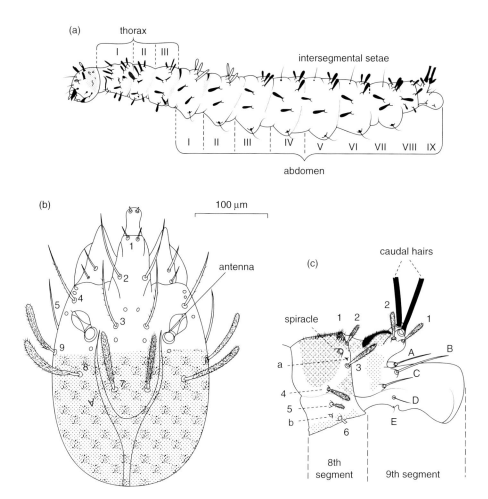

Fig. 107. (a) Lateral view of a phlebotomine larva; (b) dorsal view of head (numbers 1–9 refer to setae); (c) lateral view of terminal abdominal segments. *Source*: Abonnenc (1972).

longipalpis develop from egg to adult in 5–6 weeks. In the absence of diapauses, *P. papatasi* and *Phlebotomus caucasicus* take 6–7 weeks to develop at 25–26°C but 7–9 months if the larvae enter diapause.

5. BEHAVIOUR AND BIONOMICS

5.1 Adult behaviour

The 16-segmented antennae are pilose in both sexes, showing no sexual dimorphism. Males are, however, easily recognized by the possession of large terminalia, of which the claspers (coxite and style) and surstyle are particularly prominent. Absence of plumosity on the male antenna suggests that male phlebotomines do not form aerial swarms in which females are located by sound. Mating has been observed occurring in flight and mating dances of *Phlebotomus orientalis* occur on white surfaces at dusk. Male *P. orientalis* land and run around in all directions, stopping periodically to shake their wings before running on. Females land and behave similarly, until pairing with a male. Presumably, sexual recognition is achieved by pheromones

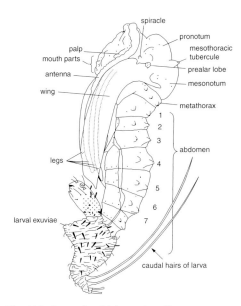

spiracle

pronotum

palp

mouth parts

mesothoracic
tubercule

antenna

prealar lobe

wing

mesonotum

metathorax

1

2

3

abdomen

legs

4

5

6

larval exuviae

7

caudal hairs of larva

Fig. 108. Pupa of a phlebotomine. *Source*: Abonnenc (1972).

emitted when the wings are shaken. Males of some species form two-dimensional swarms spaced out over the surface of the hosts on which the females feed. Male *L. longipalpis* swarm on the backs of humans and cattle, and *Phlebotomus argentipes* on the undersides of standing cattle. Female *L. longipalpis* respond to host cues more quickly and in greater numbers in the presence of the male pheromone, and *P. papatasi* by a pheromone produced by the palps and mouthparts of feeding females. The tropical rainforest *Lutzomyia vespertilionis* and *Lutzomyia ylephiletor* use tree buttresses as swarming sites. Male *L. vexator* emerge earlier than the females and rotation of the male terminalia is completed 12 h after emergence. Mating often occurs shortly after the female has fed, or while still feeding, and males inject spermatozoa directly into the spermathecal ducts, being stored in paired spermathecae of unusual appearance.

5.2 Feeding

As in other bloodsucking Nematocera, only the females are haematophagous but both sexes feed on plant juices. *Phlebotomus ariasi*

feeds on the honeydew produced by the oak-feeding aphis, *Lachanus roboris*, and both sexes of *P. papatasi* pierce leaves and stems to imbibe sap from plants at night. When *L. longipalpis* feeds on free sugar solutions, the ingested fluid passes to the crop, but when sugar solutions are imbibed by piercing a membrane, they pass directly to the midgut. Blood passes directly to the midgut, which consists of a narrow, anterior cardiac portion and a broad, sac-like posterior portion. A peritrophic membrane is secreted around the blood meal and its structure may be important in the development of the pathogenic protozoans of the genus *Leishmania*. In *P. longipes*, the peritrophic membrane develops within 24 h of feeding, reaches a maximum in 48 h, before breaking up on the third day and finally is excreted with the residue of the blood meal after 6–7 days. In arid savannah and desert areas, phlebotomines seek a favourable microclimate and may be found in the burrows of rodents or in termitaria, where females feed on the mammalian occupants, or on hosts in the close vicinity, and lay their eggs. This habit, coupled with the short flight range that is characteristic of the subfamily, leads to local concentrations of phlebotomines and the diseases they transmit, and it accounts for the focality in disease distribution.

Most species are exophilic, but a few have become endophilic in human dwellings and anthropophilic, the latter being the most important in medical and veterinary medicine (e.g. *P. papatasi* in the Mediterranean and Middle East to longitude 80° E, *Phlebotomus sergenti* in Iran, *P. argentipes* in India and *L. longipalpis* in north-east Brazil). However, some subpopulations of sand flies may prefer different habitats, such as two subpopulations of *P. ariasi*, which inhabit oak woodlands (sylvatic population) and houses (domestic population), respectively.

Most species are unable to develop eggs without a blood meal (e.g. *P. longipes*; *Lutzomyia californica*, *L. longipalpis* and *Lutzomyia trapidoi*), but autogeny has been demonstrated in species such as *Phlebotomus mascittii*, *P. papatasi*, *Lutzomyia beltrani*, *Lutzomyia lichyi*, *Lutzomyia gomezi*, *Lutzomyia shannoni* and *Lutzomyia cruciata*.

L. *longipalpis* feeds on a wide range of hosts, but shows a preference for larger animals. Chickens also represent an important attractant for phlebotomine sand flies such as L. *longipalpis* (the main vector of *Leishmania infantum*) and it has been reported that people cohabiting with chickens are about four times more likely to be infected by L. *infantum* than people living in the same area but not in close association with chickens. However, chickens are not suitable hosts for L. *infantum* under experimental conditions.

In rural Ethiopia, P. *longipes* is exophilic and a cattle feeder, but in Addis Ababa it has become endophilic, attacking people near cattle sheds. *Phlebotomus pedifer*, a sibling species of P. *longipes*, is a cave-dwelling species which, on Mt Elgon in Kenya, was found to feed largely (87%) on cattle that entered the cave, whereas most of the remaining feeds (9%) were taken from hyrax.

P. *orientalis* in Ethiopia is exophilic but markedly anthropophilic, and *Lutzomyia wellcomei* in Brazil is largely anthropophilic (65%) but also feeds regularly on rodents (25%). Another forest-dwelling phlebotomine, *Lutzomyia flaviscutellata* is zoophilic, feeding on ground-dwelling rodents of the genera *Oryzomys* and *Proechimys*. P. *caucasicus* lives in the burrows of the giant gerbil (*Rhombomys opimus*) on which it feeds.

On humans, phlebotomines feed on exposed areas of skin and these are the sites for the development of the ulcers of cutaneous leishmaniasis. Feeding is mainly nocturnal and crepuscular. In Belize, biting of ten species of *Lutzomyia* was at its highest between 18:00 and 24:00 h, and a few species continued biting until 06:00 h. The nocturnal feeding cycle of P. *argentipes* was concentrated around midnight (22:00–03:00 h) in Bengal, but in Sri Lanka it was mainly after midnight (02:00–06:00 h). In Ethiopia, the feeding of P. *orientalis* on humans began after sunset and rapidly reached a peak, from which it declined and ceased when the temperature reached 16°C, which in the highlands was rarely more than 4 h after sunset. Some species, for example P. *papatasi*, will feed during the day under shaded conditions indoors and others will feed if disturbed from their resting places. Adult L. *longipalpis*

spend the major part of their lives in animal sheds. Males spend about 2 days in a shed before leaving and are the first to colonize new sheds, to which females are attracted by a sex pheromone emitted by the males. Initially, more females are attracted to the shed than males, but a stable population is reached after about 2 weeks, with males outnumbering females.

5.3 Breeding sites

Females lay their eggs usually in organically-rich moist sites, such as in the contaminated soil of rodent or other animal burrows or shelters in more arid areas, in leaf litter on the forest floor and around the bases of forest trees in rainforests, as well as in rock piles, caves and termite mounds. The eggs of L. *longipalpis* have a soluble pheromone on their surface, which attracts gravid females to oviposit and remains active for at least 6 days, but, to be effective, more than 80 eggs have to be present. In the laboratory, extracts of commercial rabbit food, rabbit faeces and oviposition pheromone were equally attractive to ovipositing females, with significantly more eggs laid on extract combining oviposition pheromone and rabbit food. P. *papatasi* breeds in a range of sites, with greatest numbers being reared from the burrows of R. *opimus* and cattle sheds (more than 1000 individuals/site) and fewer from burrows of *Meriones erythrourus* and unoccupied storerooms (100–250/site). In Egypt, the breeding sites of *Phlebotomus langeroni* and P. *papatasi* were characterized by being rich in organic matter, having a high moisture content, a high percentage of silt and a pH of about 7.5. In Panama, L. *trapidoi* was the dominant species and favoured hillsides and streams in the vicinity of large lianas (*Orouparia* and *Sabicea*), while *Lutzomyia carrerai carrerai* was associated with hilltops and larger trees (e.g. *Anacardium*).

5.4 Seasonality and longevity

The slow rate of sand fly development limits the number of generations which may be produced each year. Palaearctic species are often bivoltine, for example early summer eggs of *P. papatasi* and *P. caucasicus* develop without diapause, but late summer eggs give rise to diapausing larvae, thus producing two generations per annum. In cooler climates, species overwinter in the larval stage, which may be a true diapause or a temperature-induced cessation of growth. Perhaps as an adaptation to survival in a hostile environment, emergence of a generation may be extended by delaying hatching of eggs and emergence of adults from pupae. The abundance of *Lutzomyia* species seems to be dependent on rainfall and temperature. The numbers of *L. flaviscutellata* were found to decline during the rain season (January–May) and then increase to reach a peak in December–January. Diapausing eggs have also been reported.

P. papatasi is expected to survive for up to six ovarian cycles under field conditions, while *P. ariasi* for one and a half ovarian cycles. Infected individuals of the latter species survive at least 29 days.

5.5 Flight range and dispersal

Phlebotomines rest by day in dark, cool, humid niches where the microclimate is favourable for survival (e.g. *L. beltrani* rests in caves). *Phlebotomus longipes* occurs in a variety of cavities, including caves, tree holes and burrows. Three closely related species of the subgenus *Paraphlebotomus* are associated with rodent burrows in the former Soviet Union. *Phlebotomus mongolensis* is found in oases where the conditions are relatively cool and humid, *P. caucasicus* in loess desert and *Phlebotomus andrejevi* in very hot sandy deserts.

Phlebotomines have a characteristic hopping flight and, if disturbed, quickly settle again a short distance away. In the open, they are very sensitive to wind speed, thus fly close to the ground and feed only under near calm conditions. Even in the forest, 80% of the blood meals of *Lutzomyia olmeca olmeca* are taken near the ground, while *L. trapidoi* rests in the lower layers by day and ascends into the canopy at night to feed on arboreal vertebrates. However, *Lutzomyia trinidadensis* takes 60%

of its feeds at 13 m and less than 5% of them at ground level.

The flight range of phlebotomines is short, usually 100–200 m, but they may disperse up to 1 km or more. Mark–release–recapture experiments showed that unengorged *P. ariasi* dispersed more than 1 km from the point of release and, 3 days after release, one female was captured 2.2 km away. Engorged females remained more local and males dispersed no more than 600 m. In a coastal desert area of North Africa, *P. papatasi* dispersed a maximum of 1.5 km. In two experiments, the average distance travelled by unfed females was greater (820 m) compared with fed females (620 m) and males (600 m). In a semi-arid valley in Colombia, *L. longipalpis* females dispersed a maximum of 960 m, but only a small percentage dispersed more than 470 m and males only averaged 95 m. When 20,000 marked *L. trapidoi* were released at ground and canopy (30 m) levels in forest, 90% of those recaptured were taken within 57 m of the point of release and only four individuals were collected at the limit of observations (200 m).

6. MEDICAL AND VETERINARY IMPORTANCE

Phlebotomines are rarely present in sufficient density to reach pest proportions and their medical and veterinary relevance is based on their role as vectors of various pathogens, the most important of which are species of *Leishmania* causing human cutaneous, visceral and mucocutaneous leishmaniases. These diseases are widely, but patchily, distributed throughout the warmer areas of the world. They are mostly zoonoses in which people become involved by entering the focus of the disease. Some species of *Leishmania* also infect dogs, and domestic foci of disease may develop in association with endophilic vectors.

Other pathogens of which phlebotomines are vectors include bacteria causing bartonellosis, a disease of humans living in certain high-altitude valleys in the Andes of South America, and sand fly or papatasi (from Italian 'pappataci') fever virus. Phlebotomines are also implicated in the transmission of vesicular stomatitis, a virus disease of cattle and horses.

6.1 Arboviruses

6.1.1 Bunyaviridae

The species of Bunyavirus are primarily mosquito-borne and tick-borne but they also include Phleboviruses transmitted mainly by phlebotomine sand flies, of which about two-thirds are transmitted by *Lutzomyia* spp. in the Neotropical region and the other third by *Phlebotomus* spp. in the Palaearctic and Afrotropical regions.

SAND FLY (PHLEBOTOMUS) FEVER (SF) VIRUSES
Eight phlebotomine-transmitted viruses are associated with this human disease, in the Old World with SFS (Sicilian strain), SFN (Neapolitan strain), Toscana (TOS) and Karimabad (KAR) viruses in southern Europe and northern Africa to eastern Asia, and in the New World with Alenquer (ALE), Candiru (CRU), Chagres (CHG) and Punta Toro (PTV) viruses in Central and South America. They typically cause an influenza-like disease with short, sharp, non-fatal fever, although a TOS strain has caused meningitis and meningoencephalitis in Italy and is known also from Spain and Portugal. Vectors include *P. papatasi*, *Phlebotomus perniciosus* and *Phlebotomus perfiliewi* in the Old World and *L. trapidoi* and *L. ylephiletor* in the New World. The viruses have an extrinsic (sand fly host) cycle of 7–10 days and appear to be maintained vertically in sand fly populations by transovarial transmission. It has been thought generally that there is not an animal reservoir, but a number of animals (e.g. sheep, gerbils and bats) have been shown to be seropositive, suggesting that a wildlife cycle may exist. There is no specific treatment for the usually relatively mild disease, and personal protection (e.g. topical repellents, bed nets) against sand fly bites is advised to prevent infection in endemic regions.

6.1.2 Rhabdoviridae

The important arboviruses in this family are the vesicular stomatitis (VS) viruses, which cause a major disease of horses, cattle and pigs in the Americas. This infection is caused by four viruses: VSI (Indiana serotype), VSNJ (New Jersey serotype), VSA (Alagoas serotype) and VSC (Cocal serotype). Although VSNJV is primarily an animal pathogen, humans can be infected. The morbidity rate in herds is usually low and there is no mortality. This condition superficially resembles foot-and-mouth disease but is much less serious. The mode of transmission during epizootics is unclear but VSV multiplies in *L. trapidoi*, which can transmit the virus transovarially to the next generation. Transovarial transmission also occurs in *L. ylephilator* but not in *Lutzomyia sanguinaria* or *L. gomezi*. There is evidence that the epizootic in cattle in Colorado in 1982 was an extension of an enzootic cycle in elk (*Cervus elephas*) and mule deer (*Odoicoleus hemionus*).

6.2 Bartonellosis

Members of the Bartonellaceae are polymorphic, often rod-shaped bacteria which are distinguished from the Anaplasmataceae by structural characteristics. *Bartonella bacilliformis* is found on or in the erythrocytes and in the cytoplasm of endothelial cells of humans, to whom it is highly pathogenic. It is also found in the phlebotomine sand fly, *Lutzomyia verrucarum*.

B. bacilliformis is restricted to certain high mountain valleys in the western and central Cordilleras of the Andes in South America and is unusual in that it produces two strikingly different human diseases: a progressive anaemia with high mortality (40%), referred to as Oroya fever, and a benign cutaneous eruption known as 'Verruga peruana'. These two conditions are referred to jointly as Carrion's disease, in memory of Carrion, who in 1885 inoculated himself with the organisms of 'Verruga peruana' and developed and died from Oroya fever, thus proving tragically and dramatically that the same organism caused both diseases.

Carrion's disease is an anthroponosis, no animal reservoir host is known and transmission is via the phlebotomine sand fly, *L. verrucarum*, which is nocturnal in habit. *B.*

bacilliformis is in the circulating blood and is picked up by *L. verrucarum* when it feeds on an infected human. The cycle of *B. bacilliformis* development in the sand fly is unknown, but the pathogen has been isolated from its gut and mouthparts. The disease is endemic in certain mountain valleys between 750 and 2750 m altitude, where it has been known to persist for 300 years. Above that height, the night temperatures are too low for *L. verrucarum* to be active and, below 750 m, the climate is too arid for its survival (although that does not explain the absence of *L. verrucarum* and Carrion's disease from well-watered riverine habitats below 750 m).

6.3 Leishmaniasis

Leishmaniasis is a disease of humans caused by infection with species of *Leishmania* in which the parasites are intracellular amastigotes in the reticuloendothelial cells. They are transmitted from host to host by the bites of phlebotomines in which the parasites are motile and extracellular. In the Old World, the vectors are species of *Phlebotomus* and, in the New World, species of *Lutzomyia*. Clinical leishmaniasis takes three main forms – visceral, cutaneous and mucocutaneous. In the New World, the diseases are zoonotic, involving reservoir hosts, while in the Old World, some are anthroponoses with no animal reservoir. The only known exception to the involvement of phlebotomids as vectors for *Leishmania* parasites seems to be the recent discovery in Australia of a *Leishmania* species infecting kangaroos that appears to be vectored by a species of Ceratopogonidae (for details see the entry on 'Biting Midges').

A range of criteria is used to separate species of *Leishmania* which are morphologically very similar, including biological, immunological, geographical, clinical, behavioural, molecular and morphological characteristics. Two subgenera infecting humans are recognized – *Leishmania* and *Viannia* – on the basis of where the protozoan develops in the gut of the phlebotomine vector (a third subgenus, *Sauroleishmania*, infects reptiles as their vertebrate hosts). Development of *Leishmania* (*Leishmania*) species is supra-

pylarian, being confined to the midgut and foregut. In *Leishmania* (*Viannia*) species, development is peripylarian, with a prolific and prolonged phase of development occurring in the hindgut (pylorus), followed by migration to the midgut and foregut.

The developmental stages of *Leishmania* in the sand fly vector and the mechanism of transmission have been the subject of some debate for many years. For the *Leishmania* (*Leishmania*) species in the Old World, when a phlebotomine feeds on an infected host, it ingests amastigotes with the blood meal, some of which will divide before becoming procyclic promastigotes, which multiply and quickly become the predominant form in the midgut. Then, two further forms of promastigotes are recognized: long, slender nectomonad promastigotes that gather at the anterior end of the blood meal, break out and move towards the stomodeal (proventricular) valve at the junction of the midgut and foregut. At this point, they transform into shorter replicative leptomonad promastigotes which are responsible for the secretion of a promastigote secretory gel (PSG). Some of the nectomonad promastigotes transform into a broad haptomonad promastigote with flagellae and attach to the cuticular lining of the stomodeal valve.

With the *Leishmania* (*Viannia*) species in the New World, the initial developmental events are similar to the above but, following replication, the procyclic promastigotes are found predominantly in the pyloric region of the hindgut, where they attach as haptomonad promastigotes to the cuticle of the hindgut, although some are thought to go forward to the anterior region of the midgut. Thereafter, they move forward to gather in the anterior midgut, secrete PSG and transform into metacyclic promastigotes (much as occurs in *Leishmania* (*Leishmania*) species).

For both subgenera, it is now thought that transmission occurs when the plug of PSG is egested (regurgitated) into the skin of the vertebrate host during the fly's attempts at feeding and the associated metacyclic promastigotes are similarly inoculated. It seems that the sand fly becomes 'blocked' by the 'gel plug' secreted by the promastigotes in the anterior midgut, thus forcing the fly to expel the 'plug' before it can feed, thereby depositing it and infective metacyclic promastigotes into the skin of the vertebrate host.

Infection rates reported from wild-caught phlebotomines have varied from 0 to 15.4%, the latter being found in *L. trapidoi*. These infections will include both pathogenic and non-pathogenic leishmaniae. Dissection of substantial numbers of *P. orientalis* in the Sudan and *P. longipes* in Ethiopia gave infection rates of 2.4 and 3.1%, respectively. Infection rates in *P. papatasi* have ranged from 0.2 to 8.7% and in *P. caucasicus* from 2.6 to 10.5%. Infection rates in *Lutzomyia* spp. are usually below 1%.

6.3.1 Visceral leishmaniasis

Visceral leishmaniasis, known also as 'kala-azar', is caused by infection with *Leishmania donovani*. Symptoms may be highly variable, depending on the individual immune responsiveness. Hepatosplenomegaly, weight loss and anaemia may occur, leading to death in untreated established cases. Vectors of *L. donovani* include *P. argentipes* in India, where humans are the reservoir, and *P. orientalis* and *Phlebotomus martini* in East Africa, where the reservoir hosts are uncertain. *L. infantum* is a zoonosis of dogs and *P. perniciosus* and *Phlebotomus neglectus* are the main vectors in the central and western Mediterranean basin. In France, the vector is *P. ariasi* and the parasite is distributed by movement of infected dogs and by *P. ariasi* itself, which has been found to disperse up to 750 m. In the Neotropical region, *L. infantum* (syn. *L. chagasi*) is a zoonosis of dogs and foxes, with the vector being *L. longipalpis*. In recent decades, there has been a resurgence of visceral leishmaniasis associated with HIV infections in southern Europe and this appears to be associated with the use of contaminated syringe needles shared for illegal drug use.

6.3.2 Cutaneous and mucocutaneous leishmaniasis

Cutaneous leishmaniasis involves nodular and ulcerative skin lesions. In the Old World, these lesions may heal completely in a few months to a few years. In the New World, such infections are more dangerous and in 3–5% of cases result in mucocutaneous leishmaniasis ('espundia') with destructive nasopharyngeal lesions, which can result in death. This development usually occurs within 2 years but can occur up to 30 years after the primary infection. Other cutaneous manifestations of infection are leishmaniasis recidivans, a chronic, drug-resistant infection found mainly in Iraq and Iran, which may persist for 20–40 years and, in India, post kala-azar dermal leishmaniasis (PKDL) in people cured of kala-azar from 2–10 years earlier.

OLD WORLD CUTANEOUS LEISHMANIASIS

Leishmania tropica is associated with dry cutaneous leishmaniasis and is an anthroponosis of urban areas. It is found predominantly in densely populated settlements from Greece, through Turkey and the Middle East, to India. Person-to-person transmission is maintained by *P. sergenti*.

Leishmania major is associated with wet cutaneous leishmaniasis and is a zoonosis of rural areas. It occurs along the North African coast, in the arid inland region of West Africa, in the Middle East and Arabian Peninsular and east and south of the Caspian and Aral Seas, where it parasitizes ground-dwelling rodents such as the giant gerbil, *Rhombomys opimus*, among which the main vector is the strongly zoophilic *P. caucasicus*. The disease is spread to humans by *P. papatasi*, which is markedly, but not exclusively, anthropophilic. *Leishmania aethiopica* is a parasite of hyraxes (*Procavia*, *Heterohyrax*, *Dendrohyrax*) transmitted by *P. longipes* and *P. pedifer*. These sand flies feed equally easily on hyraxes and cattle, and humans are bitten and become infected when they associate closely with the sand flies' main hosts.

Mechanical transmission can occur via inoculation of infected material from lesions, and this has been used in parts of South-west Asia to immunize young girls in an inconspicuous place in order to avoid future disfiguring lesions.

NEW WORLD CUTANEOUS LEISHMANIASIS

In humans, the suprapylarian leishmaniae, *Leishmania mexicana* and *Leishmania amazonensis*, produce mild cutaneous lesions; they are parasites of forest rodents and opossums and are transmitted by species of the *L. flaviscutellata* group.

L. mexicana infects the log cutters and collectors of chicle (chewing gum latex) who work in the forest for periods of about 6 months during the rainy season. The main host is the rodent *Ototylomys phyllotis* and the vector is *L. olmeca olmeca*, which is not strongly attracted to humans, except when its daytime resting places in the forest floor leaf litter are disturbed. By contrast, *L. amazonensis*, whose main hosts are species of *Proechimys*, rarely causes human infections because the vector *L. flaviscutellata* is not very anthropophilic.

Human infections with *Leishmania braziliensis* are associated with jungle activities, particularly land clearing, when the risk of infection may be very high (70–80%). It also occurs in forest remnants close to human populations. The primary cutaneous lesions may be followed by mucocutaneous disease, while the primary lesion is still active or may occur many years after the primary lesion has disappeared. There is no proven reservoir host and many species of *Lutzomyia* (e.g. *Lutzomyia spinicrassa*, *L. carrerai carrerai*, *L. welcomei* and *Lutzomyia whitmani*) have been implicated as potential vectors.

Leishmania peruviana is a member of the *L. braziliensis* complex, which causes self-healing skin lesions in humans in the high (1200–3000 m) Peruvian and Ecuadorean Andes. The condition is known as 'uta'. Dogs are probably the most likely peridomestic reservoir, but infections also occur in wild rodents (*Phyllotis andinum*, *Akodon* spp.). The most likely vectors are *Lutzomyia peruensis* and *L. verrucarum*, which have been found naturally infected.

Two other leishmanial infections associated with humans involved in forest activities are caused by *Leishmania guyanensis* and *Leishmania panamensis*. Infection rates of 60–90% have been recorded in small, exposed populations. *L. guyanensis* is rarely involved in mucocutaneous leishmaniasis, but it has a tendency to develop multiple ulcers by secondary spread via the lymphatic system from a single lesion. Infections of *L. panamensis* may produce a chain of enlarged nodes along efferent lymph channels. Its

cutaneous lesions are slow to heal, may persist for more than 10 years and infections can proceed to mucocutaneous leishmaniasis. The reservoirs for both are thought to be sloths and opossums and the vectors include *Lutzomyia umbratilis* and *Lutzomyia anduzei* for *L. guyanensis* and *L. trapidoi* and *L. ylephiletor* for *L. panamensis*.

7. PREVENTION AND CONTROL

Phlebotomines do not disperse widely and avoidance of infested localities can be a useful practical measure to reduce human leishmaniasis. In some circumstances, measures can be directed against the vector and/or the reservoir. Unfortunately, however, the breeding places of most sand fly species are not understood and defined well enough for control programmes to be directed against the larval stages. Consequently, the adult fly is often targeted in control strategies, which consist of the application of residual insecticides in resting sites such as animal shelters and houses where peridomestic species (such as *P. papatasi* and *L. longipalpis*) are present. With the exception of situations where local dogs serve as the hosts, management options for the treatment of reservoirs are generally not feasible. Environmental modification of sand fly habitats or management of parasite hosts may be useful in some circumstances, but not usually when the vectors or hosts are sylvan.

In the absence of control measures, reliance must be placed on personal protection, and the use of bed nets (untreated or treated with insecticides), insecticide coils and diffusers can be advised when such sand flies are a problem for local residents, while topical repellents may help protect visitors to problem areas.

Where the vector is endophilic and transmission occurs within households, spraying of houses with a residual insecticide can reduce human leishmaniasis. However, when spraying ceases, the emergence of new cases has been observed. Spraying of fences, external walls and peripheral vegetation may provide some level of protection in residential areas (barrier spraying).

Several insecticide formulations applied directly to dogs (using impregnated collars or spot-on formulations) have been proposed to aid the control of canine leishmaniasis caused by *L. infantum*. The use of repellents and/or insecticides, generally containing synthetic pyrethroids alone or in combination with other chemicals with repellent and/or insecticide properties, is still the most effective strategy to reduce the risk of infection in dogs in endemic areas. In some circumstances, attempts at controlling *L. infantum* infections have involved culling of stray and feral dogs. However, besides being controversial, this strategy has been proved ineffective in reducing the incidence of the human disease in Brazil.

Local control of *L. major* has been achieved in parts of Central Asia by applying insecticides and rodenticides to gerbil burrows to control both the reservoir hosts and the vectors, but the possibility of reducing human leishmaniasis by controlling the rodent reservoir needs to be evaluated carefully. While good results were obtained in the former Soviet Union by the complete destruction of *Rhombomys* colonies, where fat sand rats (*Psammomys* spp.) were the reservoir hosts, successful control was difficult and it seemed more practical to locate new human settlements away from *Psammomys* colonies.

In forest situations, reliance must be on personal protection involving the use of repellents, wearing of long-sleeved shirts and trousers, sleeping under bed nets and avoiding infested areas at the time when the vectors are active.

Treatment of infections has often used sodium stibogluconate or meglumine antimonite, but pentamidine isothionate or liposomal amphotericin B can be required in non-responding cases (particularly with mucocutaneous lesions), while miltefosine is useful for visceral cases and paromomycin is also safe and effective (particularly when given in combination with sodium stibogluconate).

SELECTED BIBLIOGRAPHY

Alencar, R.B., de Queiroz, R.G. and Barrett, T.V. (2011) Breeding sites of phlebotomine sand flies (Diptera: Psychodidae) and efficiency of extraction techniques for immature stages in terra-firme forest in Amazonas State, Brazil. *Acta Tropica* 118, 204–208.

Alexander, B. (1995) A review of bartonellosis in Ecuador and Columbia. *American Journal of Tropical Medicine and Hygiene* 52, 354–359.

Alexander, B. (2000) Sampling methods for phlebotomine sandflies. *Medical and Veterinary Entomology* 14, 109–122.

Alexander, B. and Maroli, M. (2003) Control of phlebotomine sandflies. *Medical and Veterinary Entomology* 17, 1–18.

Amora, S.S., Bevilaqua, C.M. and Feijo, F.M. (2009) Control of phlebotomine (Diptera: Psychodidae) leishmaniasis vectors. *Neotropical Entomology* 38, 303–310.

Ashford, R.W. (1996) Leishmaniasis reservoirs and their significance in control. *Clinics in Dermatology* 14, 523–532.

Bates, P.A. (2007) Transmission of *Leishmania* metacyclic promastigotes by phlebotomine sand flies. *International Journal for Parasitology* 37, 1097–1106.

Bates, P.A. and Ashford, R.W. (2006) Old World Leishmaniasis. In: Cox, F.E.G., Wakelin, D., Gillespie, S.H. and Despommier, D.D. (eds) *Topley and Wilson's Microbiology and Microbial Infections, 10th edn, Parasitology*. Hodder Arnold, London, pp. 283–312.

Bates, P.A. and Rogers, M.E. (2004) New insights into the developmental biology and transmission mechanisms of *Leishmania*. *Current Molecular Medicine* 4, 601–609.

Comer, J.A. and Tesh, R.B. (1991) Phlebotomine sand flies as vectors of vesiculovirus: a review. *Parassitologia* 33 (Suppl. 1), 143–150.

Courtenay, O., Gillingwater, K., Gomes, P.A.F., Garcez, L.M. and Davies, C.R. (2007) Deltamethrin-impregnated bednets reduce human landing rates of sandfly vector *Lutzomyia longipalpis* in Amazon households. *Medical and Veterinary Entomology* 21, 168–176.

Debboun, M. and Strickman, D. (2012) Insect repellents and associated personal protection for a reduction in human disease. *Medical and Veterinary Entomology*, doi: 10.1111/j.1365-2915.2012.01020.x

Depaquit, J., Grandadam, M., Fouque, F., Andry, P.E. and Peyrefitte, C. (2010) Arthropod-borne viruses transmitted by Phlebotomine sandflies in Europe: a review. *Eurosurveillance* 15, 40–47.

Desjeux, P. (2004) Leishmaniasis: current situation and new perspectives. *Comparative Immunology Microbiology and Infectious Diseases* 27, 305–318.

Desjeux, P. and Alvar, J. (2003) Leishmania/HIV co-infections: epidemiology in Europe. *Annals of Tropical Medicine and Parasitology* 97, Supplement 1, S3–15.

Dougall, A.M., Alexander, B., Holt, D.C., Harris, T., Sultan, A.H., Bates, P.A., *et al.* (2011) Evidence incriminating midges (Diptera: Ceratopogonidae) as potential vectors of *Leishmania* in Australia. *International Journal for Parasitology* 41, 571–579.

Dujardin, J.C., Campino, L., Canavate, C., Dedet, J.P., Gradoni, L., Soteriadou, K., *et al.* (2008) Spread of vector-borne diseases and neglect of leishmaniasis, Europe. *Emerging Infectious Diseases* 14, 1013–1018.

Elnaiem, D.A. (2011) Ecology and control of the sand fly vectors of *Leishmania donovani* in East Africa, with special emphasis on *Phlebotomus orientalis*. *Journal of Vector Ecology* 36 (Supplement 1), S23–31.

Faiman, R., Cuno, R. and Warburg, A. (2009) Control of phlebotomine sand flies with vertical fine-mesh nets. *Journal of Medical Entomology* 46, 820–831.

Faiman, R., Kirstein, O., Freund, M., Guetta, H. and Warburg, A. (2011) Exclusion of phlebotomine sand flies from inhabited areas by means of vertical mesh barriers. *Transactions of the Royal Society of Tropical Medicine and Hygiene* 105, 512–518.

Feliciangeli, M.D. (2004) Natural breeding places of phlebotomine sandflies. *Medical and Veterinary Entomology* 18, 71–80 (Corrigendum in 18, 453–454).

Freitas, V.C., Parreiras, K.P., Duarte, A.P.M., Secundino, N.F.C. and Pimento, P.F.P. (2012) Development of *Leishmania* (*Leishmania*) *infantum chagasi* in its natural sandfly vector *Lutzomyia longipalpis*. *American Journal of Tropical Medicine and Hygiene* 86, 606–612.

Garcia, A.L., Parrado, R., Rojas, E., Delgado, R.. Dujardin, J.C. and Reithinger, R. (2009) Leishmaniases in Bolivia: comprehensive review and current status. *American Journal of Tropical Medicine and Hygiene* 80, 704–711

Gavgani, A.S., Hodjati, M.H., Mohite, H. and Davies, C.R. (2002) Effect of insecticide-impregnated dog collars on incidence of zoonotic visceral leishmaniasis in Iranian children: a matched-cluster randomised trial. *Lancet* 360, 374–379.

Hide, G., Mottram, J.C., Coombs, G.H. and Holmes, P.H. (1997) *Trypanosomiasis and Leishmaniasis. Biology and Control.* CAB International, Wallingford, UK.

Ilango, K. (2010) A taxonomic reassessment of the *Phlebotomus argentipes* species complex (Diptera: Psychodidae: Phlebotominae). *Journal of Medical Entomology* 47, 1–15.

Killick-Kendrick, R. (1978) Recent advances and outstanding problems in the biology of phlebotomine sandflies. *Acta Tropica* 35, 297–313.

Killick-Kendrick, R. (1990) Phlebotomine vectors of the leishmaniases: a review. *Medical and Veterinary Entomology* 4, 1–24.

Killick-Kendrick, R. (1999) The biology and control of phlebotomine sand flies. *Clinical Dermatology* 17, 279–289.

Klowden, M.J. (ed.) (2011) Special Issue: Sand fly research and control. *Journal of Vector Ecology* 36, Supplement 1.

Lainson, R. and Rangel, E.F. (2005) *Lutzomyia longipalpis* and the eco-epidemiology of American visceral leishmaniasis, with particular reference to Brazil – a review. *Memorias do Instituto Oswaldo Cruz* 100, 811–827.

Lainson, R. and Shaw, J.J. (2006) New World Leishmaniasis. In: Cox, F.E.G., Wakelin, D., Gillespie, S.H. and Despommier, D.D. (eds) *Topley and Wilson's Microbiology and Microbial Infections, 10th edn, Parasitology*. Hodder Arnold, London, pp. 313–349.

Lambert, M., Dereure, J., El-Safi, S.H., Bucheton, B., Dessein, A., Boni, M., *et al.* (2002) The sandfly fauna in the visceral-leishmaniasis focus of Gedaref, in the Atbara-River area of eastern Sudan. *Annals of Tropical Medicine and Parasitology* 96, 631–636.

Lane, R.P. (1993) Sandflies (Phlebotominae). In: Lane, R.P. and Crosskey, R.W. (eds) *Medical Insects and Arachnids*. Chapman and Hall, London, pp. 78–109.

Lewis, D.J. (1973) Phlebotomidae and Psychodidae. In: Smith, K.G.V. (ed.) *Insects and Other Arthropods of Medical Importance*. British Museum (Natural History), London, pp. 155–179.

Lewis, D.J. (1974) The biology of the Phlebotomidae in relation to leishmaniasis. *Annual Review of Entomology* 19, 363–384.

Lewis, D.J. (1982) A taxonomic review of the genus *Phlebotomus* (Diptera: Psychodidae). *Bulletin of the British Museum of Natural History (Entomology)* 45, 121–209.

Lewis, D.J., Young, D.G., Fairchild, G.B. and Minter, D.M. (1977) Proposals for a stable classification of the phlebotomine sandflies (Diptera: Psychodidae). *Systematic Entomology* 2, 319–332.

Magill, A.J. (1995) Epidemiology of leishmaniasis. *Dermatology Clinics* 13, 505–523.

Maroli, M., Ciufolini, M.G. and Verani, P. (1993) Vertical transmission of *Toscana* virus in the sandfly *Phlebotomus perniciosus*, via the second gonotrophic cycle. *Medical and Veterinary Entomology* 7, 283–286.

Maroli, M., Mizzoni, V., Siragusa, C., D'Orazi, A. and Gradoni, L. (2001) Evidence for an impact on the incidence of canine leishmaniasis by the mass use of deltamethrin-impregnated dog collars in southern Italy. *Medical and Veterinary Entomology* 15, 358–363.

Maroli, M., Feliciangeli, M.D., Bichaud, L., Charrel, R.N. and Gradoni, L. (2012) Phlebotomine sand flies and spreading of leishmaniases and other diseases of Public Health concern. *Medical and Veterinary Entomology* in press.

Molyneux, D.H. and Ashford, R.W. (1983) *The Biology of Trypanosoma and Leishmania, Parasites of Man and Domestic Animals*. Taylor and Francis, London, 294 pp.

Molyneux, D.H. and Killick-Kendrick, R. (1987) Morphology, ultrastructure and life-cycles. In: Peters, W. and Killick-Kendrick, R. (eds) *The Leishmaniases in Biology and Medicine*, Vol I. Academic Press, London, pp. 121–176.

Monath, T.P. (ed.) (1988) *The Arboviruses: Ecology and Epidemiology*. Vol. IV. CRC Press, Boca Raton, Florida.

Monath, T.P. (ed.) (1989) *The Arboviruses: Ecology and Epidemiology*. Vol. V. CRC Press, Boca Raton, Florida.

Oliveira, A.G., Galati, E.A.B., Fernandes, C.E., Dorval M.E.C. and Brazil, R.E. (2012) Ecological aspects of phlebotomines (Diptera: Psychodidae) in endemic area of visceral leishmaniasis, Campo Grande, State of Mato Grosso do Sul, Brazil. *Journal of Medical Entomology* 49, 43–50.

Orshan, L., Szekely, D., Khalfa, Z. and Bitton, S. (2010) Distribution and seasonality of *Phlebotomus* sand flies in cutaneous leishmaniasis foci, Judean Desert, Israel. *Journal of Medical Entomology* 47, 319–328.

Otranto, D., Testini, G., Buonavoglia, C., Parisi A., Brandonisio, O., Circella, E., *et al.* (2010) Experimental and field investigations on the role of birds as hosts of *Leishmania infantum*, with emphasis on the domestic chicken. *Acta Tropica* 113, 80–83.

Oumeish, O.Y. and Parish, L.C. (ed) (1999) Leishmaniasis. *Clinics in Dermatology* 17, 247–344.

Pech-May, A., Escobedo-Ortegon, F.J., Berzunza-Cruz, M. and Rebollar-Tellez, E.A. (2010) Incrimination of four sandfly species previously unrecognized as vectors of *Leishmania* parasites in Mexico. *Medical and Veterinary Entomology* 24, 150–161.

Pinto, I.S., Andrade Filho, J.D., Santos, C.B., Falqueto, A. and Leite, Y.L.R. (2010) Phylogenetic relationships among species of *Lutzomyia*, subgenus *Lutzomyia* (Diptera: Psychodidae). *Journal of Medical Entomology* 47, 16–21.

Pratlong, F., Dereure, J., Ravel, C., Lami, P., Balard, Y., Serres, G., *et al.* (2009) Geographical distribution and epidemiological features of Old World cutaneous leishmaniasis foci, based on the isoenzyme analysis of 1048 strains. *Tropical Medicine and International Health* 14, 1071–1085.

Ready, P.D. (2011) Should sand fly taxonomy predict vectorial and ecological traits? *Journal of Vector Ecology* 36 (Supplement 1), S17–22.

Reithinger, R., Coleman, P.G., Alexander, B., Vieira, E.P., Assis, G. and Davies, C.R. (2004) Are insecticide-impregnated dog collars a feasible alternative to dog culling as a strategy for controlling canine visceral leishmaniasis in Brazil? *International Journal for Parasitology* 34, 55–62.

Reithinger, R., Brooker, S. and Kolaczinski, J.H. (2007) Visceral leishmaniasis in eastern Africa – current status. *Transactions of the Royal Society of Tropical Medicine and Hygiene* 101, 1169–1170.

Service, M.W. (ed.) (2001) *Encyclopedia of Arthropod-transmitted Infections*. CAB International, Wallingford, UK.

Sharma, U. and Singh, S. (2008) Insect vectors of *Leishmania*: distribution, physiology and their control. *Journal of Vector Borne Diseases* 45, 255–272.

Shaw, J.J. and Lainson, R. (1987) Ecology and epidemiology: New World. In: Peters, W. and Killick-Kendrick, R. (eds) *The Leishmaniases in Biology and Medicine*, Vol I. Academic Press, London, pp. 291–363.

Tabbabi, A., Ghrab, J., Aoun, K., Ready, P.D. and Bouratbine, A. (2011) Habitats of the sandfly vectors of *Leishmania tropica* and *L. major* in a mixed focus of cutaneous leishmaniasis in southeast Tunisia. *Acta Tropica* 119, 131–137.

Tarallo, V.D., Dantas-Torres, F., Lia, R.P. and Otranto, D. (2010) Phlebotomine sand fly population dynamics in a leishmaniasis endemic peri-urban area in southern Italy. *Acta Tropica* 116, 227–234.

Tesh, R.B. (1988) The genus *Phlebovirus* and its vectors. *Annual Review of Entomology* 33, 169–181.

Warburg, A. and Faiman, R. (2011) Research priorities for the control of phlebotomine sand flies. *Journal of Vector Ecology* 36 (Supplement 1), S10–16.

Ward, R.D. (1990) Some aspects of the biology of phlebotomine sandfly vectors. *Advances in Disease Vector Research* 6, 93–126.

Williams, P. (1970) Phlebotomine sandflies and leishmaniasis in British Honduras (Belize). *Transactions of the Royal Society of Tropical Medicine and Hygiene* 64, 317–364.

Scorpions (Arachnida: Scorpiones)

1. INTRODUCTION

Scorpions are widely distributed and are most common in the tropics and subtropics, and in arid zones of temperate regions. They are widely feared as dangerous venomous creatures; however, the majority of the approximately 1500 known species are not aggressive and do not present a serious medical risk to humans. There are about 50 or so species that do present a significant risk and about half of these are considered capable of causing death in humans.

2. TAXONOMY

Although the taxonomy of this group is not settled, up to 17 families have been recognized, including more than 150 reported genera and more than 1400 described species.

3. MORPHOLOGY

Scorpions are relatively large arachnids (up to 23 cm long) with eight legs and powerful chelate pedipalps (Fig. 109). They are most easily recognized by these grasping claws and the narrow, segmented tail, which ends in a large globular sting, terminating in a large curved spine. The body of a scorpion is divided into two parts: the prosoma (i.e. cephalothorax) and the opisthosoma (comprising the mesosoma, abdomen and metasoma; that is, the tail). On the prosoma are the mouthparts (chelicerae) and pedipalps (used for immobilization of the prey, defence and sensory purposes), four pairs of legs, two eyes on the top and usually two to five pairs of eyes along the front corners. The opisthosoma consists of seven mesosomal segments and

five metasomal segments; the first mesosomal (abdominal) segment bears a pair of genital opercula that cover the gonopore, the second bears the basal plate with the pectines and each of segments three to seven has a pair of spiracles that are the openings for the scorpion's respiratory organs (book lungs); the metasoma (tail) comprises five caudal segments and a sixth bearing the telson, which consists of the vesicle (which holds a pair of venom glands) and the hypodermic aculeus (the venom-injecting barb).

4. LIFE CYCLE

Scorpions are found in virtually every habitat, but they can be roughly distinguished according to their microhabitats as 'tree living' or 'ground dwelling' (including both rocky and sandy environments). Tree-living scorpions hide away during the day under bark, in tree holes and among epiphytes, while ground-dwelling scorpions live in burrows, under rocks and in rocky crevices and various wood and bark surface debris.

Most scorpions reproduce sexually. The pectines on the ventral surface are mechanoreceptors and contact chemoreceptors used by the male during mating to select a suitable place on which to deposit a spermatophore. The male then pulls the female over the spermatophore, which the female detects with her pectines and spreads her genital opercula to take in its contents.

Unlike most other arachnids, scorpions are viviparous, with the young carried about on the mother's back until they have undergone at least one moult. The size of the litter depends on the species and environmental factors, and can range from two to over a hundred scorplings (average is 10–20). The

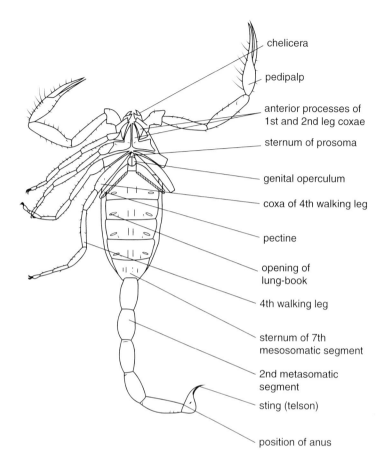

chelicera

pedipalp

anterior processes of
1st and 2nd leg coxae

sternum of prosoma

genital operculum

coxa of 4th walking leg

pectine

opening of
lung-book

4th walking leg

sternum of 7th
mesosomatic segment

2nd metasomatic
segment

sting (telson)

position of anus

Fig. 109. Ventral view of a scorpion. *Source*: Snow, K.R. (1970) *The Arachnids*. Routledge and Kegan Paul, London.

young generally resemble their parents and growth is accomplished by periodic moulting; scorpions often require five to seven moults to reach maturity (which can take from 6 months to 6 years, depending on species). Adults can live for many years (up to 25 has been reported).

5. BEHAVIOUR AND BIONOMICS

In general, scorpions are nocturnally predatory on soft-bodied arthropods and hide under shelters during the day. Some scorpions are ambush predators, remaining motionless near the entrance to their burrows and detecting the prey by contact receptors on the pedipalps.

Other scorpions are active predators and can orientate to and capture prey up to 15 cm away. The vibrations coming from the approaching prey are detected by the tarsal hairs and the slit sensillum on the two terminal segments of the legs. When a large prey item is seized in the pedipalps, the sting is brought rapidly over the top of the body and thrust into the victim, with venom being injected forcibly into the prey by muscular action (although small prey may not be stung). As with spiders, scorpions can only ingest food in a liquid form and have external digestion, with digestive juices from the gut being egested on to the food which is being torn up by the chelicerae. The liquid food is then sucked up and digested.

6. MEDICAL AND VETERINARY IMPORTANCE

All known scorpion species possess venom to kill or paralyse their prey, but it is also used as a defence against predators. In general, scorpions that have large claws are relatively harmless, while the more dangerous species have slender chelae. The large *Hadrurus arizonensis* is comparatively harmless, while the small *Centruroides exilicauda* (formerly *Centruroides sculpturatus*) is deadly. There is also variation within a genus; *C. exilicauda/sculpturatus* and *Centruroides limpidus* are harmful, while *Centruroides pantherinus* and *Centruroides vitatus* are not. Certain species of scorpion are aggressive and will attack humans deliberately, while others will attack only when threatened. Most of the dangerous species belong to the family Buthidae.

In humans, scorpion stings can produce localized and/or systemic reactions. Localized reactions may occur, from the mild, with only transitory pain and swelling, to severe, characterized by tissue destruction, haemorrhaging and necrosis at the sting site because of cytotoxins in the venom. Systemic reactions can also be variable, from mild (with persisting pain and numbness, muscle spasms, chest tightness, difficulty in swallowing and nausea), to neurologic effects (with respiratory and cardiovascular problems caused by the neurotoxins contained in the venom). When death ensues, it is usually because of cardiac or respiratory failure and it may occur within a few hours. It has been estimated that over 5000 people die each year from scorpion stings. Most stings occur because scorpions have taken shelter in shoes or clothing, or because inquisitive humans explore too casually under stones or into holes in the ground.

Scorpion venom is a complex mixture and it varies between species. The effect of the toxin depends on the species of scorpion and is independent of size. The venom of a single species of scorpion may contain one toxin which subdues insects, another that is effective against crustaceans and a third that deters mammalian predators. It has been likened to Cobra snake venom because of the similarity of the victim's response, but the two venoms are distinct. Scorpion venoms are homologous in amino acid sequences of the proteins which they include, as are the venoms of elapid snakes (the family that includes cobras). However, the amino acid sequences of the venom proteins in the two groups are quite different. They also act differently on the nervous system. Elapid toxins produce an antidepolarizing block of the end plate, while scorpion venoms depolarize different target cells.

A number of scorpion families are of concern in this context, but the two most important are the Buthidae and the Hemiscorpiidae. The Buthidae contains most of the species that are dangerously venomous to humans and other animals. The most important genera are *Androctonus* and *Leiurus* in northern Africa and western Asia, *Mesobuthus* in Asia and India, *Centruroides* in the southern USA and Mexico, *Tityus* in South America and *Lychas* in Australia. The Hemiscorpiidae includes the only non-buthid scorpion known to cause significant human mortality, *Hemiscorpius lepturus*, which is found in Iran and Iraq, and its venom contains a potent cytotoxin that causes serious tissue damage and necrosis, as well as severe systemic symptoms.

Other genera, such *Heterometrus* and *Nebo* in the family Scorpionidae in the Middle and Far East, have been associated with serious cases, and some species of *Urodacus* in the family Urodacidae in Australia can cause severe pain and pyrexia, with transient general toxicity and prostration for a day or so, but others generally appear to be relatively harmless or cause only mild and/or quickly resolving symptoms.

7. PREVENTION AND CONTROL

Maintaining masonry walls, clearing of woodpiles, stored materials and litter, and other shelter harbourage around homes, will reduce the risk of contact with scorpions and, where scorpions are common, articles of clothing and footwear should be shaken to dislodge any that might be sheltering inside (particularly if outdoors). Although chemical control of scorpions with pesticides has been used in the past, organochlorine compounds are no longer useable in most situations and, in general, pesticides are not recommended for use against scorpions. The emphasis should be on physical prevention of scorpions while entering homes and outbuildings; for example, a row of glazed ceramic tiles with slippery faces that scorpions cannot climb has been recommended for facing walls, steps and other ground-touching surfaces around homes.

Treatment of scorpion sting is largely symptomatic. Mild envenomation usually involves use of cold packs and analgesics to relieve pain. Antihistamines, steroids and sedatives are thought to be often of little benefit in uncomplicated cases. Cases with systemic reactions should receive rapid medical attention, with atropine given to relieve impacts on the parasympathetic system, while other drugs such as calcium gluconate and sodium phenobarbital can be used to relieve muscle spasm and prevent convulsions, respectively. Specially prepared antivenins can be effective if administered soon (1–2 h) after stinging.

SELECTED BIBLIOGRAPHY

Alexander, J.O.D. (1984) *Arthropods and Human Skin*. Springer-Verlag, Berlin, pp. 199–207.

Al-Sadoon, M.K. and Jarrar, B.M. (2003) Epidemiological study of scorpion stings in Saudi Arabia between 1993 and 1997. *Journal of Venomous Animals and Toxins including Tropical Diseases* 9, 1–8.

Bettini, S. (ed.) (1978) *Arthropod Venoms*. Springer-Verlag, Berlin.

Chippaux, J.P. and Goyffon, M. (2008) Epidemiology of scorpionism: a global appraisal. *Acta Tropica* 107, 71–79.

Cloudsley-Thompson, J.L. (1992) Scorpions. *Biologist* 39, 206–210.

Cloudsley-Thompson, J.L. (1993) Spiders and scorpions (Araneae and Scorpions). In: Lane, R.P. and Crosskey, R.W. (eds) *Medical Insects and Arachnids*. Chapman and Hall, London, pp. 659–682.

Couraud, F. and Jover, E. (1984) Mechanisms of action of scorpion toxins. In: Tu, A.T. (ed.) *Handbook of Natural Toxins*, Vol. II. Dekker, New York, pp. 659–678.

Fet, V., Sissom, W.D., Lowe, G. and Braunwalder, M.E. (2000) *Catalog of the Scorpions of the World (1758–1998)*. The New York Entomological Society, New York.

Isbister, G.K., Volschenk, E.S., Balit, C.R. and Harvey, M.S. (2003) Australian scorpions: a prospective study of definite stings. *Toxicon* 41, 877–883.

Isbister, G.K., Volschenk, E.S. and Seymour, J.E. (2004) Scorpion stings in Australia: five definite stings and a review. *Internal Medicine Journal* 34, 427–430.

Keegan, H.L. (1980) *Scorpions of Medical Importance*. University Press of Mississippi, Jackson, Mississippi.

Kuehn, B.M. (2011) Treatment for scorpion stings. *Journal of the American Medical Association* 306, 1315.

Luca, S.M. and Meier, J. (1995) Biology and distribution of scorpions of medical importance. In: Meier, J. and White, J. (eds) *Handbook of Clinical Toxicology of Animal Venoms and Poisons*. CRC Press, Boca Raton, Florida, pp. 205–219.

Polis, G.A. (ed.) (1990) *The Biology of Scorpions*. Stanford University Press, Stanford, California.

Southcott, R.V. (1976) Arachnidism and allied syndromes in the Australian regions. *Records of the Adelaide Children's Hospital* 1, 97–187.

Spiders (Arachnida: Araneae)

1. INTRODUCTION

Spiders are ubiquitous and regarded globally as venomous creatures. Although some spiders are highly venomous, with bites that can be life threatening, relatively fewer species are of great significance for human or animal health and several that were previously considered of medical importance have been recognized in recent years as being relatively harmless.

2. TAXONOMY

Spider taxonomy is controversial. Arguably, spiders are currently classified into three suborders: Mesothelae (primitive spiders), Mygalomorphae (spiders characterized by chelicerae that operate with a vertical motion and with two pairs of book lungs) and Araneomorphae (characterized by chelicerae that operate towards each other in a horizontal plane and with usually only one pair of book lungs). Currently, 110 families have been recognized, with more than 3800 genera and 42,000 described species; the great majority of these are ranked within the Araneomorphae.

3. MORPHOLOGY

In general, spiders are characterized by eight legs, four pairs of eyes but no antennae on a uniform prosoma (anterior portion of the body), joined by a narrow pedicel to an unsegmented opisthosoma (hind portion), which has appendages known as spinnerets that can exude silk to fashion webs of various types (Fig. 110a and b). The pedipalps are tactile, leg-like structures, shorter than the ambulatory legs. In the male, they are modified as intromittent organs. Male spiders are readily

recognized by their terminal swelling on their pedipalps. The chelicerae are two-segmented but not chelate; the distal segment is sharply pointed and bears at its tip the opening of the poison duct, whose gland may be contained in the basal segment of the chelicera or, more usually, in the anterior part of the prosoma (Fig. 110b). Some spiders are extremely small (less than 0.5 mm), but some of the largest (*Tarantula* species) may have body lengths of up to 9 cm and leg spans of more than 20 cm.

4. LIFE CYCLE

Most activity occurs during the warm seasons in temperate climates. Mating often results in the male spiders being eaten by female counterparts, with the females subsequently laying hundreds of eggs, often encased in silk cases. The young develop through their larval stages within the egg and hatch after about 2 weeks as nymphs (miniature adults but sexually immature). These spiderlings must moult (between 2 and 12 times, depending on species) to grow and, although most adults live for no more than a year or two (with males dying soon after mating, when not eaten by females), some of the larger spiders have been known to live for many years in captivity. Overwintering in temperate climates occurs, dependent on species, either as mature females, late-stage nymphs or early nymphs in the egg sac.

5. BEHAVIOUR AND BIONOMICS

Spiders are predators and the various species have many different strategies for capturing prey, which they kill with their venomous fangs. Spiders do not ingest solid food; they

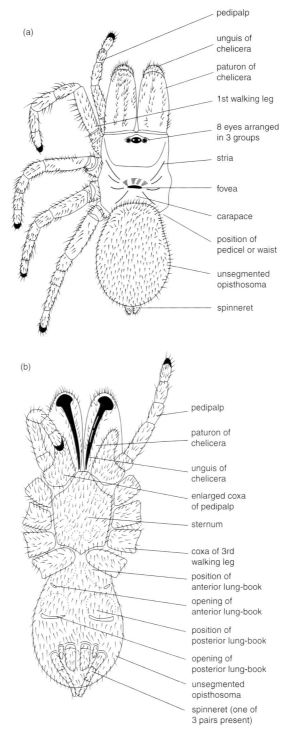

Fig. 110. (a) Dorsal and (b) ventral views of a mygalomorph spider. *Source*: Snow, K.R. (1970) *The Arachnids*. Routledge and Kegan Paul, London.

liquidize it externally via exuding digestive enzymes, and also in some species by grinding it with their pedipalps, and they then suck up the liquefied tissues.

6. MEDICAL AND VETERINARY IMPORTANCE

Most spiders are harmless to humans, being either unable to penetrate the human skin or having ineffective venom. As mentioned above, in the majority of spiders (suborder Araneomorphae), the fangs in action converge from the sides towards the midline horizontally like pincers and relatively few can effect a serious bite. However, in the Mygalomorphae, the parallel fangs strike vertically downwards and forwards and, as they are usually large and heavy spiders, they can penetrate skin more readily. Some of these larger species cause a degree of pain at least and temporary local reaction, while others can cause severe symptoms.

Spider venom is adapted to the species of prey on which the spider usually feeds, mostly invertebrates, although the mygalomorph spider, *Selonocosmia javanensis*, attacks and kills small birds. Venom components and effects vary with the different genera and species and the different animals envenomed, respectively. Venoms can be cytotoxic or neurotoxic and contain various enzymes, amino acids, polyamines, histamine and other compounds specific to a particular group of spiders or individual species.

A number of families of both suborders have previously been considered of medical importance, but only the following four are currently considered life threatening.

6.1 Ctenidae (tropical wolf spiders, armed spiders, wandering spiders)

These araneomorph spiders of the genus *Phoneutria* (e.g. *Phoneutria boliviensis*, *Phoneutria fera*, *Phoneutria nigriventer* and *Phoneutria reidyi*) are found in tropical South and Central America. They do not build webs but actively hunt at night, wandering and looking for prey among leaf litter and ground

vegetation. They are of moderate to large size (some reaching 5 cm body length), are aggressive and their bite is very painful; their venom, which includes histamine and serotonin as well as a neurotoxin, causes a range of symptoms, can lead to respiratory failure and is potentially fatal, but antivenom is available in Brazil.

6.2 Hexathelidae (venomous funnel-web tarantulas, e.g. Sydney funnel-web spider)

Of these mygalomorph spiders, the Sydney funnel-web spider, *Atrax robustus*, is usually considered the most important, as it has caused many human deaths in Australia. The common name of 'funnel-web spider' in Australia is shared by species of both genera *Atrax* and *Hadronyche*, and all species of both genera should be regarded as harmful. The females are medium-sized (3–5 cm body length) and robust spiders, forming subterranean nests in burrows and natural cavities and not commonly being encountered; however, the males wander more widely at night in search of mates and are more likely to enter houses. Arguably, these are the world's most harmful spiders (based on toxin and case studies), but severe envenomation is rare. The males are particularly aggressive and more venomous than the females; severe envenoming has occurred with six species, with at least two species (*A. robustus* and *Hadronyche formadibilis*, the tree funnel-web) known to have inflicted fatal bites to humans. The bite is extremely painful (due to the depth of fang penetration and the acidic venom) and it causes a range of neurological symptoms that can lead to coma and death – particularly in children. However, effective antivenom is available against *A. robustus* and it is effective also for the other medically important species.

6.3 Sicariidae (recluse spiders, e.g. brown recluse spider)

Araneomorph spiders of the genus *Loxosceles* produce cytotoxic venom which causes tissue destruction. The group is located principally in the tropics, but some species have been introduced widely to temperate regions in the Americas, Europe, Asia and Africa. They are usually found among ground litter, but some have become closely associated with human dwellings and other buildings. These are not large spiders, usually about 10–20 mm body length and relatively slender, lightish brown in colour with a violin-shaped marking on the prosoma. They make only irregular tangled webs and actively hunt at night. The venom is highly cytotoxic, with the enzyme sphingomyelinase destroying lipid cells and thus causing extensive damage. The initial bite may be painless and pass unnoticed, but locally extensive necrosis and ulceration can develop around the bite site, and healing can take some months. This type of reaction has been attributed to various *Loxosceles* species from different regions of the world, including *Loxosceles reclusa* in the USA and *Loxosceles rufescens* in Europe, with deaths attributed to *Loxosceles laeta* in Chile (although antivenom is available for this and other *Loxosceles* species in South America). In southern Africa, the venom of the genus *Sicarius* (e.g. six-eyed sand spider, *Sicarius hahni*) is also cytotoxic, producing necrotic wounds, and it is regarded as potentially lethal to humans.

6.4 Theridiidae (cobweb spiders, e.g. black widow, red-back spiders)

Species of *Latrodectus*, an araneomorph genus, produce neurotoxic venom (the main active component is a protein which acts on motor nerve endings) that can cause a severe human reaction with pain, swelling, nausea, sweating and neuromuscular signs (muscle weakness or paralysis). Recovery is usually achieved within 48 h and death is rare in healthy adults.

The *Latrodectus mactans* complex is a group of black spiders with a female body length of about 10 mm (males are usually 3–4 mm in length) and various red or yellow patches on the opisthosoma. They are widely distributed throughout the warmer parts of the world, being shy, retiring spiders living in irregular tangled webs, often among domestic

and peridomestic litter and trash. *L. mactans* is the black widow spider of the USA, *Latrodectus tridecimguttatus* occurs in southern Europe, *Latrodectus hasselti* is the red-back spider of Australia, *Latrodectus indistinctus* occurs in southern Africa and there are three species (including *L. mactans*) in Brazil. Effective antivenom is available against bites of the *L. mactans* group; while its use is somewhat controversial in some regions, because of concerns about allergic reactions, it is used commonly in Australia with few problems.

6.5 Other spider families of lesser importance

Other families that have been considered of medical importance but which are now recognized as relatively harmless:

6.5.1 Actinopodidae (mouse spiders)

These mygalomorph spiders are found mostly in Australia (although there is one species in Chile) and they live in burrows covered with trapdoors. They are of medium size (up to 3 cm in body length), relatively robust and generally quite dark in colour (although some males may have bright red chelicerae). The bites of several species of Australian mouse spiders have produced serious symptoms, albeit relatively rare (unlike the funnel-web, the mouse spider is far less aggressive towards humans and may often give 'dry' bites). In a review of 40 verified cases, the bites gave only minor local effects, except for a case of severe neurotoxic envenomation in a young child (but funnel-web antivenom was effective in treating the condition because of the similarity between the venoms of these groups).

6.5.2 Agelenidae (araneomorph funnel-web spiders, e.g. hobo spider)

These spiders include the purportedly venomous European hobo spider, *Tegenaria agrestis*, which is a native of Europe but has been introduced into the north-west of the USA. It is a medium-sized species (12–20 mm body length) that builds funnel-shaped webs around and in dwellings. Reports on the bite of this spider include not painful bites, where the cytolytic venom (as with *Loxosceles* spp.) causes skin necrosis and slow-healing ulcers, with possible associated systemic effects (e.g. sweating, nausea, muscle weakness, headaches and disorientation). However, this spider has not been proven (or disproven) to cause necrosis and there is considerable doubt as to its relative medical importance.

6.5.3 Lamponidae (white-tailed spiders)

These araneomorph spiders are endemic in Australasia; they are about 2 cm in length and have a white spot near the spinnerets (and sometimes others on the dorsum of the opisthosoma). They are often found in houses and have been reported frequently to be associated with necrotic skin lesions in Australia. However, a study of more than 100 cases of verified *Lampona* spp. bites has revealed only mild effects and no evidence of necrotic reactions.

6.5.4 Lycosidae (wolf spiders)

This is a worldwide group of burrowing hunting araneomorph spiders (of various sizes up to 3.5 cm) which will inflict defensive bites, including *Lycosa antibucana* in South America that has been reported to cause severe swelling and lymphangitis. Wolf spiders have been associated with necrotic lesions, but a review of more than 500 verified bites showed no evidence of dermonecrosis and only painful but otherwise insignificant injuries.

6.5.5 Miturgidae (sac spiders)

This is an araneomorph group of small (5–10 mm body length), vagrant, night-hunting, light-coloured spiders that live in silk-lined, sac-like retreats under leaf litter, stones or bark, with species of the genus *Cheiracanthium*, such as *Cheiracanthium punctorium* in Europe, *Cheiracanthium japonicum* in Japan and *Cheiracanthium mordax* in Australia, and *Cheiracanthium inclusum* and *Cheiracanthium mildei* in the New World. There are reports of bites causing pain, swelling, dizziness and other minor systemic effects such as nausea and, rarely, minor ulceration; however, there is little evidence for any necrosis resulting from unremarkable bite reactions and they are generally not considered to be a major problem anywhere.

6.5.6 Theraphosidae (tarantulas – bird-eating spiders)

These mygalomorph spiders are most commonly recognized as the very large and hairy tarantulas (2.5–10 cm in body length and 8–30 cm leg span) of various colorations. They are found in both Old and New World tropic and subtropical areas, living in silk-lined burrows and cavities. Although most are relatively harmless, some genera (e.g. *Poecilotheria* in India and *Acanthoscurria*, *Pamphobeteus* and *Phormictopus* in South America) have species that can cause severe envenomation and can be life threatening because their venoms contain highly active proteolytic enzymes and neurotoxins. Some New World tarantula species have barbed abdominal hairs, which the spider can flick off with its back legs to deter an attacker; these hairs can mechanically cause skin irritation and inflammation in the eyes, nose, mouth and respiratory passages in small animals and humans.

7. PREVENTION AND CONTROL

Avoidance of spiders or their habitats is the best way to be protected against spider bites. Removal of shelter or harbourage opportunities provided by woodpiles, loose rockwalls and accumulation of trash, etc., is advisable if there are threats of spiders locally. Pesticides can be used to reduce populations around domestic dwellings, but there is some evidence that the chemicals can excite the more robust ground-dwelling spiders and may make them potentially more harmful.

Treatment of spider bite is dependent on the species involved, but a general response should include washing the bite site, cold compresses and analgesics, and antibiotics if secondary infections occur. For necrotic lesions associated with recluse spiders and envenomation by funnel-web, wandering (armed) and widow spiders, specialist medical assistance should be sought (with a pressure bandage applied and limb immobilization

secured for funnel-web spiders). For urticaria
from tarantula hairs, oral antihistamines and
topical corticosteroids may alleviate symptoms;
in cases of ocular or respiratory involvement,
specialist medical treatment should be sought
immediately. Comments on antivenoms are
included above with the particular spider
groups.

SELECTED BIBLIOGRAPHY

Bettini, S. (ed.) (1978) *Arthropod Venoms*. Springer-Verlag, Berlin.

Coddington, J.A. and Levi, H.W. (1991) Systematics and evolution of spiders (Araneae). *Annual Review of Ecology and Systematics* 22, 565–592.

Cooke, J.A., Roth, V.D. and Miller, F.H. (1972) The urticating hairs of theraphosid spiders. *American Museum Novitates* 2498, 1–43.

da Silva, P.H., da Silveira, R.B., Appel, M.H., Mangili, O.C., Gremski, W. and Veiga, S.S. (2004) Brown spiders and loxoscelism. *Toxicon* 44, 693–709.

Duchen, L.W. and Gomez, S. (1984) Pharmacology of spider venoms. In: Tu, A.T. (ed.) *Handbook of Natural Toxins*, Vol 2. Dekker, New York, pp. 483–512.

Foelix, R.F. (1996) *Biology of Spiders*, 2nd edn. Oxford University Press, New York.

Gaver-Wainwright, M.M., Zack, R.S., Foradori, M.J. and Lavine, L.C. (2011) Misdiagnosis of spider bites: bacterial associates, mechanical pathogen transfer, and hemolytic potential of venom from the hobo spider, *Tegenaria agrestis* (Araneae: Agelenidae). *Journal of Medical Entomology* 48, 382–388.

Graudins, A., Padula, M., Broady, K.W. and Nicholson, G. (2001) Red-back spider (*Latrodectus hasselti*) antivenom prevents the toxicity of widow spider venoms. *Annals of Emergency Medicine* 37, 154–160.

Isbister, G.K. and Hirst, D. (2002) Injuries from spider spines not spider bites. *Veterinary and Human Toxicology* 44, 339–342.

Isbister, G.K. and White, J. (2004) Clinical consequences of spider bites: recent advances in our understanding. *Toxicon* 43, 477–492.

Isbister, G.K., Seymour, J.E., Gray, M.R. and Raven, R.J. (2003) Bites by spiders of the family Theraphosidae in humans and canines. *Toxicon* 41, 519–524.

Jelinek, G.A. (1997) Widow spider envenomation (latrodectism): a worldwide problem. *Wilderness and Environmental Medicine* 8, 226–231.

Lucas, S. (1988) Spiders in Brazil. *Toxicon* 26, 759–772.

Lucas, S.M., Da Silva, P.I. Jr, Bertani, R. and Cardoso, J.L.C. (1994) Mygalomorph spider bites: a report on 91 cases in the State of São Paulo, Brazil. *Toxicon* 32, 1211–1215.

Maretic, Z. (1983) Latrodectism: variations in clinical manifestations produced by *Latrodectus* species of spiders. *Toxicon* 21, 457–466.

Marques-da-Silva, E., Souza-Santos, R., Fischer, M.L. and Rubio, G.B.G. (2006) *Loxosceles* spider bites in the state of Parana, Brazil: 1993–2000. *Journal of Venomous Animals and Toxins Including Tropical Diseases* 12, 110–123.

Ori, M. (1984) Biology of and poisoning by spiders. In: Tu, A.T. (ed.) *Handbok of Natural Toxins*, Vol 2. Dekker, New York, pp. 397–440.

Platnick, N.I. (2011) *The World Spider Catalog*, version 12 (http://research.amnh.org/iz/spiders/catalog/INTRO1.html, accessed July 2011).

Southcott, R.V. (1976) Arachnidism and allied syndromes in the Australian regions. *Records of the Adelaide Children's Hospital* 1, 97–187.

Sutherland, S.K. (1990) Treatment of arachnid poisoning in Australia. *Australian Family Physician* 19, 1, 17, 50–55, 57–62.

Sutherland, S.K. and Tibbals, J. (2001a) *Australian Animal Toxins: The Creatures, Their Toxins and Care of the Poisoned Patient*. Oxford University Press, Melbourne, Australia.

Sutherland, S.K. and Tibbals, J. (2001b) The genera *Atrax* and *Hadronyche*, funnel-web spiders. In: *Australian Animal Toxins: The Creatures, Their Toxins and Care of the Poisoned Patient*. Oxford University Press, Melbourne, Australia, pp. 402–464.

Vetter, R.S. (2008) Spiders of the genus *Loxosceles* (Aranea, Sicariidea): a review of biological, medical and psychological aspects regarding envenomations. *The Journal of Arachnology* 36, 150–163.

Vetter, R.S. and Isbister, G.K. (2008) Medical aspects of spider bites. *Annual Review of Entomology* 53, 409–429.

Vetter, R.S. and Rust, M.K. (2008) Refugia preferences by the spiders *Loxosceles reclusa* and *Loxosceles laeta* (Araneae: Sicariidae). *Journal of Medical Entomology* 45, 36–41.

Williams, S.T., Khare, V.K., Johnston, G.A. and Blackall, D.P. (1995) Severe, intravascular haemolysis associated with brown recluse spider envenomation. A report of two cases and a review of the literature. *American Journal of Clinical Pathology* 104, 463–667.

Stable Flies (Diptera: Muscidae, *Stomoxys*)

1. INTRODUCTION

Stable flies (*Stomoxys*) are parasitic, blood-feeding flies (Diptera). They have biting mouthparts and both sexes are blood feeders. Although they bite humans and can be localized pests, they are not of great medical significance. None the less, they pester dogs and can have a huge economic impact on the health and productivity of cattle, being one of the most widespread and economically important pests to attack cattle.

The genus contains about 18 species, of which the most common species of importance in temperate habitats is *Stomoxys calcitrans*, the stable fly. This is now found worldwide, after being introduced into North America from Europe during the 1700s. *Stomoxys niger* and *Stomoxys sitiens* may replace *S. calcitrans* as important pests of livestock in Afrotropical and Oriental regions.

2. TAXONOMY

Within the large dipteran family Muscidae, the subfamily Stomoxyinae contains ten genera, including the blood-feeding *Stomoxys* and also *Haematobosca* and *Haematobia* (see the entry for 'Horn Flies'). Species of *Stomoxys* may be readily recognized by the possession of an elongate, sclerotized proboscis with palps that are less than half the length of the proboscis. In both *Stomoxys* and *Haematobia*, the arista carries hairs on the dorsal side only.

3. MORPHOLOGY

Stomoxys are about 7–8 mm in length and are generally grey in colour, with four longitudinal dark stripes on the thorax, and the abdomen has a grey and dark brown pollinosity, forming variable but characteristic dark spots on the II and III abdominal segments (Fig. 111). The wing's discal vein (M_{1+2}) is slightly apical forward curved (Fig. 112). The abdomen is shorter and broader than that of the house fly, *Musca domestica*, which it resembles, but the projecting proboscis is sufficiently prominent and

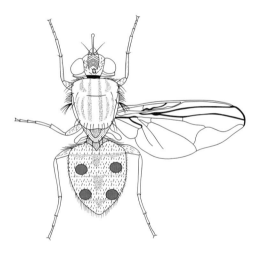

Fig. 111. *Stomoxys calcitrans.* Drawn from specimens and Greenberg (1971).

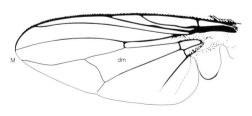

Fig. 112. The wing of *Stomoxys calcitrans.*

conspicuous to distinguish species of this genus from species of *Musca*. The palps are small and thin and only a quarter to half of the length of the proboscis. In the third-stage larva, the right mouth-hook is larger than the left, the anterior spiracles have about six openings and the posterior spiracles have three S-shaped slits surrounding a central button (Fig. 113).

4. LIFE CYCLE

The elongate, yellowish-white egg hatches into a saprophagous maggot, which undergoes three moults. The eggs are about 1 mm in length and hatch in 5–10 days, depending on temperature. The cream-coloured larvae pass through three stadia until, when fully developed at about 10 mm in length, they pupate. The puparia are brown and about 6 mm in length.

The breeding sites of *S. calcitrans* are characterized mainly by the presence of rotted urine or manure-contaminated straw, grass or leaves and shade, but larvae may be found in animal bedding, lawn cuttings, rotting vegetables and rotting stranded seaweed on beaches. The number of eggs matured by an individual female *S. calcitrans* ranges from 200 to 600, and the female scatters her eggs throughout the chosen larval medium, with pupation occurring in the drier parts.

At 26.7°C, the egg stage lasts 23 h and the three larval stages 23 h, 27 h and approximately 7 days, respectively. At 30–31°C, the pupal stage lasts 5 days. The

Fig. 113. The posterior spiracles of a larval *Stomoxys calcitrans.*

female is anautogenous, requiring several blood meals (up to 9) to complete ovarian development, and the average blood meal (25.8 mg) is three times the average body weight (8.6 mg).

5. BEHAVIOUR AND BIONOMICS

S. calcitrans is an endophilic species, closely associated with stables and human dwellings, mostly when cattle or horses are present. In summer, the adults survive for 3–4 weeks and considerably longer at cooler times of the year. Adults assemble on sunlit, light-coloured objects, from which males dart out after flying insects. Males mount females in the air or on the ground, with copulation occurring on a perch. Insolation can raise the body temperature by more than 10°C, to 22–28°C, but when the body temperature reaches 31–34°C, they seek shade. Under ordinary conditions, *S. calcitrans* can disperse up to approximately 5 km, but much further (maximum recorded 225 km) under exceptional weather conditions. Cold fronts carry large numbers of *S. calcitrans* long distances at heights of 30–60 m.

During spring and summer time, *S. calcitrans* has a bimodal diurnal pattern of feeding, locating hosts by responding to carbon dioxide and octenol emanations. Both sexes are obligatory blood feeders. Individual *S. calcitrans* may feed more than once a day, biting their host low down on the body. They attack the ankles of humans and the belly, lower body and limbs of domestic stock, particularly cattle and horses, but also the ears of dogs. They usually probe and attempt to feed on a number of hosts in rapid succession. Approximately 3 min is required for a single blood meal. After a blood meal, flies can be seen resting on sunny stable walls or fences.

Stomoxyines show strong phototactic responses to UV and blue radiation, which explains the greater attractiveness of traps using UV reflective fibreglass panels (Alsynite®). They remain largely in areas of strong sunlight and they bite mainly out of doors, although they will follow animals inside to feed. They will also enter buildings during rainy weather in autumn.

6. MEDICAL AND VETERINARY IMPORTANCE

6.1 *Stomoxys calcitrans*

The bite of stable flies is painful and, as such, they are serious pests of animals. They can be a particular problem for some dogs, as they savagely attack the points of upright ears, causing bleeding and tissue damage, which may become necrotic. In large numbers, these flies are a great source of annoyance to grazing cattle (particularly around stables and feedlots); they may reduce milk yield by 25%, or as much as 40–60%, and may reduce weight gain and feeding efficiency in stabled cattle. One analysis estimated national losses in the USA cattle industries due to stable flies to total more than US$2200 million/year. The economic threshold for *S. calcitrans* seems to be as low as 6 flies/animal. When densities exceed this threshold, the increase in growth rates of the cattle in response to fly control strategies will compensate economically for the cost of implementing the control strategies.

The salivary secretions of *S. calcitrans* may cause toxic reactions with an immuno-suppressive effect, rendering the host more susceptible to disease.

There is no evidence that the fly is important as a vector for significant human or animal pathogens, but they do serve as intermediate hosts of nematode (*Habronema* and *Draschia*) parasites of horses. Also, because they are persistent biters, often engage in interrupted feeding and feed more than once per day, they are possible mechanical vectors for a range of other blood-dwelling pathogens, such as *Trypanosoma evansi* (causing surra of equines and dogs), anthrax and *Dermatophilus congolensis*. Further, there is some experimental indication that they are capable mechanical vectors of other pathogens, including arboviruses such as West Nile virus and Rift Valley fever virus, if they feed again within a few hours of feeding on an infected host.

6.2 *Stomoxys niger*

In Africa, *S. niger* breeds in decaying plant material, with the population generally at its peak between October and December and lowest from May to July. In Zanzibar, the highest concentrations of *S. niger* have been reported in heavily forested areas on poorly drained soils. They are particularly active just after sunrise and before sunset. They fly close to the ground, with maximum numbers being caught in traps at a height of 30 cm, and there is evidence that they can locate hosts 2–5 km away.

7. PREVENTION AND CONTROL

Insecticide-impregnated ear tags, tail bands and halters, largely containing pyrethroids, together with pour-on, spot-on and spray preparations, are widely used to reduce fly annoyance in cattle and horses. Dogs' ears can be protected with petroleum jelly.

Various types of screens and electrocution grids for buildings are available to reduce fly nuisance, but the best methods of control are those aimed at improving sanitation and reducing breeding places (source reduction). For example, in stables and farms, manure should be removed or stacked in large heaps, where the heat of fermentation will kill the developing stages of flies, as well as eggs and larvae of helminths. In addition, insecticides applied to the surface of manure heaps may prove beneficial in some circumstances, but this should be used cautiously, since such applications will also kill beneficial insects.

Aerosol space sprays, residual insecticides applied to walls and ceilings and insecticide-impregnated cards and strips may all reduce fly numbers indoors. Insecticides may also be incorporated in solid or liquid fly baits, using attractants such as various sugary syrups or hydrolysed yeast and animal proteins. However, given the high rates of reproduction, high rates of dispersal and multiple generations per year, area-wide control is generally impractical.

SELECTED BIBLIOGRAPHY

Buschman, L.L. and Patterson, R.S. (1981) Assembly, mating and thermoregulating behavior of stable flies under field conditions. *Environmental Entomology* 10, 16–21.

Campbell, J.B. and Berry, I.L. (1989) Economic threshold for stable flies on confined livestock. *Miscellaneous Publications of the Entomological Society of America* 74, 18–22.

Catangui, M.A., Campbell, J.B., Thomas, G.D. and Boxler, D.J. (1997) Calculating economic injury levels for stable flies (Diptera: Muscidae) on feeding heifers. *Journal of Economic Entomology* 90, 6–10.

Crosskey, R.W. (1993) Stable-flies and horn-flies (bloodsucking Muscidae). In: Lane, R.P. and Crosskey, R.W. (eds) *Medical Insects and Arachnids*. Chapman and Hall, London, pp. 389–402.

Doyle, M.S., Swope, B.N., Hogsette, J.A., Burkhalter, K.L., Savage, H.M. and Nasci, R.S. (2011) Vector competence of the Stable Fly (Diptera: Muscidae) for West Nile Virus. *Journal of Medical Entomology* 48, 656–668.

Gilles, J., David, J.F., Tilard, E., Duvallet, G. and Pfister, K. (2008) Potential impacts of climate change on stable flies, investigated along an altitudinal gradient. *Parasitology Research* (Suppl 1) 103, S147–S159.

Greene, G.L. (1989) Seasonal population trends of adult stable flies. *Miscellaneous Publications of the Entomological Society of America* 74, 12–17.

Hogsette, J.A., Ruff, J.P. and Jones, C.J. (1989) Dispersal behaviour of stable flies (Diptera: Muscidae). *Miscellaneous Publications of the Entomological Society of America* 74, 23–32.

Holloway, M.T.P. and Phelps, R.J. (1991) The responses of *Stomoxys* spp. (Diptera: Muscidae) to traps and artificial host odours in the field. *Bulletin of Entomological Research* 81, 51–55.

Mock, D.E. and Greene, G.L. (1989) Current approaches in chemical control of stable flies. *Miscellaneous Publications of the Entomological Society of America* 74, 46–53.

Morgan, C.E., Thomas, G.D. and Hall, R.D. (1983) Annotated bibliography of the stable fly, *Stomoxys calcitrans* (L.), including references on other species belonging to the genus *Stomoxys*. *University of Missouri-Columbia Agricultural Experiment Station Research Bulletin* 1049, 1–190.

Muenworn, V., Duvallet, G., Thainchum, K., Tuntakom, S., Tanasilchayakul, S., Prabaripa, A., *et al.* (2010) Geographic distribution of Stomoxyine flies (Diptera: Muscidae) and diurnal activity of *Stomoxys calcitrans* in Thailand. *Journal of Medical Entomology* 47, 791–797.

Patterson, R.S. (1989) Biology and ecology of *Stomoxys nigra* and *S. calcitrans* on Zanzibar, Tanzania. *Miscellaneous Publications of the Entomological Society of America* 74, 2–11.

Taylor, D.B., Moon, R.D., Campbell, J.B., Berkebile, D.R., Scholl, P.J., Broce, A.B., *et al.* (2010) Dispersal of stable flies (Diptera: Muscidae) from larval development sites in a Nebraska landscape. *Environmental Entomology* 39, 1101–1110.

Taylor, D.B., Moon, R.D. and Mark, D.R. (2012) Economic impact of stable flies (Diptera: Muscidae) on dairy and beef cattle production. *Journal of Medical Entomology* 49, 198–209.

Zumpt, F. (1973) *The Stomoxyine Biting Flies of the World (Diptera: Muscidae)*. Gustav Fischer Verlag, Stuttgart, Germany.

Stomach Bot Flies (Diptera: Oestridae, Gasterophilinae)

1. INTRODUCTION

Stomach bots are common parasites of horses and donkeys; however, three species have been described from Asiatic and African rhinoceroses and three from African and Indian elephants. They infest the gastrointestinal and pharyngeal mucosa of their hosts, causing inflammation, sloughing of the tissue and ulcerations. The most important genus is *Gasterophilus*. Humans are occasionally infested with *Gasterophilus* spp., generally with a creeping dermal myiasis and only rarely with cavity (oral and ocular) infestations.

2. TAXONOMY

Within the superfamily Oestroidea, there are three major families of myiasis-producing flies: Oestridae (see also the entry for 'Nasal Bot Flies'), Calliphoridae (see the entry for 'Blow Flies') and Sarcophagidae (see the entry for 'Flesh Flies'). The Oestridae contains about 150 species of flies whose larvae cause obligate myiases and are known as the bots and warbles. There are four subfamilies of importance: Gasterophilinae, Oestrinae (see the entry for 'Nasal Bot Flies'), Hypodermatinae (see the entry for 'Warble Flies') and Cuterebrinae (see the entry for Human Bat Fly 'Torsalo Fly'). Within the Gasterophilinae, the most important genus, *Gasterophilus*, consists of eight described species, three of which have a worldwide distribution. Other, less important genera include *Gyrostigma*, which infest rhinoceroses, and *Cobboldia*, which infest African and Indian elephants.

Of the eight known species of *Gasterophilus*, six parasitize domestic horses and donkeys and two are restricted to zebras. Of these, *Gasterophilus nigricornis* has the most limited distribution (southern Asiatic part of the Palaearctic region), whereas *Gasterophilus intestinalis* is the most important and most widely distributed horse bot. Four species, *Gasterophilus nasalis*, *Gasterophilus haemorrhoidalis*, *Gasterophilus pecorum* and *Gasterophilus inermis*, parasitize both horses and zebras, and *G. nigricornis* and *Gasterophilus ternicinctus* are found only in zebras. *G. intestinalis* was originally a Palaearctic species, but it has since been introduced to many parts of the world and is now the most important horse bot in Europe, USA and Australia. The second commonest is *G. nasalis*, followed by *G. inermis*, *G. pecorum* and *G. haemorrhoidalis*, which are sympatric in some countries of the Mediterranean basin where animals are free grazing.

3. MORPHOLOGY

Bot flies are robust, dark flies of 10–15 mm in length, the postscutellum is undeveloped and the squamae are small (Fig. 114). A bulbous, greater ampulla is present below the wing base. The maxillary palps are less reduced and carry sensillae, which are considered to function as olfactory receptors and are used to seek out the gasterophilines' hosts. The body is densely covered with yellowish hairs. In the female, more abdominal segments are exposed than is usual in the Cyclorrhapha and the abdomen is characteristically recurved ventrally. The wings of adult *Gasterophilus* characteristically have no cross-vein dm-cu. The wing of *G. pecorum* is very dark, that of *G. intestinalis* has a broad transverse median band and dark areas at the end of vein iv and the wing apex (Fig. 115).

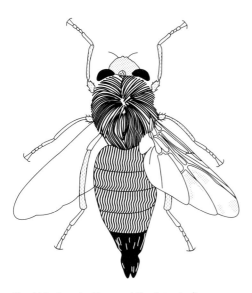

Fig. 114. Female *Gasterophilus intestinalis.*

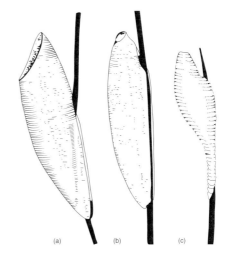

Fig. 116. Eggs of *Gasterophilus* species. (a) *G. intestinalis*; (b) *G. nasalis*; (c) *G. haemorrhoidalis*. *Source*: Cameron, A.E. (1942) *Transactions of the Highland and Agricultural Society.*

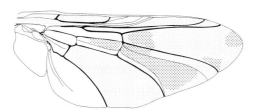

Fig. 115. Wing of *Gasterophilus intestinalis.*

Eggs are yellow in colour and attach singly to the hair of their host (Fig. 116). The larvae are an off-white colour and are armed with spines. The larva has well-developed mouth-hooks and the posterior spiracles open by three bent slits in a shallow concavity. When mature, in the stomach or passed in faeces, the larvae are cylindrical, 16–20 mm long and reddish-orange in colour, with posterior spiracles. Most *Gasterophilus* larvae have two rows of stout spines anteriorly on most segments, but in *G. nasalis* there is only one row. The spines are sharply pointed in *G. haemorrhoidalis* and blunt in *G. intestinalis.* In all species, the two posterior spiracles are united along their inner margins. Distinction of mature larvae of the various species is based on the numbers and distribution of the spines present on various segments (Fig. 117).

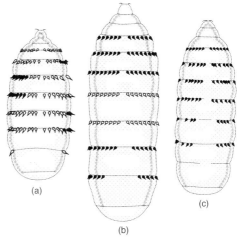

Fig. 117. Larvae of *Gasterophilus* species. (a) *G. nasalis* (16 mm); (b) *G. intestinalis* (18 mm); (c) *G. haemorrhoidalis* (18 mm). *Source*: Cameron, A.E. (1942) *Transactions of the Highland and Agricultural Society.*

4. LIFE CYCLE

In the northern temperate regions, one generation of *Gasterophilus* spp. occurs each year, but in the tropics and subtropics, multiple generations can occur each year. In a typical life cycle, the eggs are laid on the host; the first-stage larvae penetrate the tissues of the oral cavity and throat; the second and third stages are attached to the intestinal tract for several months before the mature larvae are voided in the faeces and pupate. Some of the specific variations on this pattern are described in the following.

Fecundity is roughly correlated with the size of the adult. *G. haemorrhoidalis* produces about 160 eggs, *G. nasalis* and *G. inermis* from 300 to 500, *G. intestinalis* from 400 to 700 and the largest species, *G. pecorum*, from 1300 to 2400 eggs. *G. pecorum* lays its glossy black eggs in batches of 10–15 on vegetation, mainly grass; the dark eggs of *G. haemorrhoidalis* are laid on the hairs around the lips, while the yellowish eggs of *G. nasalis*, *G. inermis* and *G. intestinalis* are laid on the intermandibular space below the jaws, on the cheeks and on the front legs, respectively. The eggs of *G. haemorrhoidalis* have a long, corrugated, stalk-like pedicel and the flanges which attach the eggs of *G. intestinalis* and *G. nasalis* to the hairs extend for half the length of the egg or its full length, respectively (Fig. 116). It is therefore possible to identify eggs of *Gasterophilus* to species level.

Attachment of the egg of *G. intestinalis* to host hairs is facilitated by the presence of an attachment groove filled with an adhesive material produced by or from the follicle cells and of lateral extensions which surround the hair completely. Eggs of *G. pecorum* and *G. intestinalis* do not hatch until they are ingested by a horse or stimulated by warmth, moisture and friction. At 25–30°C, embryonic development of *G. intestinalis* is completed within 2–4 days; however, hatching occurs after 3–6 days, resulting in a minimum period for emergence of larvae of 5–10 days. The viability of embryonated eggs is related inversely to humidity and temperature. At 10°C, eggs remain viable for 8 weeks at 100% RH (relative humidity) and for 12 weeks at 25% RH. Eggs laid late in autumn will retain the ability to hatch for several months, except in the occurrence of subzero temperatures.

Eggs of *G. inermis* hatch spontaneously and the larvae burrow into the cheek of the horse, causing a condition known as 'summer dermatitis'. Eggs of *G. nasalis* also hatch

spontaneously and the larvae migrate towards the mouth and enter the oral cavity between the lips. First-stage larvae of *G. intestinalis* burrow into the mucous membrane of the tongue. However, together with larvae of *G. nasalis*, they have been found in the alveolar space between the teeth and below the gum line, where they cause necrosis, formation of pus and difficulties in swallowing. The larvae localize in this region for up to a month and develop to second stage. Eggs of *G. haemorrhoidalis* hatch under the stimulus of moisture and burrow into the epidermis of the lip, migrating into the mouth through the subepithelial layer.

The second- and third-stage larvae are attached to the wall of the intestinal tract by well-developed mouth-hooks. The larvae of *G. pecorum* occur on the soft palate and at the basis of the tongue, with older third-stage larvae passing to the stomach; those of *G. intestinalis* are in the cardiac region of the stomach and those of *G. nasalis* occur in the pyloric region of the stomach and in the first portion of the duodenum, respectively. Larvae of *G. haemorrhoidalis* occur in the fundus of the stomach and the duodenum, with third-stage larvae reattaching to the rectal mucosa, where they may occur together with *G. inermis*. In an area where all six horse-infesting species were present, 7% of larvae were in the oral cavity, 56% in the stomach, 25% in the intestine and 12% in the rectum.

Larvae feed on tissue exudates. Although haemoglobin can be detected within larvae of *G. haemorrhoidalis*, unusually it is produced by the larva itself to assist with gas exchange. The larvae live in an environment where the oxygen tension approaches zero. They therefore produce an intracellular haemoglobin with a very high affinity for oxygen, which releases bound oxygen only at very low oxygen tensions. The haemoglobin is present in most larval tissues, but is concentrated in certain modified fat body cells of the so-called 'tracheal organ' or 'organe rouge'.

Larvae pupate in the soil and *G. intestinalis* pupae can survive for 8 weeks at 21°C and 18–20 days at 27–32°C. The most favourable conditions for survival of pupae are 29°C at 80–92% RH. No pupation occurs at 5°C. High survival of over 70% may be recorded between the post-feeding larva and adult emergence stages in several species.

5. BEHAVIOUR AND BIONOMICS

All species of *Gasterophilus* were restricted originally to the Palaearctic and Afrotropical regions, but three species, *G. nasalis*, *G. haemorrhoidalis* and *G. intestinalis*, have been introduced inadvertently into the New World.

The adults are diurnal and their activity peaks in the early afternoon in warm, sunny weather; no activity is observed on cloudy days, in strong winds or in heavy rain. Adults are short-lived and, even under favourable conditions, may live for 1 day only. Adults do not feed and the mouthparts are rudimentary.

Adult gasterophilines buzz, a habit shared by other oestrids. This is associated with endothermic heat production, which may raise the temperature of the thorax by 12°C above ambient and is the prelude to flight. Heat loss is regulated by the insulation provided by the thoracic hair and by the restriction of the haemolymph circulation to the abdomen. The optimal temperature for flight is 20–24°C, when the thoracic temperature is 31–32°C.

Mating occurs in the area surrounding either a single horse or a small group of horses, where solitary hovering males will establish and defend a territory. Mating also occurs on hilltops, where the males hover and aggressively pursue passing objects, probably in search of females. Males can hover in winds of 15–20 km/h and are active at temperatures of 19–34°C. Individuals can hover for up to 30 min without landing; males make contact with a female on the wing, couple and sink to the ground, where copulation is completed within 3–4 min.

Horses may be infested with gasterophiline larvae at any time of the year. A typical pattern of infestation has been described in Kentucky, USA. Virtually all horses were infested with *G. intestinalis* and 81% with *G. nasalis*. The average number of *G. intestinalis* varied from a low of 50 in September to a high of 229 in March, and the corresponding numbers for *G. nasalis* were 14 in September and 82 in February. Second-stage larvae of *G. intestinalis* from the previous season continued to reach the stomach of their host until April and were not voided in large numbers as post-feeding larvae until August, when the second-stage larvae of the current season which had reached the stomach in July would have developed into third-stage larvae. A similar overlap in generations has been reported for *G. nasalis*, with mature third-stage larvae from the previous year voiding from March to August and second-stage larvae of the new generation reaching the stomach in July and developing into third-stage larvae within 5–7 weeks, before the previous generation's infestation had been cleared. Hatching of the eggs of *G. nasalis* is not delayed and, consequently, in temperate regions of the northern hemisphere, infestations occur only between May and November. Eggs of *G. intestinalis* can remain viable for long periods of time, and although oviposition only occurs from early May till late October, infestations occur throughout the whole year, except for April. In Italy, the largest number of *Gasterophilus* larvae retrieved from a single host was 738 (689 *G. intestinalis* in the stomach, 39 *G. nasalis* in the duodenum and 10 *G. inermis* in the rectum). In a survey carried out in eastern Australia, 64% of the horses examined were infested by *G. intestinalis*, 19% had ulcers in the oesophageal region of the stomach, with 92% of the ulcerated stomachs associated with infestations of *G. intestinalis*. Ulcers are most common in early summer, when the deeply embedded third-stage larvae are almost fully developed.

6. MEDICAL AND VETERINARY IMPORTANCE

The presence of larvae in the buccal cavity of horses may lead to stomatitis, with ulceration of the tongue. Active tunnelling removes virtually all tissue in the path of the larvae including nerves and capillaries, leading to haemorrhage and exocytosis into the tunnels, which fill with erythrocytes mixed with macrophages, lymphocytes and some eosinophils. The tunnels may become infected with bacteria, which results in microabscesses composed of clotted erythrocytes, bacteria, disintegrating epithelial cells and large numbers of neutrophils. Cells surrounding the tunnel exhibit pyknosis and epithelial hydropic degeneration and become separated from each other. On attachment by their oral hooks to the stomach lining, larvae provoke an inflammatory reaction with the formation of funnel-shaped ulcers surrounded by a rim of hyperplastic epithelium. Heavy infestations result in chronic gastritis, loss of condition and, in rare cases, perforation and death. Burrowing of the first-stage larvae in the mouth lining, tongue and gums can produce pus pockets, loosen teeth and cause loss of appetite in the host. Larvae attached to the gastrointestinal mucosa cause inflammation and ulceration.

The adult fly can cause irritation and intense avoidance reactions when hovering around the host and laying eggs on the skin. Ovipositing females may be tenacious, laying eggs on mobile as well as stationary animals. In the presence of ovipositing gasterophilines, horses become nervous and may injure themselves. Female flies will pursue galloping horses and immediately resume oviposition when the horse stops.

Very occasionally, larvae of *Gasterophilus* spp. infest humans (mostly those who have close contact with horses). Oviposition can occur on to human body hair or infestation can occur by direct contact with hatching eggs on a horse. Experimentally, it has been shown that while the first-stage larva of *G. nasalis* cannot penetrate intact skin and that *G. intestinalis* can only penetrate damaged human skin, the first-stage larvae of the other four species of horse bots, including *G. haemorrhoidalis* and *G. pecorum*, are capable of penetrating intact human skin. The hatched larvae cause a creeping myiasis, in which the larva tunnels in the epidermis and causes considerable irritation as it advances up to 30 cm a day. The infestation may end spontaneously, but the larva may need to be excised. Human external ocular myiasis by *Gasterophilus* spp. has been reported in patients with a history of horse handling, and this is likely to be due to the accidental introduction of first-stage larvae into the eyes via the fingers. More rarely, oral myiasis and pulmonary myiasis have also been reported.

7. PREVENTION AND CONTROL

Effective control can be achieved by removal of the eggs from the host's coat, but this requires daily examination of the animal, paying particular attention to the area around the lips. Infestation of stomach bots in horses can be controlled by oral administration of organophosphate insecticides, but these generally have been replaced by broad-spectrum macrocyclic lactone compounds, such as ivermectin. In temperate areas, during winter, almost the entire *Gasterophilus* population will be present as larvae in the stomach. A single treatment during winter, therefore, should break the cycle effectively but, in certain areas where adult fly activity is prolonged by mild conditions, additional treatments may be required. There is some evidence to indicate that infestation with gasterophilines can be assessed by detecting circulating anti-*Gasterophilus* antibodies (for example, using ELISA).

Surgical removal may be required in humans with subdermal larvae, but great care should be taken during the extraction process to avoid rupturing the larva *in situ*. The administration of broad-spectrum antibiotics may help avoid secondary bacterial infection.

SELECTED BIBLIOGRAPHY

Catts, E.P. (1979) Hilltop aggregation and mating behaviour by *Gasterophilus intestinalis* (Diptera: Gasterophilidae). *Journal of Medical Entomology* 16, 461–464.

Colwell, D.D., Hall, M.J.R. and Scholl, P.J. (2006) *Oestrid Flies: Biology, Host–Parasite Relationships, Impact and Management*. CAB International, Wallingford, UK.

Colwell, D.D., Otranto, D. and Horak, I.G. (2007) Comparative scanning electron microscopy of *Gasterophilus* third instars. *Medical and Veterinary Entomology* 21, 255–264.

Drudge, J.H., Lyons, E.T., Wyant, Z.N. and Tolliver, S.C. (1975) Occurrence of second and third instars of *Gasterophilus intestinalis* and *Gasterophilus nasalis* in stomachs of horses in Kentucky. *American Journal of Veterinary Research* 36, 1585–1588.

Francesconi, F. and Lupi, O. (2012) Myiasis. *Clinical Microbiology Reviews* 25, 79–105.

Hall, M.J.R. and Wall, R. (1994) Myiasis of humans and domestic animals. *Advances in Parasitology* 35, 258–334.

Hatch, C., McCaughey, W.J. and O'Brien, J.J. (1976) The prevalence of *Gasterophilus intestinalis* and *G. nasalis* in horses in Ireland. *Veterinary Record* 98, 274–276.

Humphreys, W.F. and Reynolds, S.E. (1980) Sound production and endothermy in the horse botfly, *Gasterophilus intestinalis*. *Physiological Entomology* 5, 235–242.

McGraw, T.A. and Turiansky, G.W. (2008) Cutaneous myiasis. *Journal of the American Academy of Dermatology* 58, 907–926.

Otranto, D., Milillo, P., Capelli, G. and Colwell, D.D. (2005) Species composition of *Gasterophilus* spp. (Diptera, Oestridae) causing equine gastric myiasis in southern Italy: parasite biodiversity and risks for extinction. *Veterinary Parasitology* 10 (133), 111–118.

Rastagaev, Yu.M. (1978) Subcutaneous myiasis in man caused by larvae of the horse bot-fly. *Meditsinskaya Parazitologiya i Parazitarnye Bolezni* 47, 72–73.

Roelfstra, L., Vlimant, M., Betschart, B., Pfister, K. and Diehl, P.A. (2010) Light and electron microscopy studies of the midgut and salivary glands of second and third instars of the horse stomach bot, *Gasterophilus intestinalis*. *Medical and Veterinary Entomology* 24, 236–249.

Sukhapesna, V., Knapp, F.W., Lyons, E.T. and Drudge, J.H. (1975) Effect of temperature on embryonic development and egg hatchability of the horse bot *Gasterophilus intestinalis* (Diptera: Gasterophilidae). *Journal of Medical Entomology* 12, 391–392.

Tolliver, S.C., Lyons, E.T. and Drudge, J.H. (1974) Observations on the specific location of *Gasterophilus* spp. larvae in the mouth of the horse. *Journal of Parasitology* 60, 891–892.

Waddell, A.H. (1972) The pathogenicity of *Gasterophilus intestinalis* larvae in the stomach of the horse. *Australian Veterinary Journal* 48, 332–335.

Zumpt, F. (1965) *Myiasis in Man and Animals in the Old World*. Butterworths, London.

Ticks (Hard) (Acari: Ixodidae)

1. INTRODUCTION

The ticks (Ixodida) are relatively large acarines, all of which are blood-feeding ectoparasites of vertebrates. The order can be divided broadly into 'hard' and 'soft' ticks, based on the possession of a dorsal scutum in the Ixodidae (the 'hard' ticks), which is absent in the Argasidae (the 'soft' ticks). Hard ticks are usually relatively large and long-lived. During this time, they feed periodically, taking large blood meals, and often with long intervals off the host between each meal. Since a large proportion of the life cycle of most hard tick species occurs off the host, the habitat in which they live is of particular importance. It must contain a large enough concentration of host species to sustain the tick population and have a sufficiently high humidity for the ticks to maintain their water balance. Tick bites can directly damage humans and domestic animals, but of particular importance is the fact that hard ticks also transmit a wide range of pathogenic viral, bacterial and protozoal agents. Indeed, ticks transmit a greater variety of pathogens than any other vector taxon. Hence, although the ticks are a relatively small order, they are one of the most important groups of arthropods of medical and veterinary interest.

2. TAXONOMY

The Acari are a subclass of the Arachnida, with two superorders: Parasitiformes (Anactinotrichida) and Acariformes (Actinotrichida). The Parasitiformes possess one to four pairs of lateral stigmata posterior to the coxae of the second pair of legs and have freely movable coxae. There are four orders of Parasitiformes, two of which have few species and are of no medical or veterinary significance and two, the Mesostigmata (Gamasida) and the Ixodida (Metastigmata), which are of significant medical and veterinary importance.

The Ixodida contains about 900 species arranged in three families: the largest, the Ixodidae (hard ticks), has 14 genera and 704 species, while the Argasidae (soft ticks) has 5 genera and about 195 species. The third family, the Nuttalliellidae, comprises only a single species, *Nuttalliella namaqua*, found in the Afrotropical region (this family probably represents a basal lineage of largely extinct species which fed on reptiles, and *N. namaqua* may be considered the evolutionary link between the hard and soft ticks).

Within the Ixodidae, there are six main genera: *Amblyomma*, *Ixodes*, *Dermacentor*, *Hyalomma*, *Rhipicephalus* and *Haemaphysalis*. A number of smaller genera have also been described, such as *Margaropus*, in which have been ranked only three species found in Africa and collected from giraffe, zebra and horses, and *Bothriocroton*, of which there are seven species found in Australasia on monitor lizards, monotremes and marsupials. The family Ixodidae is often divided into Prostriata (only *Ixodes*) and Metastriata (all other ixodid genera).

3. MORPHOLOGY

The Ixodidae are large, bloodsucking Acari with a terminal movable gnathosoma in all stages. The gnathosoma consists of the basis capituli, paired four-segmented palps, paired chelicerae and a ventral median hypostome, armed with rows of backwardly directed teeth, which attach the tick securely to its host. The fourth segment of the palp is reduced and recessed on the ventral surface of the third

© R.C. Russell 2013. *The Encyclopedia of Medical and Veterinary Entomology* (R.C. Russell *et al.*)

segment. The capitulum is terminal and always visible when the tick is viewed from above. In the Ixodidae, sexual dimorphism is well developed, the dorsal scutum being small in the female and almost covering the whole of the dorsal surface in the male. When eyes are present, they are located dorsally at the sides of the scutum. The stigmata are large and posterior to the coxae of the fourth pair of legs. The pulvillus is well developed. The genital opening and the anus are located ventrally, the former being at the level of the second pair of legs and the latter a little posterior to the fourth pair of legs. Haller's

organ, a sensory pit packed with sensillae which are used in host recognition, is located dorsal distally on the tarsus of the first pair of legs (Fig. 118).

Ixodes is the largest genus in the family, with more than 243 species. They are small, inornate ticks (Fig. 118), with the gnathosoma of the female being considerably longer than that of the male. Often, the second segment of the palp is constricted at the base, creating a gap between the palp and the mouthparts. There are no eyes or festoons. The anal groove passes anteriorly to the anus and, for this reason, *Ixodes* is said to be prostriate. In

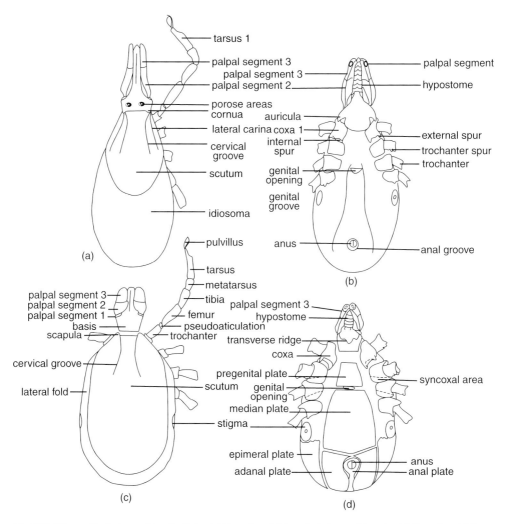

Fig. 118. Diagnostic features of *Ixodes* ticks. (a) Dorsal and (b) ventral views of female. (c) Dorsal and (d) ventral views of male. *Source*: Arthur (1965).

other genera, the anal groove is either posterior to the anus or obsolete and they are referred to as metastriate. In the male, there are seven ventral plates including a median row of three – pregenital, median and anal – a pair of adanals and a pair of epimerals. The margins of the epimerals, which are placed posterolaterally, are often indistinct.

There are 166 easily recognizable species in the genus *Haemaphysalis* (Fig. 119). They are small, inornate ticks with short mouthparts (described as brevirostrate). The basis capituli is rectangular and the base of the second palpal segment is often expanded, projecting laterally beyond the basis capituli. The second and third palpal segments taper anteriorly, so that the gnathosoma anterior to the basis capituli appears to be triangular. There are no eyes in either sex and no ventral plates in the male. Festoons are uniform, rectangular areas along the posterior margin of the body, separated by grooves and best seen in unfed specimens, as they are lost in engorged females.

The 82 species of *Rhipicephalus* are small metastriate, brevirostrate, reddish or blackish-brown ticks, which are mostly inornate (Fig. 120). The basis capituli is hexagonal dorsally

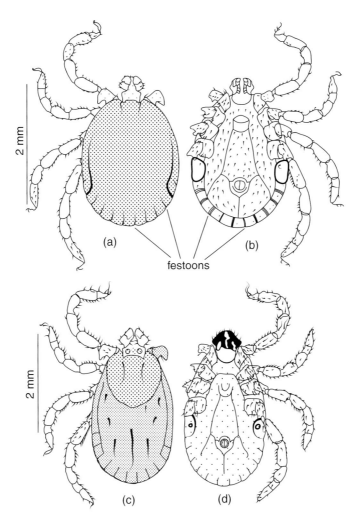

Fig. 119. *Haemaphysalis longicornis*. (a) Dorsal and (b) ventral views of male. (c) Dorsal and (d) ventral views of female. *Source*: Hoogstraal *et al.* (1968).

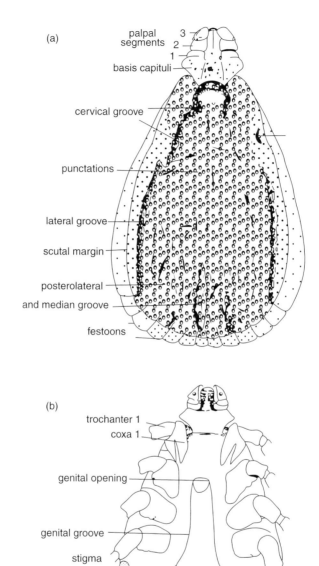

(a)

palpal segments
3
2
1

basis capituli

cervical groove

punctations

lateral groove

scutal margin

posterolateral
and median groove

festoons

(b)

trochanter 1
coxa 1

genital opening

genital groove

stigma

anus

accessory adanal shield

adanal shield

anal groove

Fig. 120. (a) Dorsal and (b) ventral views of male *Rhipicephalus sanguineus. Source*: Nuttall, G.H.F. (1911) *Ticks: A Monograph of the Ixodoidea*, Part II, p. 122. Cambridge University Press.

and eyes and festoons are present. Coxa I is bifid in both sexes. The male has adanal and accessory adanal plates on the ventral surface and, when replete, the central festoon may expand, forming a single caudal process posteriorly. The subgenus *Boophilus* (formerly considered a genus in its own right but now subsumed into the genus *Rhipicephalus*) contains five species. These are small, inornate, brevirostrate ticks in which the anal groove is obsolete. The basis capituli is hexagonal

dorsally, and there are simple eyes laterally on the scutum. Ventrally, coxa I is bifid, and there are paired adanal and accessory adanal plates flanking the anus posteriorly (Fig. 121). Replete males may develop a caudal process, which might be present even in unfed males (e.g. *Rhipicephalus* (*Boophilus*) *microplus*).

The 34 species of *Dermacentor* are medium to large, usually ornate, metastriate, brevirostrate ticks (Fig. 122). The basis capituli is rectangular dorsally and eyes and festoons are present. Coxa I is bifid in both sexes and coxa IV is greatly enlarged in the male, which has no ventral plates (Fig. 123).

The 27 species and subspecies of *Hyalomma* are medium-sized metastriate ticks with long mouthparts (i.e. longirostrate) (Fig.

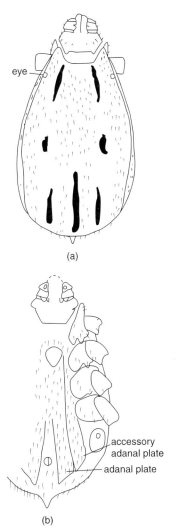

Fig. 121. (a) Dorsal and (b) ventral views of male *Boophilus microplus*. *Source*: Arthur, D.R. (1960) *Ticks: A Monograph of the Ixodoidea*, Part V, p. 209. Cambridge University Press.

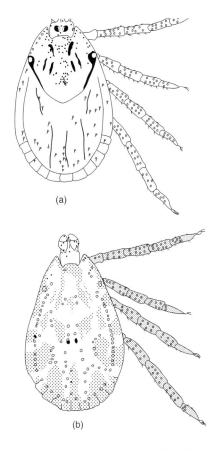

Fig. 122. *Dermacentor andersoni*. (a) Dorsal view of female; (b) dorsal view of male. *Source*: Arthur, D.R. (1962) *Ticks and Disease*. Pergamon Press, Oxford.

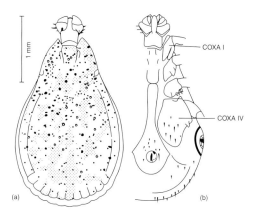

Fig. 123. (a) Dorsal and (b) ventral views of male *Dermacentor reticulatus. Source*: Arthur, D.R. (1963) *British Ticks.* Butterworth, London.

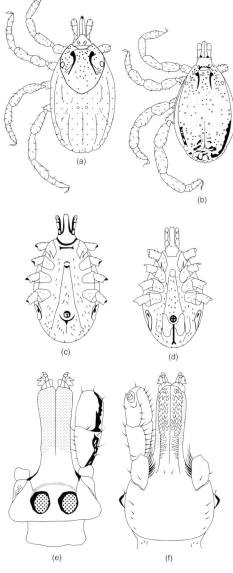

Fig. 124. (a) Dorsal and (c) ventral views of female *Hyalomma excavatum.* (b) Dorsal and (d) ventral views of male *H. excavatum.* (e) Dorsal and (f) ventral views of capitulum of female *H. aegyptium. Sources*: (a, b, c and d) Arthur, D.R. (1962) *Ticks and Disease.* Pergamon Press, Oxford. (e and f) Nuttall, G.H.F. (1911) *Ticks – A Monograph of the Ixodoidea,* Part II, p. 122. Cambridge University Press.

124). The basis capituli is subtriangular dorsally and eyes are present. Festoons and ornamentation of the scutum are variable characters which may be present or absent. The male has one pair of adanal plates and accessory adanal plates may be present. Coxa I is bifid.

The 130 species of *Amblyomma* are large, mostly ornate, metastriate, longirostrate ticks, with eyes and festoons but no adanal plates in the male (Fig. 125). Basis capituli are of variable shape and coxae bear distinct spurs, which are of taxonomic interest.

4. LIFE CYCLE (GENERAL)

4.1 Oviposition

There are four stages in the life cycle: egg, larva, nymph and adult. The female drops off its vertebrate host and seeks a sheltered situation in which to develop and lay a single large batch of eggs, after which it dies. Typically, a batch contains several hundred to several thousand brown, globular eggs (up to 5000 in *Rhipicephalus sanguineus*), and oviposition continues for many days. *Ixodes ricinus* has been recorded as depositing an egg every 3–12 min.

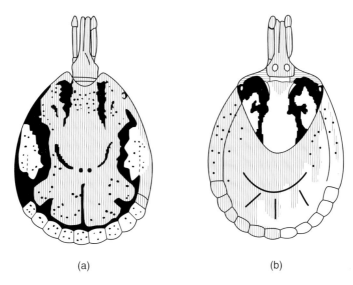

(a)   (b)

Fig. 125. Dorsal view of (a) male and (b) female *Amblyomma hebraeum. Source*: Arthur, D.R. (1962) *Ticks and Disease.* Pergamon Press, Oxford.

During oviposition, the tick bends its gnathosoma in an arc ventrally, until it is oppressed tightly to the ventral surface and the tip of the gnathosoma is near the genital opening. Gene's organ is everted from between the dorsal posterior margin of the capitulum and the anterior margin of the scutum, which has a swollen base and two short horns. During the extrusion of an egg, the lining of the vagina prolapses through the genital opening holding the egg, which is deposited between the horns of Gene's organ, and the prolapse is then retracted.

The function of Gene's organ is to apply a waterproofing wax to the egg, after which the organ is withdrawn and the egg is left on the hypostome. When the capitulum returns to its normal forwardly directed position, the egg is deposited above the female. Consequently, the egg batch is to be found above the shrivelled body of the female. If the eggs do not receive a wax coat, they shrivel and die. In *I. ricinus*, an incomplete waterproof layer is added during the egg's passage down the oviduct and waterproofing is completed by Gene's organ, which, in this species, secretes an easily spreading and penetrating soft wax. In the first few days of incubation, the wax penetrates the

outer layers of the egg to reach the inner membrane. The critical temperature at which the wax ceases to waterproof the egg is 35°C in *I. ricinus* and 44°C in *Hyalomma marginatum*. Depending on climatic conditions, eggs hatch in 2 weeks to several months, releasing a hexapod larva.

4.2 Larva

Larvae have neither stigmata nor a tracheal system and therefore water loss is solely through the cuticle. When a larva succeeds in acquiring a host, it attaches and feeds for several days, before dropping off engorged. Larvae that are not successful in finding a host move down to the more humid layers near the ground, where, at suitable humidities, they are able to take up water from the atmosphere. In this way, larvae can survive for considerable periods unfed until they can attach to a host.

4.3 Nymph and adult stages

The larva digests its blood meal and moults to become an octopod nymph, which has stigmata and a tracheal system, making it more susceptible to desiccation. The nymph feeds for 4–8 days on a host, engorging more rapidly towards the end of the period, before dropping off to find a suitable location in which to digest its meal, moult to the adult stage and attach to a new host. Mating of prostriate ticks may take place on the ground before a female adult acquires a host or on the host itself, whereas metastriate ticks mate exclusively on the host. The male feeds but does not engorge. The engorged, mated female drops off the host and hides away in a suitable location to digest the blood meal and lay an egg batch.

4.4 Life cycle variations

The majority of ixodid ticks have the type of life cycle described above and are referred to as three-host ticks. However, ticks of the genus *Margaropus* and *Rhipicephalus* subgenus *Boophilus* are one-host ticks, and engorged larvae and nymphs do not drop off the host but remain attached and moult *in situ*. The subsequent stage reattaches on the same individual host (also known as 'monotropic' ticks). A few species in other genera are also one-host ticks, including *Dermacentor nitens* (a Neotropical ear tick of horses), *Dermacentor albipictus* (the North American moose tick) and *Hyalomma scupense* in the Palaearctic region. Further, a few species have adopted a two-host cycle in which the larva and nymph occur on the same individual host, but the nymph then drops off and the adult parasitizes a different individual of the same or another host species (also known as 'ditropic' ticks). Included in the two-host ticks which parasitize domestic animals are: *Rhipicephalus evertsi* in the Afrotropical region and *Rhipicephalus bursa*, *Hy. marginatum* and *Hyalomma detritum* in the Palaearctic region.

To overcome adverse conditions, or times when the host is not present, diapause may occur at any stage of the life cycle and both pre- and post-feeding. There may be several generations in a year (e.g. *R. (Bo.) microplus*) or the life cycle may extend over 3–4 years and up to 7 years in cold subarctic latitudes.

5. BEHAVIOUR AND BIONOMICS

5.1 Host finding and attachment

Although many ixodid ticks are not host specific, they are not indiscriminate in the hosts they parasitize. A few show an extremely wide host range, and these are often of economic importance, but most occur on a limited range of hosts, which they parasitize with varying intensities, and a few do appear to be host specifc. Some hosts are commonly and heavily infested and others infrequently and lightly. Each species is adapted to its host or host range, concentrating on particular parts of the host's body and adjusting seasonal and daily activity cycles to the host's behaviour and availability.

Where there is little or no vegetation cover, ticks, for example *Hyalomma asiaticum*, detect the presence of a host by ground vibrations and emerge from their burrows and move towards the host. In vegetated areas, tick larvae seek a host by climbing vegetation and accumulate near the tips of grasses and similar plants. When a host approaches, larvae detect its presence mainly by sensory receptors in Haller's organ on the tarsus of the first pair of legs, and they quest for the host by waving those legs in the air. Sensory receptors in the anterior pit of Haller's organ respond to odours, especially of phenolic compounds, and other receptors respond to humidity, temperature and ammonia, while those in the posterior capsule respond to carbon dioxide, odours and temperature. Other receptors on the tarsus detect phenolic compounds, which are produced in large quantities by feeding females and act as sex pheromones, and others detect temperature gradients and respond to the host's body heat. Once on the host, the sensory receptors of the tarsi and the palps provide information to enable the tick to find a suitable site for attachment.

The tick pierces the skin of its host with its chelicerae and inserts the barbed hypostome to secure it to the host initially. The cheliceral digits have mechanoreceptors and sensory receptors, which respond to adenosine-5'-triphosphate (ATP) and other haemato-phagostimulants. The pattern of feeding of female *Rhipicephalus appendiculatus* involves the pharyngeal dilator muscles initiating fast sucking for 5 min, followed by 1–2 min rest, when the floor of the salivarium, the fused paired salivary ducts, is actively lowered. When the chelicerae of a feeding *Boophilus* are exposed to the blood of a resistant host, the feeding pattern deteriorates and the tick may detach.

In ticks with short mouthparts, cement secreted during the first 24 h spreads over the surface of the skin to secure it initially to the host. During the next 96 h, extra cement is secreted, which penetrates the keratinized layers of the stratum corneum, fills the lesion and attaches the tick more firmly. In ticks with longer mouthparts, there is less need of cement and it is reduced to a sheath around the mouthparts within the host.

5.2 Feeding and water elimination

During engorgement, the body weight of a tick increases by about 200 times. An unfed female *I. ricinus* weighs about 2 mg and takes in 600 mg of blood while feeding, but half to two-thirds of the water contained in the ingested blood is eliminated before feeding ceases and the final weight of the engorged tick is about 240 mg. In argasid ticks, excess water is eliminated via the coxal apparatus, but no such structure is present in ixodid ticks. Little urine is produced by the Malpighian tubules and excess fluid is eliminated by salivation, the voluminous saliva being passed back into the host.

Mammalian blood is hyposmotic to tick tissues. When *I. ricinus* feeds, for example on sheep, which have a blood chloride concentration of 0.5%, the chloride content of its haemolymph rises from 0.72% in the unfed tick to 0.88% in the fed tick. This can only be achieved by the secretion of hyposmotic saliva. The saliva contains anticoagulants, weak hydrolytic enzymes and pharmacological agents, some of which may play important roles in the feeding process and in pathogen transmission.

5.3 Water balance

The only liquid imbibed by ticks is blood, and hence it is vital to their survival off the host to conserve their water content and minimize losses. Water loss in a non-feeding tick occurs through the cuticle or through the stigmata. The critical equilibrium activity (CEA) is the threshold at which ticks neither loose nor gain water. For *Amblyomma americanum*, a species associated with woodlands, the CEA occurs at 88% RH (relative humidity) and for the more xeric *Amblyomma cajennense* at 82% RH. At 75% RH and 22–23°C, 50% of female *A. cajennense* survive for 9 months and males for 7 months, whereas adult *R. sanguineus* survive for up to 20 months under laboratory conditions (70–80% RH, 25–26°C).

Ticks have the ability to absorb water by passive and active water uptake when the moisture content of the atmosphere is above the CEA. Active uptake is achieved by the salivary glands secreting a hygroscopic fluid, which absorbs water from the atmosphere and is then imbibed by the tick. Adults have the greatest resistance to water loss and the highest rate of water imbibition. When kept at 0% RH and 25°C, unfed female *I. ricinus* may lose 17% of their body weight in 24 h. In contrast, ticks that are adapted to hotter environments, such as *Hy. marginatum*, lose considerably less water under the same conditions.

To reduce water loss from the stigmata, these complex structures can be opened and closed. Female *Amblyomma variegatum* held in an atmosphere of high carbon dioxide, when the spiracles would be kept open, lost 17 times as much water at 0% RH as at 93% RH. The water loss of larvae remained unchanged, since they had no spiracles. Ticks are able to keep water balance if completely flooded, reducing their metabolism and using oxygen stocked in air chambers below the stigmata surface. Hence, female *R. sanguineus* and *Dermacentor variabilis* can survive water immersion up to 3 and 15 days, respectively.

5.4 Pheromones, aggregation, mating and allomones

Several different pheromones are produced by ixodid ticks. They are detected by sensory receptors on Haller's organ, on the terminal segment of the palps and on the cheliceral digits. Assembly pheromones are species specific, affect all stages in the life cycle, but are not universally produced. They have been demonstrated in *Ixodes holocyclus*, a paralysis tick of Australia, and *Bothriocroton* (*Aponomma*) *concolor*, which parasitizes echidnas (monotreme mammals) in Australia. A similar pheromone is produced by the American black-legged (or deer) tick, *Ixodes scapularis*. Males of species of *Amblyomma* which parasitize large ungulates in North America and southern Africa secrete an aggregation-attachment pheromone, which attracts both adults and nymphs. Female *Amblyomma maculatum* and *Amblyomma hebraeum* will not attach to hosts unless feeding males are present. The pheromone is a blend of phenols secreted by males that have been feeding for 5 days and reaches a peak after 6–8 days. A similar period is required to complete spermatogenesis.

In *Ixodes*, mating occurs both on and off the host, but in the Metastriata, mating occurs only on the host, because both sexes need to feed before becoming sexually competent. Attractant sex pheromones are secreted by females and attract males which have fed for 3–4 days. One such pheromone, 2,6-dichlorphenol, has been found in 14 species and four genera, including *Amblyomma*, *Rhipicephalus*, *Dermacentor* and *Hyalomma*. In *Dermacentor andersoni*, the pheromone is synthesized early in adult life by the foveal gland, but it is not released until the female is feeding. In *R. appendiculatus*, both synthesis and release of the pheromone are associated with feeding. In response to the pheromone, the male detaches and orientates to the female. In *D. variabilis*, males may migrate as far as 43 cm to reach females on the opposite side of a dog host. Males of *Hyalomma dromedarii* and *Hyalomma excavatum* both secrete the same attractant sex pheromone, but in differing amounts, which enables males to find conspecific females. Discrimination in *D. variabilis* and *D. andersoni* involves two further sex pheromones, a mounting sex pheromone, which in *D. variabilis* is cholesteryl oleate, and a highly species-specific genital sex pheromone.

In *I. scapularis*, insemination is essential for the female to engorge fully, and more blood is ingested when mating and feeding occur together. In *I. ricinus* and *I. scapularis*, mating also occurs off the host in the herbage, where males are readily attracted to a pheromone emitted by virgin females. This water-soluble pheromone functions as both an assembly and sex pheromone. Unmated female *I. ricinus* of the spring population diapause over winter, but if they are mated in summer, they may not enter diapause but will become active in autumn.

During mating, the mouthparts of the male are used to stimulate the female prior to spermatophore transfer. In *Ixodes*, the hypostome and chelicerae are inserted into the genital opening of the female and in the Metastriata only the tips of the chelicerae. Males do not usually copulate before spermatids have been formed; these are transferred to the female in a spermatophore and males can copulate 20–30 times. Female *Dermacentor occidentalis* have been shown to mate up to five times. *Haemaphysalis longicornis* is unusual in that while it has sexual reproduction in the greater part of its range, in the cold parts of its distribution the species may be parthenogenetic.

Metastriate ticks (e.g. *A. americanum* and *D. variabilis*) are characterized by the presence of large wax glands on their dorsolateral surfaces. When the tick is disturbed, these glands secrete copiously a squalene-rich substance which can function as an allomone by repelling predators.

6. MEDICAL AND VETERINARY IMPORTANCE

Ixodid ticks are responsible for a toxicosis that is life threatening for both humans and other animals, and they also play an important role as vectors of a wide range of pathogens to humans and domestic animals. They are involved in the transmission of arboviruses to humans (e.g. agents of tick-borne encephalitis, Crimean-Congo haemorrhagic fever and Kyasanur Forest disease) and domestic animals (louping ill). They transmit rickettsiae to humans (e.g. agents of Rocky Mountain and Mediterranean spotted fever, ehrlichiosis and Queensland tick typhus) and to domestic animals (e.g. agents of ehrlichiosis, Q fever and anaplasmosis). They also transmit bacteria (agents of Lyme disease, relapsing fever and tularaemia) to humans and other animals and protozoa (agents of babesiosis and theileriosis) and filaria (agents of dermal filariasis) to animals.

Their most important role in a medical context is probably as vectors of agents of diseases such as Lyme disease, Rocky Mountain spotted fever and tick-borne encephalitis, while in a veterinary/economic context it is the transmission of the protozoans causing babesiosis and theileriosis to cattle, although heavy tick infestations also damage hides and cause a loss in live weight gain (estimated at 0.6 g/day for every engorged female *R. (Bo.) microplus* and 4–5 g/day for female *A. variegatum*).

6.1 Vectorial role of tick genera

6.1.1 *Ixodes*
The most important group is the Holarctic 'ricinus–persulcatus complex', which includes the North American black-legged tick, *I. scapularis*, the European sheep tick, *I. ricinus*, and the northern Palaearctic taiga tick, *Ixodes persulcatus*, which occurs from the Baltic Sea to Japan. All three are vectors of Lyme disease and *I. persulcatus* and *I. ricinus* are also the chief vectors of tick-borne encephalitis virus. *Ixodes rubicundus* and *I. holocyclus* cause paralysis in mammals in southern Africa and Australia, respectively.

6.1.2 *Haemaphysalis*
Ticks of this genus are most strongly represented in the Oriental region, where *Haemaphysalis spinigera* is the vector of the arbovirus causing Kyasanur Forest disease. *Haemaphysalis leachi*, the yellow dog tick, is a widespread parasite of carnivores, especially dogs, in the tropics and subtropics. *Haema-physalis punctata* parasitizes cattle in the Palaearctic region, as does *H. longicornis* in the Australian and Oriental regions.

6.1.3 *Rhipicephalus*
The genus reaches its greatest diversity in the Afrotropical region, where *R. appendiculatus*, the brown ear tick, is the main vector of *Theileria parva*, the causative agent of east coast fever in cattle. Other important species on domestic animals are *R. evertsi* and *Rhipicephalus simus*. The brown dog tick, *R. sanguineus*, is the most widespread ixodid species and, although commoner in the warmer parts of the world, it survives in heated buildings in urban communities in the northern hemisphere. The ticks of the subgenus *Boophilus* are one-host ticks which parasitize large mammals, especially cattle: *R. (Bo.) microplus*, the pantropical cattle tick, which occurs in the Neotropical and Afrotropical regions, with the closely related *Rhipicephalus (Bo.) australis* in the Australian region; *Rhipicephalus (Bo.) decoloratus* occurs in tropical Africa and *Rhipicephalus (Bo.) annulatus* in North America. They are vectors of babesiosis to cattle.

6.1.4 *Dermacentor*
Ticks of this genus are found in the New World and include the Rocky Mountain wood tick, *D. andersoni*, and the American dog tick, *D. variabilis*, both vectors of Rocky Mountain spotted fever and the cause of tick paralysis; the former is found in the Rocky Mountain states and the latter is abundant east of the Rocky Mountains and also occurs in limited areas on the Pacific Coast of the USA. In the Palaearctic region, species of *Dermacentor*, including *Dermacentor marginatus*, are vectors of Siberian tick typhus and *Dermacentor reticulatus* parasitizes cattle and horses.

6.1.5 *Hyalomma*
Hyalommas are tough, hardy ticks which survive where humidity is low, climatic conditions extreme, hosts rare and hiding places sparse. They probably originated in the desert regions of Kazakhstan and Iran in the Palaearctic region. *Hyalomma anatolicum* and species of the *Hy. marginatum* complex

play a major part in the transmission of the arbovirus causing Crimean-Congo haemorrhagic fever.

6.1.6 *Amblyomma*

Species of *Amblyomma* are large, ornate, metastriate, longirostrate ticks with eyes and festoons, but no adanal plates in the male. The genus is particularly well represented in the New World. Some species, particularly the Cayenne tick, *A. cajennense*, and the Neotropical yellow dog tick, *Amblyomma aureolatum*, are potential vectors of rickettsial agents to humans and animals.

6.2 Tick paralysis

More than 60 different species of ticks belonging to ten genera have been associated with a salivary toxicosis that can cause paralysis in humans and/or various other animals. The most important paralysis-causing ticks are *I. holocyclus* in eastern Australia and *I. rubicundus* in South Africa and *D. andersoni* and *D. variabilis* in North America. Of less importance are *R. evertsi* in South Africa, *D. occidentalis* and *Ixodes brunneus* in North America and *H. punctata* and *I. ricinus* in Europe.

I. holocyclus is present in moist, vegetated habitats along the eastern coast of Australia. Its principal hosts are bandicoots, but it will attach to a wide range of alternative hosts including livestock, pets and humans. The natural hosts of *I. holocyclus* are not greatly inconvenienced by the paralysing toxin, as the majority become immune to its effect. In domestic animals, however, it has been estimated that *I. holocyclus* causes 100,000 cases of tick paralysis a year and the death of 10,000 calves, young animals being more susceptible than adults. In South Africa, *I. rubicundus*, the Karoo paralysis tick, is estimated to cause about 30,000 deaths a year in livestock, mainly sheep; tick paralysis due to *R. evertsi* occurs primarily in sheep and goats and its severity is related to the number of female ticks which have engorged. Recently, tick paralysis has been associated with *R. sanguineus* infestation in young dogs, confirmed ('*ex juvantibus*' – by inference) by

early and complete recovery of some of them following acaricidal treatment and tick removal.

Typically, paralysis is associated with the feeding of an adult female tick; it is generally seen as muscular incoordination in dogs and other animals, while in humans (where children are most often affected) it can be an ascending symmetrical motor paralysis where, at the onset, paralysis affects the lower limbs ascending to the torso, upper limbs and head regions within a few hours, and respiration is affected, which can be life threatening. First symptoms usually occur 5–7 days after attachment and a single female tick may be sufficient to paralyse completely and kill an adult human. Except in the case of *I. holocyclus*, removal of the attached tick terminates the condition and allows complete recovery; however, with *I. holocyclus*, physical removal of the tick may worsen the symptoms rapidly for 1–2 days. The toxin secreted by *I. holocyclus* is known as holocyclotoxin and it has been isolated and partially characterized; an antitoxin raised in hyperimmunized dogs is available and is widely used with dogs, but also can be used with cats, sheep, goats, pigs, calves and foals.

6.3 Tick-borne pathogens

6.3.1 *Arboviruses*

Arboviruses are defined as viruses that multiply in both their vertebrate and invertebrate hosts, and the term is therefore restricted to viruses that are transmitted biologically by arthropods. Arboviruses are found in five families: the Bunyaviridae, Flaviviridae, Reoviridae, Rhabdoviridae and Togaviridae. Hard ticks are vectors of a number of bunyaviruses and flaviviruses and one reovirus of medical or veterinary significance.

CRIMEAN-CONGO HAEMORRHAGIC FEVER VIRUS (CCHFV)

This bunyavirus is enzootic in the Palaearctic, Oriental and Afrotropical regions, chiefly in steppe, savannah, semi-desert and foothill biotopes, where one or two *Hyalomma* species are the predominant ticks parasitizing domestic and wild animals. In the southern republics of the former Soviet Union, 20% of human infections result in clinical illness with mortalities of 5–10%. In South Africa, the rates may be higher, with 70% of infections resulting in clinical disease with mortalities of 35% and, in Central Asia and in the Middle East, 35–50% among clinical cases. The incubation period in humans is 2–5 days and clinical disease 7–10 days, involving fever, severe headache, myalgia, malaise, rash and internal and external bleeding from the orifices. Infection occurs through the bite of the infected tick or by crushing infected ticks in contact with the skin or from shearing tick-infested sheep.

CCHFV survives transstadially and interseasonally in several tick species and is transmitted transovarially in members of the *Hy. marginatum* complex. Several tick species belonging to different genera have been reported to be reservoirs or vectors of CCHFV. They include one-host ticks of the subgenus *Boophilus*, two-host ticks of the genus *Hyalomma*, including the *Hy. marginatum* complex, *Hy. anatolicum* and *R. bursa*. Three-host ticks of the genera *Ixodes*, *Haemaphysalis*, *Amblyomma*, *Dermacentor*, *Hyalomma* and *Rhipicephalus* serve chiefly to maintain enzootic foci of CCHFV. Circulation occurs between ticks and various wild and domestic animals, and ticks of the *Hy. marginatum* complex and *Hy. anatolicum* are especially important in causing epidemics and outbreaks of CCHFV on account of their great numbers and their aggressiveness in seeking human hosts. Epidemics occur under a combination of favourable conditions and environmental changes which favour the survival of large numbers of *Hyalomma* and their hosts.

NAIROBI SHEEP DISEASE VIRUS (NSDV)

This bunyavirus causes a severe disease of sheep and goats, with fever, collapse and diarrhoea, and mortality may reach 90%. Outbreaks have been recorded in various regions of eastern Africa. The vector is the three-host tick, *R. appendiculatus*, in which transovarial and transstadial transmission occurs. No virus or antibodies to NSDV have been found in wild ruminants or rodents and the virus appears to be restricted to sheep, goats and *R. appendiculatus*. Human infections with NSDV occur only rarely as a self-limiting, febrile illness with abdominal and back pain, headache, nausea and vomiting.

TICK-BORNE ENCEPHALITIS VIRUS (TBEV)

This flavivirus is widely distributed (as a number of different virus subtypes) across the northern Palaearctic region. At least three subtypes of TBEV are usually recognized: a European, central Siberian and a Far Eastern subtype (the latter known originally as Russian spring–summer encephalitis virus, RSSEV). The Far Eastern subtype occurs in Siberia, the southern republics of the former Soviet Union and north-eastern China, and the European subtype occurs in Europe, including Russia west of the Ural Mountains. There are several thousand cases a year in Europe and the republics of the former USSR. It has been estimated that inapparent cases outnumber clinical cases by 30 to 1. Infection begins as an influenza-like illness, with fever and non-specific symptoms, but can progress to neurological involvement, encephalitis and death. RSSEV causes a more severe disease, with 20–30% mortality compared with the mortality caused by the European subtypes (1%).

I. persulcatus is the main vector in the Far East and *I. ricinus* in the west. The viruses survive in the ticks by transstadial and transovarial transmission. A large range of small forest mammals and birds circulate the viruses and provide hosts for larvae and nymphs of the vector ticks, whereas adults parasitize medium-size to large wild and domestic mammals.

Transmission of TBEV can also occur through consuming fresh milk or cheese from infected goats or sheep. Patient management is symptomatic as there is no specific treatment or antiviral therapy, but safe and effective vaccines are available.

KYASANUR FOREST DISEASE VIRUS (KFDV)

This flavivirus was first detected in Karnataka State in India in 1957, when an epidemic in the human population was associated with an epizootic in monkeys and deaths occurred in both humans and monkeys. In humans, the incubation period is 3–8 days, with the disease sometimes having two episodes 1–2 weeks apart. There may be mild meningoencephalitis and some haemorrhagic manifestations, with 5% mortality. Isolations of virus have been made from seven species of *Haemaphysalis* ticks; the most important is *H. spinigera*, in which transstadial transmission but not transovarial transmission occurs. KFDV has been recovered from a variety of small rodents, squirrels and shrews, which are probably the maintenance hosts, with monkeys, langur (*Presbytis entellus*) and bonnet macaque (*Macaca radiata*) acting as amplifying hosts. On recovery, monkeys are immune. The emergence of KFDV as a human disease has resulted from a rapidly increasing human population having contact with the forest. Cattle grazed in and beside the forest provide additional hosts for *H. spinigera*. Meanwhile, forest clearing may cause an increased interface of forest and human agricultural activity, with a greater number of people visiting the forest area potentially being exposed to KFDV infection. There is no known treatment for this highly infectious disease, but an effective inactivated vaccine is currently in use in India. KFDV has generally been thought to be restricted to India, but many persons living in the Andaman and Nicobar Islands have been reported as being seropositive for KFDV. Variants of the virus have been identified from Saudi Arabia (e.g. Alkhurma haemorrhagic fever virus) and China.

OMSK HAEMORRHAGIC FEVER VIRUS (OHFV)

This flavivirus causes a human disease found in western Siberia, where there are two cycles of infection. In summer, the infection is tick-borne and, in winter, 'mechanical' infections occur among muskrat trappers and skinners. The virus has been recovered from *D. reticulatus*, which has a wide spectrum of hosts, including the muskrat (*Ondatra zibethicus*), which is highly sensitive to the virus. The tick transmits the virus transovarially and transstadially, and horizontally to laboratory animals during feeding. Human infection can also be acquired by handling infected muskrat carcasses or by contact with or drinking water infected by muskrats or water voles. The disease causes fever, which is commonly bimodal, and haemorrhages, but is somewhat more benign than TBEV, with only a maximum 3% mortality rate arising from the acute febrile, myalgic and haemorrhagic symptoms. There is no specific treatment and supportive management is important.

LOUPING ILL VIRUS (LIV)

This flavivirus causes an acute encephalomyelitis in sheep and is frequently fatal. It has been recorded from the British Isles (particularly Scotland and northern England) and Ireland. It is enzootic in grouse and can infect fox, hare and various other animals. The main vector is the sheep tick, *I. ricinus*, in which transstadial but (apparently) no transovarial transmission occurs. The term 'louping ill' refers specifically to the encephalitic disease in sheep in the British Isles, but a similar syndrome is recognized in various regions in Europe. A wide variety of other animals are susceptible to the virus, including humans, in which it is not fatal but can produce serious symptoms including fever, vomiting, malaise, lymphadenopathy and meningoencephalitis, with a long recovery period. There is an effective vaccine to protect sheep, and acaricides are often used to protect sheep against questing ticks.

POWASSAN VIRUS (POWV)

This flavivirus occurs principally in eastern North America, from Canada south to Pennsylvania, New York and Massachusetts, and in parts of the upper mid-west America. The virus is maintained in a cycle of forest-mammal reservoir hosts such as groundhogs (*Marmota monax*), red squirrels (*Tamiasciurus hudsonicus*) and white-footed mice (*Peromyscus leucopus*) and *Ixodes* ticks (principally *Ixodes cookei*, but also *Ixodes marxi* and *I. scapularis* – with the latter increasingly becoming the most important as a bridge vector for human infection because of its expanding range in the USA). Transovarial transmission of POWV has been demonstrated in both *I. cookei* and *I. scapularis*. Asymptomatic infections in humans are common, but infection can lead to fever, headache and signs of neurological involvement, with convulsions and coma. The fatality rate can be up to 25% and there are often severe neurological sequelae in more than 50% of survivors. With no specific antiviral treatment available, support management of patients is required.

COLORADO TICK FEVER VIRUS (CTFV)

This is a reovirus, endemic in the western USA and western Canada; it is enzootic in wild rodents, in which it causes no apparent disease. Humans are susceptible to infection; typically, the symptoms are non-specific, with fever, myalgia, headache and nausea, sometimes a rash; encephalitis and haemorrhagic signs may occur, particularly in children, but fatalities have been few. It is thought that birds and various smaller mammals (such as ground squirrels, chipmunks, deermice and woodrats) are the main reservoir hosts, but larger vertebrates may also be involved. A range of ticks has yielded isolates of CTFV but the principal vector is the three-host tick, *D. andersoni*, which can transmit the virus transovarially and transstadially. There is no specific treatment for CTFV infection, so patient management is symptomatic.

6.3.2 Rickettsiae (and related organisms)

Tick-borne spotted fevers are widely spread throughout the world, occurring in every continent and zoogeographical region. They are caused by various species of *Rickettsia* that circulate in a wide range of mammals, and sometimes birds, through the agency of ixodid ticks. The rickettsiae are transmitted transstadially and transovarially in the tick vector. They cause benign infections of the non-human host but serious, even fatal, disease in humans. In recent years, the spotted fever group rickettsiae that can cause human illness have been expanded well beyond the previous well-known group of *Rickettsia rickettsii*, *Rickettsia conorii*, *Rickettsia sibirica* and *Rickettsia australis*. For instance, what was once thought to be simply *R. conorii* in Europe, Africa and Asia has been reclassified as an assemblage of new species, including *Rickettsia africae*, *Rickettsia aeschlimanii*, *Rickettsia helvetica*, *Rickettsia massiliae*, *Rickettsia mongolotimonae*, *Rickettsia parkeri* and *Rickettsia slovaca*. Further, in recent years, novel rickettsiae have been reported from ticks collected in other areas of the world, including South America and Australia. However, as relatively little is known of the ecology of many of these, we will address below only the agents of greatest concern: *R.*

rickettsii, *R. sibirica*, *R. conorii*, *R. africae*, *R. australis*, *Rickettsia japonica* and other rickettsiales.

RICKETTSIA RICKETTSII

This agent of Rocky Mountain spotted fever (RMSF) is widely distributed within and outside the Rocky Mountains, occurring throughout most of the continental USA, with most cases originating in the south-eastern and south-central states. It is also found in south-western Canada and Central and South America, with cases reported in Mexico, Argentina, Brazil, Colombia, Costa Rica and Panama. There is an incubation period of about 1 week, with 2–3 weeks of disease, characterized by chills, headache, fever, rash and photophobia. This disease has a normal mortality of about 20%, and early and adequate medication (with tetracyclines, doxycycline or chloramphenicol) is essential to prevent fatalities.

The main human vectors of RMSF are *D. andersoni* and *D. variabilis* in the USA, and probably also in Canada, with *A. americanum* involved in Texas and Oklahoma, *A. cajennense* in Central and South America and *R. sanguineus* also in the USA, as well as in Mexico and potentially in Brazil. Other ticks may eventually be implicated in certain regions (e.g. *A. aureolatum* in Brazil). In the eastern USA, the vector is *D. variabilis*, which in the immature stages occurs on rodents and in the adult occurs on larger mammals (including humans and dogs). In the western USA, the disease is associated with fieldworkers and the vector is *D. andersoni*, of which the immature stages are indiscriminate parasites of small mammals, and the adults parasitize hares, larger wild and domestic animals and humans.

Among wild hosts, infections with *R. rickettsii* are transitory, but serological tests have demonstrated antibodies to *R. rickettsii* in many species of birds and mammals, including ground squirrels, chipmunks, rodents, leporids and carnivores. In the laboratory, infected female *D. andersoni* transmit the rickettsia to 100% of their daughters, and from them to their progeny. Nevertheless, in the field, infections of *D. andersoni* with *R. rickettsii* are much lower, less than 14%. Transstadial and transovarial transmission of the rickettsia do not reduce its virulence. Infection results from the bite of the tick, and the faeces of infected *D. andersoni* do not appear to be important in transmission.

Control of RMSF involves avoidance of likely tick-infested locations and the wearing of boots and protective clothing, especially on the lower limbs, which are most likely to pick up ixodid ticks. The protection offered by clothing can be increased by the application of a suitable repellent/toxicant, e.g. permethrin, followed by daily body examinations to remove any attached ticks. Early removal is considered to reduce the likelihood of infection.

RICKETTSIA SIBIRICA

Sometimes spelt 'siberica', *R. sibirica* is closely related to *R. rickettsii*. The two species are geographically distinct, with *R. rickettsii* being confined to the western hemisphere and *R. sibirica* to the northern Palaearctic region from Armenia in the west to eastern Siberia, China and Pakistan; however, the geographical boundaries of this rickettsia are still not known. Strains of *R. sibirica* have been recovered from at least 18 species of mammal, mostly rodents. Nine species of ixodid ticks may potentially be involved as reservoirs and vectors: four species of *Dermacentor*, three of *Haemaphysalis*, *R. sanguineus* and *Hy. asiaticum*. In *D. marginatus*, naturally or experimentally infected with *R. sibirica*, the rickettsia survives for at least 5 years or through 4 generations. The immature stages of *Dermacentor* occur on small mammals, particularly rodents, hedgehogs, hares and small carnivores, but are rare on birds (suggesting a secondary role of birds as reservoirs of *R. sibirica*). Adult *Dermacentor* parasitize medium-sized to large wild and domestic mammals.

Ticks of the genus *Dermacentor* are common and widely distributed in Eurasia, with species occupying different ecological niches. *D. marginatus* occurs in lowland and alpine steppes of western Eurasia and further west into central Europe. *Dermacentor silvarum* has an eastern distribution, extending from western Siberia to the Pacific Ocean. It infests the taiga and shrub-wormwood steppes. In the Far East, in shrub and fern marshes adjoining taiga, *Haemaphysalis concinna* is the chief vector. It gives way to *Hy. asiaticum*

in the semi-desert steppe and *Dermacentor nuttalli* in the alpine, forested and desert steppes of central and eastern Siberia.

RICKETTSIA AFRICAE

This agent of African tick bite fever was long considered a milder form of *R. conorii* in Africa, but it is now recognized as a separate species and is prevalent in *Amblyomma* ticks (particularly *A. hebraeum*) in sub-Saharan Africa. It also occurs in the West Indies, where *A. variegatum* ticks have been introduced on cattle. In humans, the disease usually exhibits fever, sometimes a rash and regional adenopathy and often features multiple eschars (a reflection of the aggressive biting behaviour of the *Amblyomma* vectors), but it is relatively mild compared with other spotted fevers and has a very low fatality rate without treatment.

RICKETTSIA CONORII

This agent of Mediterranean spotted fever (MSF), or Boutonneuse fever, has been the most ubiquitous rickettsia of the spotted fever group, previously being reported as occurring widely in southern Europe, Africa, India and the Oriental region, although many of these records are now known to be due to other rickettsiae recently separated from *R. conorii*. Current distribution is usually accepted to be around the Mediterranean and Black Seas and parts of sub-Saharan Africa, Russia and India (but further research on rickettsial strains may lead to further refinement). The popular names for the disease caused by *R. conorii* are often formed from the geographical location of infections (e.g. Israeli spotted fever, Marseilles fever).

R. conorii can cause acute suffering in humans; the disease presents as an acute febrile illness, with regional adenopathy, a disseminated maculopapular rash and an eschar at the bite site. The infection is generally less severe than *R. rickettsii*, and antibiotic treatment is usually effective in maintaining a benign course, but some cases are severe and fatalities have resulted. Without treatment, mortality is reported to be about 2%. The normal route of transmission of *R. conorii* to humans is through the bite of the tick, but infection may also result from contamination of eye and nasal mucosa from crushed ticks or their faeces. It is transmitted principally by the dog tick, *R. sanguineus*. Although the vertebrate reservoirs for *R. conorii* have not been well defined, and it has been suggested that ticks serve as both vectors and reservoirs for most of the spotted fever rickettsia, antibodies to *R. conorii* have been detected in a range of domestic and wild animals in various regions. In many countries, dogs are suspected of being a major reservoir and dogs have been shown experimentally to be capable of infecting *R. sanguineus* ticks.

RICKETTSIA AUSTRALIS

This agent of Queensland tick typhus (QTT) is found in coastal eastern Australia from northern Queensland to Victoria. Several species of marsupials, including bandicoots, are suspected to be reservoir hosts and, on circumstantial evidence, the probable principal vector is *I. holocyclus*, a widely distributed and unusually indiscriminate feeder, attacking almost any bird or mammal. The clinical appearance of QTT is similar to that of other spotted fever group rickettsial infections, with sudden onset of fever, headache and muscular pain, a maculopapular rash, influenza-like symptoms, lymph node tenderness and an eschar at the bite site. The disease is generally considered to be only mild to moderately severe, and antibiotic (doxycycline, azithromycin, ciprofloxacin) management is usually effective, but one fatal case has been reported.

A number of other tick-associated rickettsiae have been found in Australia in recent decades, including *Rickettsia honei* as the agent responsible for Flinders Island spotted fever (and linked to the reptilian tick, *Bothriocroton hydrosauri* (formerly *Aponomma hydrosauri*)) and *R. honei* strain *marmionii* as the agent of the variant Flinders Island spotted fever, but relatively little is known of their respective ecologies or clinical impacts.

RICKETTSIA JAPONICA

This agent of Japanese, or Oriental, spotted fever is not known outside Japan, but hundreds of cases have been reported, mainly from the coast of the central and south-west regions. Infections are relatively typical of milder spotted fever infections (no fatal cases have been reported), with fever, headache and an eschar, and there is an extensive macular rash that becomes petechial after a few days. The rickettsia has been detected in several ticks, with *Haemaphysalis flava*, *H. longicornis* and *Ixodes ovatus* suspected as vectors (and perhaps reservoirs).

EHRLICHIA CANIS

This agent of canine monocytic ehrlichiosis (CME) is an important cause of disease of dogs worldwide, which is usually mild but meningitis can be a common feature, as are ocular symptoms such as retinal dysfunction. Also, in certain breeds (e.g. German shepherds), it can produce a severe haemorrhagic condition known as tropical canine pancytopenia. *E. canis* can last for up to 5 years in dogs without clinical expression. The vector is the dog tick, *R. sanguineus*. Treatment of acute infections with doxycycline is usually effective.

EHRLICHIA CHAFFEENSIS

This agent of human monocytic ehrlichiosis, transmitted by *A. americanum*, is a parasite of white-tailed deer (*Odocoileus virginianus*) in the south-eastern and south-central USA. In recent decades, *Ehrlichia chaffeensis*, and also *Ehrlichia ewingii*, have been reported as causing a human ehrlichiosis, a usually mild (but occasionally severe and rarely fatal) disease that can present as a non-specific acute fever with headaches, muscle and joint pain and sometimes a rash, and, frequently, central nervous system disorder. *E. ewingii*, similarly transmitted by *A. americanum* in the southern USA, causes an illness in dogs that may act as a reservoir for human infections.

EHRLICHIA RUMINANTIUM

This pathogen (formerly known as *Cowdria ruminantium*) is the agent of heartwater disease in cattle, sheep and goats in southern Africa, where it is regarded as the most important tick-borne disease in the region. It also occurs in wild ruminants, including blesbok (*Damahscus albifrons*), wildebeest (*Connochaetes gnu*) and springbok (*Antidorcas marsupialis*), which may become symptomless carriers. Peracute cases show only high fever, prostration and convulsions, ending in death. In acute cases, the disease lasts about 6 days with 50–90% mortality. *Ehrlichia ruminantium* is found in the Afrotropical region and has been introduced into some islands in the West Indies. Ticks become infected by feeding on a range of ruminant and non-ruminant animals with acute infections; transstadial transmission is documented. Twelve species of *Amblyomma* are known to be capable of transmitting *E. ruminantium*, of which *A. variegatum* and *A. hebraeum* are the most important and widely distributed vectors in Africa (although *Amblyomma gemma* and *Amblyomma lepidum* are thought to be important vectors in semi-arid areas in eastern Africa).

ANAPLASMA PHAGOCYTOPHILUM

This pathogen (formerly *Ehrlichia phagocytophila*) is the agent of tick-borne fever (TBF) in ruminants and of human granulocytic anaplasmosis (HGA), formerly known as human granulocytic ehrlichiosis (HGE). *Ehrlichia equi*, an agent of disease in horses and some other animals, is now considered a subjective synonym of *E. phagocytophila* and therefore is also now known as *Anaplasma phagocytophilum*. This rickettsial agent is transmitted by *I. ricinus* in Europe and *I. scapularis* and *Ixodes pacificus* in North America.

It causes a mild disease in cattle and sheep in the UK, Ireland, Scandinavia, Sweden and Spain but, more importantly, it increases the susceptibility of lambs to staphylococcal infection and louping ill. In Norway, infected lambs had lower live weights in autumn. *A. phagocytophilum* may persist in the bloodstream of sheep for life, forming a major source of infection in an enzootic area. The mortality rate is negligible in cattle, the main losses being due to abortion when pregnant cows become infected. Related organisms occur in Africa and India.

In humans, *A. phagocytophilum* was first recognized causing a human infection following bites by *I. scapularis* in the north-eastern and mid-west USA, and *I. pacificus* in California. Symptoms include fever, headache, leucopenia and thrombocytopenia, but absence of skin rash. Now known from many temperate regions of the world, transmission in Europe involves *I. ricinus*. Reservoir hosts appear to be small mammals, particularly rodents (such as the white-footed mouse, *P. leucopus*, in the north-east of the USA), and the pathogen is transmitted transstadially in the tick vectors. Treatment is generally with doxycycline, although some other tetracyclines can be effective.

ANAPLASMA MARGINALE

This agent of bovine anaplasmosis is a pathogen of cattle in the tropics and subtropics, occurring in sub-Saharan Africa, Australia, the Mediterranean, Asia, the former Soviet Union and North and South America. It causes severe debility, anaemia, jaundice and abortion in adult cattle. Young animals are relatively resistant to infection but, in cattle more than 3 years old, a peracute condition can develop and death may occur within 24 h. When susceptible animals move into an infected area, or the vector population expands into a previously disease-free area, the morbidity rate can be very high and the mortality may be 50% or more. Recovered animals require prolonged convalescence and will continue to be infected for the rest of their lives.

Anaplasma is transmitted biologically by ticks and mechanically by various bloodsucking flies, particularly tabanids, and also eye flies (chloropids). Many species of ticks have been shown capable of transmitting the pathogen; *R. (Bo.) decoloratus* and *R. simus* are vectors in Africa, *R. (Bo.) microplus* is important in Australia and, in the USA, *D. albipictus*, *D. andersoni*, *D. occidentalis* and *D. variabilis* are considered to be vectors. Transstadial (but not transovarial) transmission occurs. *Anaplasma marginale* has a wide range of other hosts, including zebu cattle, water buffalo, African antelopes, American deer and camels. The African buffalo (*Syncerus caffer*) is refractory to infection.

The related *Anaplasma centrale* causes mild, inapparent disease in cattle and is found in Australia, South America, South-east Asia and the Middle East. *Anaplasma ovis*, which occurs in Africa, in the Mediterranean region, the former Soviet Union and the USA, usually produces subclinical infections in sheep and goats, although a severe anaemia may develop (especially in goats suffering from a concurrent disease).

COXIELLA BURNETII

This agent of Q fever belongs to a monotypic genus and it has been proposed that it is not related to other Rickettsieae. It has a worldwide distribution and was described originally from Australia, where it is circulated among bandicoots by *Haemaphysalis humerosa* and *I. holocyclus*, and among kangaroos by *Amblyomma triguttatum*. However, cattle, sheep and goats can be important reservoirs and most human infection is by contact with infected tissues and by inhalation, most commonly in people associated with domestic animals and particularly with slaughterhouses. Q fever is an acute, self-limiting disease of 3–6 days duration, with fever, severe headache and pneumonia occurring in about 60% of cases and hepatitis in a third of cases. Most cases are inapparent or unrecognized. Infection produces a solid but non-sterile immunity and may produce endocarditis, a serious condition, 2–20 years later. Treatment with doxycycline and tetracycline is effective and there is a vaccine available in Australia for workers at high risk of exposure in slaughterhouses.

Coxiella burnetii is enzootic in cattle, sheep and goats, which shed large numbers of the rickettsiae at parturition in the placenta and fetal fluids. It is normally regarded as being non-pathogenic for domestic livestock but there are records of very heavy infections causing abortion in sheep and goats; it is widely disseminated among wild mammals and birds and has been isolated from domesticated camels, buffaloes and horses. Natural infections have been found in species from eight genera of Ixodidae, and various ticks likely play a role in transmitting *C. burnetii* between wild and domestic animals. Indeed, a wide range of animals (including dogs, foxes, hares, moose, racoons, deer and birds) has been found to be infected.

6.3.3 Bacteria

BORRELIA BURGDORFERI S.L.

Borrelia burgdorferi s.l., including *Borrelia afzelli* and *Borrelia garinii* and other regional spirochaete species, are the agents of Lyme disease. Named from the town of Lyme in Connecticut, USA, where the first cases were described, Lyme disease is now recognized as widespread in temperate climates of the northern hemisphere, in both the Old and New World, but it is not definitively known from anywhere in the southern hemisphere (despite various reports).

The causative agent, *B. burgdorferi* s.l., is now known to be a genetically diverse complex, with at least ten genospecies, the most common of which are *B. burgdorferi* s.s. from North America and *B. afzelii* and *B. garinii* from Eurasia. Others include *Borrelia valaisiana*, *Borrelia bissettii*, *Borrelia japonica*, *Borrelia lusitaniae*, *Borrelia andersonii*, *Borrelia turdae*, *Borrelia sinica* and *Borrelia tanuki*, although some of these have only rarely or never been associated with disease in humans.

In general, three stages of disease have been recognized: localized with flu-like symptoms and a characteristic skin rash occurring 1–3 weeks after infection; disseminated disease with neurological and/or cardiac symptoms usually weeks to months later; and a persistent stage with arthritic manifestations weeks to years later. However, the various genospecies have been associated with different clinical presentation; for example, *B. burgdorferi* s.s. with arthritis, *B. afzelii* with dermatological symptoms and *B. garinii* with neurological symptoms.

The reservoir hosts for the spirochaetes are, in general terms, small mammals (mostly rodents) and birds (with reservoir competence being dependent on spirochaete species). The principal vectors are members of the *Ixodes* 'ricinus–persulcatus complex', which includes the North American *I. scapularis* in the north-east and upper mid-west of the USA, *I. pacificus* in the far west USA, *I. ricinus* in Europe and *I. persulcatus* in Eurasia. The tick vectors acquire spirochaetes from vertebrate hosts, and the spirochaetes colonize the gut and transstadial transmission occurs. On host feeding by the next stage of the tick, the spirochaetes multiply in the new blood in the midgut and, over several days, penetrate the gut epithelium and migrate to the salivary glands, for delivery to a new host (although transovarial transmission has been reported for *B. burgdorferi* in the 'ricinus–persulcatus' group of ticks, there remains doubt as to whether it actually occurs and it has been proposed that the spirochaetes being detected in unfed larvae are not *B. burgdorferi* but another spirochaete). A tick needs to feed for several hours (usually more than 48 h) to transmit *B. burgdorferi* successfully and, on humans, adult ticks are likely to be found and removed within 2 days of attachment, thus before achieving transmission. The much smaller nymphs are more likely to be overlooked and so are considered to be the main source of human infections (although with *I. persulcatus* it has been proposed that the adult female seems to be the stage that mostly transmits to humans).

Various other *Ixodes* species are involved in the ecological cycles of *Borrelia* species and may be important in maintaining sylvan transmission cycles, but many do not bite humans and thus are not considered to be major vectors in the important endemic regions. In eastern North America, *I. scapularis* is the primary vector, whereas in western North America, the primary vector is *I. pacificus*, which occurs from California north to British Columbia and eastwards to Nevada, Utah and Idaho. In Eurasia, the primary vector, *I. ricinus*, occurs in a broad band of Europe eastwards to the Ural Mountains and south to the Mediterranean, particularly in deciduous woodland, where there is high humidity (>80% RH), or in meadows and moorland with high rainfall.

Avoidance of Lyme disease involves the same anti-tick precautions for protection against *R. rickettsii*, i.e. avoidance of tick habitats, wearing of protective clothing treated with a tick repellent/toxicant and careful inspections for ticks on the body after exposure in tick-infested areas. Human infections are treated with antibiotics, and the drugs of choice vary with the symptoms presented, but amoxicillin, doxycycline, ceftriaxone, cefotaxime, ceftnaxone and penicillin have been recommended, depending on clinical circumstances. For a few years, a vaccine against *B. burgdorferi* s.s. (but not cross-protective against the Eurasian spirochetes) was available in North America, but it was withdrawn from the market.

BORRELIA THEILERI

The agent of bovine borreliosis occurs in Africa, India, Indonesia, Australia and South America. It causes a mild disease in cattle and a febrile disease in horses. The vectors are species of the subgenus *Boophilus* (particularly *R. (Bo.) annulatus*, *R. (Bo.) decoloratus* and *R. (Bo.) microplus*), although in Africa it is also associated with *R. evertsi*.

Rhicephalus (Bo.) spp. are one-host ticks and can only be vectors if there is transovarial transmission. *Borrelia theileri* invades the haemocoel of the tick and infects the central ganglion and ovary, from which rates of transovarial transmission as high as 80% have been recorded; however, larval ticks seem to be incapable of transmitting *B. theileri*. Following transstadial transmission, in feeding nymphs and adults there is massive multiplication of *B. theileri* in the haemocytes and release into the haemolymph for transmission to new hosts.

FRANCISELLA TULARENSIS

This agent of tularaemia is a Gram-negative, obligately aerobic, pleomorphic, non-motile coccobacillus which causes disease in humans and many other warm-blooded animals, including sheep, horses, pigs, cattle and birds. Tularaemia is ubiquitous in the northern hemisphere between latitudes 30° and 71°N, occurring in Europe (mostly eastern and northern), Russia, Japan, Canada, USA and Mexico. At least three subspecies of *Francisella tularensis* have been recognized as causes of disease in humans possibly related to ticks: *Francisella tularensis* subsp. *tularensis* (causing type A tularaemia), *Francisella tularensis* subsp. *holarctica* (causing type B tularaemia) and *Francisella tularensis* subsp. *novicida* (a relatively non-virulent strain formerly known as *Francisella novicida*). Of the two subspecies most commonly associated with ticks and human disease, *F. tularensis* subsp. *tularensis* is the more virulent agent and is found predominantly in North America, while *F. tularensis* subsp. *holarctica*, which causes a milder disease, is endemic in both the Palaearctic and Nearctic regions of the northern hemisphere (but has been reported also from the southern hemisphere, e.g.

Australia). *F. novicida* is known only from a few immunocompromised cases in North America and a fourth subspecies, *Francisella tularensis* subsp. *mediasiatica*, has been reported from central Asia, but little is known about it.

F. tularensis can be transmitted from host to host by a variety of routes, including bloodsucking arthropods, water, food and inhalation. In addition, the organism possesses the potent property of being able to penetrate unbroken skin and therefore can be transmitted by contact with infected material. It usually produces a marked reaction at the site of entry, which in 70–80% of cases becomes an ulcer. Human tularaemia is an acute, febrile, infectious zoonotic disease with an incubation period of 3–4 days; typically, an ulcer forms at the inoculation site, there is glandular involvement and lymphadenopathy. The introduction of as few as ten organisms subcutaneously can cause disease in non-vaccinated individuals, and mortality may be as high as 35% of untreated septicaemic cases. Treatment with streptomycin has reduced the mortality rate in humans to virtually zero.

Nearly 300 species of wild mammals and birds and more than 50 species of arthropods have been found naturally infected. In North America, *F. t. tularensis* occurs particularly in lagomorphs and wild rodents and is transmitted by the bites of ticks (*A. americanum*, *D. andersoni*, *D. variabilis* and *Haemaphysalis leporispalustris*) and tabanid horse flies. *F. t. palaearctica* has been associated with lemmings and mosquitoes in Sweden and hares and voles in Russia, with *D. nuttalli* as the vector.

Infections of *F. tularensis* in ticks are long lasting. The organism multiplies in the midgut epithelium and haemolymph of the tick and is transferred when the tick is feeding, but transovarial transmission has also been demonstrated in *D. andersoni* and other ixodid ticks in North America. The ability of *F. tularensis* to survive away from its vertebrate host favours mechanical transmission by bloodsucking flies, especially tabanids (e.g. *Chrysops discalis* is recognized as being a major mechanical vector of *F. tularensis* in the USA).

Tularaemia is most severe in sheep, in which the morbidity rate in North America may be as high as 40%, with a mortality of 50% especially in young animals. In horses, there is fever and foals are affected more seriously than older animals; in swine, tularaemia causes fever in piglets but is latent in adult pigs.

6.3.4 Protozoa (also described as the 'piroplasms')

BABESIOSES

Approximately 100 species of *Babesia* have been described from nine orders of mammals, with the greatest numbers being found in rodents, carnivores and ruminants (mainly livestock, in which they cause severe disease in cattle). *Babesia* species are usually divided into two groups: the large and the small species groups. The small one includes *Babesia bovis* and the large group includes *Babesia bigemina*. In studies of babesiosis throughout the world, attention has been concentrated mainly on *B. bovis* and its boophilid tick vectors and also on other species infecting cattle, including *B. bigemina*, *Babesia divergens* and *Babesia major*. Less information is available on equine (*Babesia caballi*) and canine (*Babesia gibsoni*, *Babesia canis*) babesioses.

Multiplication of *Babesia* spp. involves schizogony in the vertebrate host, gamogony in the intestinal cells of the tick, with fusion of gametes followed by sporogony in the salivary glands and other organs. All species invade the erythrocytes of their vertebrate host, but in *Babesia equi* (i.e. *Theileria equi*) and *Babesia microti*, sporozoites introduced by the tick initially invade the lymphocytes, in which schizogony occurs. The infected lymphocytes burst, releasing motile merozoites, which invade erythrocytes. The merozoites undergo rapid reproduction and, in the process, destroy the erythrocytes in which they develop, causing severe anaemia in the host. Some merozoites do not divide but develop into ovoid or spherical gamonts, which develop no further until ingested by the tick.

Ixodid ticks acquire infection with *Babesia* by the alimentary route when they feed on an infective host, or transovarially via the female parent. Vertical transmission from generation to generation (without the ingestion of additional babesia) occurs with *Babesia ovis* in *R. bursa*, whereas *B. bovis* and *B. bigemina* have only transovarial transmission in *R. (Bo.) microplus*. Infections within the tick are transmitted transstadially, and this is the only mode of transmission for *B. microti*.

With *B. canis*, *B. ovis* and other species of *Babesia*, infections are usually acquired by the adult female tick, although *R. sanguineus* can transmit *Babesia vogeli* in the adult stage, having become infected in the preceding nymphal stage. The three-host ticks *I. ricinus*, *H. punctata* and *R. sanguineus* acquire, respectively, *B. divergens*, *B. major* and *B. vogeli* in the adult stage and can transmit the parasite in all stages of the next generation, although larvae of *R. sanguineus* only transmit when present in large numbers.

R. bursa acquires infection with *B. ovis* in the last 4 h before detachment and, in this two-host tick, transmission is by the adult stage of the succeeding generation. In *R. bursa*, where both alimentary and vertical infections with *B. ovis* occur, alimentary infections produce higher infection rates than from vertical transmission.

BABESIA IN CATTLE

The most widespread species are *B. bovis* and *B. bigemina*, which are widely distributed between 32°S and 40°N, where they are responsible for serious losses in Latin America and Asia. In the USA, the main parasite was *B. bigemina*, which was brought under control by the eradication of the vector *R. (Bo.) annulatus*. In Australia, *B. bovis* is more important than *B. bigemina*, although both species occur and are transmitted by the same vector, *R. (Bo.) microplus*. Other species are involved: *Babesia jakimovi* in Russia, *Babesia ovata* in Japan and *Babesia occultans* in South Africa.

Susceptible cattle exposed to babesial infection can suffer a morbidity rate of 90% and an almost equally high mortality rate. After an incubation period of 2–3 weeks, babesial infections cause fever, profound anaemia, haemoglobinuria (hence 'redwater' being a common name for the disease) and either

death may occur within 24 h or the disease last for 3 weeks. Survivors become carriers for a variable period of time, usually about 6 months, followed by a further 6 months of sterile immunity, which may differ significantly according to individual or population immunological responsiveness and species of *Babesia* causing the infection. Repeated infections can make the immunity permanent, but in the absence of further infections, cattle become susceptible again after 1 year. All races of cattle are equally susceptible to *B. bigemina*, but zebu and Afrikaner cattle have a higher resistance to *B. bovis* than European breeds. Indian (*Bos indicus*) and zebu-type cattle are relatively free from the disease, because of their resistance to heavy tick infestations. In enzootic areas, calves receive passive immunity from maternal antibodies in the colostrum, which is protective for about 11 weeks. The greatest infection rate occurs in animals from 6 to 12 months of age and is uncommon in animals older than 5 years, but the severity of the disease increases with age. Cattle that have recovered from infections are immune to developing disease in response to a homologous challenge, but may suffer subclinical superinfection and may develop clinical disease in response to a heterologous challenge.

Babesia bovis

B. bovis (syn. *Babesia argentina*, *Babesia berbera*, *Babesia colchica*) is widely distributed throughout the world, occurring in Europe, Central and South America, Africa, Asia and Australia. In southern Europe, the vector is *R. bursa*, while in tropical and subtropical areas, the vectors are species of the subgenus *Boophilus*. *R. (Bo.) microplus* is widely distributed in the tropics, while in Africa, *R. (Bo.) decoloratus* is an important vector. *R. (Bo.) microplus* was considered to be the only boophilid in Australia, but recent evidence revalidated *R. (Bo.) australis*, which is now considered to be the local species. Infection is acquired by the female tick, which transmits the parasite transovarially to its larvae, and these represent the infective stage. The larval tick injects sporozoites, which invade the erythrocytes, and the severity of the reaction is a function of the parasitaemia reached in the host. Infected erythrocytes clump together and block capillaries, causing brain damage and anoxia of internal organs.

The duration of a single infection of *B. bovis* is of the order of 18 months to 4 years, with a fluctuating parasitaemia presenting cycles at 3- to 8-week intervals. The fluctuations are due to change in antigenic type, of which more than 100 have been recognized. On passage through the tick, the strain of *B. bovis* reverts to its basic antigen, but this also is variable. In the more natural situation where the infected animal is exposed to repeated infections (i.e. superinfections), the parasitaemia rises smoothly from zero to a maximum in 1–2 years and then declines.

Babesia bigemina

This has a similar distribution to that of *B. bovis* and it is also found in the southern former Soviet Union. Important vectors of *B. bigemina* are *R. (Bo.) microplus* and *R. evertsi*, which acquire the parasite when the adult female is feeding and transmit it to the succeeding nymphal stage and adult. *B. bigemina* can be transmitted transovarially through several generations of *R. (Bo.) microplus*. Clinical disease is produced when the parasitaemia exceeds 1%, but, unlike the most pathogenic *B. bovis*, there is no clumping of infected red cells and therefore no blockage of capillaries in organs, such as the brain. The duration of a single infection of *B. bigemina* is of the order of 6–12 months. Because *R. (Bo.) microplus* is a vector of several pathogens of cattle, the clinical picture can be complicated by synergistic pathogenicity caused by various agents, which simultaneously infect the same individual (e.g. *B. bigemina*, *B. bovis* and *A. marginale*). Strains of *B. bigemina* exist, with the African strain being highly pathogenic and the Australian strain considerably less so.

Babesia divergens

B. divergens (syn. *Babesia caucasica*, *Babesia occidentalis*, *Babesia karelica*) is a small parasite which causes disease and death in cattle in Europe and North Africa. It can produce high parasitaemias, but the infected erythrocytes do not clump and therefore there is no cerebral involvement. In western and central Europe, the vector is *I. ricinus*. The parasite is acquired by the adult tick and transmitted transovarially to the next generation. Differently from *B. bovis* or *B. bigemina*, infection occasionally may continue into the next (F_2) generation and, since the life cycle of *I. ricinus* extends for 3 years, the parasite may survive in the tick for up to 4 years.

Babesia major

B. major is a large species found in cattle in Europe and the Middle East. The vector is *H. punctata*, in which the parasite acquired by the female tick is transmitted transovarially and probably passed in the succeeding adult stage. Although in infections the erythrocytes clump, few fatalities occur in cattle. Infections in American bison introduced into south-east England were very severe, with 50% mortality before drugs were used to cure the survivors.

Babesia jakimovi

B. jakimovi, a large *Babesia* whose natural host is the Siberian roe deer (*Capreolus capreolus*), causes a severe and often fatal disease of cattle, and also infects reindeer and elk. Species of *Ixodes* are vectors in the field, and the laboratory evidences suggest that *I. ricinus* transmit *B. jakimovi* transovarially to the succeeding generation.

BABESIA IN SHEEP

Two species of *Babesia* occur in sheep, a large species, *Babesia motasi*, and a small one, *B. ovis*. Some strains of *B. motasi* are infective for sheep only and others for both sheep and goats. *B. motasi* is a significant pathogen on its own and it participates in synergistic pathogenicity with *Theileria* and *Ehrlichia* species. The distribution of *B. motasi* coincides with that of various species of *Haemaphysalis* in Europe, Africa and Asia, and that of *B. ovis* coincides with that of *R. bursa* and *R. evertsi* likewise in Europe, Africa and Asia. When adult sheep become infected, they can act as carriers for more than 2 years. An attenuated vaccine has produced solid immunity in sheep.

BABESIA IN HORSES

Formerly, a large species, *B. caballi*, and a small one, *B. equi*, were said to parasitize equines and cause equine piroplasmosis, also known as 'biliary fever'. The disease occurs in parts of Europe, Africa and Asia, and parts of the Americas. However, *B. equi* is now recognized as a species of *Theileria* and is discussed further below under theilerioses.

B. caballi is known from Europe, Africa, Asia and the Americas. A number of species of three genera of ixodid ticks (i.e. *Dermacentor*, *Hyalomma* and *Rhipicephalus*) have been identified as vectors of *B. caballi*, with transstadial transmission shown for four species and transovarial for nine species.

The incubation period for *B. caballi* is 10–30 days, with infections often being inapparent and intrauterine infection rare. The symptoms are fever, anorexia, malaise and weight loss and, in neglected horses, severe anaemia. Treatment with diminazene diaceturate is effective in clearing *B. caballi*. Chemical control of ticks is probably feasible under highly intensive management systems. In free-ranging systems in endemic areas, foals should be exposed to natural infection, allowing them to develop immunity without clinical disease while protected by natural non-specific resistance, thus resulting in an endemic stable disease situation.

BABESIA IN PIGS

Pigs support a large, *Babesia trautmanni*, and a small, *Babesia perroncitoi*, species. *B. trautmanni* is present in southern Europe, the former Soviet Union and parts of Africa. It may cause a severe condition, with 60–65% of the erythrocytes being parasitized. Infections are commoner in pigs aged 4–6 months. Wild pigs, warthog and bushpig are likely significant reservoirs of this parasite. *B. perroncitoi* has a very limited distribution in the Sudan, Sardinia and Italy, and may be a significant pathogen of pigs. The vectors of both species are suspected rather than known.

BABESIA IN DOGS

The large species in dogs are *B. canis*, *B. vogeli* and *Babesia rossi*, and the small species comprise three genetically and clinically distinct species, *B. gibsoni* and *Babesia conradae* and a *Babesia microti*-like piroplasm named *Theileria annae*. *B. canis* causes mild to severe disease and is the predominant species in temperate regions of Europe, with *B. vogeli*, the least virulent species, being present in Europe, Africa, Asia, Australia and

North and South America, and *B. rossi*, notoriously the most virulent species, being reported in western, eastern and southern Africa. *B. gibsoni* is present in five continents (being common in Asia, Africa and Europe), while *B. conradae* is known from the western USA and *T. annae* from Spain.

Fatal cases due to anoxia occur when parasitaemias are of about 40–45%, whereas parasitaemias of 2–14% are usually benign. Foxes and jackals act as reservoirs of *B. gibsoni*. In South-east Asia, the vector of *B. gibsoni* is *H. longicornis*. In warmer countries, *B. vogeli* is transmitted by *R. sanguineus*, whereas *B. rossi* is transmitted by *H. leachi* and, in temperate regions of Europe, *B. canis* is transmitted by *D. marginatus* and *D. reticulatus*. Transovarian transmission of *B. canis* has been observed through five generations.

In France, infection with *Babesia* spp. has a bimodal pattern: a spring–summer peak among domestic dogs, with *R. sanguineus* as the vector of *B. vogeli*, and an autumn–winter peak among hunting dogs, when *Dermacentor* spp. transmits *B. canis*.

B. canis shows synergistic pathogenicity with *E. canis*, with *B. canis* destroying the erythrocytes and *E. canis* impeding red cell production. Infected red cells clump together and block capillaries in the brain, leading to death. Young puppies are highly susceptible to *B. canis*.

BABESIA IN HUMANS

B. divergens of cattle in Europe and *B. microti* (which it has been proposed should be called *Theileria microti* but is retained here for this account) of rodents in Europe and North America can infect humans, causing a malaria-like disease, following transmission by *I. ricinus* in Europe and *I. scapularis* in eastern North America. Generally, human babesioses are a relatively mild, often asymptomatic disease, except when the host is immunosuppressed or has been splenectomized, in which case it can have a fatal outcome. The severity of the disease also varies with the age of the patient, with elderly patients tending to develop severe illness. Treatment involves oral quinine and intravenous clindamycin. The epidemiology of human babesiosis is similar to that of Lyme disease. *B. microti*, a parasite of the common white-footed mouse, *P. leucopus*, is transmitted by *I. scapularis*, whose populations are dependent on deer as hosts. The larval ticks acquire *B. microti* by feeding on infected white-footed mice (*P. leucopus*) or some species of shrew. The gametocytes ingested by larvae invade the salivary glands and pass transstadially to the nymphal stage, which transmit sporozoites while feeding. While *I. ricinus* transmits *B. divergens* transovarially, this does not occur with *B. microti* in *I. scapularis*.

CONTROL OF BABESIOSIS

Effective drugs are available to treat the disease in cattle, early treatment being essential to prevent the animal dying from anaemia, and these drugs are amicarbalide, quinuronium sulfate, diminazene aceturate, imidocarb dipropionate, trypan blue, pentamidine and phenamidine, although the first two have been withdrawn because of a manufacturing safety issue and the third, which is widely used in the tropics, has been withdrawn from Europe for marketing reasons. The anti-theilerial drugs, parvaquone and buparvaquone, are also reputed to be effective for babesiosis. National registration authorities have restricted access to some of these drugs in certain countries, as some (e.g. the diamidine derivative, diminazine) are associated with a high rate of toxic side effects. These drugs can delay the progress of clinical signs but only rarely achieve a true eradication of the infection.

Vaccination with living parasites and imidocarb has to be controlled carefully to ensure that the drug does not prevent the parasite eliciting an immune response from the host, leaving the animal unprotected. An attenuated vaccine has given excellent results and has the advantage of being less virulent and non-infective to ticks. In Europe, commercial vaccines are available for canine babesiosis.

Controlling babesiosis by controlling the tick vector should aim to protect animals from tick bites or at least reduce the level of tick infestation, minimizing tick damage and allowing all animals to become infected early in life, when they are less susceptible. Tick and

Babesia spp. reinfections throughout the lives of the animals help to maintain acquired immunity (enzootic stability). Tick control that is 'too successful' prevents animals acquiring infection and immunity, risking a more damaging outbreak later.

THEILERIOSES

Two genera of the Theileriidae, *Theileria* and *Cytauxozoon*, are recognized as relevant in the current context. *Theileria* are particularly important parasites of cattle and five species of *Theileria* are found in cattle in various parts of the world, with the most important being *T. parva* and *Theileria annulata*. *Theileria lestoquardi* (syn. *Theileria hirci*) causes malignant theileriosis in sheep and goats, while *Theileria ovis* and *Theileria separata* cause mild theilerioses in small ruminants and sheep, respectively. Further, *B. equi* has been recognized as a species of *Theileria* in horses (causing biliary fever or piroplasmosis) and is included here (however, although it has been proposed that *B. microti* should be renamed *Theileria microti*, that rodent organism has been included with the babesioses for this account). Species of *Cytauxzoon*, particularly *Cytauxzoon felis*, have been reported as causing serious disease and death in wild and domestic cats.

THEILERIA IN CATTLE

An infective tick introduces sporozoites after it has been feeding for 3–5 days, during which time kinetes have invaded its salivary glands and produced infective sporozoites. Within 10 min, the sporozoites have invaded lymphoid cells of the host and, after 72 h, they begin to multiply by binary fission, producing large schizonts containing 13–50 nuclei. These macroschizonts are 'Koch's blue bodies'. They stimulate the infective lymphocytes to divide, producing infected daughter cells (uninfected lymphocytes do not divide). When only one or two schizonts are present in a lymphocyte, they become spherical and relatively large (microschizonts) and they multiply to produce merozoites. The merozoites are found in erythrocytes beginning 8 (*T. annulata*) or 13 days (*T. parva*) after infection. Up to 90% of the erythrocytes may be infected with *T. annulata*. Commonly, two different forms of merozoites are seen – comma shaped and spherical – dividing by binary fission and leading to the destruction of the erythrocytes, causing anaemia in the host. Spherical merozoites are probably gamonts which do not develop further until ingested by a tick.

The pathogenicity of a *Theileria* species is related to the density of schizonts in lymphocytes and piroplasms in erythrocytes. *T. parva*, *T. annulata* and *T. lestoquardi* produce numerous schizonts and piroplasms and are very pathogenic. *Theileria orientalis*, *T. ovis* and *Theileria mutans* produce few schizonts, but may cause varying degrees of anaemia when piroplasms are abundant. Schizonts of the benign parasites *Theileria velifera* and *T. separata* have not been described and only scanty infections of the erythrocytes are found.

In the tick, sexual stages develop from spherical merozoites 2–4 days after cessation of feeding. Ray bodies are found together with spherical 'macrogametes'. The ray bodies are considered to be microgamonts, which develop into uninucleated gamete-like stages. Syngamy of gametes occurs 6 days after feeding, giving rise to a zygote. Six to 24 days later, the spherical zygote becomes a motile kinete, found in the intestinal cells of the tick. After the tick has moulted and attached to a host, the motile kinetes leave the intestinal cells and invade the cells of the salivary glands (cell types d and e of the type III acinus), but not those of other organs. The parasite becomes polymorphic, with nuclear division producing thousands of small cytomeres. Five days after attachment, cytomere production is complete and they develop into ovoid sporozoites (almost 50,000 sporozoites/host cell).

Theileria parva

T. parva is a parasite of the African buffalo (*Syncerus caffer*), causes classic 'east coast fever' (ECF) of East and central Africa. Previously, 'corridor disease' (a more serious illness) of East and southern Africa was said to be caused by *T. parva lawrencei*, and 'January disease' or 'Zimbabwean theileriosis' (a milder illness) by *T. parva bovis*. However, these were not sound taxa and the three subspecies are now known to be genetically identical, with the variations in incidence, pathogenicity and morphology simply reflecting the antigenic diversity of *T. parva*.

T. parva occurs in eastern, central and southern Africa. It is lethal to European (*B. taurus*) and zebu (*B. indicus*) cattle and to the water buffalo (*Bubalus bubalis*). The main vector is the brown ear tick, *R. appendiculatus*, and in the more arid regions of eastern Africa it is *Rhipicephalus zambeziensis*. In Angola, where *R. appendiculatus* is absent, and in Zaire, the vector is *Rhipicephalus duttoni*.

In cattle, fever is dependent on the density of schizonts in the host, and its time of onset is dependent on the number of sporozoites injected. The incubation period is 1–3 weeks. The disease is characterized by high fever, catastrophic lymphoblastosis, which destroys vital organs and impairs the immune system, emaciation and death in 7–10 days. Formation of merozoites is time dependent and, at all dosages, piroplasms appear in erythrocytes in 13 days, being the prepatent period independent of the number of piroplasms. At death, 50% of the erythrocytes may contain piroplasms. With a virulent strain of *T. parva*, single tick infections can kill over 90% of susceptible cattle. In susceptible hosts, imported cattle or previously unexposed indigenous stock, *T. parva* can cause 90–100% mortality, and even zebu cattle, which have a natural immunity to the disease, suffer a 5% mortality in calves. Cattle that survive have a solid immunity in which piroplasms are not visible.

Transmission of *T. parva* is exclusively transstadial. Infections acquired by the larva or nymph are transmitted in the succeeding nymph or adult stage. There is considerable variation in the proportion of feeding ticks which become infected and in the intensity of infection in individual ticks. One factor that influences this is the parasitaemia of the host. In one series of experiments, the overall infection rate among ticks was 35%; however, a significantly higher percentage (61%) was infected when the parasitaemia of the host was 41–50% than at parasitaemias of 6–40% and <5%, when the respective infection rates were 33 and 27%. There is no close relationship between the intensity of infection in the salivary glands of the tick and the level of parasitaemia in the host animal. When nymphal and larval *R. appendiculatus* are fed on the same host, the infection rate in the resulting adults is higher than in the nymph, but sporozoites are produced more rapidly in nymphs than in adults, appearing in 2 days compared to 4 days in adult *R. appendiculatus*. Infectivity with *T. parva* is lost after 11 months, even though the tick may survive for 1–2 years.

There is a relationship between tick burden and the intensity of ECF that is meaningful at low infestation rates. An average of five adult ticks per animal (two or three per ear) will sustain enzooticity and one to four per head will induce epizooticity in which heavy losses of adult cattle can occur, while an average of less than one per individual animal can allow sporadic outbreaks. In ECF-free areas, infestations of adult *R. appendiculatus* average one adult for every five or more beasts.

Theileria annulata

The agent causing tropical theileriosis or Mediterranean coast fever (not particularly appropriate terms) in a region extending from southern Europe and northern Africa in the west via the Middle East, Russian Federation and central Asia to northern China. Although somewhat less pathogenic than *T. parva*, *T. annulata* may be economically the more important species because of its wider distribution in the world. In enzootic areas, European cattle may suffer mortalities of up to 40–80% and even zebu cattle, which are much more resistant to *Theileria* than imported cattle breeds, may still suffer 15–20% mortality, with deaths occurring mainly in calves. The disease involves hypertrophy of the lymphoid tissue, fever, weight loss and haemolytic anaemia, with more than 30% of the erythrocytes being parasitized. Recovered cattle have a good, persistent immunity of the premunity type in which there is persistent parasitaemia.

The vectors of *T. annulata* are various species of *Hyalomma*, most of which have only one generation per year. Cattle which have recovered from infection with *T. annulata* continue to harbour parasites in the circulating blood and ticks easily become infected, with infection rates of up to 64% being recorded in field-collected *Hy. anatolicum*.

T. annulata is transmitted by three-host ticks such as *Hy. anatolicum*, two-host ticks

such as *Hy. detritum*, and even by *Hy. scupense*, a one-host tick in which the adult remains on the host over the winter period and transmits the disease by moving from one host to another, when cattle are in close contact. The development of *T. annulata* in *Hy. excavatum* involves the formation of microgametes and macrogametes, with four microgametes being formed from each microgamont; these are formed in the first 96 h and from day 5 after repletion zygotes appear in the epithelial cells of the gut and grow steadily to day 12. The zygotes transform into kinetes, which at first move within the epithelial cells, then are found in the haemolymph from day 17, reach the salivary glands 18 days after repletion and transform into fission bodies. Infected host cells become greatly enlarged and the parasite divides several times before sporozoites are formed and released into the saliva. This takes from 2 to 7 days, depending on whether they are young ticks or have been starved for up to 6 months, but when ticks have been starved for more than 6 months, no sporozoites may develop during their feeding period.

CONTROL OF BOVINE THEILERIOSIS

Halofuginone given orally or intramuscularly and intramuscular parvaquone are effective treatments. Cattle can be immunized by the 'infection–treatment method', which involves injecting *T. parva* sporozoites into an animal and controlling the resulting infection by the administration of parvaquone. Controlling the disease by controlling the vector is an expensive and demanding operation. It is necessary to treat cattle weekly and the cost of acaricides can be prohibitive.

THEILERIA IN HORSES
Theileria equi

T. equi (formerly *B. equi*) causes equine biliary fever (also called equine piroplasmosis) in horses. The disease occurs in parts of southern Europe, Africa, Asia and parts of the Americas. A number of species of three genera of ixodid ticks, *Dermacentor*, *Hyalomma* and *Rhipicephalus*, have been identified as vectors and transstadial transmission has been demonstrated in nine species.

Infections with *T. equi* develop into clinical cases after an incubation period of 12–14 days, and intrauterine infection of the fetus is a complication which causes serious losses. The symptoms are fever, anorexia, malaise and weight loss and, in neglected horses, severe anaemia. *T. equi* produces a parasitaemia of 1–5% and occasionally to more than 20%.

CYTAUXZOON IN CATS

C. felis (and perhaps other *Cytauxzoon* species) causes a serious disease in felids. The initial tissue phase of infection occurs with schizonts within macrophages lining blood vessels; this is followed by an erythrocytic phase, and development of the schizogonous form is responsible for the severe and fatal disease. In domestic cats, there can be a rapid course of illness, with death occurring in fewer than 5 days, following anaemia, depression, anorexia, vomiting, jaundice, splenomegaly, hepatomegaly and high fever, with hypothermia typically developing just prior to death. The disease has been described from regions of the eastern USA, where the reservoir host appears to be the wild felid bobcat (*Lynx rufus*) and ticks are considered the vectors (transmission of *C. felis* by *A. americanum* has been demonstrated experimentally). There are also reports of *Cytauxzoon* infections in cats from Europe (Spain, France and Italy), Asia (Mongolia) and South America (Brazil), but there is some doubt about the relationships of those organisms with *C. felis* and no information on local natural hosts or vectors.

6.3.5 Nematodes (filarioids)

Ticks are recognized as vectors of filarioids of the genus *Cercopithifilaria* (28 species), which parasitize primates, ungulates, rodents, carnivores and marsupials. *Cercopithifilaria* adults are usually localized beneath cutaneous tissues, whereas microfilariae are always in the dermis, the infestation being transmitted by ixodid ticks. Besides the first species known to infest dogs, namely *Cercopithifilaria grassii* (formerly known as *Dipetalonema grassii*) and *Cercopithifilaria bainae*, recently, microfilariae of a different, yet unnamed species (*Cercopithifilaria* sp. I) have been characterized and the occurrence of this nematode has overlapped with the distribution of its

vector (the brown dog tick, *R. sanguineus*), with prevalence rates reaching up to 21.6% in dogs from southern Europe. Histological evidence indicated that microfilariae of *Cercopithifilaria* sp. I might induce erythematous and papular dermatitis in infested dogs.

7. PREVENTION AND CONTROL

Due to the huge veterinary and economic importance of ticks, and the disease-causing pathogens they transmit, a significant effort has been and continues to be expended on their control and prevention. Control strategies for the prevention of veterinary disease range from the use of chemical acaricides and pheromones to non-chemical methods such as habitat modification and host removal. Tick management for the prevention of human infection involves avoidance practices, personal protection measures with clothing and topical repellents, self-inspections and antibiotic treaments and vaccines (where applicable).

7.1 Acaricide use

Chemical acaricides are applied to a host either by various methods such as total immersion in a dipping bath or in the form of a spray, shower, spot-on, slow-release ear tags for livestock, collars for dogs or bait stations that allow a self-application of acaricide (e.g. for deer). A wide variety of formulations of organophosphate and pyrethroid insecticides (e.g. permethrin, deltamethrin, flumethrin) are available for application. Macrocyclic lactones or closantel given by the parenteral route have also been shown to be a useful aid in the control of ticks. Topical acaricidal compounds, such as fipronil (phenylpyrazole), imidacloprid (chloronicotinyl, pyridylmethylamine), selamectin (macrocyclic-lactone) and organophosphates (e.g. malathion, ronnel, chlorpyrifos, fenthion, dichlorvos, cythoate, diazinon, propetamphos, phosmet) and carbamates (e.g. carbaryl, isoprocarb) can be used to kill ticks on the host. There is some suggestion that insect growth regulators such as methoprene may also be used, although the evidence for this is unclear. The efficiency of acaricides can be enhanced by using them in appropriate combinations or combining them with tick pheromones. The tick sex pheromone, 2,6-dichlorophenol, combined with the acaricide, propoxur, has suppressed populations of *D. variabilis* on dogs.

The long-term control of three-host hard ticks is geared to the period required for the adult female stage to become fully engorged, which varies from 4 to 10 days, according to the species. If an animal is treated with an acaricide which has a residual effect of, say, 3 days, it will be at least 7 days before any fully engorged female reappears following treatment (i.e. 3 days' residual effect plus a minimum of 4 days for engorgement). Weekly treatment during the tick season should therefore kill the adult female ticks before they are engorged, except in cases of very severe challenge, when the treatment interval has to be reduced to 4 or 5 days. Theoretically, weekly treatment should also control the larvae and nymphs, but in several areas, the peak infestations of larvae and nymphs occur at different seasons to the adult females and the duration of the treatment season has to be extended.

Acaricides can be applied directly to the tick's microhabitat. To reach and kill the ticks (which are usually covered by leaf litter or vegetation or hidden in crevices), the chemicals must be applied strategically with respect to space and time. Vapour treatments have adverse environmental impacts and so are usually used only in kennels, barns or around houses. Granular formulations of organophosphate, carbamate and pyrethroid (deltamethrin or bifenthrin) insecticides have been shown to be particularly effective in controlling populations of *I. scapularis* nymphs on residential properties in the north-east USA with a single spring application.

Permethrin- or fipronil-impregnated materials placed strategically in the field for pathogen reservoir rodents to collect for nesting have been shown to be relatively effective in reducing both the numbers of larval ticks on the mice and the density of infected host-seeking nymphs in some locations, but they have not been widely effective. An alternative strategy targeting the pathogen in the reservoir

host has also been tried in Lyme disease risk areas, with antibiotics being placed in rodent baits; however, this technology is still undergoing development and environmental assessment, and its future realization is uncertain.

With continual use of acaricides, tolerance and resistance inevitably arise and the development of resistance to acaricides is a major problem in many parts of the world. The development of resistance can be delayed by appropriate product rotation.

Humans can obtain some protection against ticks by the application of repellents such as diethyl toluamide (DEET) or pyrethrins to the skin, or DEET or pyrethroids (e.g. permethrin) to clothing, but this approach has short protective time and there is no commercially available repellent for livestock.

Based on their mode of action, the formulations of ectoparasiticides available on the market may prevent feeding (e.g. avoiding the beginning of blood feeding or disrupt contact between the arthropod parasite and the host) or cause the death of the arthropod parasite after a certain period since the beginning of feeding. The speed of tick kill and the residual activity are the most important factors, with a desired characteristic for acaricidal products being an efficacy of greater than 90% within 48 h after treatment and prevention of reinfestation. A more rapid acaricidal effect also enhances the prevention of pathogen transmission in treated animals.

7.2 Tick management and control

For livestock, traditional control methods such as burning of cattle pastures are still used in some areas and are generally practised during a dry period before rains, when ticks are inactive. This technique is still a most useful one in extensive range conditions, and provided it is used after seeding of the grasses has taken place, regeneration of the pastures will occur rapidly following the onset of rains. Cultivation of land and, in some areas, improved drainage help to reduce the prevalence of tick populations and can be used where more intensive systems of agriculture prevail. Pasture 'spelling' in which domestic livestock are removed from pastures for a period of time has been used in semi-extensive or extensive areas. However, pasture spelling is likely to be ineffective for three-host ticks because their immature stages occur on wild hosts.

Anti-tick vaccines are another alternative to acaricides. A commercially available vaccine, TickGARD®, has been developed against *R. (Bo.) microplus/australis* in Australia. The vaccine destroys the cells lining the tick's gut, producing gaps in the gut wall which allow blood to leak into the haemocoel. It reduces tick fertility by up to 70%, females producing fewer eggs with lower hatch rates and less viable larvae. The primary injection should be given when the infestation on cattle is low and booster doses given every 6–10 weeks. The vaccine is against a 'concealed' antigen, the tick's midgut cells, to which cattle are never exposed, and hence the need for booster injections. For optimal response, cattle should be in nutritionally good condition. The advantages of vaccination are a reduction in the frequency of insecticidal treatment and the absence of residues in milk and meat.

A combination of these strategies can be implemented together in large-scale eradication programmes. However, there are only a small number of cases where complete area-wide eradication of ticks has been successful. One of the most well-known examples was the eradication of *R. (Bo.) annulatus* and *R. (Bo.) microplus* in the USA by 1960. Failure to eradicate a tick population often occurs because of reinvasion and repopulation. For example, attempts to eradicate *R. (Bo.) microplus/australis* from Australia and Papua New Guinea failed, as the reinvasion of ticks in eradication zones could not be prevented.

One critical issue to be considered in attempting eradication is that removal of the vector tick followed by its reintroduction can be more damaging, because the cattle will not acquire immunity. Cattle infected early in life develop immunity without developing clinical disease and maintain their immunity throughout adult life. Non-immune adult cattle develop clinical disease. Hence, one approach may be to maintain tick infestation at a level which confers immunity on young animals so that there is no clinical disease and, at the same time, infestation is not heavy enough to cause direct damage. In Australia, this result can be achieved by limiting numbers of engorged female *R. (Bo.) microplus/australis* to about ten per animal per day.

For protection of humans from tick-borne diseases, vegetation management is also an appropriate practice to reduce tick populations by removing the shelter (e.g. short brush, ground cover, leaf litter or other vegetation) that maintains the higher humidity conditions required for their survival. Wearing appropriate protective clothing (long-sleeved shirt tucked into trousers and long trousers tucked into socks) increases the opportunities for ticks to be noticed and removed and decreases the risks of tick attaching to exposed skin; further, application of topical repellents (e.g. DEET (diethyl toluamide) or permethrin) to clothing and exposed skin effectively decrease the risk of tick bites.

SELECTED BIBLIOGRAPHY

Allsopp, B.A. (2010) Natural history of *Ehrlichia ruminantium*. *Veterinary Parasitology* 167, 123–135.

Aubry, P. and Geale, D.W. (2011) A review of bovine anaplasmosis. *Transboundary and Emerging Diseases* 58, 1–30.

Barnett, S.F. (1977) Theileria. In: Kreier, J.P. (ed.) *Parasitic Protozoa*, Vol IV. Academic Press, New York, pp. 77–113.

Bennett, C.E. (1995) Ticks and Lyme disease. *Advances in Parasitology* 36, 343–405.

Black, W.C. and Piesman, J. (1994) Phylogeny of hard- and soft-tick taxa (Acari: Ixodida) based on mitochondrial 16S rDNA sequences. *Proceeding of the National Academy of Sciences* 91, 10034–10038.

Black, W.C., Klompen, J.S. and Keirans, J.E. (1997) Phylogenetic relationships among tick subfamilies (Ixodida: Ixodidae: Argasidae) based on the 18S nuclear rDNA gene. *Molecular Phylogenetics and Evolution* 71, 29–44.

Bock, R.E. and De Vos, A.J. (2001) Immunity following use of Australian tick fever vaccine: a review of the evidence. *Australian Veterinary Journal* 79, 832–839.

Bock, R., Jackson, L., De Vos, A. and Jorgenson, W. (2004) Babesiosis of cattle. *Parasitology* 129, S247–269.

Brackney, M.M., Marfin, A.A., Staples, J.E., Stallones, L., Keefe, T., Black, W.C., *et al.* (2010) Epidemiology of Colorado tick fever in Montana, Utah, and Wyoming, 1995–2003. *Vector-Borne and Zoonotic Diseases* 10, 381–385.

Brianti, E., Otranto, D., Dantas-Torres, F., Weigl, S., Latrofa, M.S., Gaglio, G., *et al.* (2012) *Rhipicephalus sanguineus* (Ixodida, Ixodidae) as intermediate host of a canine neglected filarial species with dermal microfilariae. *Veterinary Parasitology* 10, 330–337.

Burri, C., Bastic, V., Maeder, G., Patalas, E. and Gern, L. (2011) Microclimate and the zoonotic cycle of tick-borne encephalitis virus in Switzerland. *Journal of Medical Entomology* 48, 615–627.

Carli, E., Trotta, M., Chinelli, R., Drigo, M., Sinigoi, L., Tosolini, P., *et al.* (2012) *Cytauxzoon* sp. infection in the first endemic focus described in domestic cats in Europe. *Veterinary Parasitology* 183, 343–352.

Carroll, J.E., Allen, P.C., Hill, D.E., Pound, J.M., Miller, J.A. and George, J.E. (2002) Control of *Ixodes scapularis* and *Amblyomma americanum* through use of the '4 poster' treatment device on deer in Maryland. *Experimental and Applied Acarology* 28, 289–296.

Castellaw, A.H., Showers, J., Goddard, J., Chenney, E.F. and Varela-Stokes, A.S. (2010) Detection of vector-borne agents in lone star ticks, *Amblyomma americanum* (Acari: Ixodidae), from Mississippi. *Journal of Medical Entomology* 47, 473–476.

Chapman, A.S., Murphy, S.M., Demma, L.J., Holman, R.C., Curns, A.T., *et al.* (2006) Rocky Mountain spotted fever in the United States, 1997–2002. *Vector-Borne and Zoonotic Diseases* 6, 170–178.

Dahlgren, F.S., Mandel, E.J., Krebs, J.W., Massung, R.F. and McQuiston, J.H. (2011) Increasing incidence of *Ehrlichia chaffeensis* and *Anaplasma phagocytophilum* in the United States, 2000–2007. *American Journal of Tropical Medicine and Hygiene* 85, 124–131.

Dantas-Torres, F. (2007) Rocky Mountain spotted fever. *The Lancet Infectious Diseases* 7, 724–732.

Dantas-Torres, F. (2008) The brown dog tick, *Rhipicephalus sanguineus* (Latreille, 1806) (Acari: Ixodidae): from taxonomy to control. *Veterinary Parasitology* 152, 173–185.

Dantas-Torres, F, Giannelli, A. and Otranto, D. (2012) Starvation and overwinter do not affect the reproductive fitness of *Rhipicephalus sanguineus*. *Veterinary Parasitology* 185, 260–264.

Davies, F.G. (1997) Nairobi sheep disease. *Parassitologia* 39, 95–98.

Debboun, M. and Strickman, D. (2012) Insect repellents and associated personal protection for a reduction in human disease. *Medical and Veterinary Entomology*, doi: 10.1111/j.1365-2915.2012.01020.x

Deblinger, R.D. and Rimmer, D.W. (1991) Efficacy of a permethrin-based acaricide to reduce the abundance of *Ixodes dammini* (Acari: Ixodidae). *Journal of Medical Entomology* 28, 708–711.

De Waal, D.T. (1992) Equine piroplasmosis: a review. *British Veterinary Journal* 148, 6–13.

Dobson, S.J. and Barker, S.C. (1999) Phylogeny of the hard ticks (Ixodidae) inferred from 18S rRNA indicates the genus *Aponomma* is paraphyletic. *Molecular Phylogenetics and Evolution* 11, 288–295.

Dolan, M.C., Maupin, G.O., Schneider, B.S., Denatale, C., Hamon, N., Cole, C., *et al.* (2004) Control of immature *Ixodes scapularis* (Acari: Ixodidae) on rodent reservoirs of *Borrelia burgdorferi* in a residential community of southeastern Connecticut. *Journal of Medical Entomology* 41, 1043–1054.

Dolan, M.C., Schulze, T.L., Jordan, R.A., Dietrich, G., Schulze, C.J., Hojgaard, A., *et al.* (2011) Elimination of *Borrelia burgdorferi* and *Anaplasma phagocytophilum* in rodent reservoirs and *Ixodes scapularis* ticks using a doxycycline hyclate-laden bait. *American Journal of Tropical Medicine and Hygiene* 85, 1114–1120.

Doudier, B., Olano, J., Parola, P. and Brouqui, P. (2010) Factors contributing to emergence of *Ehrlichia* and *Anaplasma* spp. as human pathogens. *Veterinary Parasitology* 167, 149–154.

Dumler, J.S., Barbet, A.F., Bekker, C.P., Dasch, G.A., Palmer, G.H., Ray, S.C., *et al.* (2001) Reorganization of genera in the families Rickettsiaceae and Anaplasmataceae in the order *Rickettsiales*: unification of some species of *Ehrlichia* with *Anaplasma*, *Cowdria* with *Ehrlichia*, and *Ehrlichia* with *Neorickettsia*, descriptions of five new species combinations and designation of *Ehrlichia equi* and 'HGE agent' as subjective synonyms of *Ehrlichia phagocytophila*. *International Journal of Systematic and Evolutionary Microbiology* 51, 2145–2165.

Dworkin, M.S., Shoemaker, P.C. and Anderson, D.E. (1999) Tick paralysis: 33 human cases in Washington state, 1946–1996. *Clinical Infectious Diseases* 29, 1435–1439.

Eisen, R.J., Lane, R.S., Fritz, C.L. and Eisen, L. (2006) Spatial patterns of Lyme disease risk in California based on disease incidence data and modeling of vector-tick exposure. *American Journal of Tropical Medicine and Hygiene* 75, 669–676.

Eisen, R.J., Piesman, J., Zielinski-Gutierrez, E. and Eisen, L. (2012) What do we need to know about disease ecology to prevent Lyme disease in the northeastern United States? *Journal of Medical Entomology* 49, 11–22.

Ergonul, O. (2006) Crimean-Congo haemorrhagic fever. *Lancet Infectious Diseases* 6, 203–214.

Estrada-Pena, A. and Jongejan, F. (1999) Ticks feeding on humans: a review of records on human-biting Ixodoidea with special reference to pathogen transmission. *Experimental and Applied Acarology* 23, 685–715.

Estrada-Pena, A., Venzal, J.M., Nava, S., Mangold, A., Guglielmone, A.A., Labruna, M.B. and de la Fuente, J. (2012) Reinstatement of *Rhipicephalus (Boophilus) australis* (Acari: Ixodidae) with rediscription of the adult and larval stages. *Journal of Medical Entomology* 49, 794–802.

Estrada-Pena, A., Venzal, J.M. and Sanchez Acedo, C. (2006) The tick *Ixodes ricinus*: distribution and climate preferences in the western Palaearctic. *Medical and Veterinary Entomology* 20, 189–197.

Fish, D. and Childs, J.E. (2009) Community-based prevention of Lyme disease and other tick-borne diseases through topical application of acaricide to white-tailed deer: background and rationale. *Vector-Borne and Zoonotic Diseases* 9, 357–364.

Foley, J.E. and Nieto, N.C. (2010) Tularemia. *Veterinary Microbiology* 140, 332–338.

Franke, J., Hildebrandt, A., Meier, F., Straube, E. and Dorn, W. (2011) Prevalence of Lyme disease agents and several emerging pathogens in questing ticks from the German Baltic coast. *Journal of Medical Entomology* 48, 441–444.

Fuente, J. de la, Estrada-Pena, A., Venzal, J.M., Kocan, K.M. and Sonenshine, D.E. (2008) Overview: ticks as vectors of pathogens that cause disease in humans and animals. *Frontiers in Bioscience* 13, 6938–6946.

George, J.E., Pound, J.M. and Davey, R.B. (2004) Chemical control of ticks on cattle and the resistance of these parasites to acaricides. *Parasitology* 129, S353–366.

Giannelli, A., Dantas-Torres, F. and Otranto, D. (2012) Underwater survival of *Rhipicephalus sanguineus* (Acari: Ixodidae). *Experimental and Applied Acarology* 57, 171–178.

Ginsberg, H.S. and Stafford, K.C. III (2005) Management of tick and tick-borne diseases. In: Goodman, J.L., Dennis, D.T. and Sonenshine, D.E. (eds) *Tick-Borne Diseases of Humans*. ASM Press, Washington DC, pp. 65–86.

Gothe, R., Kunze, K. and Hoogstraal, H. (1979) The mechanism of pathogenicity in the tick paralyses. *Journal of Medical Entomology* 16, 357–369.

Gothe, R., Gold, Y. and Bezuidenhout, J.D. (1986) Investigations into the paralysis-inducing ability of *Rhipicephalus evertsi mimeticus* and that of hybrids between this subspecies and *Rhipicephalus evertsi evertsi*. *Onderstepoort Journal of Veterinary Research* 53, 25–29.

Graham, O.H. and Hourrigan, J.L. (1977) Eradication programs for the arthropod parasites of livestock. *Journal of Medical Entomology* 13, 629–658.

Grattan-Smith, P.J., Morris, J.G., Johnston, H.M., Yiannikas, C., Malik, R., Russell, R.C., *et al.* (1997) Clinical and neurophysiological features of tick paralysis. *Brain* 120, 1975–1987.

Gray, J.S. (1987) Mating and behavioural diapause in *Ixodes ricinus* L. *Experimental and Applied Acarology* 3, 61–71.

Gray, J., Zintl, A., Hildebrandt, A., Hunfeld, K.-P. and Weiss, L. (2010) Zoonotic babesiosis: overview of the disease and novel aspects of pathogen identity. *Ticks and Tick-borne Diseases* 1, 3–10.

Guglielmone, A.A., Beati, L., Barros-Battesti, D.M., Labruna, M.B., Nava, S., Venzal, J.M., *et al.* (2006) Ticks (Ixodidae) on humans in South America. *Experimental and Applied Acarology* 40, 83–100.

Guglielmone, A.A., Robbins, R.G., Apanaskevich, D.A., Petney, T.N., Estrada-Pena, A., Horak, I.G., *et al.* (2010) The Argasidae, Ixodidae and Nuttalliellidae (Acari: Ixodida) of the world: a list of valid species names. *Zootaxa* 2528, 1–28.

Gyuranecz, M., Rigo, K., Dan, A., Foldvari, G., Makrai, L., Denes, B., *et al.* (2011) Investigation of the ecology of *Francisella tularensis* during an inter-epizootic period. *Vector-Borne and Zoonotic Diseases* 11, 1031–1035.

Hall, R. and Baylis, H.A. (1993) Tropical theileriosis. *Parasitology Today* 9, 310–312.

Hall-Mendelin, S., Craig, S.B., Hall, R.A., O'Donoghue, P., Atwell, R.B., Tulsiani, S.M., *et al.* (2011) Tick paralysis in Australia caused by *Ixodes holocyclus* Neumann. *Annals of Tropical Medicine and Parasitology* 105, 95–106.

Hayes, E.B. (2005) Tularemia. In: Goodman, J.L., Dennis, D.T. and Sonenshine, D.E. (eds) *Tick-borne Diseases of Humans*. ASM Press, Washington DC, pp. 207–217.

Heyman, P., Cochez, C., Hofhuis, A., van der Giessen, J., Sprong, H., Porter, S.R., *et al.* (2010) A clear and present danger: tick-borne diseases in Europe. *Expert Review of Anti-infective Therapy* 8, 33–50.

Hoogstraal, H. (1966) Ticks in relation to human diseases caused by viruses. *Annual Review of Entomology* 11, 261–308.

Hoogstraal, H. (1967) Ticks in relation to human diseases caused by *Rickettsia* species. *Annual Review of Entomology* 12, 377–420.

Hoogstraal, H. (1985) Argasid and nuttalliellid ticks as parasites and vectors. *Advances in Parasitology* 24, 135–238.

Hopla, C.E. (1974) The ecology of tularemia. *Advances in Veterinary Science and Comparative Medicine* 18, 25–53.

Hubálek, Z., Halouzka, J. and Juřicová, Z. (2003) Longitudinal surveillance of the tick *Ixodes ricinus* for borreliae. *Medical and Veterinary Entomology* 17, 46–51.

Irwin, P.J. (2009) Canine babesiosis: from molecular taxonomy to control. *Parasites and Vectors* 26, 2 Suppl 1, S4.

Jin, H., Wei, F., Liu, Q. and Qian, J. (2012) Epidemiology and control of human granulocytic anaplasmosis: a systematic review. *Vector-Borne and Zoonotic Diseases* 4, 269–273.

Johnson, R.C., Kodner, C., Jarnefeld, J., Eck, D.K. and Xu, Y. (2011) Agents of human anaplasmosis and Lyme disease at Camp Ripley, Minnesota. *Vector-Borne and Zoonotic Diseases* 11, 1529–1534.

Jongejan, F. and Uilenberg, G. (2004) The global importance of ticks. *Parasitology* 129, S3–14.

Jouda, F., Perret, J.-L. and Gern, L. (2004) *Ixodes ricinus* density, and distribution and prevalence of *Borrelia burgdorferi* sensu lato infection along an altitudinal gradient. *Journal of Medical Entomology* 41, 162–169.

Joyner, L.P. and Donnelly, J. (1979) The epidemiology of babesial infections. *Advances in Parasitology* 17, 115–140.

Kakoma, I. and Mehlhorn, H. (1994) Babesia of domestic animals. In: Kreier, J.P. (ed.) *Parasitic Protozoa*, 2nd edn, Vol 7. Academic Press, San Diego, California, pp. 141–216.

Kaufman, W.R. (1979) Control of salivary fluid secretion in ixodid ticks. In: Rodriguez, J.G. (ed.) *Recent Advances in Acarology*, Vol 1. Academic Press, New York, pp. 357–363.

Keim, P., Johansson, A. and Wagner, D.M. (2007) Molecular epidemiology, evolution, and ecology of Francisella. *Annals of the New York Academy of Sciences* 1105, 30–66.

Keirans, J.E. and Robbins, R.G. (1999) A world checklist of genera, subgenera, and species of ticks (Acari: Ixodida) published from 1973–1997. *Journal of Vector Ecology* 24, 115–129.

Keirans, J.E., Clifford, C.M., Hoogstraal, H. and Easton, E.R. (1976) Discovery of *Nuttalliella namaqua* Bedford (Acarina: Ixodoidea: Nuttalliellidae) in Tanzania and re-description of the female based on scanning electron microscopy. *Annals of the Entomological Society of America* 69, 926–932.

Klompen, J.S.H., Black, W.C., Keirans, J.E. and Oliver, J.H. Jr (1996) Evolution of ticks. *Annual Review of Entomology* 41, 141–162.

Kocan, K.M. (2000) Anaplasmosis control: past, present, and future. *Proceedings of the New York Academy of Science* 916, 501–509.

Kocan, K.M., de la Fuente, J., Blouin, E.F., Coetzee, J.F. and Ewing, S.A. (2010) The natural history of *Anaplasma marginale*. *Veterinary Parasitology* 167, 95–107.

Lane, R.S., Piesman, J. and Burgdorfer, W. (1991) Lyme borreliosis: relation of its causative agent to its vectors and hosts in North America and Europe. *Annual Review of Entomology* 36, 587–609.

Lees, A.D. (1946a) Chloride regulation and the function of the coxal glands in ticks. *Parasitology* 37, 172–184.

Lees, A.D. (1946b) The water balance in *Ixodes ricinus* L. and certain other species of ticks. *Parasitology* 37, 1–20.

Levin, M.L., Killmaster, L.F. and Zemtsova, G.E. (2012) Domestic dogs (*Canis familiaris*) as reservoir hosts for *Rickettsia conorii*. *Vector-Borne and Zoonotic Diseases* 12, 28–33.

Lucas, J.M.S. (1954) Fatal anaemia in poultry caused by heavy tick infestation. *Veterinary Record* 66, 573–574.

Lysyk, T.J. (2010) Tick paralysis caused by *Dermacentor andersoni* (Acari: Ixodidae) is a heritable trait. *Journal of Medical Entomology* 47, 210–214.

Lysyk, T.J., Veira, D.M. and Majak, W. (2009) Cattle can develop immunity to paralysis caused by *Dermacentor andersoni*. *Journal of Medical Entomology* 46, 358–366.

McCosker, P.J. (1981) The global importance of babesiosis. In: Ristic, M. and Kreier, J.P. (eds) *Babesiosis*. Academic Press, New York, pp. 1–24.

McDade, J.E. (1990) Ehrlichiosis – a disease of animals and humans. *Journal of Infectious Diseases* 161, 609–617.

McLean, R.G., Shriner, R.B., Pokorny, K.S. and Bowen, G.S. (1989) The ecology of Colorado tick fever in Rocky Mountain National park in 1974. III. Habitats supporting the virus. *American Journal of Tropical Medicine and Hygiene* 40, 86–93.

McQuiston, J.H., Paddock, C.D., Holman, R.C. and Childs, J.E. (1999) The human ehrlichiosis in the United States. *Emerging Infectious Diseases* 5, 635–642.

Mahoney, D.F. (1994) The development of control methods for tick fevers of cattle in Australia. *Australian Veterinary Journal* 71, 283–289.

Mantke, O.D., Escadafal, C., Niedrig, M. and Pfeffer, M. (2011) Tick-borne encephalitis in Europe, 2007 to 2009. *Eurosurveillance* 16, 7–18.

Margos, G., Vollmer, S.A., Ogden, N.H. and Fish, D. (2011) Population genetics, taxonomy, phylogeny and evolution of *Borrelia burgdorferi* sensu lato. *Infection Genetics and Evolution* 11, 1545–1563.

Maupin, G.O., Fish, D., Zultowsky, J., Campos, E.G. and Piesman, J. (1991) Landscape ecology of Lyme disease in a residential area of Westchestyr County, New York. *American Journal of Epidemiology* 133, 1105–1113.

Mehlhorn, H., Schein, E. and Ahmed, J.S. (1994) Theileria. In: Kreier, J.P. (ed.) *Parasitic Protozoa*, 2nd edn, Vol 7. Academic Press, San Diego, California, pp. 217–304.

Monath, T.P. (ed.) (1988) *The Arboviruses: Ecology and Epidemiology*, Vols. I–IV. CRC Press, Boca Raton, Florida.

Mukhebi, A.W. (1992) Economic impact of theileriosis and its control in Africa. In: Norval, R.A.I., Perry, B.D. and Young, A.S. (eds) *The Epidemiology of Theileriosis in Africa*. Academic Press, London, pp. 379–403.

Mwase, E.Y., Pegram, R.G. and Mather, T.N. (1990) New strategies for controlling ticks. In: Curtis, C.F. (ed.) *Appropriate Technology in Vector Control*. CRC Press, Boca Raton, Florida, pp. 94–102.

Needham, G.R. and Teel, P.D. (1991) Off-host physiological ecology of Ixodid ticks. *Annual Review of Entomology* 36, 659–681.

Nolan, J., Roulston, W.J. and Schnitzerling, H.J. (1979) The potential of some synthetic pyrethroids for control of the cattle tick (*Boophilus microplus*). *Australian Veterinary Journal* 55, 463–466.

Nolan, J., Schnitzerling, H.J. and Bird, P. (1985) The use of ivermectin to cleanse tick infested cattle. *Australian Veterinary Journal* 62, 386–388.

Norval, R.A.I., Lawrence, J.A., Young, A.S., Perry, B.D., Dolan, T.T. and Scott, J.B. (1991) *Theileria parva* influence of vector, parasite and host relationships on the epidemiology of theilenosis in southern Africa. *Parasitology* 102, 347–356.

Ogden, N.H., Cripps, P., Davidson, C.C., Owen, G., Parry, J.M., Timms, B.J., *et al.* (2000) Ixodid tick species attaching to domestic dogs and cats in Great Britain and Ireland. *Medical and Veterinary Entomology* 14, 332–338.

Oliver, J.H.J. (1989) Biology and systematics of ticks (Acari: Ixodida). *Annual Review of Ecology and Systematics* 20, 397–430.

Otranto, D. and Wall, R. (2008) New strategies for the control of arthropod vectors of disease in dogs and cats. *Medical and Veterinary Entomology* 22(4), 291–302.

Otranto, D., Dantas-Torres, F. and Breitschwerdt, E.B. (2009) Managing canine vector-borne diseases of zoonotic concern: part one. *Trends in Parasitology* 25, 157–163.

Otranto, D., Brianti, E., Latrofa, M.S., Annoscia, G., Weigl, S., Lia, R.P., *et al.* (2012a) On a *Cercopithifilaria* sp. transmitted by *Rhipicephalus sanguineus*: a neglected, but widespread filarioid of dogs. *Parasites and Vectors* 5, 1.

Otranto, D., Dantas-Torres, F., Tarallo, V.D., Ramos, R.A., Stanneck, D., Baneth, G., *et al.* (2012b) Apparent tick paralysis by *Rhipicephalus sanguineus* (Acari: Ixodidae) in dogs. *Veterinary Parasitology*. [Epub ahead of print]

Parola, P. and Raoult, D. (2001) Ticks and tickborne bacterial disease in humans: an emerging infectious threat. *Clinical Infectious Diseases* 32, 897–928.

Parola, P., Paddock, C.D. and Raoult, D. (2005) Tick-borne rickettsioses around the world: emerging diseases challenging old concepts. *Clinical Microbiology Reviews* 18, 719–756.

Pattnaik, P. (2006) Kyasanur forest disease: an epidemiological view in India. *Reviews in Medical Virology* 16, 151–165.

Pegram, R.G., Lemche, J., Chizyuka, H.G.B., Sutherst, R.W., Floyd, R.B., Kerr, J.D., *et al.* (1989) Effect of tick control on liveweight gain of cattle in central Zambia. *Medical and Veterinary Entomology* 3, 313–320.

Pegram, R.G., Wilson, D.D. and Hansen, J.W. (2000) Past and present national tick control programs. Why they succeed or fail. *Annals of the New York Academy of Science* 916, 546–554.

Piesman, J. and Dolan, M.C. (2002) Protection against Lyme disease spirochete transmission provided by prompt removal of nymphal *Ixodes scapularis* (Acari: Ixodidae). *Journal of Medical Entomology* 39, 509–512.

Piesman, J. and Eisen, L. (2008) Prevention of tick-borne diseases. *Annual Review of Entomology* 53, 323–343.

Porter, R., Norman, R. and Gilbert, L. (2011) Controlling tick-borne diseases through domestic animal management: a theoretical approach. *Theoretical Ecology* 4, 321–339.

Porterfield, J.S. (1989) *Andrewes' Viruses of Vertebrates*. Balliere-Tindall, London.

Randolph, S.E. (1993) Climate, satellite imagery and the seasonal abundance of the tick *Rhipicephalus appendiculatus* in southern Africa: a new perspective. *Medical and Veterinary Entomology* 7, 243–258.

Randolf, S.E. (2001) The shifting landscape of tick-borne zoonoses: tick-borne encephalitis and Lyme borreliosis in Europe. *Philosophical Transactions of the Royal Society London Series B* 356, 1045–1056.

Randolf, S.E. (2004) Tick ecology: processes and patterns behind epidemiological risk posed by ixodid ticks as vectors. *Parasitology* 129, S37–65.

Randolph, S.E. and Sumilo, D. (2007) Tick-borne encephalitis in Europe: dynamics of changing risk. In: Takken, W. and Knols, B.G.J. (eds) *Emerging Pests and Vector-Borne Diseases in Europe. Ecology and Control of Vector-Borne Diseases*, 1, pp. 187–206.

Randolph, S.E., Gern, L. and Nuttall, P.A. (1996) Co-feeding ticks: epidemiological significance for tick-borne pathogen transmission. *Parasitology Today* 12, 472–479.

Raoult, D. and Marrie, T. (1995) Q fever. *Clinical and Infectious Diseases* 20, 489–496.

Reese, S.M., Dietrich, G., Dolan, M.C., Sheldon, S.W., Piesman, J., Petersen, J.M., *et al.* (2010) Transmission dynamics of *Francisella tularensis* subspecies and clades by nymphal *Dermacentor variabilis* (Acari: Ixodidae). *American Journal of Tropical Medicine and Hygiene* 83, 645–652.

Reese, S.M., Petersen, J.M., Sheldon, S.W., Dolan, M.C., Dietrich, G., Piesman, J., *et al.* (2011) Transmission efficiency of *Francisella tularensis* by adult American dog ticks (Acari: Ixodidae). *Journal of Medical Entomology* 48, 884–890.

Reichard, M.V., Edwards, A.C., Meinkoth, J.H., Snider, T.A., Meinkoth, K.R., Heinz, R.E., et al. (2010) Confirmation of *Amblyomma americanum* (Acari: Ixodidae) as a vector for *Cytauxzoon felis* (Piroplasmorida: Theileriidae) to domestic cats. *Journal of Medical Entomology* 47, 890–896.

Rizzoli, A., Hauffe, H.C., Carpi, G., Vourch, G.I., Neteler, M. and Rosa, R. (2011) Lyme borreliosis in Europe. *Eurosurveillance* 16, 2–9.

Ruebush, T.K. (1991) Babesiosis. In: Strickland, G.T. (ed.) *Hunter's Tropical Medicine*. Saunders, Philadelphia, Pennsylvania, pp. 655–658.

Russell, R.C. (2001) The medical significance of acari in Australia. In: Halliday, R.B., Walter, D.E., Proctor, H.C., Norton, R.A. and Colloff, M.J. (eds) *Acarology: Proceedings of the 10th International Congress*. CSIRO Publishing, Melbourne, Australia, pp. 535–546.

Ruzek, D., Yakimenko, V.V., Karan, L.S. and Tkachev, S.E. (2010) Omsk haemorrhagic fever. *Lancet* 376, 2104–2113.

Schoeman, J.P. (2009) Canine babesiosis. *Onderstepoort Journal of Veterinary Research* 76, 59–66.

Schulze, T.L., Jordan, R.A., Hung, R.W., Taylor, R.C., Markowski, D. and Chomsky, M.S. (2001) Efficacy of granular deltamethrin against *Ixodes scapularis* and *Amblyomma americanum* (Acari: Ixodidae) nymphs. *Journal of Medical Entomology* 38, 344–346.

Selmi, M., Martello, E., Bertolotti, L., Bisanzio, D. and Tomassone, L. (2009) *Rickettsia slovaca* and *Rickettsia raoultii* in *Dermacentor marginatus* ticks collected on wild boars in Tuscany, Italy. *Journal of Medical Entomology* 46, 1490–1493.

Short, N.J. and Norval, R.A.I. (1981) The seasonal activity of *Rhipicephalus appendiculatus* Neumann 1901 (Acarina: Ixodidae) in the highveld of Zimbabwe Rhodesia. *Journal of Parasitology* 67, 77–84.

Smith, R.D. and Rogers, A.B. (1998) *Borrelia theileri*: a review. *Journal of Spirochetal and Tick-borne Diseases* 5, 63–68.

Smith, R.P. (2011) Ticks: the vectors of Lyme disease. In: Halperin, J.J. (ed.) *Lyme Disease: An Evidence-Based Approach. Advances in Molecular and Cellular Microbiology*, 20, pp. 1–28.

Solberg, V.B., Miller, J.A., Hadfield, T., Burge, R., Schech, J.M. and Pound, J.M. (2003) Control of *Ixodes scapularis* (Acari: Ixodidae) with topical self-application of permethrin by white-tailed deer inhabiting NASA, Beltsville, Maryland. *Journal of Vector Ecology* 28, 117–134.

Sonenshine, D.E. (1986) Tick pheromones. *Current Topics in Vector Research* 2, 225–263.

Sonenshine, D.E. (1991) *Biology of Ticks*, Vol 1. Oxford University Press, New York.

Sonenshine, D.E. (1993) *Biology of Ticks*, Vol 2. Oxford University Press, New York.

Spickett, A.M. and Heyne, H. (1988) A survey of Karoo tick paralysis in South Africa. *Onderstepoort Journal of Veterinary Research* 55, 89–92.

Steelman, C.D. (1976) Effects of external and internal arthropod parasites on domestic livestock production. *Annual Review of Entomology* 21, 155–178.

Steere, A.C. (1989) Lyme disease. *New England Journal of Medicine* 321, 586–596.

Stone, B.F. (1988) Tick paralysis, particularly involving *Ixodes holocyclus* and other *Ixodes* species. *Advances in Disease Vector Research* 5, 61–85.

Sutherst, R.W., Dallwitz, M.J., Utech, K.B.W. and Kerr, J.D. (1978) Aspects of host finding by the cattle tick, *Boophilus microplus. Australian Journal of Zoology* 26, 159–174.

Sutherst, R.W., Norton, G.A., Barlow, N.D., Conway, G.R., Birley, M. and Comins, H.N. (1979a) An analysis of management strategies for cattle tick (*Boophilus microplus*) control in Australia. *Journal of Applied Ecology* 16, 359–382.

Sutherst, R.W., Utech, K.B.W., Kerr, J.D. and Wharton, R.H. (1979b) Density dependent mortality of the tick, *Boophilus microplus*, on cattle – further observations. *Journal of Applied Ecology* 16, 397–403.

Sutherst, R.W., Maywald, G.F., Kerr, J.D. and Stegeman, D.A. (1983) The effect of cattle tick (*Boophilus microplus*) on the growth of *Bos indicus* X *B. taurus* steers. *Australian Journal of Agricultural Research* 34, 317–327.

Teel, P.D., Ketchum, H.R., Mock, D.E., Wright, R.E. and Strey, O.F. (2010) The Gulf Coast tick: a review of the life history, ecology, distribution, and emergence as an arthropod of medical and veterinary importance. *Journal of Medical Entomology* 47, 707–722.

Telford, S.R., Gorenflot, A., Brasseur, P. and Spielman, A. (1993) Babesial infections in humans and wildlife. In: Kreier, J.P. (ed.) *Parasitic Protozoa*, Vol 5. Academic Press, San Diego, California, pp. 1–47.

Thorner, A.R., Walker, D.H. and Petri, W.A.J. (1998) Rocky Mountain spotted fever. *Clinical Infectious Diseases* 27, 1353–1360.

Troughton, D.R. and Levin, M.L. (2007) Life cycles of seven Ixodid tick species (Acari: Ixodidae) under standardized laboratory conditions. *Journal of Medical Entomology* 44, 732–740.

Uilenberg, G. (2006) *Babesia* – a historical overview. *Veterinary Parasitology* 138, 3–10.

Waladde, S.M. (1987) Receptors involved in host location and feeding in ticks. *Insect Science and Its Application* 8, 643–647.

Waladde, S.M. and Rice, M.J. (1982) The sensory basis of tick feeding behaviour. *Current Themes in Tropical Science* 1, 71–118.

Walker, D.H. and Dumler, J.S. (1996) Emergence of the ehrlichioses as human health problems. *Emerging Infectious Diseases* 2, 18–29.

Webster, M.C., Fisara, P. and Sargent, R.M. (2011) Long-term efficacy of a deltamethrin-impregnated collar for the control of the Australian paralysis tick, *Ixodes holocyclus*, on dogs. *Australian Veterinary Journal* 89, 439–443.

Wikel, S.K. (1996) Host immunity to ticks. *Annual Review of Entomology* 41, 1–22.

Willadsen, P. (1980) Immunity to ticks. *Advances in Parasitology* 18, 293–313.

Wisseman, C.L. (1991) Rickettsial infections. In: Strickland, G.T. (ed.) *Hunter's Tropical Medicine*. W.B. Saunders, Philadelphia, Pennsylvania, pp. 256–286.

Woldehiwet, Z. (2010) The natural history of *Anaplasma phagocytophilum. Veterinary Parasitology* 167, 108–122.

Yabsley, M.J. (2010) Natural history of *Ehrlichia chaffeensis*: vertebrate hosts and tick vectors from the United States and evidence for endemic transmission in other countries. *Veterinary Parasitology* 167, 136–148.

Yoder, J.A., Pollack, R.J. and Spielman, A. (1993) An ant-diversionary secretion of ticks: first demonstration of an acarine allomone. *Journal of Insect Physiology* 39, 429–435.

Ticks (Soft) (Acari: Argasidae)

1. INTRODUCTION

The ticks (Ixodida) are relatively large acarines, all of which are blood-feeding ectoparasites of vertebrates. The order can be divided broadly into 'hard' and 'soft' ticks, based on the possession of a dorsal scutum in the Ixodidae (the 'hard' ticks), which is absent in the Argasidae (the 'soft' ticks). In the Argasidae, the body is leathery and unsclerotized, with a textured surface. Soft ticks differ from hard ticks in many aspects of their biology; in particular, they are generally nocturnal and visit the host frequently to take relatively small blood meals. Argasids are more common in deserts or dry conditions and, in contrast to the hard ticks, argasid ticks tend to live in close proximity to their hosts: in chicken coops, pigsties, pigeon lofts, bird's nests, animal burrows or dens. In these restricted and sheltered habitats, the hazards associated with host finding are reduced and more frequent feeding becomes possible. While ixodid ticks are those mainly responsible for transmission of the pathogens of greatest medical and veterinary concern, argasid ticks are also important as parasites and vectors of various pathogens, including viruses (e.g. African swine fever virus), bacteria (e.g. *Borrelia duttonii*) and rickettsia (e.g. *Aegyptianella pullorum*), and their bites also may cause severe allergic reactions.

2. TAXONOMY

The Acari are a subclass of the Arachnida with two superorders: Parasitiformes (Anactinotrichida) and Acariformes (Actinotrichida). The Parasitiformes possess one to four pairs of lateral stigmata posterior to the coxae of the second pair of legs and have freely movable coxae. There are four orders of Parasitiformes, two of which have few species and are of no medical or veterinary significance, and two, the Mesostigmata (Gamasida) and the Ixodida (Metastigmata), which are of significant medical and veterinary importance. The Ixodida contains 900 recognized species arranged in three families: the largest, the Ixodidae (hard ticks), includes 14 genera and 704 species, while the Argasidae (soft ticks) includes 5 genera and 195 recognized species. The third family, the Nuttalliellidae, comprises only a single species, *Nuttalliella namaqua*, found in the Afrotropical region (this family probably represents a basal lineage of largely extinct species which fed on reptiles, and *N. namaqua* may be considered the evolutionary link between the hard and soft ticks).

Within the Argasidae, three genera, *Argas*, *Ornithodoros* and *Otobius*, contain species of medical and/or veterinary importance, while a fourth genus, *Antricola*, includes species that are associated exclusively with bats.

Sixty-one species have been described in the genus *Argas*, and they are allocated to seven subgenera which are structurally and biologically distinct. Two subgenera, *Argas* and *Persicargas*, parasitize birds; other subgenera are associated with bats and a small number of other mammals, while *Argas* (*Microargas*) *transversus* is a permanent ectoparasite of the Galapagos giant tortoise (*Geochelone elephantopus*).

There are 114 species of *Ornithodoros*, of which three species, *Ornithodoros moubata*, *Ornithodoros lahorensis* and *Ornithodoros savignyi*, have become associated with people and/or their domestic animals. The taxonomic position of the two or more known strains of *O. moubata* is not satisfactorily resolved and the term *O. moubata* will be used to cover both the hut-dwelling strain, which feeds on people and chickens, and the strain living in burrows and feeding on the occupants – warthogs, antbears and porcupines.

Only two species of *Otobius* have been described, *Otobius megnini* from cattle and *Otobius lagophilus* from cottontail and jack rabbits in western North America.

3. MORPHOLOGY

Argasids are tough, leathery ticks with a textured and unsclerotized dorsal surface. None the less, larvae of most argasids have a dorsal scutum, which is important for taxonomic purposes. The cuticle in unfed ticks may be marked characteristically with grooves or folds. There is little differentiation between the sexes. In nymphs and adults, the gnathosoma is not usually visible from the dorsal view, being located ventrally in a recess, known as the camerostome (Fig. 126). The fourth segment of the palp is similar in size to the other three. When eyes are present, they are lateral in position in folds above the legs. The stigmata are small and placed anteriorly to the coxae of the fourth pair of legs (Fig. 127). The pad-like pulvillus between the claws is either absent or rudimentary in adults and nymphs, but may be well developed in larvae.

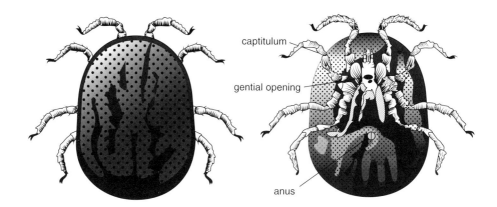

Fig. 126. *Ornithodoros moubata*. Dorsal (left) and ventral (right) views of female. *Source*: Castellani, A. and Chalmers, A.J. (1913) *A Manual of Tropical Medicine*. Bailliere Tindall, London.

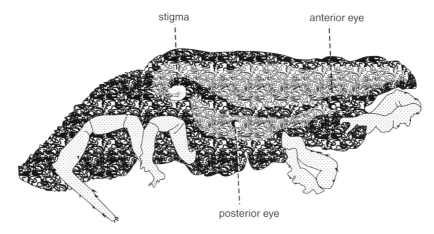

Fig. 127. *Ornithodoros savignyi*. Lateral view. *Source*: Patton, W.S. and Evans, A.M. (1929) *Insects, Ticks, Mites and Venomous Animals. Part 1: Medical*. H.R. Grubb, Croydon.

In *Argas*, the margin of the body is distinctly flattened and usually structurally different from the dorsal surface. The flattened margin remains distinct even when the tick is fully fed. There is usually a lateral sutural line present (Fig. 128). Eyes are absent. All stages of *Argas* are found in the resting places of the birds and bats that they parasitize. In *Ornithodoros* and *Otobius*, there is no lateral sutural line and no distinct margin to the body. In *Ornithodoros*, the integument bears mammillae and, in *Otobius*, the integument is spiny in the nymph and granulated in the adult.

4. LIFE CYCLE

Argasid ticks typically have a multi-host developmental cycle. Most argasid ticks, except larvae of certain *Argas* and *Ornithodoros*

species, feed rapidly (15–30 min), then drop off the host and shelter in cracks and crevices in the environment, where they moult to become first-stage nymphs. When a new host becomes available, nymphs climb rapidly on to the host, feed, drop off and again hide in sheltered spots in the environment, where they moult. In some species, up to seven nymphal moults occur before the final adult stage is reached; however, the exact number of nymphal stages varies, even within the same species. The number may depend on several factors, such as blood volume ingested during the blood meals and sex of the adult ticks (i.e. males usually emerge sooner than females, thus requiring a smaller number of nymphal stages). Adults mate off the host and feed several times. The adult female lays batches of 400–500 eggs after each blood meal (i.e. multiple gonotrophic cycles).

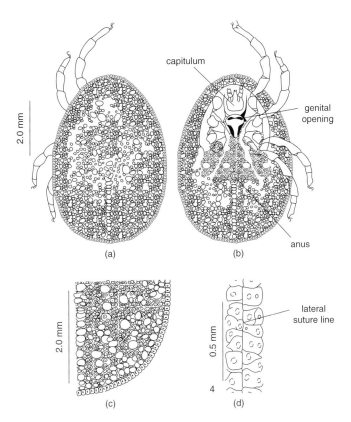

Fig. 128. *Argas* (*Persicargas*) *walkerae*. (a) Dorsal and (b) ventral views of female. (c) Female dorsal integument (posterior quadrant) and (d) lateral integument in the same area. *Source*: Kaiser and Hoogstraal (1969).

The high resistance to starvation during the development of argasids, as well as the multiple nymphal stages, gives soft ticks a characteristically much longer life cycle (many years) than that of hard ticks. Diapause is a major factor regulating the time needed for development of many argasid species that must survive in empty burrows or nests for several months until their hosts return or new hosts arrive.

5. BEHAVIOUR AND BIONOMICS

5.1 *Argas*

Argas persicus is of considerable veterinary importance as the most widespread argasid tick feeding on poultry. It originates in the Palaearctic region, where it occurs on domestic poultry and wild birds, but it has been introduced by humans with their poultry into all other zoogeographical regions, with the possible exception of the Neotropical region, where a close related species, *Argas miniatus*, occurs.

The unfed adult *A. persicus* is pale yellow in colour, becoming darker when fed. In outline, the body is oval to pear-shaped, being broadest behind the legs at about the level of the anus. Females measure 7–10 mm × 5–6 mm and males are slightly smaller, measuring 4–5 mm × 2.5–3 mm. The quadrangular cells at the body margin are not striated.

In southern Africa, the main argasid parasitizing poultry is *Argas walkerae* (Fig. 128); *Argas arboreus* parasitizes Ciconiiformes (i.e. herons, egrets, storks, ibises) in the Afrotropical region; *Argas robertsi* fills the same niche in the Australian and Oriental regions. In Queensland, *A. robertsi* also parasitizes chickens, and may do so elsewhere in the region, but the commoner chicken argasid is *A. persicus*. Although very similar, the validity of three of these four closely related species has been confirmed by cross-breeding in the laboratory. Few eggs and no progeny are produced when *A. arboreus*, *A. walkerae* and *A. persicus* are cross-mated, proving that these three species are isolated reproductively. *Argas sanchezi* and *Argas radiatus*, in addition to *A. persicus*, attack poultry in the

southern USA, along the Gulf of Mexico and the Mexican border.

Females deposit yellowish-brown, spherical eggs in cracks and crevices of poultry houses, where they hatch to produce a larva, which has a more or less circular outline, subterminal mouthparts visible from above and no stigmata. Larvae of *A. arboreus* have a simple respiratory system opening to the exterior through slit-like ostia above coxae I and II. The larva attaches itself to chicken, particularly under the wings, where it feeds for about a week. It then falls off, becoming inactive before moulting to a nymph. There are two to four nymphal stages before the adult. Females locate their host primarily by its odour and, secondarily, by its radiant heat. Digestion of blood is in three stages – haemolysis, 'rapid' digestion involving endocytosis and 'slow' digestion in which the stored blood meal is consumed gradually.

The adults produce an aggregation pheromone which brings the sexes together for mating. Usually, the female feeds before mating and the coxal fluid is excreted after the tick has fed (and not during feeding, as in *Ornithodoros*). During mating, the male inserts a spermatophore into the female genital opening. Eggs are matured and laid over a period of several days after a blood meal and not in a single batch. The median time for survival of unfed larvae of *A. persicus* is about 20 days. Depending on the availability of hosts, there may be up to ten generations a year, but the life cycle could extend to 2 years if hosts are available sporadically.

Desiccation is a major threat to the survival of unfed ticks; at 96% RH (relative humidity), survival may be almost 100% over more than 100 days and 50% at 0% RH.

5.2 *Ornithodoros moubata*

O. moubata, also known as the *O. moubata* complex (Fig. 126), is widely distributed throughout East Africa and northern South Africa, extending into the drier parts of central Africa. It lives in cracks in walls and in the earth floors of huts. During the day, it hides away in dark locations, including the possessions of the occupants. Consequently,

O. moubata has been spread by people as they move from one area to another.

The eggs hatch in about 8 days at 30°C and give rise to larvae, which do not feed but remain motionless until they moult into nymphs 4 days later. The nymphs feed on blood, taking 20–25 min to acquire a full meal. After an interval, the first nymphal stage moults into the second stage, which feeds and repeats the process. The number of nymphal stages is variable, with adult males being produced after four nymphal stages and females after five. Mating, which occurs after the female has fed, stimulates ovarian development, with oviposition occurring 10–15 days later. Full digestion of the blood meal and egg development can be delayed for many months if the female is not mated.

Feeding involves periods of active suction, when saliva, containing an effective anticoagulant, is introduced into the host, and periods of rest when blood is stored in the midgut and its diverticula. There is no open connection between the midgut and the rectum and no faeces are passed. The Malpighian tubules open into the hindgut, where their excretory products are deposited. The tick excretes a great deal of the watery component of blood via the coxal apparatus, which opens just behind the coxae of the first pair of legs. Excretion of coxal fluid begins about 15 min after the start of feeding and continues during feeding and for about an hour afterwards. The coxal apparatus can excrete 30 times its own volume in 20 min. Over a considerable period of time, a female will lay several batches of comparatively large, spherical (0.9 mm in diameter), glistening golden yellow eggs. During oviposition, the tick bends its head towards the genital opening and extrudes Gene's organ from the base of the gnathosoma. Each egg is 'handled' by the Gene's organ, which coats the egg with a waxy, waterproof layer, which is vital for its development and survival.

Guanine has been shown to act as a non-specific, non-volatile, persistent assembly pheromone which acts by contact and is responded to by other argasids (e.g. *A. persicus*, *Ornithodoros porcinus*) and some ixodids (e.g. *Amblyomma cohaerens*, *Rhipicephalus appendiculatus*). Female *Ornithodoros* secrete a non-specific pheromone in their coxal fluid, which is most active 4–6 days after the female has fed and evokes courtship behaviour in sexually active males. Aggregations of *O. moubata* develop in response to pheromones produced by both sexes. Males respond more readily to the female pheromone than do females to the male pheromone. The receptors are located on the fourth segment of the palps. The function of the pheromone is to bring the sexes together for mating and for food location, and hence starved ticks show an enhanced response to pheromones which are not species specific (e.g. *O. moubata* is attracted to *Ornithodoros tholozani* and *A. persicus*).

In mating, the male crawls beneath the female so that their ventral surfaces are in contact and the male uses its mouthparts to dilate the female genital opening. A bulb-shaped spermatophore appears at the genital opening of the male and is introduced into the female opening by the mouthparts of the male. Later, when the sperm are mature, the spermatophore ruptures and they are released into the uterus. A female may mate more than once.

In common with other argasid ticks, *O. moubata* has considerable powers of survival from starvation and desiccation. A crystalline cuticular wax coats the body and this has a melting point of 63°C (while that of *O. savignyi* is 75°C). It is also protected by an overlying cement layer, together preventing water loss through the cuticle. Unmated and unfed adult female *O. moubata* have survived for more than 3 months when kept at 32°C and at 0% RH. When fed once before being starved and maintained under a more favourable humidity (85% RH), there were differences in survival between strains of *O. moubata*. In three different populations, 60% survived for 9, 18 and 56 months. Survival is favoured by the fact that *O. moubata* is able to take up water through the spiracles when exposed to humid air (95% RH). Hence, in the absence of hosts, the tick is able to survive for long periods, even up to 5 years under suitable conditions of humidity. Such resistance to starvation enables foci of *O. moubata* to persist in the absence of hosts for a long period. Coupled with the ability to feed on alternate hosts, foci can be regarded, for practical purposes, as permanent.

The source of the blood meal can have a considerable effect on fecundity, and female *O. moubata* have been found to produce nearly twice as many eggs when they feed on porcine compared with bovine blood. Further, nymphs reared on porcine blood developed more quickly and reached the adult stage in fewer stages than nymphs fed on bovine blood.

A female may produce up to eight egg batches, but the number of eggs in each batch declines steadily from about 140 in the first batch to 33 in the eighth batch. In the absence of repeated mating, the percentage of females ovipositing declines rapidly to less than 30% after the third batch of eggs. If females mate after every blood meal, the proportion which oviposits remains high. On average, a female will lay about 500 eggs in her lifetime.

5.3 *Ornithodoros savignyi*

O. savignyi is known as the 'eyed tampan' (in contrast to *O. moubata*, the 'eyeless tampan'), because it possesses two pairs of simple eyes in the folds above coxa I and between coxae III and IV (Fig. 127). It is also known as the 'sand tampan', because it buries itself in sandy and loose clay soils, under trees, near wells and shady spots frequented by domestic stock – particularly camels, cattle, mules – and people, on which it feeds. It does not occur in huts. On standing cattle, it feeds on the legs just above the hooves. The numbers of *O. savignyi* in infested localities can reach plague proportions.

The biology and life cycle of *O. savignyi* are very similar to that of *O. moubata*, but it has a much wider geographical distribution – occurring in the arid parts of Africa, Arabia, India and Sri Lanka. *O. savignyi* is not known to transmit any pathogen, but camels and cattle suffer greatly from exposure to large numbers and may even die from the volume of blood lost.

5.4 *Ornithodoros lahorensis*

O. lahorensis is a serious pest of sheep, cattle and camels in Asia, the southern republics of the former Soviet Union and south-east Europe, from sea level to 2900 m. When larvae find a host, they remain on it for 3–6 weeks, engorging four times and moulting three times. The engorged nymph detaches and drops to the ground, where it moults into an adult. Given the opportunity, the adult will feed rapidly on another host, after which females will deposit batches of 300–500 eggs. This species is facultatively autogenous and can mature two batches of eggs without a blood meal. Unfed adults can live for 18 years and larvae for 1 year.

5.5 *Ornithodoros porcinus*

In a study of animal burrows in East Africa, over 40% of them were found to be infested by *O. porcinus*. The estimated numbers of ticks ranged from a few up to 250,000. The most important environmental conditions for tick survival are a neutral soil and a favourable temperature, with an optimum at 24°C. Vegetation around the burrow conceals the occupants from predators and favours the presence of suitable hosts for the tick. The optimum altitude for *O. porcinus* is 900–1500 m above sea level, but above 1900 m is not favourable.

5.6 *Otobius megnini*

O. megnini (Fig. 129), the spinose ear tick, originated in the Americas, from where it has been introduced into southern Africa and India. It is mainly a parasite of cattle and horses but has been recorded from a range of hosts in North America, including donkeys, sheep, goats, dogs, cats, deer and rabbits. In India, it has been recorded from cattle, sheep and humans. This tick is associated with stables and animal shelters.

The female tick lays her eggs in cracks and crevices 1 or 2 m above the ground in the walls of animal shelters. This behaviour ensures that the emerging larva is at a height to transfer easily to the bodies of large, stabled domestic animals. The eggs hatch in 11 days in summer and in 3–8 weeks under cooler conditions. The eggs are small, oval and reddish in colour. A hexapod larva 0.5 mm in length emerges from the egg. Its terminal gnathosoma is very

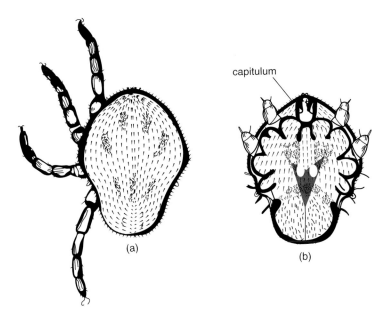

Fig. 129. *Otobius megnini.* (a) Dorsal and (b) ventral views of partially fed nymph. *Source*: Cooley and Kohls (1944).

long, accounting for more than one-third of the length of the unfed larva. There are two pairs of ocellus-like eyes present dorsally. The larva enters an ear of its host, where it engorges and may attain a length of 4 mm; the engorgement takes 5–10 days and is followed by a quiescent period before the larva moults to become an octopod nymph.

There are two octopod nymphal stages, each characterized by having the gnathosoma on the ventral surface and a spiny integument, from which the tick acquires its common name, 'spinose ear tick'. The nymphs reattach to the skin lining the ear, suck blood and remain in the ear for an unusually long time; most leave the host within 5 weeks, but they may remain for several months. The fully-fed nymph measures up to 8 mm. Nymphs drop from the host and creep into cracks and crevices in walls and woodwork, under stones or under the bark of trees, usually low down, where they develop into adults. The body of the adult is fiddle-shaped, being constricted posterior to the fourth pair of legs. The adult does not feed and its hypostome is poorly developed and without teeth.

The female can wait up to 18 months to be impregnated and then lays up to 1500 eggs in small batches over a period of a few weeks to several months. This species has considerable powers of survival in the absence of hosts and it can persist in empty cattle sheds and stables for more than 2 years. Unfed larvae usually survive for less than a month, but under favourable circumstances can last as long as 4 months. Infestations of *O. megnini* cannot be eradicated easily by vacating premises, except for excessively long periods.

O. megnini is more common in hot and arid areas and it is not present in wet areas. In South America, it has been found at altitudes of up to 2600 m above sea level. At this altitude in Bolivia, nymphs of *O. megnini* were present all year-round in the ears of dairy cattle and, during the rainy season, clusters of 150 nymphs and larvae could be found under the tails of a small percentage (<15%) of cattle. These ticks do not transmit pathogens but do considerable damage to the ears, eardrums and auricular nerves by their feeding. The ear can be choked with ticks, wax and other debris and the eardrum be ruptured, favouring secondary infections. Badly infested calves, sheep and goats may die and loss of condition in infested animals is common.

6. MEDICAL AND VETERINARY IMPORTANCE

The medical and veterinary importance of Argasid ticks is associated with the local lesions (skin rash, pruritus, mucosal hyperaemia) and systemic reactions (e.g. IgE-mediated anaphylaxis, tick toxicosis) that their bites can cause to humans and animals, and in addition, they act as vectors of a number of pathogenic organisms.

6.1 *Argas* species

Adult *Argas*, particularly the engorged females, are easily seen on the skin, commonly beneath the wings of chickens and turkeys, and also of wild birds. With high infestation burdens, egg laying decreases and may stop altogether as a result of the infestation. Argasids are most active in the warm dry season, causing emaciation, weakness, slow growth and anaemia, which may be fatal. However, the ticks only feed for a limited period.

Inflammation and raised areas will be present from tick bites and small granulomatous reactions (consisting of a mixed inflammatory cell response with fibrosis) may form at the site of tick bites. They are rarely found in commercial caged layer operations, but are more common in simpler poultry houses. The feeding of *A. persicus* and *A. walkerae* larvae can cause a condition known as fowl paralysis, since they attach and feed for days on animals. It is known that *A. reflexus* may cause IgE-mediated anaphylaxis in human patients. *A. persicus*, and probably the other related chicken-feeding species, transmits two important pathogens to poultry, *Borrelia anserina* and *A. pullorum*.

6.1.1 *Borrelia anserina*
B. anserina is a spirochaete that is pathogenic for geese, ducks, turkeys, pheasants, canaries, chickens and grouse, causing an acute condition with fever, cyanosis and diarrhoea, and mortality is normally high in domestic chickens. Treatment is usually based on antibiotics and the application of acaricides to reduce tick populations. Avian borreliosis occurs in southern Europe, the Middle East, Africa, India, Australia, South America and the western USA. The vectors of *B. anserina* are species of *Argas* (subgenus *Persicargas*), including *A. persicus* and *A. arboreus*. The development of *B. anserina* in argasid ticks involves the ingested borreliae penetrating the midgut and appearing in the haemocoel, in which they multiply and invade certain tissues, including the central nerve mass, salivary glands and gonads. Heavy infections of *B. anserina* develop and both transstadial and transovarial transmission occur. However, there are two other common Egyptian bird-parasitizing argasids, *Argas streptopelia* and *Argas hermanni*, that resist infection and show only limited transstadial and no transovarial transmission.

6.1.2 *Aegyptianella pullorum*
A. pullorum is a rickettsia that infects chickens, geese, ducks and quail, and the ostrich has been found infected naturally. Pigeons and turkeys are refractory to infection and guinea fowl is of uncertain susceptibility. A range of wild birds has been infected experimentally, but natural infections of wild birds need to be investigated in the field before deciding that these are infections with *A. pullorum*. The effect of infection varies with the age of the bird. Fatalities occur in chickens up to the age of 4 weeks, with mortality declining rapidly over that period. Poultry infected after the age of 12 weeks develop a low persistent parasitaemia. Fowls that have recovered clinically may remain infective to argasid ticks for up to 18 months. Treatment is typically effective with tetracycline and other broad-spectrum antibiotics. The infection is found in southern Europe, the Middle East, Africa and the Indian subcontinent. The vectors of *A. pullorum* are ticks of the subgenus *Persicargas* of the genus *Argas*, in particular *A. persicus* and *A. walkerae*. Transmission in the argasid is both transstadial and transovarial, although only a small percentage of larvae are infected by the transovarial route, which is regarded as being of little epizootiological importance. In *A. walkerae*, *A. pullorum* develops by intensive multiplication in the intestinal epithelium,

haemocytes and salivary glands, completing its cycle by 30 days. After an infected feed, the parasites are to be found in the intestinal epithelium after 24 h and, at the end of 14 days, the intestinal cells are heavily parasitized. After 2–3 weeks, the parasites appear in the haemocytes and multiply rapidly until the end of the 4th week, when the parasites occur in the salivary glands. This development can take place in all developmental stages of the tick, which remains infected for life, with infections even persisting for 2 years in starved nymphs and female ticks. *A. pullorum* is introduced into its bird host with the saliva of the feeding tick.

6.2 *Ornithodoros* species

Ticks of the genus *Ornithodoros* are known to cause local lesions and systemic illness generally referred to as 'tick toxicosis'. They also act as vectors of bacterial and viral pathogens.

6.2.1 *Borrelia* spirochaetes

These cause relapsing fevers (commonly known as 'tick-borne relapsing fever' or 'endemic relapsing fever') in humans. The relapsing fevers have a wide distribution, being found in Africa, Asia and the Americas, with assorted tick–*Borrelia* vector–pathogen associations and vertebrate hosts, mostly rodents, in the different regions. However, the most important of the associations is that of ticks of the *O. moubata* complex and *B. duttonii* in eastern, central and southern Africa. Here, the disease has been thought to be an anthroponosis, with humans being the only known host and *O. moubata* being the vector. However, recently there has been evidence suggesting it is a zoonosis, as it is in other parts of the world, where rodents and their *Ornithodoros* ticks (e.g. *Ornithodoros erraticus* in North Africa, *O. tholozani* in Asia, *Ornithodoros hermsi*, *Ornithodoros parkeri* and *Ornithodoros turicata* in the western and south-western USA and *Ornithodoros rudis* and *Ornithodoros talaje* in Central and South America) are involved in transmission cycles.

When *O. moubata* feeds on a person (or other animals) infected with *B. duttonii*, it takes in borreliae with the blood into the midgut. The borreliae then penetrate the midgut and enter the haemocoel, where they multiply and invade various organs, including the central nerve mass, the coxal glands and the ovaries. Only in nymphal ticks do the salivary glands become heavily infected, and therefore only nymphal ticks can transmit *B. duttonii* by bite. In adult ticks, the salivary glands are virtually free of spirochaetes and they transmit *B. duttonii* via infected coxal fluid, with the borreliae entering the new host through skin abrasions or by direct penetration of the unbroken skin.

Infection in humans results in a fever with a recurring (relapsing) pattern every few days; headache, myalgia and a petechial rash may occur, among other symptoms, and it can be a serious disease in young children. A treatment regime of penicillin followed by tetracycline is usually effective. In *O. moubata*, *B. duttonii* is passed both transstadially and transovarially, with filial infection rates being as high as 90%. *B. duttonii* can be transmitted transovarially through at least five generations, and hence centres of endemic relapsing fever tend to persist for long periods of time. Transmission of *B. duttonii* to humans is associated with populations of *O. moubata*, which are synanthropic and have habits comparable to those of bedbugs, multiplying in cracks and crevices of human habitations, from which they emerge at night to feed on the sleeping occupants. Survival of the disease in a location is aided by the long time that individual ticks may live.

Borrelia Coriaceae This is an undescribed bacterial pathogen whose aetiology is uncertain. It is likely to be transmitted by the soft tick *Ornithodoros coriaceus*, causing 'epizootic bovine abortion' (EBA). This is a severe condition of rangeland cattle in the western USA. The infection causes late-term abortions in beef cattle and deer.

6.2.2 Arboviruses

AFRICAN SWINE FEVER VIRUS (ASFV)

Currently the only member of the family Asfarviridae, this causes a lethal haemorrhagic disease in domesticated pigs, with mortality often 100% and following quickly after the onset of fever. It was first reported from eastern and southern Africa, associated with warthogs and bushpigs, in which the infections were subclinical and the vector was *O. porcinus*. In the 1950s, the virus was exported to Europe, where it caused major problems following local transmission by *O. erraticus*, and later it arrived in Central and South America, where different species of *Ornithodoros* have been shown to be competent experimental vectors. In Europe, *O. erraticus* is able to act as a reservoir of ASFV for at least 5 years after the removal of infectious hosts.

Experimentally, the virus passes from one tick generation to another by transovarian transmission. However, the level of viraemia needed to infect the tick has never been detected in warthogs, raising the possibility of another host being involved or of special conditions being required for warthogs to produce a sufficiently intense infecting viraemia. Apart from tick transmission, the virus can be passed between domestic pigs by close contact and fomites, and contaminated foodstuffs. It is widely considered to be the most important epidemic disease affecting the pig industry, and there is no vaccine available and no treatment for the infection.

7. PREVENTION AND CONTROL

Argasid ticks, which exist in and around animal housing, poultry houses and enclosures, can be controlled by the application of an acaricide to their environment, coupled with treatment of the population on the host. However, since unfed adults can survive hidden in cracks and crevices of the environment for up to 4 years, while unfed larvae can survive for several months, control measures must concentrate on treating all possible resting places with an approved insecticide and must be repeated regularly over a long period to kill ticks that

were too well hidden to be affected by earlier treatments.

Treatments may be effected using acaricidal sprays or emulsions containing organophosphates and pyrethroids. All niches and crevices in affected buildings should be sprayed, and nesting boxes and perches in poultry houses should be painted with acaricides. At the same time as premises are treated, birds should be dusted with a suitable acaricide or, in the case of larger animals, sprayed or dipped at monthly intervals. In poultry houses, all new birds should be treated prior to introduction into an existing flock. Control of argasid ticks can be assisted by the elimination of cracks in walls and perches, which provide shelter to the free-living stages.

Control of *Ornithodorus* ticks affecting humans can be achieved with the application of residual insecticides in and around houses, with particular focus on the numerous cracks and crevices where they may be hiding within the dwelling; such control has been shown to reduce the incidence of tick-borne relapsing fever in children in Africa. The use of sleeping nets and topical repellents can also be protective against the ticks at night, and nightlights can deter the ticks from entering sleeping rooms. Where rodent reservoirs are involved in transmission cycles, rodent control (see the entry on 'Fleas') should be undertaken.

Control of *O. megnini* on cattle can be achieved by the application of carbamate, pyrethroid and organophosphate biocides applied directly to the ears of the infected animals and/or by the application of ear tags containing amitraz (formamidine), permethrin or a combination of both. Infestation of horses by *O. megnini* has been controlled successfully by the administration of fipronil (phenylpirazol) into the ears of the animals in Australia. Acaricides administered orally (sulfur, famphur, eprinomectin) or by injection (ivermectin, doramectin) have failed to control infestations in cattle.

SELECTED BIBLIOGRAPHY

Bates, P. (2012) Ticks (Ixodida). In: Bates, P. (ed.) *External Parasites of Small Ruminants. A Practical Guide to Their Prevention and Control.* CAB International, Wallingford, UK, pp. 49–61.

Beelitz, P. and Gothe, R. (1991) Investigations on the host seeking and finding of *Argas* (*Persicargas*) *walkerae* (Ixodidae: Argasidae). *Parasitology Research* 77, 622–628.

Boinas, F.S., Wilson, A.J., Hutchings, G.H., Martins, C. and Dixon, L.J. (2011) The persistence of African swine fever virus in field-infected *Ornithodoros erraticus* during the ASF endemic period in Portugal. *PloS One* 6, e20383.

Burgdorfer, W. and Hayes, S.F. (1990) Vector–spirochaete relationships in louse-borne and tick-borne borrelioses with emphasis on Lyme disease. *Advances in Disease Vector Research* 6, 127–150.

Chellappa, D.J. (1973) Notes on spinose ear tick infestations in man and domestic animals in India and its control. *Review of Applied Entomology* 63, 2902.

Dantas-Torres, F., Venzal, J.M., Bernardi, L.F., Ferreira, R.L., Onofrio, V.C., Marcili, A., *et al.* (2012) Description of a new species of bat-associated argasid tick (Acari: Argasidae) from Brazil. *Journal of Parasitology* 98, 36–45.

Dautel, H. and Knulle, W. (1998) The influence of physiological age of *Argas reflexus* larvae (Acari: Argasidae) and of temperature and photoperiod on induction and duration of diapauses. *Oecologia* 113, 46–52.

Estrada-Pena, A. and Jongejan, F. (1999) Ticks feeding on humans: a review of records on human-biting Ixodoidea with special reference to pathogen transmission. *Experimental and Applied Acarology* 23, 685–715.

Estrada-Pena, A., Mangold, A.J., Nava, S., Venzal, J.M., Labruna, M. and Guglielmone, A.A. (2010) A review of the systematics of the tick family Argasidae (Ixodida). *Acarologia* 50, 317–333.

Galun, R., Sternberg, S. and Mango, C. (1978) Effects of host species on feeding behaviour and reproduction of soft ticks (Acari: Argasidae). *Bulletin of Entomological Research* 68, 153–157.

Geigy, R. (1968) Relapsing fevers. In: Weinman, D. and Ristic, M. (eds) *Infectious Blood Diseases of Man and Animals*, Vol. II. Academic Press, New York, pp. 175–216.

Guglielmone, A.A., Robbins, R.G., Apanaskevich, D.A., Petney, T.N., Estrada-Pena, A., Horak, I.G., *et al.* (2010) The Argasidae, Ixodidae and Nuttalliellidae (Acari: Ixodida) of the world: a list of valid species names. *Zootaxa* 2528, 1–28.

Hefnawy, T., Bishara, S.I. and Bassal, T.T.M. (1975) Biochemical and physiological studies of certain ticks (Ixodoidea): effects of relative humidity and starvation on the water balance and behaviour of adult *Argas* (*Persicargas*) *arboreus* (Argasidae). *Experimental Parasitology* 38, 14–19.

Hoogstraal, H. (1985) Argasid and nuttelhellid ticks as parasites and vectors. *Advances in Parasitology* 24, 135–238.

Hoogstraal, H., Clifford, C.M., Keirans, J.E. and Wassef, H.Y. (1979) Recent developments in biomedical knowledge of *Argas* ticks (Ixodoidea: Argasidae). In: Rodriguez, J.G. (ed.) *Recent Advances in Acarology*, Vol. II. Academic Press, New York, pp. 269–278.

Jori, F. and Bastos, A.D.S. (2009) Role of wild suids in the epidemiology of African swine fever. *EcoHealth* 6, 296–310.

Khalil, G.M. (1979) The subgenus *Persicargas* (Ixodoidea; Argasidae: *Argas*). 31. The life cycle of *A.* (*P.*) *persicus* in the laboratory. *Journal of Medical Entomology* 16, 200–206.

Kleiboeker, S.B., Burrage, T.G., Scoles, G.A., Fish, D. and Rock, D.L. (1998) African swine fever virus infection in the argasid host, *Ornithodoros porcinus porcinus*. *Journal of Virology* 72, 1711–1724.

Lopez, J.E., McCoy, B.N., Krajacich, N.J. and Schwan, T.G. (2011) Acquisition and subsequent transmission of *Borrelia hermsii* by the soft tick *Ornithodoros hermsi*. *Journal of Medical Entomology* 48, 891–495.

McCall, P.J., Hume, J.C.C., Motshegwa, K., Pignateli, P., Talbert, A. and Kisinza,W. (2007) Does tick-borne relapsing fever have an animal reservoir in East Africa? *Vector-Borne and Zoonotic Diseases* 7, 659–666.

Mango, C.K.A. and Galun, R. (1977) *Ornithodoros moubata*: breeding *in vitro*. *Experimental Parasitology* 42, 282–288.

Mans, B.J., de Klerk, D., Pienaar, R. and Latif, A.A. (2011) *Nuttalliella namaqua*: a living fossil and closest relative the ancestral tick lineage: implications for the evolution of blood-feeding in ticks. *PloS One* 6, e23675.

Monath, T.P. (1988) *The Arboviruses: Ecology and Epidemiology*, Vols. I–II. CRC Press, Boca Raton, Florida.

Montasser, A.A. (2010) The fowl tick, *Argas (Persicargas) persicus* (Ixodoidea: Argasidae): description of the egg and redescription of the larva by scanning electron microscopy. *Experimental and Applied Acarology* 52, 343–361.

Montasser, A.A., Marzouk, A.S., El-Alfy, S.H. and Baioumy, A.A. (2011) Efficacy of abamectin against the fowl tick, *Argas (Persicargas) persicus* (Oken, 1818) (Ixodoidea: Argasidae). *Parasitology Research* 109, 1113–1123.

Nava, S., Mangold, A.J. and Guglielmone, A.A. (2009) Field and laboratory studies in a Neotropical population of the spinose ear tick, *Otobius megnini*. *Medical and Veterinary Entomology* 23, 1–5.

Oliver, J.H. (1989) Biology and systematics of ticks (Acari: Ixodoidea). *Annual Review of Ecology and Systematics* 20, 397–430.

Otieno, D.A., Hassanall, A., Obenchain, F.D., Sternberg, A. and Galun, R. (1985) Identification of guanine as an assembly pheromone of ticks. *Insect Science and its Application* 6, 667–670.

Peirce, M.A. (1974) Distribution and ecology of *Ornithodoros moubata porcinus* Walton (Acarina) in animal burrows in East Africa. *Bulletin of Entomological Research* 64, 605–619.

Phillips, J.S. and Adeyeye, O.A. (1996) Reproductive bionomics of the soft tick, *Ornithodoros turicata* (Acari: Argasidae). *Experimental and Applied Acarology* 20, 369–380.

Pini, A. and Hurter, L.R. (1975) African swine fever: an epizootiological review with special reference to the South African situation. *Journal of the South African Veterinary Association* 46, 227–232.

Robinson, G.G. (1942) The mechanism of insemination in the argasid tick, *Ornithodorus moubata* Murray. *Parasitology* 34, 195–198.

Russell, R.C. (2001) The medical significance of Acari in Australia. In: Halliday, R.B., Walter, D.E., Proctor, H.C., Norton, R.A. and Colloff, M.J. (eds) *Proceedings of the 10th International Congress of Acarology*. CSIRO Publishing, Melbourne, Australia, pp. 535–546.

Schwan, T.G., Raffel, S.J., Schrumpf, M.E., Webster, L.S., Marques, A.R., Spano, R., *et al.* (2009) Tick-borne relapsing fever and *Borrelia hermsii*, Los Angeles County, California, USA. *Emerging Infectious Diseases* 15, 1026–1031.

Sonenshine, D.E. (1991) *Biology of Ticks*, Vol 1. Oxford University Press, New York.

Sonenshine, D.E. (1993) *Biology of Ticks*, Vol 2. Oxford University Press, New York.

Talbert, A., Nyange, A. and Molteni, F. (1998) Spraying tick-infested houses with lambda-cyhalothrin reduces the incidence of tick-borne relapsing fever in children under five years old. *Transactions of the Royal Society of Tropical Medicine and Hygiene* 92, 251–253.

Tsetse Flies (Diptera: Glossinidae)

1. INTRODUCTION

Tsetse flies, Glossina species, are distributed throughout over 10 million km² of the sub-Saharan Afrotropical region, involving 37 countries. In the past, tsetse have occurred in the Nearctic, since four species of fossil Glossina have been found in beds of Oligocene age in Colorado, USA; however, no established populations now exist outside sub-Saharan Africa.

In terms of tsetse distributions, four subdivisions of the Afrotropical region have been recognized: western Africa, which is approximately the area north of the Gulf of Guinea; central Africa, which extends from western Africa eastwards to the Sudan and south to Angola and includes the densely forested belt west of the western Rift Valley; eastern Africa, which extends from the Sudan to Tanzania and borders the Indian Ocean; and southern Africa, which includes Africa south of the Democratic Republic of Congo (formerly known as Zaire) and Tanzania, south of 10°S.

Both sexes are haematophagous and can inflict painful bites, and a number of species are actual or potential vectors of the trypanosomes causing 'sleeping sickness' and 'nagana' in humans and domestic animals, respectively, in several parts of Africa.

2. TAXONOMY

Only one genus, Glossina, is included in the family Glossinidae. The genus Glossina contains 31 living taxa, 23 species and 8 subspecies. The species are assigned to three subgenera (Glossina, Nemorhina and Austenina), which are also known as the morsitans, palpalis and fusca groups, named after the commonest species in each group.

The status of the various subspecies is unresolved, but there is evidence of genetic incompatibility causing sterility when subspecies of Glossina morsitans are cross-mated.

The subgenus Glossina contains five species, of which two, G. morsitans and Glossina pallidipes, are of major economic importance, and Glossina swynnertoni and Glossina austeni are of local significance. They are commonly found in savannah woodland and evergreen thickets, except for G. austeni, which is restricted to coastal forest and relict forest. The three subspecies of G. morsitans (morsitans, centralis and submorsitans) range widely in tropical Africa. Glossina longipalpis is found in western and central Africa. G. pallidipes occurs in the Democratic Republic of Congo and eastern and southern Africa, G. swynnertoni in eastern Africa and G. austeni in eastern and southern Africa.

The subgenus Nemorhina also contains five species, of which three, Glossina palpalis, Glossina fuscipes and Glossina tachinoides, occur in riverine and lakeside habitats and are particularly important as vectors of human trypanosomiasis. This subgenus occurs mainly in western and central Africa, with G. tachinoides extending into eastern Africa and Glossina pallicera occurring in Angola. G. fuscipes is found mainly in central and eastern Africa, but also occurs in Chad, Angola and Zambia.

The 13 species of the subgenus Austenina are, except for Glossina brevipalpis and Glossina longipennis, forest dwellers found mainly in western and central Africa, with lesser representation in eastern Africa and Angola. G. longipennis is found only in eastern Africa and G. longipalpis in eastern and southern Africa. These species have little contact with humans or livestock and are of little economic importance, except when cattle are moved into forested areas, as in western Uganda, for example.

3. MORPHOLOGY

Tsetse flies (Glossinidae) are readily identifiable. They are medium to large brown flies with a long, forwardly directed proboscis, sheathed by equally long palps, measuring 6–14 mm in length, excluding the proboscis (Fig. 130). The antenna has the typical cyclorrhaphan structure with an elongated third segment, but it is distinctive in that the rays of the arista are feathered and are only present on the dorsal side (Fig. 131). When viewed laterally, the third segment bears a fringe of hairs, the length of which is a specific character. The eyes are dichoptic in both sexes. The wings are folded scissor-like at rest and extend a short distance beyond the end of the abdomen. The wing venation is characterized by the discal cell being 'hatchet' shaped (Fig. 132). The abdomen of the male bears posteriorly on its ventral surface a button-like structure, the folded male terminalia.

Species of the subgenus *Glossina* are small to medium-sized tsetse, 6–11 mm in length, with distinct bands on the abdomen (except in *G. austeni*), the distal segments of the hind tarsus are dark dorsally and the male claspers are swollen distally. Species of the subgenus *Nemorhina* are of similar size (i.e. 6–11 mm in length), with the dorsum of the abdomen dark brown (except in *G. tachinoides*, which has a banded abdomen of the *morsitans* type), all hind tarsi are dark dorsally and male claspers are not swollen distally but joined by a membrane. In contrast, species of *Austenina* are large tsetse (11–14 mm in length), with

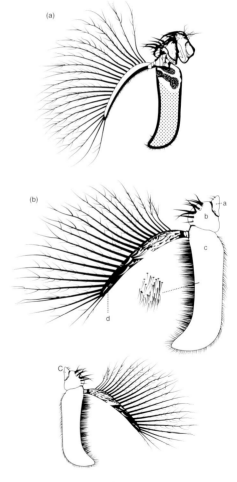

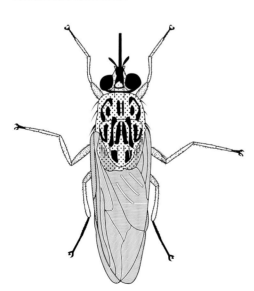

Fig. 130. Female *Glossina fuscipes* viewed from above resting on a vertical surface. *Source*: Buxton (1955).

Fig. 131. Antennae of *Glossina*. (a) *G. palpalis*; (b) *G. nigrofusca*; (c) *G. morsitans*. a, b, c = first, second and third antennal segments; d = arista. *Source*: Buxton (1955).

Fig. 132. Wing of *Glossina* species.

strong bristles on the anepimeron (ptero-pleuron) as well as on the katepisternum (sternopleuron) and in which male claspers are neither swollen distally nor joined by a membrane. In the other subgenera, *Glossina* and *Nemorhina*, there are no strong bristles on the anepimeron.

4. LIFE CYCLE

Tsetse flies are viviparous, the female producing fully grown larvae. Each ovary is composed of two polytrophic ovarioles. Ovaries and ovarioles produce oocytes alternately, beginning with the right ovary. Since the relict body left in the ovariole after ovulation can be easily recognized, it is possible to estimate the (physiological) age of tsetse flies over the first four cycles by observing the condition of the ovaries and the contents of the uterus. With care, the technique can be extended to enable the age of female flies to be determined for up to 80 days.

4.1 Larvae

With free access to a host for feeding, a female will have a fully developed egg ready for ovulation 7–9 days after emergence. Subsequent ovulations occur within an hour of parturition and therefore occur at 9- to 10-day intervals. After ovulation, the egg and then the larva are retained in the common oviduct, described as the 'uterus'. Here, the larva is nourished by secretions of the uterine (accessory) gland. To meet the growing larva's needs, the uterine gland goes through a cycle during which its cells increase in volume 100-fold. The maximum size of the cells is

reached about two-thirds of the way through pregnancy, just before the larva reaches the third stage. The secretion is composed mainly of acidic lipids at first, but changes later to protein.

The larva passes through three larval stadia and a fully-grown third-stage larva at deposition weighs more than its female parent. At 25°C, the egg stage lasts 4 days and the three larval stages 1, 1.5 and 2.5 days, respectively, in a 9-day developmental cycle (all times quoted refer to 25°C unless stated otherwise.)

First- and second-stage larvae breathe through posterior spiracles, which become elaborated into polypneustic lobes in the third stage (Fig. 133a). In each lobe there are three air chambers which open via numerous supernumerary stigmata.

4.2 Pupae

The polypneustic lobes darken and harden 24–48 h before the larva is deposited. In *G. m. centralis*, parturition appears to have a circadian basis, with larvae being deposited in the late afternoon. After larviposition, the larva is able to burrow into friable soil by peristaltic contractions of the body. The length of time during which the larvae are free-living is influenced by the time of deposition, the presence or absence of light and the nature of the substratum. The onset of pupation is accelerated by darkness and mechanical stimulation of a particulate substrate. The larva is negatively phototactic, positively thigmotactic and seeks a high humidity but moves away from free water. It burrows rapidly into the substratum and pupation occurs within 1–5 h of deposition. During barrelling and contraction, an acidic fluid is discharged from the anus and rapidly coats the puparium. Its function is uncertain, but it may include a pheromone attracting females to larviposit in the same site, or be a defensive secretion against predators and pathogens.

The puparium, measuring 3–8 mm, darkens rapidly, becoming light brown within 10 min and dark brown after several hours. Pupal apolysis occurs 2–3 days later, when the pupal cuticle is formed. At first, the pupa is cryptocephalic, but it becomes phanerocephalic

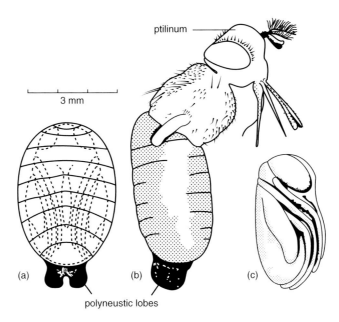

ptilinum

3 mm

(a)   (b)   (c)

polyneustic lobes

Fig. 133. Pupa and puparium of *Glossina morsitans*. (a) Puparium, showing the pupa by dotted lines; (b) adult fly emerging from puparium; (c) pupa removed from puparium. *Source*: Buxton (1955).

5 days later, when the appendages are everted. Two days later, the pupal/adult apolysis occurs, but 3 more weeks will elapse before adult development is complete and the adult emerges (Fig. 133b). The polypneustic lobes cease to function after pupal formation. The pupal spiracles are located in the same position as the adult prothoracic spiracles and they continue to function until adult emergence.

At 16, 25 and 32°C, pupation lasts for 100, 30 and 20 days, respectively. The relationship is linear from 18 to 27°C; however, above 27°C, development is slower than expected and thermal stress becomes evident, with exposure of 1 h to 40°C producing death of the pupa. In general, pupae are found in areas of bare sandy soil, holes in the ground, under logs, under fallen leaves and bark and in tufts of grass. However, in the hot season, the microclimate can become too hot for the survival of pupae and they are deposited in areas that are more sheltered and more humid.

4.3 Adult

The adult emerges using its ptilinum (Fig. 133b) to force off the cap from the puparium and then to burrow upwards to the surface. The wings are expanded, but for the first few days of its life the adult is soft and is said to be 'teneral'. During this phase, the endocuticle is secreted, the exocuticle hardened and the flight muscles developed. In *G. pallidipes*, the residual dry weight of the thorax increases linearly with age to the end of the teneral stage. Non-inseminated females are rare in the field and most females are mated at the time of their first blood meal. Newly emerged teneral flies possess low energy reserves, particularly if their pupae have been exposed to high temperatures, and the mortality of teneral flies is highest in the first week of adult life.

5. BEHAVIOUR AND BIONOMICS

5.1 Mating

Male *G. m. morsitans* respond visually to a target fly at a maximum distance of 50 cm, the cue for response being the rate at which the approaching object affects more than one ommatidium. In the field, aggregations of males, known as following swarms, are associated with large, slowly moving, conspicuous objects. In this situation, mating pairs of *G. m. morsitans* and *G. longipennis* may be seen, but mating pairs of *G. pallidipes* are rare. It is uncertain whether mating is initiated in the air or whether the male follows the female until it lands. Final recognition is achieved by a non-volatile, species-specific contact phero-mone, which appears in the cuticular waxes of a female 2 days before emergence and remains for the rest of her life. Males detect the pheromone through chemoreceptors present on the tarsi and tibiae of their legs.

5.2 Feeding and digestion

Both sexes are haematophagous and, when they feed, blood passes into the crop. Blood is transferred from the crop to the midgut, as required, by the coordinated action of valves and the posterior section of the foregut, which functions as an 'oesophageal pump'. In the midgut, the blood is enclosed in the peritrophic membrane, secreted at the proventriculus. Female *G. pallidipes* feed every 3 days and, when pregnant, feed earlier in the day and are more active on day 5 of pregnancy when the larva is in the first stage.

Meal size is probably regulated by stretch receptors in the ventral wall of the abdomen, which becomes greatly distended. Freshly fed *G. morsitans* and *G. pallidipes*, collected in the field, have been estimated to have ingested 37–76 mg of blood and in both species females took larger meals than males. The blood meal is reduced rapidly in volume in the anterior portion of the midgut by excretion of water and, within 30 min, all the blood stored in the crop will have been transferred to the midgut and reduced by half. Digestive enzymes are produced in the middle section of the midgut and include a powerful protease. Absorption of the products of digestion occurs in the posterior section of the midgut. Digestion is 60% complete within 24 h of a meal and 90% complete within 48 h.

5.3 Energy production

Three stages of the feeding cycle of *Glossina* are known. A lipogenic phase, lasting about 24 h, follows immediately after a blood meal and is accompanied by a rapid rise in oxygen consumption. During this phase, excess nitrogen is excreted and lipids synthesized; this reaches a peak 12 h after the feed. The first lipolytic phase begins on the second day after feeding and is marked by an increase in proline, mobilization of the lipid reserves and a reduced respiratory rate. The second lipolytic phase begins on the third to fourth day, when digestion is complete. The rate of lipid mobilization is reduced and the concentration of proline declines. The insect now suffers nitrogen depletion.

In flight, proline can only supply the short-term energy needs of the tsetse fly. It is converted into alanine in the flight muscles and the reverse process (alanine to proline) occurs in the abdominal fat body utilizing available lipids. This process limits continuous flight to about 2 min in *Glossina*. After 2–3 min of continuous flight, the wingbeat falls from 220 to 180 cycles/s and is incapable of lifting a load. Overall, female *G. m. morsitans* spend on average only 5 min/day in flight, and the more active males about 15 min in the hot season and over 30 min in the cold season. Teneral flies have a low wingbeat frequency because their flight muscles are underdeveloped and, consequently, they cannot lift heavy loads and they ingest smaller meals. The greater part of meals taken early in life is used to develop the thoracic musculature and the build-up of fat reserves is undertaken later.

5.4 Activity cycles

Tsetse have a bimodal rhythm of activity, generally with morning and evening peaks separated by a decline in activity at noon. However, activity patterns may vary between species: *G. pallidipes* displays increased feeding activity throughout the day, reaching a peak in the evening; *G. f. fuscipes* is characterized by a midday peak, with low activity in the morning and in the late afternoon; *G. brevipalpis* has a peak in activity before dawn and after sunset. Such rhythms are modified by environmental conditions, particularly temperature and light intensity. At 26°C, *G. m. morsitans* is positively phototactic, but at 34–38°C it seeks the dark, and the threshold for this response is lowered by 2.2°C for a tenfold increase in illumination. In the hot season, *G. m. morsitans* is active early in the day, becoming inactive when the temperature reaches 32°C; in the cool season, activity begins later in the day, and in the wet season, activity is distributed more evenly throughout the day. *G. longipalpis* is inactive below 22°C, while *G. pallidipes* feeds at temperatures between 18 and 32°C.

Under experimental conditions, *G. m. morsitans* shows a bimodal rhythm of response to a visual stimulus, with peak activity in the morning and afternoon being five times that at midday. The response increases steadily with time since the last feed, until flies become moribund after 5 days' starvation. Visual responsiveness correlates most closely with the fly's total body weight.

5.5 Host location behaviour

Host location involves the use of visual and chemosensory stimuli. Odour cues are of primary importance for activation and long-distance source detection, whereas visual cues become important at shorter distances from the host. Using odour cues, such as carbon dioxide, acetone and octenol that occur in animal breath and phenols that occur in urine, tsetse can locate hosts from distances of over 50 m. Carbon dioxide is thought to activate resting tsetse and phenols increase upwind flight towards the source. At field concentrations, acetone and octenol do not activate resting tsetse but increase their responsiveness to visual stimuli.

In odour plumes in the field, it has been shown that *G. pallidipes* and *G. m. morsitans* fly in a general upwind direction at a height of less than 50 cm above the ground in shallow curves, 1–2 m in diameter, across the odour plume at speeds in excess of 4 m/s. Within 2–3 m of leaving a plume, they make an in-flight turn and an upwind turn, when they re-enter the odour plume. Tsetse use visual cues to maintain their direction while flying. In general, short bursts of flight lasting 20–30 s are followed by up to 90 s of inactivity.

Resting tsetse are not attracted to a stationary host but in the field, about one-third

will respond to ox odour or a moving visual stimulus. In addition to movement, host recognition involves shape and colour. The more conspicuous an object is, the greater its attraction, and blue, black or white objects are more attractive than yellow or green. Tsetse are able to detect UV reflectance and this property of an object influences orientation and landing responses. Tsetse flies do not land immediately on reaching an object but encircle it, and even then they do not necessarily land. It is estimated that two-thirds of *G. palpalis* attracted to a black screen will leave without landing on it. The addition of host odours to a visually attractive object increases the landing rate greatly.

It is important to appreciate that species differ in their response to hosts, as may different populations of the same species. Similarly, individual responses to olfactory and visual stimuli are dependent on various endogenous cycles, for example nutritional depletion, circadian rhythms and, for females, the stage of pregnancy.

5.6 Host selection

Most species of tsetse have a favoured host or host range. For example, *G. swynnertoni* and *G. morsitans* were shown to feed on warthog, giraffe and buffalo in an area where zebra, wildebeest and impala formed 80% of the available hosts. Only one of the identified blood meals had been taken from the more abundant mammals. This may be due in part to the reaction of the host to the presence of the flies, because *G. morsitans* and *G. pallidipes* are thought to avoid mammals such as baboons, impala and humans, which resist attack, and to feed on relatively quiescent animals, such as pigs and cattle. Species of *Nemorhina* are opportunist feeders and frequently feed on reptiles, including crocodiles, and *G. f. fuscipes* feeds readily on monitor lizards (*Varanus* spp.).

The probability that tsetse will feed on less-favoured hosts is dependent on host availability, age and nutritional depletion. Teneral flies need a blood meal relatively quickly and young *G. morsitans* will feed on humans, whereas older flies are less likely to do so. Non-teneral flies feeding on people have smaller energy reserves than those feeding on cattle, which, in turn, have smaller energy reserves than those in swarms following a host.

A hypothesis for the evolution of tsetse flies has been proposed based on their feeding habits. It is suggested that the genus probably arose in the Mesozoic, when it would have fed on large reptiles. This ancestral habit is still reflected in the *Nemorhina*, which inhabit riverine and lacustrine environments and feed largely on reptiles. With the decline of reptiles in the Tertiary, it is postulated that tsetse flies adapted to feed on early mammals, which would have included the Suidae, or pigs, and the ancestors of large mammals, such as hippopotami, that predominated in the Eocene and Oligocene. Pigs were, and still largely are, inhabitants of forest, to which the subgenus *Austenina* is largely restricted. The ancestors of the subgenus *Glossina* moved into the savannah with the ancestors of the savannah-dwelling warthog. Ruminants such as antelope, giraffe and buffalo only became dominant in the Pliocene, and the shift of tsetse to these hosts is relatively recent. Evidence for this is to be found in their response to infection with trypanosomes.

5.7 Population dynamics

The first larviposition will be made by an adult female at about 20 days, followed by subsequent larvipositions at 10-day intervals. Equal numbers of each sex emerge; therefore, a female must live for at least 30 days to maintain the population. It has been estimated that the average longevity may be 6 weeks for males and 14 weeks for females, although this may be shorter in the dry season and longer in the wet season. However, marked female *G. palpalis* have been recaptured 6 months after having been caught initially in the wild.

The low reproduction rate of tsetse flies implies that populations of *Glossina* will be relatively steady, changing only slowly with time. The sustained maximum recorded rate of increase of tsetse in the field would permit a 20-fold increase in population size in 10 months and a 60-fold increase in less than 14 months. Bimonthly catches of *G. swynnertoni*

in Tanzania showed a 20-fold range over 10 years, and the annual mean catch showed the same variation over 20 years. Monthly catches of *G. palpalis* in northern Nigeria over 24 years showed a 60-fold range.

5.8 Resting places

Given the high energetic costs of flight, flies spend most of their time resting. The daytime and nocturnal resting places are different. *G. m. morsitans* rests on branches, boles and rot holes of trees by day and at night on twigs and leaves. This may be an adaptation to avoid nocturnal predators. By day, tsetse can see an approaching predator and by night detect the vibration caused by a predator on an easily disturbed leaf or twig. *G. pallidipes* occupies similar resting places but nearer the ground, i.e. mostly below 3 m, whereas *G. morsitans* extends up to 6 m and on occasions to 12 m. Both species show a change in resting site during the day in the hot season, when they move from branches to rot holes and boles. Biochemical evidence has been produced to indicate that, in the field, *G. m. morsitans* lives at temperatures 2–6 degrees below the average ambient temperature.

Another response to harsh conditions is for tsetse to rest nearer the ground. In the Cameroons, the behaviour of *G. tachinoides* is temperature dependent. Above 30°C, it rests nearer the ground and when the temperature exceeds 35°C, it moves to a cooler site. In intense heat, it seeks the shelter of dense stands of *Mimosa nigra*. This response has an effect on the efficacy of control measures. The severity of the hot dry season in West Africa increases with distance from the coast. In the more severe climate, it is sufficient to treat vegetation with insecticides to a height of 1.5 m to control *G. m. submorsitans*, whereas nearer the coast treatment must be applied up to 3.6 m.

5.9 Dispersal

A male *G. m. morsitans* may fly a distance of 3.3 km, flying at 11 km/h for 18 min. However, such a flight is not made in one direction but is composed of a series of short flights, each occupying about 5 s. For male *G. m. morsitans*, the average length of a step was determined to be 15.9 m and the number of steps per day as 208. The effect of such a large number of short movements is to disperse male *G. f. fuscipes* 338 m/day and the average dispersal of 13 observations on four species, mostly *G. morsitans*, was estimated to be 252 m/day. These are average numbers and, in the first week of life, male *G. morsitans* disperse only a short distance; this rises steadily as the flies age, to reach a peak in the fourth to sixth weeks, after which dispersion declines rapidly in older flies.

Such modest dispersal is consistent with the observed long-term expansion of *G. morsitans* in Nigeria, where its spread along the cattle routes has been calculated at 5 km/year, or about 100 m/week. This is about half the rate of dispersal (180 m/week) observed for individual flies. Nevertheless, when continued over many years, tsetse have considerable potential as invaders of suitable uninfested areas. In a trapping-out experiment against *G. palpalis* in a village on the Ivory Coast, for a period of 11 weeks, the daily catch of flies hardly changed but the characteristics of the captured females changed; it was evident that the removal of the original population by trapping was compensated for rapidly by the immigration of females from the bush.

6. MEDICAL AND VETERINARY IMPORTANCE

Although the bites of tsetse flies are very painful and cause marked irritation, their main significance is in the transmission of human African trypanosomiasis (HAT), known as 'sleeping sickness', and cattle trypanosomiasis, called 'nagana'. Flies become infected with protozoan trypanosome parasites during feeding on a vertebrate host and these then undergo multiplication within the fly. The fly is then infective to other hosts during subsequent feeding. Wild hosts may be asymptomatic carriers and act as reservoirs of disease. Since the infective (metacyclic) form of the trypanosome for the tsetse is also the infective form for the vertebrate hosts (including

humans), mechanical transmission may occur from cross-host biting within a short period via mouthpart contamination, but the prevalence of this is unknown.

Two subspecies of *Trypanosoma brucei*, *Trypanosoma brucei gambiense* and *Trypanosoma brucei rhodesiense*, are transmitted by tsetse flies of the subgenera *Nemorhina* and *Glossina*, respectively, and cause sleeping sickness in people in western, central and eastern Africa. Devastating epidemics of sleeping sickness have occurred in the past, and transmission has increased in some areas in recent decades. Animal trypanosomiases are caused by several species of *Trypanosoma* (e.g. *Trypanosoma brucei brucei*, *Trypanosoma congolense*, *Trypanosoma simiae*, *Trypanosoma suis* and *Trypanosoma vivax*) and are associated with tsetse flies of the subgenera *Glossina* and *Austenina*. Wherever there are tsetse flies, it is impossible to keep cattle without regular chemotherapy. Trypanosomiasis has inhibited the development of cattle-raising industries in much of tropical Africa and, even today, cattle ranching is impractical in much of Africa's savannah lands.

6.1 Life cycle of *Trypanosoma brucei*

In the vertebrate host, infective metacyclic forms injected by a feeding *Glossina* develop into rapidly dividing, long slender (LS) trypomastigotes; at a later stage, there is a switch to an intermediate (I) form and then to a non-dividing, short stumpy (SS) form lacking free flagellum which can be ingested by the tsetse vector. Trypanosomes pass with the ingested blood into the fly's crop, from which they are passed via the proventriculus into the midgut, where they are contained within the peritrophic membrane, more exactly, in the endoperitrophic space. In the midgut, the I and SS forms differentiate into procyclic trypomastigotes, which develop and divide extensively. The trypanosomes move into the ectoperitrophic space via the open free end of the membrane or by penetrating the membrane. This process takes about 2 weeks after an infective feed. In the ectoperitrophic space, the trypanosomes move forward to the proventriculus, where they pass through the soft, newly secreted peritrophic membrane to become free in the oesophagus and move down the food channel to the opening of the salivary duct in the hypopharynx. They pass along the salivary duct to reach the salivary glands, where they develop into epimastigotes, which attach to the epithelial cells of the salivary glands and multiply enormously. Extensive division takes place and finally metacyclic stages form that can be injected into a susceptible vertebrate host. A recent study has suggested that in *Trypanosoma*-infected tsetse flies, the feeding behaviour is prolonged as a result of changes in the saliva composition, therefore promoting the transmission of the protozoan parasite to the vertebrate host.

The development cycle of *T. congolense* is similar in part to that of *T. brucei*. In *G. m. morsitans*, *T. congolense* penetrates the peritrophic membrane in the central region of the midgut and, 7 days after an infective feed, heavy infections are found in both the endopentrophic and ectoperitrophic spaces; by 21 days, epimastigotes are found attached to the food canal by their flagellae, and a week later free forms are detected in the hypopharynx. Ingested trypomastigotes of *T. vivax* attach to the walls of the food channel, pharynx and oesophagus by means of hemidesmosomes formed between their flagellum and the chitinous walls of the surrounding structures. Attached trypomastigotes multiply and differentiate into epimastigotes and finally into premetacyclic forms which detach and migrate to the hypopharynx, where they mature into infective metacyclics.

The completion of the *T. vivax* life cycle is much faster (5–13 days) than those of *T. congolense* (2–3 weeks) and *T. brucei* (3–5 weeks); the cycle of *T. suis* takes about 4 weeks.

In natural populations of *Glossina* spp., there is considerable variation in infection rates; for example, from 0.2% in *G. fuscipes martinii* in Zambia to 76.6% in *G. m. morsitans* in Nigeria. The highest infection rates were of *T. vivax* and the lowest of *T. brucei*, with infections of *T. congolense* being intermediate in frequency. Susceptibility to

infection with *T. brucei* is a function of age of the vector. Unless a tsetse feeds on an infected host very soon after adult emergence, it is unlikely to become infective. The highest percentages of infection are established in flies that feed on an infected host within 48 h (preferably 24 h) of emergence (i.e. first feed) but, even then, infectivity is unlikely to exceed 10%.

The dependence of infectivity on age is related to the development of the peritrophic membrane, which is not secreted until after adult emergence. Flies that feed early in adult life have shorter membranes, in which case the trypomastigotes reach the ectoperitrophic space more easily.

6.2 Human African trypanosomiasis (sleeping sickness)

In Africa, human trypanosomiasis is present from 14°N to 29°S, involving 37 countries and some 10 million km^2, with more than 200 foci of the disease. At the turn of the 20th century, human trypanosomiasis caused half a million deaths in the Democratic Republic of Congo and half that number around Lake Victoria, East Africa. By the early 1950s, the disease was under control and human African trypanosomiasis had been virtually 'eradicated' by 1965. However, with civil and political unrest in many of the regions, it has re-emerged due to poor surveillance and problems related to access to effective chemotherapeutics, plus the cost to the various health budgets of HIV AIDS, and there is active transmission in at least 20 countries. There has been ongoing activity with epidemics in recent decades and there are estimates that tens of thousands of human lives and millions of livestock are lost annually, with up to 500,000 people infected and more than 60 million threatened.

Two forms of human sleeping sickness are recognized. Gambian sleeping sickness is caused by *T. b. gambiense* and Rhodesian sleeping sickness by *T. b. rhodesiense*. The organisms are morphologically indistinguishable but, epidemiologically, *T. b. gambiense* and *T. b. rhodesiense* are geographically and ecologically distinct and produce clinically different human diseases.

A chancre (ulcer) commonly develops at the site of an infective bite after 5–15 days. Within 1–3 weeks of being infected, trypanosomes can be found in the haemolymphatic system, producing lymphadenopathy and splenomegaly. This is followed by the trypanosomes insidiously invading the central nervous system, a development which occurs in weeks to months after infection by *T. b. rhodesiense* or in months to years with *T. b. gambiense*. The victim shows increasing indifference, lassitude and daytime somnolence, leading to death within 9 months (*T. b. rhodesiense*) or 4 years (*T. b. gambiense*).

T. b. gambiense occurs in West Africa and in western and northern central Africa and *T. b. rhodesiense* occurs in central and East Africa. *T. b. gambiense* is associated with the *Nemorhina* subgenus of *Glossina* and its vectors are the riverine and lacustrine tsetse *G. palpalis*, *G. fuscipes* and *G. tachinoides*, whereas the vectors of *T. b. rhodesiense* are the savannah species of *Glossina*: *G. morsitans*, *G. pallidipes* and *G. swynnertoni*.

6.2.1 Gambiense sleeping sickness

Gambiense sleeping sickness had been regarded as an anthroponosis, involving transmission from one person to another, but there is evidence that domestic pigs may play an active role as reservoirs of *T. b. gambiense*. In some forest and forest/savannah villages in West Africa where domestic pigs are kept, peridomestic populations of *G. palpalis* and *G. tachinoides* have become established, feeding on pigs and humans and, with the pigs acting as hosts of *T. gambiense*, the villages become foci of human trypanosomiasis.

Gambiense sleeping sickness is transmitted when there is close association between tsetse and humans, as occurs, for example, in the dry season when people and *G. palpalis* concentrate around waterholes. Other hazardous situations occur when palm cutters work in raffia beds (*Raphia sudanica*), the haunt of *G. tachinoides*, or fishermen land their catch on a lake shore frequented by *G. fuscipes*.

6.2.2 Rhodesiense sleeping sickness

This disease is an anthropozoonosis in which the natural cycle is from animal to animal and humans are an unimportant, accidental intrusion. *T. rhodesiense* is readily infective to humans, laboratory animals and wild and domestic animals. Strains pathogenic to humans have been isolated from bushbuck and some other animals, including donkeys. People become infected when their activities, as game warden, hunter or tourist (or entomologist), take them into the habitat of game animals and *Glossina* (*Glossina*) tsetse flies (even though humans are not the flies' preferred host).

6.3 Trypanosomiases of domestic animals

Virtually all domestic stock animals in Africa suffer from trypanosomiasis. The common name for the trypanosome infections in cattle and horses is 'nagana' and the three main pathogens are *T. brucei*, *T. congolense* and *T. vivax*. It has been estimated that infection leads to approximately US$35 million being spent on trypanocidal drugs to maintain livestock free of the disease and US$1.2 billion is lost due to the loss of meat and milk production and costs related to disease and vector control.

Economically, the most important trypanosome infection of domestic animals is in cattle. The most important species in cattle are *T. vivax* and *T. congolense*. In enzootic areas, infection rates can be over 60%, and in outbreaks, infections with *T. vivax* may reach 70%. After an incubation period of 8–20 days, acute infections can cause death within several weeks and in chronic cases infection persists for months or years, with the infected animals becoming carriers. Infected animals suffer intermittent fever, anaemia, are emaciated but have good appetites and are reproductively impaired.

6.3.1 *Trypanosoma vivax*

T. vivax is the main cause of trypanosomiasis in domestic animals, especially cattle, to which it maintains high virulence, particularly in West Africa. It is also pathogenic to sheep, goats, horses and camels, but not to cats or dogs. The main vectors are species of the subgenus *Glossina*.

6.3.2 *Trypanosoma congolense*

T. congolense produces severe disease in cattle, sheep and goats and chronic infections in them and camels and horses. It is transmitted by members of the subgenus *Glossina* and *G. (A.) brevipalpis* and *G. (A.) longipennis*.

6.3.3 *Trypanosoma brucei*

T. brucei is particularly pathogenic to equines and causes severe disease in sheep, goats, cats and dogs. It produces only a benign infection in cattle, but cattle and game animals are the main reservoirs. *T. brucei* is transmitted especially by the subgenus *Glossina* and other species.

6.3.4 *Trypanosoma simiae*

T. simiae occurs naturally in warthogs in East and central Africa, where it is highly pathogenic to domestic pigs, with death occurring within hours of the parasites first appearing in the circulating blood, but cattle are resistant. In view of the high feeding rates of *Glossina* species on suids, this trypanosome is transmitted by five species of *Austenina*, four *Glossina* and two *Nemorhina*.

6.3.5 *Trypanosoma suis*

T. suis is only rarely reported as a pathogen of swine and most domestic animals seem to be refractory to *T. suis*. Its natural hosts are warthogs, bushpigs and forest hogs; it is transmitted by *G. (A.) brevipalpis* and *Glossina (A.) vanhoofi*.

6.3.6 *Trypanosoma evansi*

T. evansi affects camels and horses, resulting in a disease known as 'surra', and is transmitted mechanically by bloodsucking flies other than tsetse, principally stable flies and horse flies.

7. PREVENTION AND CONTROL

7.1 Treatment of infection

Treatment of human infections is difficult, since the available drugs are complicated to administer and may cause severe adverse reactions. At present, there is no vaccine on the market and chemoprophylaxis, with pentamidine, has been used to provide high-risk protection against *T. b. gambiense* only. In humans, early infections with *T. b. gambiense* are treated with pentamidine as the first-choice drug. Suramin is the first choice against *T. b. rhodesiense* and it is generally avoided against *T. b. gambiense* because of potential allergic reactions in regions where *Onchocerca* infections occur. However, both drugs have poor penetration of the central nervous system (CNS) and, in those cases with CNS involvement, melarsoprol is the drug of choice against late infections and relapses in infection caused by both subspecies (although resistance has been appearing in some countries). Dexamethasone and diazepam are recommended for the management of encephalopathic syndrome. Eflornithine is the only new drug that has been registered for human treatment in the past 50 years. While it is not very effective against *T. b. rhodesiense*, it is recommended against late-stage *T. b. gambiense* infection, but it is relatively costly and therefore largely unaffordable for populations of developing countries. In recent years, pharmaceutical companies have donated stocks of effective drugs, which have become more deliverable through the World Health Organization and several non-governmental organizations.

For livestock, hormidium has been widely used in the treatment of trypanosome-infected ruminants, but drug resistance has reduced the usefulness of hormidium and quinapyramine. Currently, diminazene and isometamidium are most widely used because they have no cross-resistance. The latter drug is also employed as chemoprophylactic. Equines and camels are usually treated with quinapyramine. The use of drugs as prophylaxis against trypanosomiasis may result in producing drug-resistant strains of trypanosomes. Nevertheless, some native breeds of cattle are trypanotolerant (e.g. N'Dama and West African shorthorn), becoming infected but not developing anaemia or any other sign of acute illness (e.g. fever). Prevention of disease is achieved by keeping livestock in tsetse-free areas or by maintaining local control of the vector.

7.2 Vector control

7.2.1 Population sampling

Tsetse are generally present in low density, with population sizes estimated to range from 4 to 18 tsetse/ha. Traditionally, the density of tsetse flies in an area has been estimated by catches from a stationary animal, usually an ox, or during a 'fly round', when a person (with or without an animal) walks along a fixed circuit, stopping at intervals to collect the tsetse flies attracted to the mobile host. A disadvantage of this method is that humans are not attractive hosts to G. morsitans and G. pallidipes; in fact, evidence suggests that some tsetse flies are actually repelled by humans.

Although field populations of tsetse contain more females than males, paradoxically, on mobile human or animal baits more males than females are captured, the percentage of males being as high as 98%. These males are commonly presumed to be a 'mating swarm', following the host to intercept and mate with females as they visit to feed. Teneral flies and females within 48 h of parturition are also under-represented in catches, as they are relatively inactive.

Electric traps that electrocute and kill attracted flies are superior to fly rounds in that they can be operated in the absence of humans. Present-day traps incorporate colour, electric grids and bait in the form of a selection of carbon dioxide, acetone, octenol and phenols. They need both to attract and to capture tsetse flies. Plain flat traps are made more effective by being flanked by an electric grid, which catches tsetse as they circle the central visual stimulus. A range of non-return traps, such as the Biconical trap, has been developed to attract and capture tsetse. These may be baited with acetone and cow urine, and baited traps catch more representative samples than unbaited traps.

The powerful attraction of the odour of cattle to female G. m. morsitans and both sexes of G. pallidipes has been shown clearly by the use of different numbers of cattle in underground pits, where their odour can be vented above ground in the vicinity of a trap. This technique has been used to demonstrate that the number of tsetse caught is related linearly to the weight of livestock in the pit trap; in one study, 7000 tsetse flies were caught within 3 h (attracted by the odour of 11,500 kg of livestock) and large catches continued to be made for 60 days. This indicates that well-designed baited traps could replace the aerial and ground application of insecticides to control G. m. morsitans and G. pallidipes.

7.2.2 Tsetse management

There are a number of different approaches that have been taken to avoid or control tsetse populations. Protection from the flies can be afforded by appropriate clothing, although the flies are more attracted to dark colours and can bite through thin clothes. Topical repellents may be used, although these are not particularly effective. Notably, since flies are attracted to large moving objects, travelling in vehicles may increase the risk of being bitten, unless the vehicle is fly proof.

Indirect control strategies, such as the removal of resting and oviposition sites and killing wild animal reservoirs, were used extensively in the past. These involved the cutting and burning of large areas of bushy vegetation and game hunting. Such environmentally damaging approaches are no longer considered appropriate.

The approach more commonly used nowadays involves treatment of the habitat with insecticides (e.g. pyrethroids). These are normally released via aerial spraying or by ground spraying using backpacks or trucks. Such treatments may be focused on specific areas, for example riverine habitats in West Africa to control G. palpalis and G. tachinoides. Restricted spraying is more difficult to apply to the more widespread savannah species such as G. morsitans. However, large areas (6000 km^2) have been treated successfully in Botswana and Zimbabwe by the application of low dosages of non-residual insecticides using fixed-wing aircraft (e.g. endosulfan at l0 g/ha). Five or six applications, at intervals of 20 days, produced good levels of control, but reinvasion occurred from the surrounding untreated area. Multiple treatments with insecticides are also likely to have environmental impacts.

More recently, it has been shown that the use of insecticide-treated cattle may represent a cost-effective method for controlling tsetse, but its impact might be compromised by the patchy distribution of livestock. In particular, the restricted application of spray or pour-on formulations of insecticide (e.g. deltamethrin) to the lower quarters and the ears, known to be the preferential attack sites of tsetse flies on the cattle host, has led to an effective control of human and/or bovine trypanosomiasis in several African countries (e.g. Burkina Faso, Uganda, Zimbabwe). Requiring a lower dose of insecticide, such an approach proves to be more sustainable from both an economical and environmental standpoint.

Due to the relatively low population densities of tsetse and their low rates of reproduction, the use of odour-baited traps or insecticide-treated screens may be highly effective in managing their populations. A wide range of different trap designs has been created, incorporating known visual and olfactory attractants. Many of the more recent effective screens comprise simple cloth pegged between sticks, with block patterns of black sometimes combined with dark blue panels. These attract the flies, and when the screens are given a spray coating of a residual insecticide, they provide a lethal surface for any flies that rest on them. Recent studies on G. f. fuscipes in Kenya have shown that small (0.25 × 0.25 m) insecticide-treated screens are more cost-effective than the larger screens (≥1.0 × 1.0 m) currently used to control tsetse.

In the Zambezi Valley (southern Africa), odour-baited black screens at a density of 3–5 km^2 were able, in 6 months, to reduce populations of G. m. morsitans and G. pallidipes to less than 0.01% at the centre of the 600 km^2 management block. Similar success was achieved in Kenya against G. pallidipes (98–99% reduction) over 100 km^2. Often, traps and screens can be constructed relatively easily and inexpensively by farmers and treated with insecticides locally, thus reducing the need for external intervention. However, trap deployment and

maintenance can be difficult in remote areas with poor infrastructures. Recent work has shown that tree trunks painted with dark colour designs and similarly sprayed with insecticide can also serve as lethal resting sites for flies. In comparison to large-scale insecticide use, these methods are safer for the environment, relatively inexpensive and provide a more selective method of control.

The release of sterile males (known as the sterile insect technique, or SIT) has been widely tested against tsetse populations. In this approach, flies are reared in the laboratory, sterilized in the pupal stage using non-lethal doses of radiation and then male adults are released. They then mate with wild females, rendering them infertile. However, the technique is expensive, logistically complex and its application usually results in only a temporary reduction in the tsetse population, often as a result of immigration from outside the release area. Therefore, SIT may be used more effectively as part of an integrated management programme, involving also odour-baited targets along with insecticide-treated cattle. None the less, the successful eradication of a G. austeni population from Zanzibar using SIT has led several African countries to incorporate this control strategy in their national tsetse control programmes.

There is currently a coordinated effort (the Pan African Tsetse and Trypanosomiasis Eradication Campaign, PATTEC), under the aegis of the African Union, that is endeavouring to combat tsetse and trypanosomiasis infections collaboratively, using a range of approaches. Reports indicate that since 2007, Botswana and Namibia have been tsetse free and Malawi, Rwanda and Equatorial Guinea are expected to be tsetse free by 2013; eight countries have ongoing control projects, three countries have test projects and 19 countries have developed plans.

SELECTED BIBLIOGRAPHY

Abd-Alla, A.M., Parker, A.G., Vreysen, M.J. and Bergoin, M. (2011) Tsetse salivary gland hypertrophy virus: hope or hindrance for tsetse control? *PLoS Neglected Tropical Diseases* 5(8), e1220.

Allsopp, R. (1984) Control of tsetse flies (Diptera: Glossinidae) using insecticides: a review and future prospects. *Bulletin of Entomological Research* 74, 1–23.

Bales, J.D. (1991) African trypanosomiasis. In: Strickland, G.T. (ed.) *Hunter's Tropical Medicine*. Saunders, Philadelphia, Pennsylvania, pp. 617–628.

Bouyer, J., Stachurski, F., Gouro, A.S. and Lancelot, R. (2009) Control of bovine trypanosomosis by restricted application of insecticides to cattle using footbaths. *Veterinary Parasitology* 161, 187–193.

Brady, J., Gibson, G.A. and Packer, M.J. (1989) Odour movement, wind direction, and the problem of host-finding by tsetse flies. *Physiological Entomology* 14, 369–380.

Brun, R., Blum, J., Chappuis, F. and Burri, C. (2010) Human African trypanosomiasis. *Lancet* 375, 148–159.

Bursell, E. and Taylor, P. (1980) An energy budget for *Glossina* (Diptera: Glossinidae). *Bulletin of Entomological Research* 70, 187–196.

Bursell, E., Gough, A.J.E., Beevor, P.S., Cork, A., Hall, D.R. and Vale, G.A. (1988) Identification of components of cattle urine attractive to tsetse flies, *Glossina* spp. (Diptera: Glossinidae). *Bulletin of Entomological Research* 78, 281–291.

Colvin, J. and Gibson, G. (1992) Host-seeking behavior and management of tsetse. *Annual Review of Entomology* 37, 21–40.

Davis, S., Aksoy, S. and Galvani, A. (2011) A global sensitivity analysis for African sleeping sickness. *Parasitology* 138, 516–526.

Dransfield, R.D., Brightwell, R., Kilu, J., Chaudhury, M.F. and Adabie, D.A. (1989) Size and mortality rates of *Glossina pallidipes* in the semi-arid zone of southwestern Kenya. *Medical and Veterinary Entomology* 3, 83–95.

Dransfield, R.D., Brightwell, R., Kyorku, C. and Williams, B. (1990) Control of tsetse fly (Diptera: Glossinidae) populations using traps at Nguruman, south-west Kenya. *Bulletin of Entomological Research* 80, 265–276.

Eisler, M.C. (1996) Pharmacokinetics of the chemoprophylactic and chemotherapeutic trypanocidal drug isometamidium chloride (Samorin) in cattle. *Drug Metabolism and Disposition* 24, 1355–1361.

Green, C.H. (1990) The effect of colour on the numbers, age and nutritional status of *Glossina tachinoides* (Diptera: Glossinidae) attracted to targets. *Physiological Entomology* 15, 317–329.

Hargrove, J.W. (1981) Tsetse dispersal reconsidered. *Journal of Animal Ecology* 50, 351–373.

Hargrove, J.W. (1991) Ovarian ages of tsetse flies (Diptera: Glossinidae) caught from mobile and stationary baits in the presence and absence of humans. *Bulletin of Entomological Research* 81, 43–50.

Hargrove, J.W. and Coates, T.W. (1990) Metabolic rates of tsetse flies in the field as measured by the excretion of injected caesium. *Physiological Entomology* 15, 157–166.

Hide, G., Mottram, J.C., Coombs, G.H. and Holmes, P.H. (1997) *Trypanosomiasis and Leishmaniasis. Biology and Control*. CAB International, Wallingford, UK.

Ilemobade, A.A. (2009) Tsetse and trypanosomosis in Africa: the challenges, the opportunities. *Onderstepoort Journal of Veterinary Research* 76, 35–40.

Kuzoe, F.A.S. (1993) Current situation of African trypanosomiasis. *Acta Tropica* 54, 153–162.

Langley, P.A. (1977) Physiology of tsetse flies (*Glossina* spp.) (Diptera: Glossinidae): a review. *Bulletin of Entomological Research* 67, 523–574.

Leak, S.G.A. (1999) *Tsetse Biology and Ecology: Their Role in the Epidemiology and Control of Trypanosomiasis*. CAB International, Wallingford, UK.

Lindh, J.M., Torr, S.J., Vale, G.A. and Lehane, M.J. (2009) Improving the cost-effectiveness of artificial visual baits for controlling the tsetse fly *Glossina fuscipes fuscipes*. *PLoS Neglected Tropical Diseases* 3, e474.

Logan-Henfrey, L.L., Gardiner, P.R. and Mahmoud, M.M. (1992) Animal trypanosomiasis in sub-Saharan Africa. In: Kreier, J.P. and Baker, J.R. (eds) *Parasitic Protozoa*, Vol 11. Academic Press, San Diego, California, pp. 157–276.

Lutumba, P., Robays, J., Bilenge, C.M.M.B., Mesu, V.K.B.K., Moliso, D., Declerq, J., *et al.* (2005) Trypanosomiasis control, Democratic Republic of Congo, 1993–2003. *Emerging Infectious Diseases* 11, 1382–1388.

Malele, I., Nyingilili, H. and Msangi, A. (2011) Factors defining the distribution limit of tsetse infestation and the implication for livestock sector in Tanzania. *African Journal of Agricultural Research* 6, 2341–2347.

Maudlin, I. (2006) African trypanosomiasis. *Annals of Tropical Medicine and Parasitology* 100, 679–701.

Molyneux, D.H. (1980) Animal reservoirs and residual 'foci' of *Trypanosoma brucei gambiense* sleeping sickness in West Africa. *Insect Science and Its Application* 1, 59–63.

Molyneux, D.H. and Ashford, R.W. (1983) *The Biology of Trypanosoma and Leishmania, Parasites of Man and Domestic Animals.* Taylor and Francis, London, 294 pp.

Molyneux, D.H., Baldry, D.A.T. and Fairhurst, C. (1979) Tsetse movement in wind fields: possible epidemiological and entomological implications for trypanosomiasis and its control. *Acta Tropica* 36, 53–65.

Mwanakasale, V. and Songolo, P. (2011) Disappearance of some human African trypanosomiasis transmission foci in Zambia in the absence of a tsetse fly and trypanosomiasis control program over a period of forty years. *Transactions of the Royal Society of Tropical Medicine and Hygiene* 105, 167–172.

Njitchouang, G.R., Njiokou, F., Nana-Djeunga, H., Asonganyi, T., Fewou-Moundipa, P., Cuny, G., *et al.* (2011) A new transmission risk index for human African trypanosomiasis and its application in the identification of sites of high transmission of sleeping sickness in the Fontem focus of southwest Cameroon. *Medical and Veterinary Entomology* 25, 289–296.

Randolph, S.E., Dransfield, R.D. and Rogers, D.J. (1989) Effect of host odours on trap composition of *Glossina pallidipes* in Kenya. *Medical and Veterinary Entomology* 3, 297–306.

Randolph, S.E., Rogers, D.J. and Kilu, J. (1991) The feeding behaviour, activity and trap-ability of wild *Glossina pallidipes* in relation to their pregnancy cycle. *Medical and Veterinary Entomology* 5, 335–350.

Roditi, I. and Lehane, M.J. (2008) Interactions between trypanosomes and tsetse flies. *Current Opinion in Microbiology* 11, 345–351.

Rogers, D.J., Randolph, S.E. and Kuzoe, F.A.S. (1984) Local variation in the population dynamics of *Glossina palpalis palpalis* (Robineau-Desvoidy) (Diptera: Glossinidae). I. Natural population regulation. *Bulletin of Entomological Research* 74, 403–423.

Seed, J.R. and Hall, J.E. (1992) Trypanosomes causing disease in man in Africa. In: Kreier, J.P. and Baker, J.R. (eds) *Parasitic Protozoa*, Vol II. Academic Press, San Diego, California, pp. 85–155.

Shaw, A.P.M. (2009) Assessing the economics of animal trypanosomosis in Africa – history and current perspectives. *Onderstepoort Journal of Veterinary Research* 76, 27–32.

Simarro, P.P., Cecchi, G., Franco, J.R., Paone, M., Fevre, E.M., Diarra, A., *et al.* (2011) Risk for human african trypanosomiasis, Central Africa, 2000–2009. *Emerging Infectious Diseases* 17, 2322–2327.

Takken, W., Oladunmade, M.A., Dengwat, L., Feldmann, H.U., Onah, J.A., Tenabe, S.O., *et al.* (1986) The eradication of *Glossina palpalis palpalis* (Robineau-Desvoidy) (Diptera: Glossinidae) using traps, insecticide-impregnated targets and the sterile insect technique in central Nigeria. *Bulletin of Entomological Research* 76, 275–286.

Torr, S.J. (1988) Behaviour of tsetse flies (*Glossina*) in host odour plumes in the field. *Physiological Entomology* 13, 467–478.

Torr, S.J. (1989) The host-orientated behaviour of tsetse flies (*Glossina*): the interaction of visual and olfactory stimuli. *Physiological Entomology* 14, 325–340.

Torr, S.J., Maudlin, I. and Vale, G.A. (2007a) Less is more: restricted application of insecticide to cattle to improve the cost and efficacy of tsetse control. *Medical and Veterinary Entomology* 21, 53–64.

Torr, S.J., Prior, A., Wilson, P.J. and Schofield, S. (2007b) Is there safety in numbers? The effect of cattle herding on biting risk from tsetse flies. *Medical and Veterinary Entomology* 21, 301–311.

Torr, S.J., Chamisa, A., Vale, G.A., Lehane, M.J. and Lindh, J.M. (2011a) Responses of tsetse flies, *Glossina morsitans morsitans* and *Glossina pallidipes*, to baits of various size. *Medical and Veterinary Entomology* 25, 365–369.

Torr, S.J., Mangwiro, T.N.C. and Hall, D.R. (2011b) Shoo fly, don't bother me! Efficacy of traditional methods of protecting cattle from tsetse. *Medical and Veterinary Entomology* 25, 192–201.

Usman, S.B., Babatunde, O.O., Oladipo, K.J., Felix, L.A.G., Gutt, B.G. and Dongkum, C. (2008) Epidemiological survey of animal trypanosomiasis in Kaltungo local government area Gombe State Nigeria. *Journal of Protozoology Research* 18, 96–105.

Vale, G.A. (1974) The responses of tsetse flies (Diptera: Glossinidae) to mobile and stationary baits. *Bulletin of Entomological Research* 64, 545–588.

Vale, G.A. (1977) The flight of tsetse flies (Diptera: Glossinidae) to and from a stationary ox. *Bulletin of Entomological Research* 67, 297–303.

Vale, G.A. (1993) Development of baits for tsetse flies (Diptera: Glossinidae) in Zimbabwe. *Journal of Medical Entomology* 30, 831–842.

Vale, G.A. (2009) Prospects for controlling trypanosomosis. *Onderstepoort Journal of Veterinary Research* 76, 41–45.

Vale, G.A. and Torr, S.J. (2005) User-friendly models of the costs and efficacy of tsetse control: application to sterilizing and insecticidal techniques. *Medical and Veterinary Entomology* 19, 293–305.

Vale, G.A., Hargrove, J.W., Cockbill, G.F. and Phelps, R.J. (1986) Field trials of baits to control populations of *Glossina morsitans morsitans* Westwood and *G. pallidipes* Austen (Diptera: Glossinidae). *Bulletin of Entomological Research* 76, 179–193.

Vale, G.A., Lovemore, D.F., Flint, S. and Cockbill, G.F. (1988) Odour-baited targets to control tsetse flies *Glossina* spp. (Diptera: Glossinidae), in Zimbabwe. *Bulletin of Entomological Research* 78, 31–49.

Van Den Abbeele, J., Caljon, G., De Ridder, K., De Baetselier, P. and Coosemans, M. (2010) *Trypanosoma brucei* modifies the tsetse salivary composition, altering the fly feeding behavior that favors parasite transmission. *PLoS Pathogens* 6(6), e1000926.

Wall, R. and Langley, P.A. (1993) The mating behaviour of tsetse flies (*Glossina*): a review. *Physiological Entomology* 18, 211–218.

Warnes, M.L. (1989) Responses of the tsetse fly, *Glossina pallidipes*, to ox odour, carbon dioxide and a visual stimulus in the laboratory. *Entomologia Experimentalis et Applicata* 50, 245–253.

Tumbu Flies (Diptera: Calliphoridae)

1. INTRODUCTION

The testaceous Calliphorid, *Cordylobia anthropophaga*, known as the tumbu, mango or putzi fly, is found in sub-Saharan Africa, where it causes myiasis in rodents, dogs and even humans. Its larvae form a characteristic warble-like cyst just below or in the skin on their hosts, within which they develop. A second African species within the genus, *Cordylobia rodhaini*, found in the moister parts of tropical Africa, largely parasitizes antelope and rodents and, rarely, also humans.

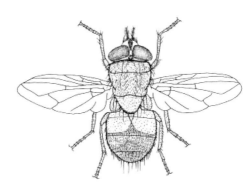

Fig. 134. *Female Cordylobia anthropophaga. Source: redrawn from James (1947).*

2. TAXONOMY

Within the calypterate Diptera, superfamily Oestroidea, the large family Calliphoridae is composed of over 1000 species divided between 150 genera. At least 80 species have been recorded as causing myiasis of vertebrates and most are known as blow flies (see the entry for 'Blow Flies and Screw-worm Flies'). Within the genus *Cordylobia*, *C. anthropophaga* is the most important species, while *C. rodhaini* is less widespread.

developed mouthparts. The arista of the antenna has setae on both sides. The thoracic squamae are without setae and the stem vein of the wing is without bristles.

A mature maggot is covered incompletely with a dense set of characteristic, small, single-toothed spines that point in a backward direction. A mature maggot can reach up to 15 mm in length and has posterior spiracles with three sinuous slits within a weakly sclerotized peritreme (Fig. 135).

3. MORPHOLOGY

Both sexes of adult tumbu fly are large, stout flies, up to 12 mm in length (Fig. 134). The body is not metallic coloured like most blow flies but is a dull yellow-brown or red-brown in colour, with two dorsal thoracic stripes, which are broad and variable. They have large, fully

4. LIFE CYCLE

Females oviposit batches of 100–300 eggs on dry sandy ground soiled by urine and excreta. The eggs hatch in 24–48 h and the larva remains buried in the sand until responding to vibrations, heat and carbon dioxide, which could signify the arrival of a host. The first-

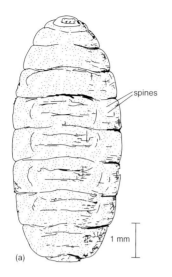

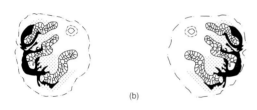

Fig. 135. Third-stage larva of *Cordylobia anthropophaga*. (a) Ventral view; (b) posterior spiracles of larva. *Source*: larva from Bertram, D.S. (1938) *Annals of Tropical Medicine* 32, 433. Spiracles drawn from various sources.

stage larvae then start to search for a host. On a suitable host, the larva will penetrate the unbroken skin and bury itself in less than a minute. Larvae penetrate the feet, genitals, tail and axillary regions of their host. The larva develops in a boil-like swelling in the skin. The swelling has an opening through which the larva breathes using its posterior spiracles. The three larval stages are completed in 8 days and then the larva leaves the host and pupates in the ground.

5. BEHAVIOUR AND BIONOMICS

Adult *C. anthropophaga* are diurnal, with peaks of activity in the early morning and late afternoon, between which times they seek shade. They feed on fermenting fruit and carrion. The maggots develop most successfully in a wide range of rodents, particularly rats. However, domestic dogs are also an important reservoir. It is assumed that rodents are the most primitive host for these flies, which have, over time, become secondarily adapted to other species, such as dogs and humans. Females are shade loving and will oviposit in dry sand, on contaminated dirt and on soiled or inefficiently washed clothing or bedding, and hence larvae are often distributed over the areas of the body covered by clothing. One female can lay up to 500 eggs in her short lifespan of 2–3 weeks. The boil-like swelling caused by larvae of *C. anthropophaga* causes considerable discomfort as the larva increases in size and there is a copious exudate of serum, blood and larval faeces.

6. MEDICAL AND VETERINARY IMPORTANCE

Cordylobia infections have been reported from many parts of the world (e.g. USA, the Middle East, Europe and Australia) in travellers from Africa. In the endemic regions, such as the tropical regions of Africa south of the Sahara Desert, myiasis caused by *C. anthropophaga* is common, but mild infestations do not usually cause clinical problems. Large infestations can lead to swelling and oedema (particularly when larvae are in close proximity to each other), and in severe cases when larvae penetrate deep into the tissues, the damage caused can be fatal.

The tumbu fly maggots infest a wide range of different hosts, with the domestic dog being a particularly important reservoir. The development and survival of the larvae varies depending on its host species, with development success being highest when infecting native rodents. Characteristics of the host, such as the thickness of the skin, influence their susceptibility to invasion, with thin-skinned animals being more vulnerable to infection than thick-skinned species. *C. rodhaini* infests humans less frequently.

7. PREVENTION AND CONTROL

Avoidance of ground contact with the larvae at prime oviposition sites is the best method of prevention. The danger of infestation from clothing can be avoided by drying clothes in full sunlight out of contact with the ground and ironing the clothing to kill any eggs or larvae before storing in covered receptacles.

For infestations, surgical removal of the maggot is usually unnecessary and occluding the breathing hole with a thick substance such as paraffin or other liquid/jelly/ointment (bacon has also been used effectively) will usually force the maggot to emerge far enough to be grasped with forceps. Application of a macrocyclic lactone, such as ivermectin, to kill the larva can be effective, but if they are not then removed surgically, there is a risk of secondary infection.

SELECTED BIBLIOGRAPHY

Colwell, D.D., Hall, M.J.R. and Scholl, P.J. (2006) *Oestrid Flies: Biology, Host–Parasite Relationships, Impact and Management*. CAB International, Wallingford, UK.

Francesconi, F. and Lupi, O. (2012) Myiasis. *Clinical Microbiology Reviews* 25, 79–105.

Geary, M.J., Hudson, B.J., Russell, R.C. and Hardy, A. (1999) Exotic myiasis with Lund's fly (*Cordylobia rodhaini*). *Medical Journal of Australia* 171, 654–655.

Hendrix, C.M., King-Jackson, D.A., Wilson, M., Blagburn, B.L. and Lindsay, D.S. (1995) Furunculoid myiasis in a dog caused by *Cordylobia anthrophaga*. *Journal of the American Veterinary Medicine Association* 207, 1187–1189.

Logar, J., Soba, B. and Parac, Z. (2006) Cutaneous myiasis caused by *Cordylobia anthropophaga*. *Wiener Klinische Wochenschrift* 118, 180–182.

McGraw, T.A., Timothy, A. and Turiansky, G.W. (2008) Cutaneous myiasis. *Journal of the American Academy of Dermatology* 58, 907–926.

Warble Flies/Old World Skin Bot Flies
(Diptera: Oestridae: Hypodermatinae)

1. INTRODUCTION

Flies of the genus *Hypoderma* are dermal parasites of Artiodactyla, Lagomorpha and Rodentia. Two species, *Hypoderma bovis* and *Hypoderma lineatum*, are important parasites of cattle and are widespread in the northern hemisphere between the latitudes of 25° and 60°N. They are not believed to have established themselves widely in the southern hemisphere. Human infestations have been reported, mostly as subdermal migratory and internal ocular myiasis.

2. TAXONOMY

Within the superfamily Oestroidea, family Oestridae, the subfamily Hypodermatinae contains six species of veterinary importance within the genus *Hypoderma*. Two species, *H. bovis* and *H. lineatum*, are parasites primarily of cattle, whereas *Hypoderma diana*, *Hypoderma actaeon*, *Hypoderma tarandi* and *Hypoderma sinense* affect roe deer, red deer, reindeer and yak, respectively. *H. bovis* is known as the northern cattle grub and *H. lineatum* as the common cattle grub.

3. MORPHOLOGY

The cattle-infesting species are bumblebee-like flies (Fig. 136) with reddish-yellow hairs at the posterior extremity of the abdomen. In *H. bovis*, the hairs anterior to the transverse suture of the thorax are variously described as whitish-yellow or reddish-yellow and contrast with those posterior to the suture, which are black. In *H. lineatum*, the hairs on the scutum

are white and yellow, with a predominance of white anteriorly and yellow posteriorly. The adults have rudimental mouthparts.

The mature third-stage larva is about 30 mm in length, with a convex ventral surface and a flat dorsal surface. Most of the body segments carry on their ventral surfaces an anterior row of larger, backwardly directed spines and a posterior band of smaller, forwardly directed spines. The spines are less developed on the dorsal surface (Fig. 137). In *H. lineatum*, the spiracles are flat, crescent shaped and with a considerable gap between the arms of the crescent surrounding the button. In *H. bovis*, the posterior spiracles are funnel shaped, with a much smaller gap between the arms than occurs in *H. lineatum*

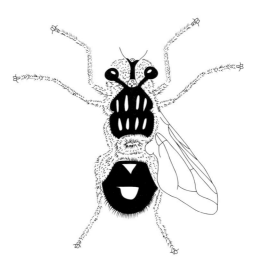

Fig. 136. Female *Hypoderma lineatum* (with left wing removed). Source: redrawn from Cameron, A.E. (1942) *Insect Pests of 1942. Transactions of the Highland and Agricultural Society.*

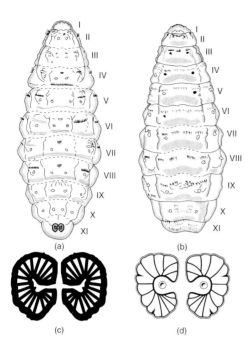

Fig. 137. Third-stage larva of *Hypoderma lineatum*. (a) Dorsal view; (b) ventral view; enlarged posterior spiracles of (c) *Hypoderma bovis* and (d) *H. lineatum*. *Source*: (a) and (b) from Cameron, A.E. (1937) *Transactions of the Highland and Agricultural Society*. (c) and (d) from James (1947).

(Fig. 137). Third-stage larvae are identified on the basis of spinulation on the tenth larval segment (dorsal view) and peritreme features (Fig. 137).

4. LIFE CYCLE

The females emerge with a fully developed complement of 300–650 eggs. The female attaches its eggs firmly to the host's hair. The base of the egg is connected to an attachment organ by a flexible petiole. The attachment organ has a central groove filled with adhesive material and a pair of adhesive-coated lateral flanges, which nearly meet to surround the hair. The adhesive material solidifies and attaches the egg firmly to the hair, and the flexible petiole enables the egg to adjust its alignment and reduce mechanical stress.

H. bovis lays its eggs while flying or when walking over the host, causing cattle to show violent avoidance reactions (gadding). It attaches a single egg per hair on the rump and upper parts of the hindleg. In contrast, ovipositing *H. lineatum* do not disturb the host and lay 7–10 eggs in line along the shaft of a single hair on the legs and lower parts of resting and standing cattle. The eggs of both species hatch within about 4 days and the first-stage larvae crawl down the hair and penetrate the skin using proteolytic enzymes (mostly collagenases) secreted from the blind midgut. The precise route taken by migrating larvae in the host is still uncertain. Several months elapse before the larvae reach their final localization on the back of the host.

Migrating larvae of *H. lineatum* are found within the wall of the oesophagus, whereas those of *H. bovis* are in the epidural fat of the spinal canal (most larvae being found in the region of the lumbar and posterior thoracic vertebrae). It has been suggested that larvae reach the spinal canal by migrating along nerve trunks or through muscles. Although it is accepted that most larvae of *H. bovis* migrate via the spinal canal, this may not be the only migration route, as the number of larvae in this region is usually smaller than on the back.

Early in the year, first-stage larvae move to their final site on the host's back, in an area 25 cm from either side of the midline from shoulder to tail, and subcutaneous warbles (from which the flies get their common name 'warble fly') are formed. The larva is now about 15 mm long but still in the first stage, and moults into the second stage soon after reaching the back. The larvae release collagenolytic enzymes accumulated in their gut which pierce the skin and form a hole through which the larvae breathe and then develop into third-stage larvae.

After from 4 to 11 weeks, the yellowish-brown mature larva forces its way through the breathing hole in the skin and drops to the ground, where it moves around actively seeking shelter. Mature larvae and pupae can survive cool to sub-zero temperatures. The duration of the pupal stage varies from 3 to 10 weeks depending on external conditions.

Where both species co-occur, adult *H. lineatum* usually appear about 4 weeks earlier than adult *H. bovis*.

5. BEHAVIOUR AND BIONOMICS

The distribution of flies of the genus *Hypoderma*, such as species affecting cattle (*H. bovis* and *H. lineatum*), has been reduced greatly over the past few years as a consequence of eradication campaigns in many countries.

There is one generation a year and, in general, adult *H. lineatum* occur on the wing from late March to the end of May. First-stage larvae undertake migration from April until September and can be found in the oesophagus from September to March, with the peak from November to January. *H. lineatum* warbles appear on the host's back from January to April, with peaks in February and March, and mature larvae leave the host and pupate from March to early May. Development and migration of *H. bovis* occurs slightly later, with adults being present from June to mid-September; larvae migrate from June to November, reach the spinal canal from November to May (main period December–March), warbles appear from March to July and pupae occur from May to August. Knowledge of this cycle is crucial for timely serological diagnoses of infestation (ELISA) throughout September–November (when warbles are still undetectable by palpation) and subsequent treatment strategies.

Adults are active on sunny days when the temperature is above 18°C. Being unable to feed, they survive between 3 and 5 days only. In the presence of suitable hosts, adults do not fly very far. However, marked flies have been recovered 300 m from the release point after 95 min, and laboratory flight mill studies indicate that *H. lineatum* is capable of flying up to 16 km. Adults emerge early in the day (07:30–08:30 h) and mate within 1 h of emergence.

H. diana is a parasite of deer and is present in a wide variety of habitats, overlapping the territory of its hosts. It is distributed throughout Europe and Asia, from 30° to 60°N, living in several different ecological zones such as mixed, deciduous and coniferous forests, wooded steppes and wetlands. The adult fly is most active in May and June, particularly on warm, sunny days. The main factors influencing the flight and oviposition of female flies are air temperature and light, with adult flies largely active around midday. As in other species, the intensity of infection and prevalence are higher in younger animals, possibly as a consequence of the lack of acquired resistance. When the larvae reach the skin on the host's back, large, soft, painful swellings of up to 3 mm diameter develop. The larvae lie in cysts containing yellow purulent fluid.

H. tarandi, known as the reindeer warble fly, resembles other species of the same genus and it is widely distributed in reindeer/caribou habitats of arctic and subarctic regions of northern Europe, North America and Russia. They are active in July and August, each female laying between 500 and 700 eggs, which are attached to the downy undercoat rather than the outer hair, preferentially on the flanks, legs and rump. After approximately 6 days, the eggs hatch and the larvae then burrow into and under the skin. Unlike other *Hypoderma* spp., the first-stage larva of *H. tarandi* migrates directly to the back in the subcutaneous connective tissue via the spine. When the larva comes to rest around September–October, a swelling is created where it feeds. The main consequence of the infestation is the damage to hides, which may result in substantial economic losses. An increasing number of human cases in people associated closely with reindeer/caribou herds have been recorded in many regions of the northern hemisphere.

6. MEDICAL AND VETERINARY IMPORTANCE

Economic losses due to *Hypoderma* activity and infestation occur for a number of reasons. Disturbance caused by actively ovipositing flies, especially *H. bovis*, reduces weight gain and may produce losses of 10–15% in milk production. The passage of larvae under the skin results in the formation of jelly-like tracks in muscle, which have to be removed at the abattoir.

In some areas of China, prevalence of infestation by *H. bovis* and *H. sinense* may reach up to 98–100%, with maximum intensities exceeding 400 warbles/animal, greatly impairing animal production. Larval death may result in the collapse of the infested animal, as a consequence of anaphylactic reactions to toxins released by dead larvae. In order to prevent these reactions, systemic insecticides should not be administered between mid-November and mid-March, when larvae are localized in the spinal canal or oesophagus. In addition, the breathing holes formed by the larvae damage the hide and reduce its value for subsequent use as leather.

Human infestations have been reported, mostly as subdermal migratory myiasis due to migrating first-stage larvae, although oral and ocular infestations (which can be severe, with destruction of the eye) have been reported, as have meningitis and skin allergies, but they are seldom fatal. Humans are accidental hosts for *Hypoderma* larvae, which cannot complete their development in this host.

Mostly, infestation occurs in people who are associated closely with cattle (through eggs being laid on human body hairs or when human skin comes in contact with newly hatched larvae on host animals). In poor social settings of China, human infestation by larval *Hypoderma* spp. has been reported to occur in up to 7% of cattle and yak farmers. The infection is usually self-limited and, while surgical removal of the larvae is effective for subdermal infestations, it is not feasible in cases of deeper invasion.

7. PREVENTION AND CONTROL

Hypoderma are highly susceptible to systemically active organophosphorus insecticides and to the macrocyclic lactones, such as ivermectin, even if used at minimum dosage. The organophosphorus preparations are applied as 'pour-ons' to the backs of cattle and are absorbed systemically from there; macrocyclic lactones can also be given by subcutaneous injection or pour-on. A single annual treatment is usually recommended, preferably in September, October or November, prior to the larvae of *H. bovis* reaching the spinal canal, therefore preventing spinal damage due to the disintegration of dead larvae. Treatment during spring when the larvae have left their resting sites and migrated under the skin of the back, although effective in achieving control of the infestation, is not recommended since the breathing third-stage larva has then perforated the hide. Successful eradication schemes have been undertaken in some areas of Europe such as the UK, Eire, Denmark and the Netherlands. However, the risk of reintroduction is a constant threat. Macrocyclic lactones administered with the feed also may be used to eliminate oestrids from farmed deer. Other forms of insecticide application include power dusters, dust bags, sprays and back rubbers.

SELECTED BIBLIOGRAPHY

Beesley, W.N. (1961) Observations on the development of *Hypoderma lineatum* De Villiers (Diptera: Oestridae) in the bovine host. *Annals of Tropical Medicine and Parasitology* 55, 18–24.

Beesley, W.N. (1977) Practical relationships between the biology and control of cattle grubs. *Veterinary Parasitology* 3, 251–257.

Boulard, C., Alvinerie, M., Argenté, G., Languille, J., Paget, L. and Petit, E. (2008) A successful, sustainable and low cost control-programme for bovine hypodermosis in France. *Veterinary Parasitology* 25(158), 1–10.

Colwell, D.D., Hall, M.J.R. and Scholl, P.J. (2006) *Oestrid Flies: Biology, Host–Parasite Relationships, Impact and Management*. CAB International, Wallingford, UK.

Colwell, D.D., Otranto, D. and Stevens, J.R. (2009) Oestrid flies: eradication and extinction versus biodiversity. *Trends in Parasitology* 25, 500–504.

Francesconi, F. and Lupi, O. (2012) Myiasis. *Clinical Microbiology Reviews* 25, 79–105.

Hadlow, W.J., Ward, J.K. and Krinsky, W.L. (1977) Intracranial myiasis by *Hypoderma bovis* (Linnaeus) in a horse. *Cornell Veterinarian* 67, 272–281.

Hassan, M.U., Khan, M.N., Abubakar, M., Waheed, H.M., Iqbal, Z. and Hussain, M. (2010) Bovine hypodermosis – a global aspect. *Tropical Animal Health and Production* 42, 1615–1625.

Klein, K.K., Fleming, C.S., Colwell, D.D. and Scholl, P.J. (1990) Economic analysis of an integrated approach to cattle grub (*Hypoderma* spp.) control. *Canadian Journal of Agricultural Economics* 38, 159–173.

Lysyk, T.J., Colwell, D.D. and Baron, R.W. (1991) A model for estimating abundance of cattle grub (Diptera: Oestridae) from the proportion of uninfested cattle as determined by serology. *Medical and Veterinary Entomology* 5, 253–258.

Otranto, D., Colwell, D.D., Traversa, D. and Stevens, J.R. (2003) Species identification of *Hypoderma* affecting domestic and wild ruminants by morphological and molecular characterization. *Medical and Veterinary Entomology* 17, 316–325.

Otranto, D., Traversa, D., Colwell, D.D., Guan, G., Giangaspero, A., Boulard, C., *et al.* (2004) A third species of *Hypoderma* (Diptera: Oestridae) affecting cattle and yaks in China: molecular and morphological evidence. *Journal of Parasitology* 90, 958–965.

Otranto, D., Paradies, P., Testini, G., Lia, R.P. Giangaspero, A., Traversa, D., *et al.* (2006) First description of the endogenous life cycle of *Hypoderma sinense* affecting yaks and cattle in China. *Medical and Veterinary Entomology* 20, 325–328.

Scholl, P.J. (1993) Biology and control of cattle grubs. *Annual Review of Entomology* 39, 53–70.

Tommeras, B.A., Nilssen, A.C. and Wibe, A. (1996) The two reindeer parasites, *Hypoderma tarandi* and *Cephenemyia trompe* (Oestridae), have evolved similar olfactory receptor abilities to volatiles from their common host. *Chemoecology* 7, 1–7.

Wilson, G.W.C. (1986) Control of warble fly in Great Britain and the European community. *Veterinary Record* 118, 653–656.

Index